Navigating Tuberculosis

Foundations and Frontiers in Enzymology

Series Editor
Professor Munishwar Nath Gupta

Foundations and Frontiers in Enzymology

Navigating Tuberculosis

Insights into Host-Pathogen Protein Interactions

Edited by

Mohammad Shariq
School of Science, Department of Life Sciences, GITAM University, Hyderabad, Telangana, India

Seyed Ehtesham Hasnain
School of Basic Sciences & Research, Sharda University, Greater Noida, Uttar Pradesh, India

Nasreen Zafar Ehtesham
Sharda University, Greater Noida, Uttar Pradesh, India

Series Editor

Munishwar Nath Gupta

ELSEVIER

ACADEMIC PRESS
An imprint of Elsevier

Academic Press is an imprint of Elsevier
125 London Wall, London EC2Y 5AS, United Kingdom
50 Hampshire Street, 5th Floor, Cambridge, MA 02139, United States

ISBN: 978-0-443-27735-1

For information on all Academic Press publications visit our website at https://www.elsevier.com/books-and-journals

Publisher: Megan Ball
Acquisitions Editor: Ali Afzal-Khan
Editorial Project Manager: Rahul Basak
Production Project Manager: Punithavathy Govindaradjane
Cover Designer: Raman Kumar

Typeset by TNQ Tech

This book is dedicated to all scientists worldwide who relentlessly pursue lasting solutions to eradicate tuberculosis and alleviate its burden on humanity.

Contents

Contributors

Farhan Ahmed
GITAM School of Life Sciences, Gandhi Institute of Technology and Management (GITAM) University, Hyderabad Campus, Hyderabad, Telangana, India

Yashika Ahuja
Department of Life Science, School of Bioscience and Technology, Sharda University, Greater Noida, Uttar Pradesh, India

Anwar Alam
Department of Life Science, School of Bioscience and Technology, Sharda University, Greater Noida, Uttar Pradesh, India; Department of Biotechnology, School of Bioscience and Technology, Sharda University, Greater Noida, Uttar Pradesh, India; DST-FIST, Sharda University, Greater Noida, Uttar Pradesh, India; Department of Life Science, School of Basic Sciences and Research, Sharda University, Greater Noida, Uttar Pradesh, India

Waseem Ali
Department of Life Science, School of Bioscience and Technology, Sharda University, Greater Noida, Uttar Pradesh, India; Department of Life Science, School of Basic Sciences and Research, Sharda University, Greater Noida, Uttar Pradesh, India

Onaiza Ansari
Department of Biotechnology, School of Chemical and Life Sciences, Jamia Hamdard, New Delhi, India

Faizan Ansari
Department of Molecular Medicine, School of Interdisciplinary Sciences and Technology, Jamia Hamdard, New Delhi, India

Sharmistha Banerjee
Department of Biochemistry, School of Life Sciences, University of Hyderabad, Hyderabad, Telangana, India

Amit Chakraborty
CSIR-Centre for Cellular and Molecular Biology, Hyderabad, Telangana, India; Academy of Scientific and Innovative Research (AcSIR), Ghaziabad, Uttar Pradesh, India

Korak Chakraborty
CSIR—Centre for Cellular and Molecular Biology, Hyderabad, Telangana, India; Academy of Scientific and Innovative Research (AcSIR), Ghaziabad, Uttar Pradesh, India

Kashish Choudhary
Department of Molecular Medicine, School of Interdisciplinary Sciences and Technology, Jamia Hamdard, New Delhi, India

Sajal Dey
BRIC-Centre for DNA Fingerprinting and Diagnostics (CDFD), Hyderabad, Telangana, India; Graduate Studies, Manipal Academy of Higher Education, Manipal, Karnataka, India

R. Divya
Immuno-genomics & Systems Biology Laboratory, BRIC-Institute of Life Sciences, Bhubaneswar, Odisha, India

Nasreen Z. Ehtesham
Department of Life Science, School of Bioscience and Technology, Sharda University, Greater Noida, Uttar Pradesh, India; Department of Life Science, School of Basic Sciences and Research, Sharda University, Greater Noida, Uttar Pradesh, India

Haleema Fayaz
Department of Life Science, School of Bioscience and Technology, Sharda University, Greater Noida, Uttar Pradesh, India

Sonam Grover
Department of Molecular Medicine, School of Interdisciplinary Sciences and Technology, Jamia Hamdard, New Delhi, India

Muhammad H. Gul
Division of Pulmonary, Critical Care & Sleep, Department of Internal Medicine, University of Nebraska Medical Center, Omaha, NE, United States

Sarah Hamid
Department of Molecular Medicine, School of Interdisciplinary Sciences and Technology, Jamia Hamdard, New Delhi, India

Seyed E. Hasnain
Department of Life Science, School of Bioscience and Technology, Sharda University, Greater Noida, Uttar Pradesh, India; Department of Biochemical Engineering and Biotechnology, IIT Delhi, New Delhi, India; Department of Life Science, School of Basic Sciences and Research, Sharda University, Greater Noida, Uttar Pradesh, India

Zin Mar Htun
Division of Pulmonary, Critical Care & Sleep, Department of Internal Medicine, University of Nebraska Medical Center, Omaha, NE, United States

Ishan
Department of Molecular Medicine, School of Interdisciplinary Sciences and Technology, Jamia Hamdard, New Delhi, India

Sana Jamily
Department of Molecular Medicine, School of Interdisciplinary Sciences and Technology, Jamia Hamdard, New Delhi, India

Yogita Kapoor
CSIR-Centre for Cellular and Molecular Biology, Hyderabad, Telangana, India; Academy of Scientific and Innovative Research (AcSIR), Ghaziabad, Uttar Pradesh, India; Ragon Institute of MGH, MIT and Harvard University, Cambridge, MA, United States

Dilpreet Kaur
Department of Life Science, School of Bioscience and Technology, Sharda University, Greater Noida, Uttar Pradesh, India

Vikas Kumar
Department of Biotechnology, Indira Gandhi National Tribal University, Amarkantak, Madhya Pradesh, India

Kunal Malik
Department of Molecular Medicine, School of Interdisciplinary Sciences and Technology, Jamia Hamdard, New Delhi, India

Vandana Maurya
BRIC-Centre for DNA Fingerprinting and Diagnostics (CDFD), Hyderabad, Telangana, India; Graduate Studies, Regional Center for Biotechnology, Faridabad, Haryana, India

Nerges Mistry
The Foundation for Medical Research, Mumbai, Maharashtra, India

Alladi Mohan
Department of Medicine, Sri Venkateswara Institute of Medical Sciences, Tirupati, Andhra Pradesh, India

Krishnaveni Mohareer
Department of Biochemistry, School of Life Sciences, University of Hyderabad, Hyderabad, Telangana, India

Sangita Mukhopadhyay
BRIC-Centre for DNA Fingerprinting and Diagnostics (CDFD), Hyderabad, Telangana, India

Anam Mursaleen
Department of Biotechnology, School of Chemical and Life Sciences, Jamia Hamdard, New Delhi, India

Vinay Kumar Nandicoori
CSIR-Centre for Cellular and Molecular Biology, Hyderabad, Telangana, India; Academy of Scientific and Innovative Research (AcSIR), Ghaziabad, Uttar Pradesh, India; BRIC-National Institute of Immunology, New Delhi, India

Sourya Prakash Nayak
Immuno-genomics & Systems Biology Laboratory, BRIC-Institute of Life Sciences, Bhubaneswar, Odisha, India

Niteen Pathak
BRIC-Centre for DNA Fingerprinting and Diagnostics (CDFD), Hyderabad, Telangana, India

Vandana Pawar
Department of Biotechnology, Indira Gandhi National Tribal University, Amarkantak, Madhya Pradesh, India

Shivangi Prandiyal
Department of Life Science, School of Bioscience and Technology, Sharda University, Greater Noida, Uttar Pradesh, India

Subhasish Prusty
Immuno-genomics & Systems Biology Laboratory, BRIC-Institute of Life Sciences, Bhubaneswar, Odisha, India

Sunil K. Raghav
Immuno-genomics & Systems Biology Laboratory, BRIC-Institute of Life Sciences, Bhubaneswar, Odisha, India

Tirumalai R. Raghunand
CSIR—Centre for Cellular and Molecular Biology, Hyderabad, Telangana, India; Academy of Scientific and Innovative Research (AcSIR), Ghaziabad, Uttar Pradesh, India

Nidhi Rani
Molecular Biophysics Unit, Indian Institute of Science, Bangalore, Karnataka, India; Jawaharlal Nehru Centre for Advanced Scientific Research, Bangalore, Karnataka, India

Syeda Rushna
Department of Biotechnology, School of Chemical and Life Sciences, Jamia Hamdard, New Delhi, India

Ruxana T. Sadikot
Division of Pulmonary, Critical Care & Sleep, Department of Internal Medicine, University of Nebraska Medical Center, Omaha, NE, United States; VA Nebraska Western Iowa Health Care System, Omaha, NE, United States

Mohd Shariq
GITAM School of Life Sciences, Gandhi Institute of Technology and Management (GITAM) University, Hyderabad Campus, Hyderabad, Telangana, India

Surendra Kumar Sharma
Department of Molecular Medicine, Jamia Hamdard (Deemed to Be University), New Delhi, India; Faculty of Medicine, Datta Meghe Institute of Medical Sciences (Deemed to Be University), Wardha, Maharashtra, India

Javaid Ahmad Sheikh
Department of Biotechnology, School of Chemical and Life Sciences, Jamia Hamdard, New Delhi, India

Gauri Shrivastava
Department of Life Science, School of Bioscience and Technology, Sharda University, Greater Noida, Uttar Pradesh, India

Kalpana Sriraman
The Foundation for Medical Research, Mumbai, Maharashtra, India

Avadhesha Surolia
Molecular Biophysics Unit, Indian Institute of Science, Bangalore, Karnataka, India; Dr. Reddy's Institute of Life Science, Hyderabad, Telangana, India

Diksha Thakuri
Department of Life Science, School of Bioscience and Technology, Sharda University, Greater Noida, Uttar Pradesh, India

Vishwanath Upadhyay
Department of Molecular Medicine, Jamia Hamdard (Deemed to Be University), New Delhi, India; Department of Life Science, Sharda University, Noida, Uttar Pradesh, India

Preface

Tuberculosis remains one of the world's most persistent and devastating infectious diseases, profoundly affecting millions of lives each year. The urgent need to understand the intricate interplay between the host immune system and *Mycobacterium tuberculosis* (*M. tb*) has not only motivated the contributors of this volume but has also galvanized scientific communities worldwide. With increasing recognition of the complexity and multidimensional nature of TB pathogenesis, this book seeks to bridge critical knowledge gaps and inspire new directions in host-pathogen research. Our purpose in assembling this volume is to provide a comprehensive, up-to-date resource that synthesizes current advances in molecular, cellular, and immunological mechanisms governing TB infection. This compilation brings together leading experts who have contributed pivotal breakthroughs, from the identification of novel proteins impacting host immunity to the discovery of sophisticated pathogen survival tactics. By carefully mapping the connections between host defenses, pathogen virulence strategies, and therapeutic opportunities, we aim to foster deeper understanding and spark innovative approaches in TB research.

The primary audience for this book includes basic scientists, graduate students, and researchers actively engaged in host-pathogen interaction studies at the molecular and cellular levels, especially within microbiology, immunology, molecular biology, and infectious diseases. Additionally, we expect the volume to serve a broader readership, including doctoral fellows, postdoctoral researchers, and professionals in public health and epidemiology, who seek insight into the fundamental principles underlying microbial pathogenesis and translational strategies for combating TB. The context for this book is rooted in a global health landscape where tuberculosis poses an enduring threat, particularly in settings with limited resources. Significant advances in technologies, from genomics and proteomics to systems biology, have transformed our perspective of how *M. tb* manipulates host processes. The chapters are organized to reflect the latest findings and tackle pressing scientific questions: How do host proteins mediate immunity? What molecular tactics does *M. tb* employ to evade defenses? How do mechanisms like autophagy, apoptosis, and programmed cell death shape disease outcomes?

Drawing on our collective expertise, we have curated chapters by renowned scientists spanning diverse geographical regions and institutional backgrounds. This diversity is reflected in the multidisciplinary approach of the book, integrating insights across cellular biology, immunology, microbiology, and translational research. The editorial team's longstanding commitment to TB research, evidenced by decades of scholarship, mentorship, and institutional leadership, ensures both rigor and relevance in the book's content. The structure of this volume follows a logical progression. The first chapters deliver foundational information on virulence mechanisms, host adaptation, and immune signaling in TB. Subsequent sections delve into pathogen strategies, host countermeasures, and key protein factors, followed by chapters on autophagy, metabolic reprogramming, interferon signaling, drug resistance, and future therapeutic avenues. Each chapter blends technical detail with broad scientific context to promote clarity and accessibility.

Our methodology in compiling this book drew on literature synthesis, expert contributions, and collaborative discussions with international and national leaders. Stringent editorial review and peer input ensured accuracy, currency, and scholarly integrity. The contributors, listed in the LOC, reflect a global and multidisciplinary community committed to advancing TB science. We gratefully acknowledge the contributions, guidance, and inspiration provided by colleagues, academic institutions, and funding bodies. This collaborative effort would not have been possible without their support.

We invite readers to engage with the volume as both a reference and a catalyst for further research. It is our hope that the book will empower scientists to tackle unresolved challenges in TB pathogenesis, drive progress toward new diagnostics and therapies, and ultimately contribute to achieving the global aim of eradicating tuberculosis.

Acknowledgments

The editors extend their deepest gratitude to the numerous individuals and institutions whose support and encouragement made this volume possible. The conception and successful realization of *Navigating Tuberculosis: Insights into Host-Pathogen Protein Interactions* have been shaped by the collective efforts of many distinguished scientists, colleagues, and organizations committed to advancing the understanding of tuberculosis pathogenesis. We are profoundly grateful to our host institutions, GITAM University, Hyderabad; Sharda University, Greater Noida; and the Indian Institute of Technology, Delhi, for their continued encouragement, academic freedom, and infrastructural support that nurtured the development of this scholarly work. Their commitment to fostering excellence in scientific research has been instrumental in enabling us to bring this ambitious project to fruition. We sincerely acknowledge the Council of Scientific and Industrial Research (CSIR), the Department of Biotechnology (DBT), the Indian Council of Medical Research (ICMR), and the Department of Science and Technology (DST), Government of India, for their sustained investment in tuberculosis research. The generous funding and visionary programs from these agencies have played a crucial role in catalyzing innovative studies on host-pathogen biology across the country.

The editors wish to express their heartfelt appreciation to all chapter contributors, eminent scientists from leading institutions in India and abroad, whose intellectual rigor and scientific insight have enriched this volume. Their willingness to share their expertise and time, despite demanding professional commitments, stands as a testament to the collaborative spirit that drives scientific advancement. We are indebted to our mentors, collaborators, and colleagues, past and present, who have guided and inspired our scientific journeys. Their thoughtful discussions, constructive feedback, and constant encouragement have profoundly shaped our understanding of infection biology and molecular pathogenesis. We also acknowledge the contributions of our research groups and students, whose curiosity, diligence, and enthusiasm for discovery have continually motivated us to pursue new questions at the interface of microbiology and immunology. A special note of appreciation is due to Elsevier's editorial and production teams for their professionalism, patience, and unwavering support throughout the proposal and editorial process. Their expertise in scientific publishing ensured that the vision of this book was transformed into a coherent and accessible resource for the global research community. Finally, we express our gratitude to our families for their unconditional support, patience, and understanding throughout the demanding process of editing this volume. Their encouragement remains the quiet strength behind our academic endeavors.

This book is dedicated to the scientific community worldwide, particularly to the researchers, clinicians, and students striving to unravel the complexities of tuberculosis and translate knowledge into solutions that advance global health. May this volume serve as both a source of insight and a catalyst for future discoveries in the relentless pursuit of a world free from tuberculosis.

One-Sentence Keywords

1. ***Mycobacterium tuberculosis* pathogenesis** — exploring molecular and cellular mechanisms driving TB infection.
2. **Host-pathogen protein interactions** — decoding the molecular dialogue between *M. tb* and human immune cells.
3. **PE/PPE protein family** — investigating unique virulence determinants shaping *M. tb* persistence.
4. **Autophagy and immune evasion** — uncovering how *M. tb* manipulates host autophagic defenses.
5. **Phagosome-lysosome fusion** — mechanisms of disruption and restoration in TB infection.
6. **Macrophage-pathogen dynamics** — front-line host responses against intracellular mycobacteria.
7. **Innate and adaptive immunity in tuberculosis** - cellular coordination and immune modulation.
8. **Metabolic reprogramming in infected macrophages** — linking immunometabolism to bacterial survival.
9. **Programmed cell death in TB** — apoptosis, pyroptosis, and necroptosis as host defense mechanisms.
10. **ESX-1 type VII secretion system** — molecular machinery underlying virulence and immune modulation.
11. **Host-directed therapies (HDTs)** — novel strategies leveraging immune pathways to control TB.
12. **Molecular and structural biology of virulence factors** — from gene clusters to functional phenotypes.
13. **Translational immunology and vaccine design** — bridging mechanistic insights to clinical application.
14. **Global tuberculosis control and elimination** — aligning molecular discoveries with the WHO's End TB goals.

Introduction

Tuberculosis (TB), caused by *Mycobacterium tuberculosis* (*M. tb*), remains one of the world's most devastating infectious diseases, claiming millions of lives every year and posing an enduring challenge to public health. Despite the availability of chemotherapeutic regimens and vaccines, TB continues to thrive in both developing and developed regions, largely due to the pathogen's exceptional adaptability, its ability to persist in latent forms, and the emergence of multidrug-resistant and extensively drug-resistant strains. These clinical realities reveal the need to deepen our understanding of the molecular and immunological interactions that define host-pathogen equilibrium and dictate disease outcomes. The study of TB pathogenesis has undergone a paradigm shift over the past 2 decades, from descriptive microbiology to integrative systems biology. With the advent of high-throughput genomics, proteomics, and immunomics, the scientific community now appreciates *M. tb* not merely as a static pathogen but as a dynamic entity capable of sensing and reshaping host responses. The crosstalk between mycobacterial proteins and host signaling pathways determines infection establishment, immune evasion, and, ultimately, disease progression or resolution. This intricate molecular dialogue forms the central theme of this book.

Navigating Tuberculosis: Insights into Host-Pathogen Protein Interactions aims to present a comprehensive synthesis of current knowledge on the molecular interplay between *M. tb* and its host. The book brings together contributions from distinguished scientists whose expertise spans molecular microbiology, immunology, biochemistry, and translational medicine. It provides readers with an in-depth exploration of how *M. tb* manipulates host cellular mechanisms, ranging from autophagy, apoptosis, and phagosomal maturation to antigen presentation and cytokine signaling, to sustain its survival and persistence. The scope of this volume extends beyond a traditional review of TB biology. It integrates discoveries across molecular mechanisms, host immune dynamics, and therapeutic interventions into a coherent framework that reflects the multi-dimensional nature of TB infection. The text offers a balanced perspective: while early chapters elucidate fundamental mechanisms of virulence and immune evasion, later sections focus on host defense strategies, regulatory circuits, and emerging concepts in host-directed therapy (HDT). The final chapters bridge molecular insights with clinical translation, discussing novel therapeutic targets, vaccine strategies, and the potential of systems-based approaches to redefine TB control.

A distinctive feature of this book is its emphasis on **protein-protein interactions (PPIs)** as the molecular foundation of host-pathogen communication. By examining these interfaces, the book uncovers how specific mycobacterial effectors hijack host pathways, subvert immune surveillance, or exploit metabolic vulnerabilities to establish infection. Through detailed examples and integrated analyses, readers are guided to appreciate how these molecular events shape cellular outcomes and influence therapeutic opportunities. Another strength lies in the book's **interdisciplinary approach**. Contributors have integrated data from in silico modeling, biochemical characterization, omics analyses, and infection models to provide a holistic view of TB biology. This approach not only broadens the scientific context but also demonstrates how molecular discoveries can inform translational strategies, particularly in the development of next-generation drugs and vaccines.

The organization of the book is designed as a logical progression, from the pathogen's molecular toolkit to the host's defensive orchestration and, finally, to therapeutic exploitation. Each chapter is self-contained yet thematically connected, allowing readers to navigate between mechanistic insights and broader biological implications. For researchers and students, this volume serves both as a **reference resource** and a **conceptual guide**. For clinicians and translational scientists, it offers a molecular perspective that complements clinical practice and drug development efforts. Ultimately, this book aspires to stimulate new thinking, foster collaboration across disciplines, and inspire innovative research at the interface of microbiology and immunology. In uniting diverse voices within a single volume, *Navigating Tuberculosis* captures the current momentum of TB research, one that is transitioning from observing the pathogen to decoding its molecular strategies and, ultimately, leveraging this understanding toward therapeutic innovation. It is our hope that this book will serve as both a comprehensive resource and a springboard for future discoveries that advance the global effort to control and eventually eliminate TB.

SECTION

Mycobacterial virulence and host interaction

I

CHAPTER

1

Mechanisms of *Mycobacterium tuberculosis* virulence: The ESX-1 locus and host countermeasures

Syeda Rushna[1], Onaiza Ansari[1], Anam Mursaleen[1], Farhan Ahmed[2], Javaid Ahmad Sheikh[1] and Mohd Shariq[2]

[1]*Department of Biotechnology, School of Chemical and Life Sciences, Jamia Hamdard, New Delhi, India;* [2]*GITAM School of Life Sciences, Gandhi Institute of Technology and Management (GITAM) University, Hyderabad Campus, Hyderabad, Telangana, India*

Highlights

1. Tuberculosis caused 10.8 million infections and 1.25 million deaths in 2023, reaffirming its global health burden.
2. The ESX-1 secretion system is central to *M. tb* virulence, enabling phagosomal rupture and cytosolic access.
3. EsxA (ESAT-6) and EsxB (CFP-10), supported by EspA, EspB, EspC, and PE/PPE proteins, orchestrate membrane damage.
4. Phagosomal rupture exposes bacterial components to cytosolic sensors, triggering cGAS-STING signaling, inflammasome activation, and pyroptosis.
5. Host countermeasures include galectin-8 (GAL-8) mediated xenophagy, ESCRT-driven membrane repair, cytosolic serpins, and acidification control.
6. These responses both restrict bacterial growth and, when dysregulated, drive immunopathology and tissue destruction.
7. Targeting ESX-1 machinery and reinforcing host repair pathways offer promising antivirulence and host-directed therapies.

Abbreviations

AAA+ ATPases Associated with diverse cellular Activities
AIM2 Absent In Melanoma 2
ASC Apoptosis-associated speck-like protein containing a CARD
ATPase Adenosine Triphosphatase
BCG Bacillus Calmette–Guérin
CARD Caspase Activation and Recruitment Domain
Casp1 Caspase-1
CFP-10 Culture Filtrate Protein 10
CFP Culture Filtrate Protein

Navigating Tuberculosis. https://doi.org/10.1016/B978-0-443-27735-1.00007-1

cGAMP cyclic GMP–AMP
cGAS Cyclic GMP–AMP synthase
CHMP4 Charged Multivesicular Body Protein 4
DNA Deoxyribonucleic Acid
Ecc ESX-conserved component
EM Electron Microscopy
ER Endoplasmic Reticulum
ESAT-6 Early Secretory Antigenic Target 6 (EsxA)
ESCRT Endosomal Sorting Complexes Required for Transport
Esp ESX Secretion-Associated Protein
ESX Type VII Secretion System (1, 3 or others)
FRET Fluorescence Resonance Energy Transfer
GAL-3, 8, 9 Galectin-3, Galectin-8, Galectin-9
GSDMD Gasdermin D
HRS Hepatocyte growth factor-regulated tyrosine kinase substrate
HTH Helix–Turn–Helix
IFN-β Interferon beta
IFN-γ Interferon gamma
IFN Interferon
IL-18 Interleukin-18
IL-1Ra Interleukin-1 Receptor antagonist
IL-1β Interleukin-1 beta
IL Interleukin
iPSC Induced Pluripotent Stem Cell KasA/KasB; Ketoacyl synthase A/B
LAM Lipoarabinomannan
LC3 Microtubule-Associated Protein 1A/1B-Light Chain 3
MAVS Mitochondrial Antiviral-Signaling Protein
MCV Mycobacteria-Containing Vacuole
MDR-TB Multidrug-Resistant Tuberculosis
MTBC Mycobacterium tuberculosis complex
mTOR Mechanistic Target of Rapamycin
MycP1 Mycosin Protease 1
NK Natural Killer cell
NLR NOD-like Receptor
NLRP3 NOD-, LRR-, and pyrin domain-containing protein 3
Nramp-1 Natural Resistance-Associated Macrophage Protein 1
Nα N-terminal alpha (acetylation)
PAMP Pathogen-Associated Molecular Pattern
PE/PPE Pro-Glu/Pro-Pro-Glu Protein families
pH Potential of Hydrogen
PhoPR/MprAB Two-component bacterial regulatory systems
pks13 Polyketide synthase 13
PPE68 Pro-Pro-Glu Protein 68
RD1 Region of Difference 1
RIG-I Retinoic acid Inducible Gene I
RNA Ribonucleic Acid
Rv Reference to specific coding sequences in M. tuberculosis
Scp2 Sterol Carrier Protein 2
SecA2 Accessory Sec secretion pathway

Spi2A Serpina3g/Spi2A (mouse cytosolic serpin)
SQSTM1/p62 Sequestosome 1
STING Stimulator of Interferon Genes
T7SS Type VII Secretion System
TAX1BP1 Tax1 Binding Protein 1
TB Tuberculosis
TBK1 TANK-Binding Kinase 1
Tim-3 T cell immunoglobulin and mucin-domain containing-3
TNF-α Tumor Necrosis Factor alpha
TPR (Canonical) Tetratricopeptide Repeat D1/D2/D3; ATPase domains of EccC1
TPR Tetratricopeptide Repeat
Tsg101 Tumor Susceptibility Gene 101
Ub Ubiquitin
Vps4 Vacuolar Protein Sorting 4
WT Wild-Type
XDR-TB Extensively Drug-Resistant Tuberculosis
Xenophagy Selective Autophagy targeting pathogens
Δ Deletion (mutant, e.g., NLRP3−/−)

1. Introduction

Tuberculosis (TB), caused by *Mycobacterium tuberculosis* (*M. tb*), continues to pose one of the greatest global health challenges. After several years in which coronavirus disease 2019 (COVID-19) dominated infectious disease mortality, TB has once again re-emerged as the world's deadliest infectious agent. According to the World Health Organization's (WHO) 2024 Global TB Report, an estimated 10.8 million people developed active TB in 2023, and approximately 1.25 million individuals lost their lives to the disease, reaffirming TB's position as the leading cause of death from a single infectious pathogen worldwide [1]. Despite intensified control efforts, the decline in TB incidence has been inconsistent. For instance, India reported a 17.7% decline in incidence since 2015, yet globally the distribution of cases remains heavily skewed. Five high-burden countries, India (26%), Indonesia (10%), China (6.8%), the Philippines (6.8%), and Pakistan (6.3%), accounted for more than half of all global cases in 2023 [1]. This uneven progress demonstrates the urgent need to deepen our understanding of the molecular mechanisms of pathogenesis and the host defense strategies that counter them.

The threat posed by TB is further amplified by the rise of multidrug-resistant TB (MDR-TB). In 2023 alone, nearly 400,000 individuals developed MDR-TB, representing about 3.7% of all new TB infections [1]. These drug-resistant forms are considerably more difficult and costly to treat, requiring longer regimens with toxic second-line drugs. The persistent reservoir of latent infection adds another layer of complexity: nearly 25% of the global population is estimated to harbor *M. tb* in an asymptomatic state [2]. While most latent infections remain under immune control, a subset progresses to active disease, particularly in individuals with weakened immunity. This vast reservoir, combined with the emergence of drug-resistant strains, highlights the fragile equilibrium between host defense and pathogen survival and illustrates why TB remains such a formidable adversary.

At the heart of TB infection lies a battle within the macrophage phagosome. The phagosome is a membrane-bound compartment where macrophages normally neutralize invading microbes. Under physiological conditions, the phagosome undergoes a tightly regulated maturation process. It fuses with late endosomes and lysosomes, acidifies, and acquires hydrolytic enzymes, creating a degradative environment hostile to pathogens. This maturation is a cornerstone of innate immunity, designed to sequester and destroy intracellular microbes [3]. However, *M. tb* has evolved highly sophisticated mechanisms to arrest phagosome maturation, thereby creating a protective intracellular niche. Following uptake by alveolar macrophages, the bacterium resides in an arrested early phagosome that fails to acquire acidic and hydrolytic lysosomal markers. Instead, this compartment retains access to early endosomal contents and nutrients, forming a permissive environment for bacterial persistence [3].

Several mycobacterial factors contribute to this blockade of phagosomal maturation. Cell envelope lipids such as lipoarabinomannan (LAM) disrupt host phosphatidylinositol-3 kinase signaling, preventing phagosome-lysosome fusion. Similarly, secreted effectors including the phosphatase SapM and various kinases interfere with host vesicular trafficking and block acidification [3]. By residing in this stalled compartment, *M. tb* avoids exposure to lysosomal hydrolases, reactive oxygen species (ROS), and antimicrobial peptides, thus evading the host's most potent defense mechanisms [4]. This adaptation is one of the reasons for the extraordinary success of *M. tb* as a long-term intracellular pathogen.

Among the diverse virulence factors of *M. tb*, the ESX-1 secretion system stands out as particularly central. The ESX-1 system, also known as a T7SS, is encoded by the RD1 locus and is indispensable for virulence. It mediates phagosomal membrane disruption, bacterial escape into the cytosol, immune modulation, and pathogen dissemination [5]. The significance of ESX-1 was first highlighted by studies of the attenuated *Mycobacterium bovis* Bacillus Calmette-Guérin (BCG) vaccine strain, which lacks the RD1 locus and therefore ESX-1 activity, rendering it non-pathogenic [6]. ESX-1 exports several effectors, most notably EsxA (ESAT-6, a 6 kDa early secretory antigenic target) and EsxB (CFP-10, a 10 kDa culture filtrate protein), which are secreted as a heterodimeric complex. EsxA functions as a membrane-disrupting effector, while EsxB serves as a stabilizing chaperone. Under acidic conditions, the EsxA/EsxB complex dissociates, allowing EsxA to damage host membranes [7]. Other ESX-1 substrates, including EspA, EspB, and EspC, further augment virulence by promoting membrane interactions, facilitating secretion, and modulating host immune responses.

Recent evidence suggests that ESX-1-mediated membrane damage is not the result of classical pore formation but instead occurs through contact-dependent disruption, distinguishing it from other well-characterized pore-forming toxins [8]. This unique mode of action shows the sophistication of the ESX-1 system. By breaching the phagosomal membrane, ESX-1 not only enables bacterial survival but also exposes mycobacterial DNA, RNA, and proteins to cytosolic sensors, thereby activating host immune pathways such as cyclic GMP-AMP synthase-stimulator of interferon response cGAMP interactor (cGAS-STING) and NOD-like receptor (NLR) inflammasomes. These responses drive production of type I interferons (IFNs), IL-1β, and IL-18, and can induce pyroptosis, a lytic form of cell death. Thus, ESX-1 functions as both a virulence mechanism and a trigger of immune surveillance, reflecting the dynamic host-pathogen interplay that defines TB pathogenesis.

Understanding the ESX-1 secretion system and its effector proteins is therefore crucial for unraveling the molecular basis of TB pathogenesis and identifying promising therapeutic targets. In this chapter, we first dissect the structural organization of ESX-1 and the functions of its core effectors. We then examine how ESX-1 enables *M. tb* to breach phagosomal membranes and trigger cytosolic immune responses, including nucleic acid sensing, inflammasome activation, and pyroptosis. Next, we highlight the host countermeasures, such as galectin (GAL)-mediated xenophagy, endosomal sorting complexes required for transport (ESCRT)-dependent membrane repair, cytosolic serpins, and phagosomal acidification, that attempt to neutralize ESX-1 activity. Finally, we discuss therapeutic implications, including direct inhibition of ESX-1, reinforcement of host defense mechanisms, and modulation of inflammatory signaling. By integrating recent advances in structural biology, cell biology, and immunology, this chapter aims to provide a comprehensive overview of ESX-1-mediated phagosomal damage and its essential role in shaping the outcome of TB infection.

2. Structural components of the ESX-1 system

The ESX-1 secretion system is a highly conserved and specialized T7SS in *M. tb* [9,10]. It is indispensable for bacterial virulence, driving host-pathogen interactions, immune modulation, and the subversion of host defenses. The genetic basis of ESX-1 lies within the RD1 locus, which is present in virulent *M. tb* strains but absent in attenuated strains, including the *M. bovis* BCG vaccine strain. Spanning approximately 9.5 kilobases, RD1 encodes nine essential genes required for the structural and functional integrity of ESX-1. The loss of RD1 in attenuated mycobacteria implicates its critical role in virulence and was instrumental in the development of live attenuated TB vaccines [11–13].

The RD1 locus encodes both structural and effector proteins that form the secretion machinery. These include Rv3871 (eccCb1), Rv3872 (phoP), Rv3873 (phoR), Rv3874 (pknF), Rv3875 (esxA/ESAT-6), Rv3876 (esxB/CFP-10), Rv3877 (eccD1), Rv3878 (eccE1), and Rv3879c (espA) [14]. Together, these proteins constitute a membrane-embedded, multi-subunit complex adapted to the unique lipid-rich cell envelope of mycobacteria. The ESX-1 machinery comprises three major elements: (i) an inner membrane complex, (ii) cytoplasmic components, and (iii) effector substrates. These components act in concert to assemble a highly organized secretion system that exports virulence factors to the host cell environment [5].

2.1 Inner membrane complex proteins

The structural foundation of ESX-1 lies in its inner membrane complex, composed of four conserved ESX components (ECC): EccB1, EccC1, EccD1, and EccE1. Together, these proteins span the inner membrane and assemble into a channel that forms the core of the secretion apparatus (Fig. 1.1). This inner membrane complex not only anchors ESX-1 within the mycobacterial cell envelope but also provides the structural framework required for translocation of effector proteins [15].

2.2 EccB1: the structural scaffold of the ESX-1 system

EccB1, encoded by Rv3869, is the first identified ECC protein and functions as a structural scaffold of the ESX-1 system. It is localized primarily in the periplasm, where it provides an anchoring framework for the secretion machinery [16]. Structurally, EccB1 contains two domains, a central domain and a periplasmic domain, distinguished by four repeat regions that form a quasi-two-fold symmetry. This arrangement suggests a stabilizing role, ensuring that the ESX-1 complex remains securely anchored in the periplasmic compartment.

EccB1 also contains an ATP-binding site, indicating that it may assemble into a hexamer and function as an ATPase. Such activity is a common feature of periplasmic proteins involved in transport and is thought to provide the energy required for ESX-1-mediated substrate translocation [17].

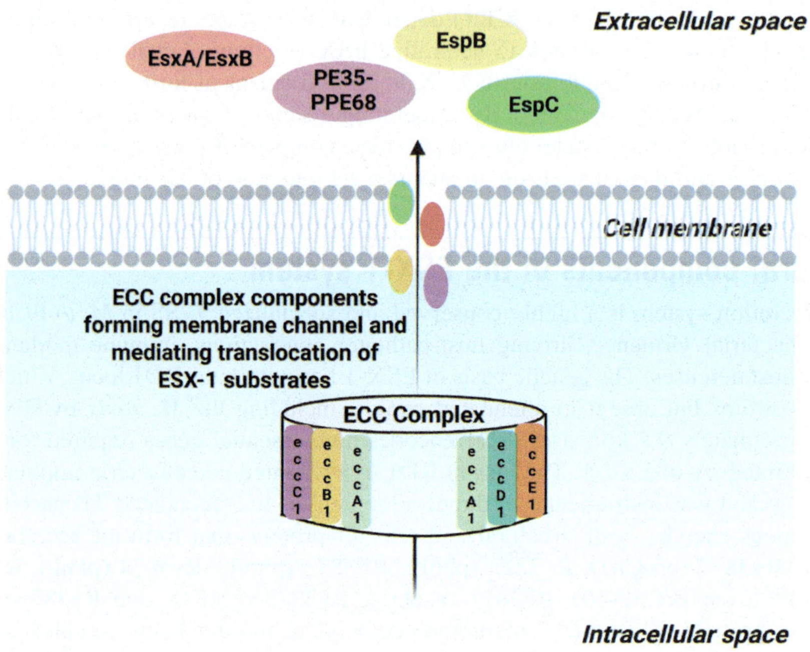

FIGURE 1.1 Schematic representation of the ESX-1 secretion machinery in *Mycobacterium tuberculosis*.

The membrane-integrated ECC complex (EccB1, EccC1, EccD1, EccA1, and EccE1) forms a channel for ESX-1 substrate translocation. Secreted proteins such as EsxA/EsxB, PE35-PPE68, EspB, and EspC participate in virulence through phagosomal membrane disruption and host immune modulation.

Crystallographic analyses have further revealed a flexible, membrane-distal domain (A2) with oscillatory movements. This dynamic feature is proposed to coordinate substrate transport, facilitating the release of effectors through the ESX-1 channel. By coupling structural anchoring with ATP-driven conformational dynamics, EccB1 serves as a central player in ESX-1 function. Its dual role in stabilizing the secretion system and energizing substrate translocation reveals its importance in sustaining the virulence of *M. tb* [18].

2.3 EccC1: the ATP-driven translocase of the ESX-1 system

EccC1, the second ESX-conserved component, is composed of two subunits, EccCa1 (Rv3870) and EccCb1 (Rv3871). It functions as an ATP-dependent translocase that powers the secretion of major virulence determinants, most notably the EsxA/EsxB heterodimer, across the mycobacterial envelope. This export is indispensable for *M. tb* to modulate host immunity and establish infection. Structurally, EccC1 contains three ATPase domains (D1, D2, and D3) belonging to the FtsK/SpoIIIE family, arranged sequentially along the C-terminal region. Among them, the D3 domain is critical for substrate recognition, specifically engaging the C-terminal region of EsxB, a key step in the secretion of the

EsxA/EsxB complex [19–21]. Structural studies reveal that EccC1 assembles into a hexameric ring, an oligomeric state essential for energy-dependent translocation and effector release [22]. Through this arrangement, EccC1 serves as the motor of ESX-1, coupling ATP hydrolysis with substrate recognition and export. Its central role demonstrates the dependence of *M. tb* virulence on a finely tuned, energy-driven secretion process.

2.4 EccD1: the transmembrane conduit of the ESX-1 system

EccD1, encoded by Rv3877, is a polytopic transmembrane protein that anchors ESX-1 to the mycobacterial inner membrane. It contains an ~110-amino acid N-terminal ubiquitin (Ub)-like domain, followed by a 30-amino acid linker and 11 predicted C-terminal α-helical transmembrane segments of the YukD family. The N-terminal domain projects into the cytoplasm, while the helices span the inner membrane, forming a barrel-like structure that enables the passage of ESX-1 substrates such as EsxA and EsxB into the periplasm [16].

The Ub-like domain dimerizes in a head-to-tail configuration, creating transmembrane pores. The dimer surface features a negatively charged groove (Asp49, Glu45, Glu60, Glu57, Asp61, and Asp50), a unique electrostatic property of ESX-1 that may attract positively charged secretion partners or regulate pore gating [23]. Comparative genomics highlights the importance of EccD1: it is conserved across pathogenic mycobacteria, such as *M. marinum*, but disrupted in attenuated strains like *M. bovis* BCG due to RD1 deletions [24,25]. Loss of eccD1 abolishes ESX-1 secretion and severely compromises bacterial virulence [26,27]. Thus, EccD1 acts as the transmembrane conduit of ESX-1, forming the essential gateway for substrate transport while preserving the stability of the secretion complex.

2.5 EccE1: the membrane anchor of the ESX-1 system

EccE1, encoded by Rv3882c, is predicted to contain 12 transmembrane helices that integrate into the inner membrane, stabilizing the ESX-1 secretion complex [28]. Its polar localization reflects the spatial organization of ESX-1 and is coordinated with other components to ensure correct assembly of the machinery. Loss of eccE1 destabilizes the ESX-1 apparatus, leading to reduced levels of associated proteins including EccB1, EccCa1, and EccD1. This disruption markedly impairs the secretion of substrates such as EsxA, EsxB, EspA, and EspC, which then accumulate in the cytoplasm rather than being exported. Consequently, phagosomal rupture activity is diminished, and virulence is severely attenuated in both in vitro and in vivo models. Interestingly, EspB secretion remains unaffected in eccE1 mutants, suggesting that EccE1 selectively governs the export of specific substrates [23]. Collectively, these findings establish EccE1 as the membrane anchor and stabilizer of ESX-1, ensuring complex integrity and efficient secretion of virulence factors required for pathogenesis.

2.6 Cytoplasmic ATPase—EccA1

EccA1, encoded by Rv3868, is a cytosolic AAA + ATPase of the CbxX/CfqX family, essential for substrate recognition and secretion through ESX-1 [29,30]. Structurally, EccA1 exhibits a two-domain architecture: an N-terminal domain with six tetratricopeptide repeats (TPRs) and a C-terminal ATPase domain. The TPRs are predicted to mediate substrate recognition, although this process deviates from classical TPR-ligand interactions due to a β-finger insertion within the binding groove [29,30]. The C-terminal domain constitutes the catalytic core, harboring conserved motifs characteristic of AAA - + ATPases, Walker A, Walker B, and the arginine finger (Arg-429). These motifs facilitate ATP binding and hydrolysis, driving conformational changes that promote substrate delivery to the membrane-associated ESX-1 complex [30]. This domain also mediates oligomerization and enzymatic activity, contributing to the thermostability of EccA1 and supporting bacterial survival under host-induced stress [29] (Fig. 1.1).

Beyond its direct role in ESX-1 secretion, EccA1 influences mycolic acid biosynthesis through interactions with lipid synthesis enzymes including Pks13, KasA, and KasB. This regulatory function is strictly ATPase-dependent. Catalytically impaired mutants, such as E398Q, fail to restore EsxA secretion, hemolysis, and mycolic acid synthesis, leading to attenuated intracellular survival and reduced virulence in macrophages and zebrafish models [31]. Thus, EccA1 links secretion dynamics with lipid metabolism, reinforcing its role as a multifunctional regulator of *M. tb* pathogenesis.

3. Regulatory components of the ESX-1 system

3.1 EspR

EspR, encoded by Rv3849, is a dimeric DNA-binding protein that functions as a nucleoid-associated transcriptional regulator and a master activator of ESX-1 in *M. tb.* Each monomer consists of an N-terminal helix-turn-helix (HTH) motif responsible for DNA recognition, and a C-terminal domain required for dimerization [32–34]. EspR positively regulates the espACD operon (Rv3616c–Rv3614c), which encodes EspA, EspC, and EspD, proteins critical for ESX-1 function. Early reports suggested that EspR might itself be secreted by ESX-1, creating a negative feedback loop. However, subsequent studies demonstrated no detectable secretion in vitro, indicating that its principal role is intracellular transcriptional regulation [35]. EspR expression is controlled by the two-component systems MprAB and PhoPR, which directly activate the espR promoter. Mutations in the MprA-binding site significantly reduce espR transcription, leading to decreased EspA and EspD mRNA and diminished EspR and EspA protein abundance [36]. This regulatory hierarchy highlights EspR as a key node connecting two-component signaling (PhoPR/MprAB) to ESX-1 activation. In summary, EspR acts as a transcriptional activator that links environmental sensing to the regulation of virulence-associated secretion.

3.2 EspN

EspN, a putative HTH transcription factor, was recently identified in *M. marinum* and is conserved in the *M. tb* complex (MTBC) [37]. EspN expression is repressed under non-infectious conditions by the transcriptional repressor EspM, but during infection, EspN becomes essential for ESX-1 function. Functionally, EspN enhances ESX-1 activity by activating *whiB6* and several ESX-1-associated genes, including *espE*, *eccA1*, and other secretion components. This creates a positive feedback loop that amplifies secretion activity during host infection [37]. The importance of EspN is shown by mutational studies: *M. marinum* strains lacking EspN exhibit reduced hemolytic activity, defective macrophage lysis, impaired intracellular replication, and diminished bacterial burden in zebrafish infection models. These phenotypes closely parallel those of ESX-1-deficient mutants, highlighting EspN as an essential infection-induced regulator [38].

4. ESX-1-mediated molecular mechanisms of virulence

4.1 Phagosome escape and cytosolic access

A hallmark of *M. tb* virulence is its ability to breach the phagosomal membrane and translocate into the host cytosol. This capacity is largely orchestrated by the ESX-1 secretion system and its repertoire of effectors, which collectively destabilize the phagosomal compartment, promote bacterial escape, and initiate a complex cascade of host immune responses. Among these, the heterodimeric complex formed by EsxA and EsxB is the most thoroughly characterized, while accessory proteins such as EspA, EspB, EspC, and the PE/PPE family function as facilitators that reinforce secretion and virulence. Together, these factors define the unique ability of virulent strains of *M. tb* to persist and replicate within macrophages.

4.2 EsxA and EsxB: the central drivers of phagosomal rupture

EsxA-6, encoded by *Rv3875*, is a 95-amino-acid protein that forms a stable 1:1 heterodimer with EsxB, encoded by *Rv3874*, and the two represent the most critical virulence determinants of the ESX-1 system. Structurally, EsxA possesses a central HTH motif flanked by flexible N- and C-terminal arms. The two major helices, each approximately 50 Å long, are of a size consistent with bilayer thickness, making them ideally suited for insertion into host membranes [39].

A defining property of EsxA is its pH-dependent activity. Under acidic conditions, EsxA undergoes Nα-acetylation at Thr-2, which facilitates its dissociation from EsxA and triggers its conformational activation [40–42]. Acidification also enhances the α-helicity of its C-terminal region, allowing EsxA to insert into phospholipid membranes. Once inserted, EsxA oligomerizes into higher-order structures that perforate host membranes, compromising phagosomal integrity and permitting bacterial escape into the cytosol [43] (Fig. 1.2). Notably, this pore-forming property is unique to EsxA; EsxB itself lacks cytolytic activity but stabilizes EsxA during folding, secretion, and translocation [44]. The functional importance of EsxA has been repeatedly validated in infection models. Wild-type *M. tb* induces strong recruitment of galectin-3 (GAL-3) to damaged phagosomal membranes, a marker of membrane rupture, whereas ESX-1-deficient mutants fail to elicit comparable GAL-3 accumulation [45]. This demonstrates EsxA as the principal effector of phagosomal rupture.

Although non-cytolytic, EsxB plays an equally indispensable role. Acting as a molecular chaperone, it forms a tight heterodimer with EsxA that is specifically recognized by the ESX-1 machinery. Deletion of *esxB* abolishes EsxA secretion, producing phenotypes indistinguishable from ESX-1-null mutants, including impaired phagosomal rupture and defective hemolysis. Thus, EsxB safeguards the conformational stability and secretion competence of EsxA, functioning as the essential cofactor that empowers its pore-forming activity. Together, EsxA and EsxB constitute the primary executors of phagosomal escape and the central axis of ESX-1-dependent virulence.

4.3 Accessory ESX-1 effectors reinforcing virulence

While EsxA and EsxB are the direct mediators of membrane disruption, additional ESX-1-associated proteins play essential supporting roles by ensuring efficient secretion, stabilizing effector function, and modulating host cell processes. EspA, encoded by *Rv3616c* within the *espACD* operon adjacent to

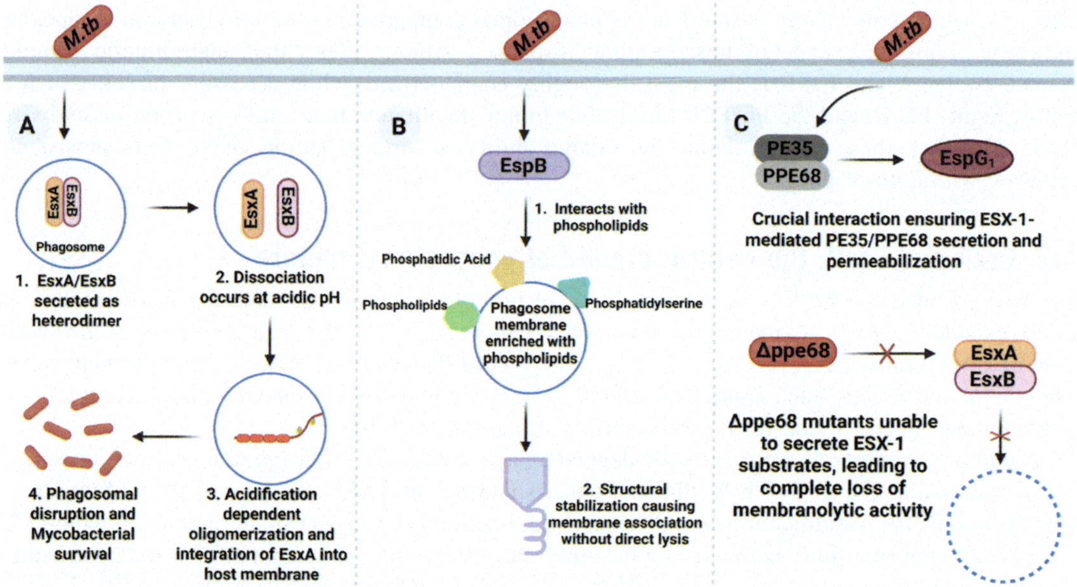

FIGURE 1.2 Representation of the functional components of the ESX-1 system that contribute to phagosomal disruption.

(A) EsxA/EsxB are secreted as a heterodimer, dissociate upon exposure to acidic pH, and EsxA inserts into the host membrane to mediate phagosomal rupture. (B) EspB phospholipid binding stabilizes membrane binding without direct lysis. (C) PE35/PPE68 interaction is necessary for the secretion of substrate. Deletion of PPE68 eliminates secretion of EsxA/EsxB, eliminating membranolytic activity.

the RD1 locus, is indispensable for secretion of EsxA and EsxB. EspA is predicted to form a flexible, four-helix bundle-like structure and can form disulfide-linked dimers via conserved cysteine residues. Although disruption of these bonds does not entirely abolish ESX-1 activity, it significantly reduces virulence, highlighting EspA's unique contribution. Mutations such as W55R, G57R, F50R, and K62A impair EsxA/B secretion, leading to defective phagosomal rupture and attenuated infection [46]. By coupling substrate export with bacterial survival, EspA serves as a secretion facilitator tightly linked to mycobacterial persistence.

EspB (*Rv3881c*) represents another key accessory effector. After secretion, EspB undergoes C-terminal cleavage by the protease MycP1, a step that promotes its oligomerization into pH-dependent heptameric ring structures. These oligomers are large enough to accommodate ESX-1 substrates, suggesting that EspB may function as a scaffold for secretion or a mediator of host-membrane interactions. EspB binds anionic phospholipids such as phosphatidic acid and phosphatidylserine, enriched in phagosomal and mitochondrial membranes, thereby anchoring itself to host organelles. While EspB does not directly lyse membranes, it interferes with phagosomal maturation and destabilizes mitochondrial integrity, facilitating intracellular persistence. Importantly, EspB contains both PE and PPE domains fused into a single polypeptide and achieves full functionality only after secretion, representing a novel paradigm of secretion-dependent activation among ESX-1 effectors [47–49].

EspC (*Rv3615c*) forms an essential complex with EspA, reinforcing ESX-1-mediated secretion. A remarkable feature of EspC is its ability to self-assemble into filamentous surface structures, potentially forming conduits for the export of EsxA and EsxB. Mutational disruption of its C-terminal residues destabilizes these filaments and compromises EsxA/B secretion [50–52]. Strains lacking EspC show impaired phagosomal rupture and attenuated virulence, confirming its indirect but indispensable contribution to ESX-1-driven pathogenesis.

Among the PE/PPE proteins, the PE35-PPE68 heterodimer is the most prominent ESX-1-associated pair. Encoded upstream of *esxB* and *esxA* within RD1, this complex interacts specifically with the chaperone EspG1, which stabilizes the heterodimer and ensures its secretion. This partnership is highly selective, as EspG1 does not interact with PE/PPE proteins from other ESX systems. The PE35-PPE68 complex is essential for EsxA secretion, and loss of PPE68 in *M. marinum* completely abolishes EsxA/B export and hemolytic activity. Complementation with the *ppe68-pe35* operon restores both secretion and virulence, highlighting the central role of this heterodimer in ESX-1–dependent membrane permeabilization [53–55]. Collectively, these accessory proteins form a supportive network that amplifies the activity of EsxA/EsxB, stabilizes secretion, and diversifies the impact of ESX-1 on host cell physiology (Fig. 1.2).

4.4 Mechanisms of phagosomal rupture

The outcome of these coordinated activities is the disruption of phagosomal integrity and the liberation of bacilli into the cytosol. Soon after phagocytosis, virulent *M. tb* strains begin damaging phagosomal membranes, whereas ESX-1-deficient strains such as *M. bovis* BCG remain trapped within intact phagolysosomes, where they are efficiently neutralized [56]. Recent studies have illuminated two complementary models of EsxA-mediated disruption. In one model, EsxA inserts into host membranes and induces pore formation. Beyond simple pores, EsxA also drives vesiculation and tubulation, remodeling lipid bilayers into buds and tubular extensions. Over 24 hours, these vesicles undergo fission coinciding with EsxA fibrillation into polymeric networks, which exert radial and tangential forces that destabilize membranes. This fibril-mediated vesiculation explains the time-dependent kinetics of *M. tb* escape, typically observed 1–2 days post-infection [6]. An alternative but not mutually exclusive contact-dependent model suggests that ESX-1 mediates direct tearing of membranes at the bacteria-host interface rather than forming discrete pores [57]. This mechanism highlights ESX-1 as a specialized secretion apparatus that delivers effectors with spatial precision at the point of bacterial engagement.

4.5 Biological consequences of phagosomal escape

The ability of *M. tb* to rupture phagosomal membranes confers multiple advantages. First, it allows evasion of lysosomal degradation, shielding the bacillus from hydrolytic enzymes and antimicrobial peptides. Second, access to the nutrient-rich cytosol provides a more favorable replicative niche. Third, cytosolic escape facilitates cell-to-cell spread, either through lysis of infected macrophages or by exploiting host membranes in a "Trojan horse" strategy for dissemination. However, this strategy also carries risks for the pathogen. Phagosomal rupture exposes bacterial DNA, RNA, and other pathogen-associated molecular patterns (PAMPs) to cytosolic sensors such as cGAS, STING, and NLRs. These surveillance pathways initiate potent innate immune responses, including type I IFN induction, inflammasome activation, and pyroptosis [58]. Thus, ESX-1-mediated escape represents a double-edged sword: while it promotes bacterial survival and replication, it also alerts host immunity and shapes the inflammatory landscape of TB.

4.6 Cytosolic DNA/RNA sensing and type I IFN induction

A central consequence of ESX-1-mediated phagosomal rupture is the leakage of microbial nucleic acids into the cytosol, which activates host nucleic acid-sensing pathways. Among these, the cGAS-STING axis is the most prominent. Under basal conditions, the cytosol is virtually devoid of DNA, making its appearance a danger signal indicative of infection or catastrophic stress. Following phagosomal damage, *M. tb* DNA escapes into the cytosol, where it is recognized by cGAS. Upon binding double-stranded DNA, cGAS generates the second messenger cyclic GMP-AMP (cGAMP), which activates STING at the endoplasmic reticulum. STING in turn recruits and activates TANK binding kinase 1 (TBK1), leading to interferon-regulatory factor 3 (IRF3) phosphorylation and transcriptional induction of type I IFNs, notably IFN-β [58].

Genetic and biochemical studies have validated this pathway in TB. In macrophages infected with *M. tb*, deletion of cGAS or STING markedly reduces IFN-β production [58,59]. cGAS-deficient macrophages completely fail to mount type I IFN responses, and direct physical binding of cGAS to *M. tb*-derived DNA confirms the bacterial genome as the ligand [58]. Importantly, ESX-1 is indispensable for this process: mutants lacking ESX-1 do not elicit robust type I IFN responses because DNA remains confined within intact phagosomes [58]. These findings establish phagosomal rupture as the essential gateway that exposes microbial DNA to cytosolic surveillance (Fig. 1.3).

Interestingly, cytosolic DNA detection is not unique to *M. tb*. Similar mechanisms are employed during infections with *Legionella pneumophila*, suggesting that DNA sensing is a conserved immune strategy against vacuolar pathogens. However, the outcome in TB is more complex, reflecting a delicate balance between protective and pathological roles of cGAS-STING signaling. On the beneficial side, cGAS-STING activation not only drives IFN-β production but also promotes autophagy. Cytosolic DNA sensing marks damaged phagosomes for autophagic clearance, an intrinsic defense pathway that restricts bacterial growth. Consistently, cGAS-deficient cells exhibit impaired autophagy and uncontrolled *M. tb* proliferation [58]. This illustrates how nucleic acid sensing simultaneously triggers antimicrobial effector mechanisms and inflammatory signaling.

However, excessive type I IFN production can be detrimental. Clinical transcriptomic studies reveal a strong IFN signature in TB patients, correlating with disease severity. Experimental models confirm this: in mice, sustained type I IFN production increases susceptibility to *M. tb* by antagonizing IL-1, a cytokine indispensable for protective immunity [60]. Mechanistically, type I IFN induces IL-1 receptor antagonist (IL-1Ra), which neutralizes IL-1 signaling. Strikingly, genetic ablation of type I IFN signaling or partial reduction of IL-1Ra restores host resistance, demonstrating how overactive IFN responses compromise protective immunity [61]. Thus, the cGAS-STING axis embodies a

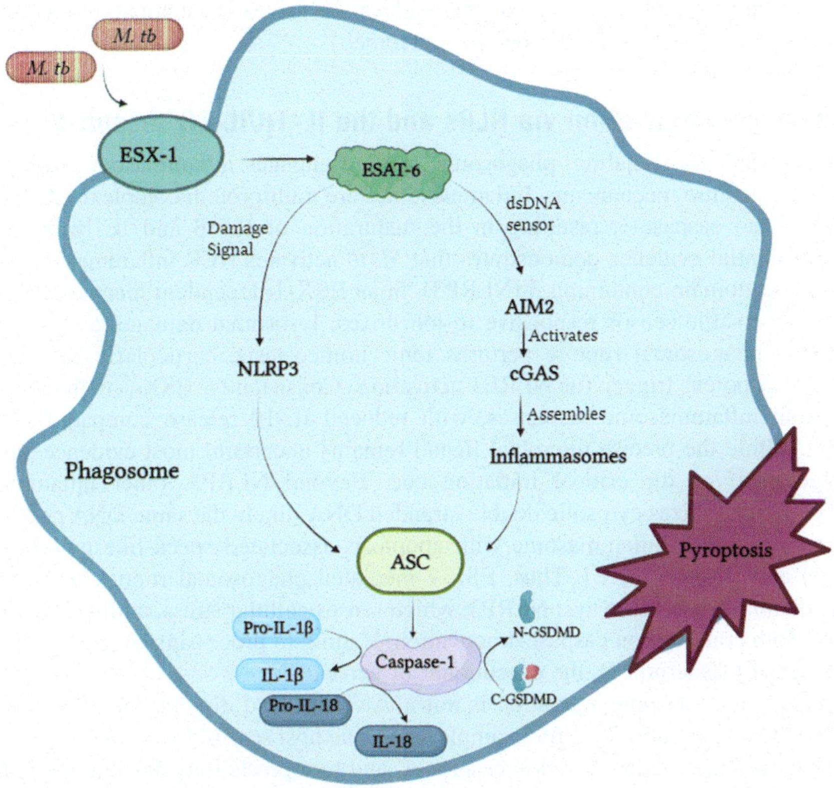

FIGURE 1.3 Diagrammatic representation of esx-1-mediated phagosomal rupture. The bacterial DNA escapes into the cytosol, where it is sensed by AIM2 and cGAS.

NLRP3, on the other hand, senses the stress and damage signal. These sensors initiate the assembly of inflammasomes, leading to recruitment of adaptor protein ASC, which in turn leads to caspase-1 activation, cytokine maturation (IL-1β, IL-18), and cleavage of the N-terminal domain of Gasdermin D, leading to pyroptotic cell death, a defensive yet inflammatory response.

double-edged sword: moderate activation promotes autophagy and defense, whereas hyperactivation exacerbates pathology by suppressing IL-1-mediated immunity.

Beyond DNA, ESX-1 activity also facilitates the translocation of bacterial RNA into the cytosol. This occurs via both the ESX-1 secretion system and the SecA2 export pathway. Once in the cytosol, bacterial RNA is sensed by RIG-I–like receptors (RIG-I and MDA5), which signal through the mitochondrial adaptor mitochondrial antiviral signaling protein (MAVS) to amplify type I IFN responses. This RNA-sensing pathway synergizes with DNA sensing to intensify IFN production. Importantly, mice deficient in MAVS or IRF3 exhibit reduced pathology, highlighting the role of excessive nucleic acid sensing in TB immunopathogenesis [62]. In summary, cytosolic nucleic acid sensing is a cornerstone of ESX-1-driven host-pathogen interaction. By hijacking antiviral surveillance pathways, *M. tb* forces the host to mount an IFN-dominated response. Moderate signaling promotes bacterial restriction via autophagy, but excessive activation drives immunopathology, illustrating the precarious balance that determines TB disease outcome.

4.7 Inflammasome activation via NLRs and the IL-1β/IL-18 response

Parallel to cGAS-STING signaling, phagosomal rupture engages inflammasome pathways, another class of cytosolic defense mechanisms. Inflammasomes are multiprotein complexes that detect danger signals and activate caspase-1, resulting in the maturation of IL-1β and IL-18 and induction of pyroptosis. Substantial evidence demonstrates that *M. tb* activates NLR inflammasomes, particularly NLR family pyrin domain containing 3 (NLRP3), in an ESX-1-dependent manner [63].

NLRP3 is a versatile sensor responsive to ion fluxes, lysosomal damage, and metabolic stress. ESX-1-mediated phagosomal rupture perturbs ionic homeostasis, particularly through K^+ efflux, which serves as a potent trigger for NLRP3 activation. Consistently, ESX-1-deficient *M. tb* strains elicit only weak inflammasome responses, with reduced IL-1β release compared with wild-type infection [63]. While the precise microbial ligand remains uncertain, most evidence supports membrane damage itself as the critical initiating cue. Beyond NLRP3, other inflammasomes also contribute. AIM2 recognizes cytosolic double-stranded DNA, likely the same DNA pool that activates cGAS, and assembles an inflammasome with apoptosis-associated speck-like protein containing a CARD (ASC) and caspase-1 [63]. Thus, ESX-1-mediated phagosomal rupture activates two complementary inflammasome pathways: NLRP3, which senses cellular stress, and AIM2, which directly detects DNA. Both converge on caspase-1 activation, leading to processing of pro–IL-1β and pro–IL-18 and cleavage of Gasdermin D, the executioner of pyroptosis.

Interestingly, inflammasome activation is not always initiated directly by ESX-1 effectors. The ESX-1–independent protein EST12, for example, binds the host scaffold protein receptor for activated C kinase 1 (RACK1), activating NLRP3, caspase-1, and pyroptosis [64]. Intriguingly, *M. tb* mutants lacking EST12 exhibit increased virulence, suggesting that inflammasome-driven IL-1β production restricts bacterial growth and contributes to host protection [65]. From the host perspective, IL-1β is a central protective cytokine in TB. It enhances phagolysosomal fusion, counterbalances excessive type I IFN signaling, and helps organize granulomas. Experimental models reinforce this: mice deficient in IL-1 receptor, caspase-1, or NLRP3 exhibit uncontrolled bacterial growth and increased susceptibility [66]. Human patients with impaired IL-1 signaling similarly develop severe TB. IL-18, though less studied, is also important, as it promotes IFN-γ production by NK and T cells, reinforcing the Th1 response essential for TB control.

However, inflammasome activity is also a double-edged sword. Excessive IL-1β release and pyroptosis contribute to necrosis within granulomas, caseation, and cavity formation, processes that damage host tissue while facilitating bacterial transmission. *M. tb* has therefore evolved mechanisms to temper inflammasome activity. For instance, the secreted phosphatase PtpB interferes with inflammasome signaling by dephosphorylating phosphoinositide's, preventing Gasdermin D pore formation and pyroptosis [67]. Remarkably, macrophages infected with PtpB-expressing bacilli still activate caspase-1 and process IL-1β, but cytokine release occurs via Gasdermin D–independent mechanisms. Similarly, ESX-3-secreted EsxH targets the ESCRT repair machinery, disrupting antigen presentation and altering immune responses [68]. Although not a direct inflammasome inhibitor, EsxH exemplifies the broader *M. tb* strategy of manipulating membrane damage responses to its advantage. In sum, inflammasome activation is a critical checkpoint in TB immunity. By detecting phagosomal rupture,

NLRP3 and AIM2 trigger protective IL-1β and IL-18 responses, but dysregulated activity drives pathology. *M. tb*, in turn, fine-tunes inflammasome outcomes through effectors that prevent excessive host cell death while sustaining inflammation conducive to transmission.

4.8 Pyroptosis induction following ESX-1-mediated damage

Pyroptosis is a lytic and highly inflammatory form of programmed cell death triggered downstream of inflammasome activation. Unlike apoptosis, which proceeds silently, pyroptosis releases cellular contents, proinflammatory cytokines, and viable bacteria into the extracellular environment. In *M. tb*-infected macrophages, inflammasome activation results in caspase-1-mediated cleavage of Gasdermin D. The liberated N-terminal fragment inserts into the plasma membrane, oligomerizes, and forms large pores that disrupt ionic gradients, induce osmotic swelling, and culminate in cell rupture [63]. This process serves dual purposes. Early during infection, pyroptosis benefits the host by eliminating heavily infected macrophages, depriving *M. tb* of intracellular niches, and exposing bacilli to extracellular immune defenses. It also promotes granuloma formation by releasing chemotactic mediators that recruit additional immune cells. Consistently, *M. tb* mutants defective in inducing pyroptosis are more virulent in mice, suggesting the protective role of this pathway [65].

Yet, as infection progresses, excessive pyroptosis becomes pathological. Chronic and unregulated macrophage lysis contributes to necrotic granulomas, caseation, and cavitation, processes that destroy lung tissue but also enhance bacterial spread. Necrotic lesions in TB often display molecular signatures of pyroptosis and necroptosis rather than apoptosis [69]. In this setting, *M. tb* exploits pyroptosis-driven necrosis: host lipids and debris provide nutrients, and tissue destruction opens routes for transmission. To fine-tune this balance, *M. tb* actively modulates pyroptosis. PtpB inhibits Gasdermin D pore formation, delaying macrophage lysis and prolonging the intracellular niche [63,64]. Similarly, EsxH interferes with ESCRT-mediated membrane repair, skewing outcomes toward sustained inflammation [68]. These strategies reflect a bacterial calculus: suppress pyroptosis when it threatens replication yet exploit it when tissue destruction favors dissemination.

The extent of pyroptosis varies across strains and hosts. Virulent *M. tb* typically induces moderate inflammasome activity, whereas *M. marinum* and hypervirulent *M. tb* strains cause extensive cell death. Host factors, such as ESCRT repair efficiency, further influence this outcome. When repair succeeds, NLRP3 activation is dampened, but when *M. tb* blocks ESCRT, danger signals persist, and pyroptosis is prolonged. In summary, pyroptosis is both protector and executioner in TB. Initiated by ESX-1-mediated phagosomal damage, it constrains bacterial growth early but contributes to immunopathology and transmission when excessive. Effective TB control hinges on balancing this pathway, sufficient to eliminate infected cells yet restrained to prevent catastrophic tissue necrosis.

5. Host immune responses and counteractive mechanisms

The ESX-1 T7SS of *M. tb* represents one of the most intensively studied virulence determinants. By perforating phagosomal membranes, ESX-1 allows bacterial DNA, RNA, and effector proteins to access the host cytosol, thereby activating a broad spectrum of innate immune pathways. These responses, though initially protective, often yield paradoxical consequences: while they can restrict bacterial growth, they also drive inflammation, tissue destruction, and disease progression. The following sections synthesize current understanding of the major host pathways counteracting ESX-1-mediated damage, highlighting both their defensive capacity and their exploitation by *M. tb*.

5.1 Galectin-8 sensing and xenophagy

Galectins (GALs), a family of β-galactoside-binding lectins, have long been recognized for extracellular functions in adhesion and signaling [70–72]. However, in the cytosol they act as critical sentinels of membrane integrity. When endosomes or phagosomes rupture, luminal glycans normally hidden from the cytoplasm become exposed. GALs bind these glycans and thereby flag damaged compartments for immune recognition [73–76]. A variety of intracellular pathogens, including *Salmonella enterica*, *Shigella flexneri*, *Listeria monocytogenes*, *Legionella pneumophila*, and *Yersinia pseudotuberculosis*, trigger GAL accumulation at vacuolar lesions, suggesting their conserved role as sensors of pathogen-induced membrane damage [75–78]. During *M. tb* infection, GAL-3, 8, and 9 are recruited to mycobacteria-containing vacuoles (MCVs). Yet for years their contribution remained poorly defined, leading many to regard them merely as markers of phagosomal rupture rather than active effectors [56,57,79–83].

Recent work has clarified this picture by establishing GAL-8 as an indispensable mediator of selective autophagy (xenophagy) during TB. In macrophages, GAL-8 recruitment occurs at ESX-1-dependent lesions. CRISPR/Cas9 knockout studies revealed that loss of GAL-8, but not GAL-3 or GAL-9, markedly reduces recruitment of ubiquitin, sequestosome 1 (SQSTM1/p62), and microtubule-associated protein 1 light chain 3 (LC3) to damaged MCVs. This impairment severely compromises xenophagic clearance and permits enhanced bacterial survival [45]. Conversely, GAL-8 overexpression amplifies LC3 recruitment and restricts intracellular growth, implicating its non-redundant role. Mechanistically, GAL-8 directly binds the selective autophagy adaptor Tax1 binding protein 1 (TAX1BP1), which in turn recruits autophagic membranes. Importantly, TAX1BP1 engagement is specific to GAL-8; neither GAL-3 nor GAL-9 can substitute. Moreover, recruitment of TAX1BP1 requires both GAL-8 and functional ESX-1, firmly linking phagosomal rupture to autophagic defense (Fig. 1.4).

While GAL-9 contributes indirectly by signaling through the T-cell receptor Tim-3 to enhance IL-1β production [84,85], and GAL-3 marks damaged vacuoles without driving clearance, it is GAL-8 that provides the molecular bridge between lesion detection and xenophagic response. This discovery addresses a long-standing gap in the field and positions GAL-8 as a central checkpoint in host resistance. Yet, *M. tb* often circumvents xenophagy. By interfering with ubiquitination and autophagy initiation, ESX-1 effectors reduce the efficiency of GAL-8, dependent clearance [86,87]. The outcome is a precarious balance: sufficient GAL-8 activity constrains bacilli, but pathogen-mediated interference frequently tips the scales toward persistence.

5.2 ESCRT-mediated phagosomal repair

Alongside xenophagy, the ESCRT system constitutes another critical line of defense against ESX-1 damage. ESCRT proteins are typically involved in multivesicular body formation and membrane scission. Yet at disrupted phagosomes, ESCRT-III components, including tumor susceptibility 101

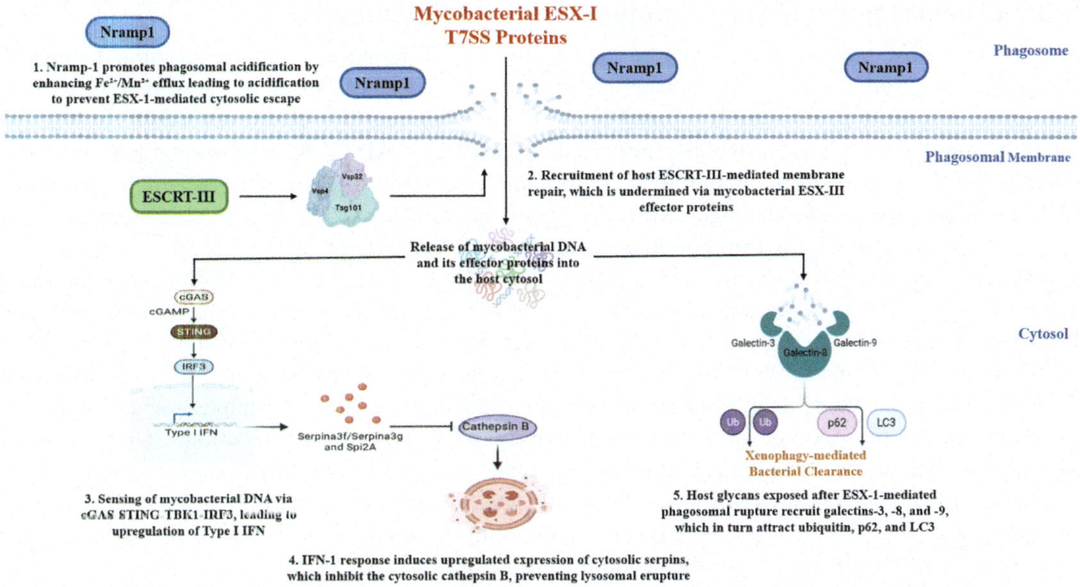

FIGURE 1.4 Systematic representation of multilayered defenses employed by the host against *M. tb*.

ESX-1-mediated phagosomal disruption, including ESCRT-III-driven membrane repair (antagonized by ESX-3 effectors), and cGAS-STING-IRF3 signaling that induces type I IFN and cytosolic serpins to inhibit cathepsin B-mediated cell death. GAL-8 senses glycan exposure upon membrane rupture, recruiting TAX1BP1 and autophagy machinery (Ub, p62, LC3) to facilitate xenophagic clearance. Concurrently, Nramp-1-mediated phagosomal acidification limits ESX-1-dependent membrane permeabilization.

(Tsg101), charged multivesicular body protein 4 (CHMP4), and vacuolar protein sorting 4 homolog A (Vps4), rapidly assemble at sites of rupture. Their coordinated activity reseals membranes, prevents cytosolic leakage of bacterial products, and restores compartmental integrity [88]. High-resolution live imaging shows that ESCRT recruitment occurs within minutes of membrane damage, providing a rapid containment mechanism. Autophagic membranes are often recruited concurrently, either to patch lesions or fully engulf compromised vacuoles, routing them toward lysosomal degradation. This dual system of ESCRT and autophagy illustrates the redundancy of host defenses against ESX-1 (Fig. 1.4).

However, *M. tb* deploys countermeasures. ESX-3-secreted effectors such as EsxG and EsxH directly disrupt ESCRT function by targeting receptor trafficking at the lysosome [56]. EsxH binds the ESCRT-0 component serine/arginine-rich splicing factor 5 (SRS5/HRS), impeding antigen presentation and vesicular repair. This interference prolongs phagosomal disruption, allowing bacilli to exploit cytosolic nutrients and avoid degradation. Thus, ESCRT-mediated repair exemplifies the host-pathogen arms race: host cells mobilize rapid resealing machinery, while *M. tb* effectors evolve to sabotage the same process. The balance between successful repair and bacterial interference decisively shapes infection outcome.

5.3 Inflammasome pathways (protective vs. pathological)

One of the most consequential host responses to ESX-1 damage is inflammasome activation. Inflammasomes are cytosolic multiprotein complexes that detect stress or microbial molecules, leading to caspase-1 activation, IL-1β and IL-18 maturation, and pyroptotic cell death. In TB, two inflammasome sensors are particularly important: NLRP3 and AIM2. NLRP3 responds to perturbations of cellular homeostasis, including K^+ efflux and lysosomal rupture, both of which accompany ESX-1-induced membrane damage. AIM2, by contrast, directly recognizes double-stranded DNA released into the cytosol, the same pool of bacterial DNA that activates cGAS [63].

Experimental knockouts (NLRP3$^{-/-}$, Casp1$^{-/-}$, GSDMD$^{-/-}$) confirm that ESX-1-mediated injury activates the NLRP3 inflammasome, leading to caspase-1 cleavage and Gasdermin D, mediated pore formation at the plasma membrane. This culminates in pyroptosis, a lytic form of cell death characterized by IL-1β and IL-18 release [79]. Importantly, this pathway does not require complete bacterial escape; mechanical perturbation of the plasma membrane alone suffices to trigger inflammasome assembly.

From a host perspective, inflammasome activity provides strong protection. IL-1β enhances phagolysosomal fusion, antagonizes maladaptive type I IFN signaling, and supports granuloma organization [66]. IL-18 promotes IFN-γ release by natural killer (NK) and T cells, reinforcing Th1 immunity. Indeed, mice deficient in inflammasome components or IL-1 signaling are highly susceptible to TB, and human genetic studies echo this vulnerability.

Yet inflammasomes also fuel pathology. Excessive IL-1β and pyroptosis drive necrosis within granulomas, caseation, and cavitation, hallmarks of advanced TB that facilitate bacterial transmission. This duality is compounded by pathogen strategies: ESX-1-independent effectors such as EST12 trigger inflammasome activation [64,65], while others like PtpB suppress Gasdermin D pore formation [67], effectively tuning host responses to bacterial advantage. Thus, inflammasomes represent a Janus-faced defense: essential for early control, yet complicit in tissue destruction when dysregulated. The central question remains how to harness protective inflammasome signaling without unleashing pathological inflammation.

5.4 Cytosolic serpins and lysosomal protease control

A less appreciated but crucial host defense involves cytosolic serpins, protease inhibitors induced by type I IFN signaling. While type I IFNs are often detrimental in TB by antagonizing IL-1, they also upregulate clade A3 serpins, notably Serpina3f and Serpina3g (Spi2A), which play protective roles [89–92]. Mechanistically, ESX-1 activity destabilizes lysosomes, releasing cathepsin B, a potent cysteine protease. Once in the cytosol, cathepsin B retains activity at neutral pH and drives host cell lytic death [93,94]. Spi2A binds and inhibits cytosolic cathepsin B, protecting infected macrophages from premature lysis [95]. This countermeasure ensures that macrophages remain viable long enough to orchestrate antimicrobial functions.

The protective roles of Spi2A extend beyond infection. By neutralizing cathepsin B, Spi2A regulates erythropoiesis, memory CD8$^+$ T cell survival, and the maintenance of granulocytes and B cell progenitors [96, 97–99]. In TB, however, the balance is delicate: while serpins defend against cathepsin-driven cytotoxicity, they are induced by the very type I IFN pathways that otherwise compromise immunity. This paradox underscores the complexity of host-directed therapy: targeting IFN signaling wholesale may suppress detrimental effects but inadvertently reduce beneficial serpin responses. Thus, cytosolic serpins exemplify a specialized adaptation of host immunity, where protease inhibition mitigates the collateral damage of lysosomal rupture, but at the cost of dependence on maladaptive IFN signaling.

5.5 Phagosomal acidification control

Finally, host cells constrain ESX-1 activity through phagosomal acidification. The natural resistance-associated macrophage protein 1 (Nramp-1), a proton-dependent divalent metal transporter, enhances acidification by effluxing Fe^{2+} and Mn^{2+} ions [100]. Acidic phagosomes stabilize membranes, activate hydrolytic enzymes, and limit ESX-1-mediated rupture. Experimental data demonstrate that Nramp-1–competent macrophages exhibit markedly reduced phagosomal escape compared with Nramp-1-deficient cells. Pharmacological inhibition of acidification with bafilomycin A1 reproduces the permissive phenotype, confirming acidic pH as a barrier to cytosolic access. Mechanistically, Nramp-1-driven ion transport not only deprives bacteria of essential metals but also establishes an environment hostile to ESX-1-mediated destabilization. These insights extend to in vivo infection. Using β-lactamase-based fluorescence resonance energy transfer (FRET) reporters, researchers have shown that *M. tb* accesses the cytosol of lung and spleen phagocytes during natural infection, challenging the long-held paradigm of TB as a vacuole-confined pathogen. Nevertheless, the frequency of cytosolic access is significantly shaped by Nramp-1 expression, reinforcing its role as a gatekeeper of niche selection.

The implications are profound. Reinforcing phagosomal acidification could serve as a host-directed therapeutic strategy, blocking cytosolic escape and limiting dissemination. Moreover, enhancing vacuolar integrity may improve vaccine efficacy by keeping antigens in compartments favorable for presentation. The ESX-1 secretion system epitomizes the evolutionary arms race between *M. tb* and its host. By rupturing phagosomes, *M. tb* gains cytosolic access and manipulates host signaling, but this same breach triggers a diverse array of countermeasures: GAL-8-mediated xenophagy, ESCRT-dependent repair, inflammasome signaling, serpin-mediated protease control, and Nramp-1, driven acidification. Each pathway offers protection, yet each is vulnerable to pathogen subversion or prone to pathological consequences when dysregulated. Understanding how these mechanisms intersect, when they succeed, when they fail, and how *M. tb* exploits their weaknesses, is central to designing host-directed therapies. Rather than targeting a single pathway, effective interventions will likely require balancing protective responses (autophagy, inflammasome priming, serpins, acidification) while curbing pathological inflammation and tissue damage. Only by mastering this balance can we tilt the arms race in favor of the host.

6. Therapeutic implications

The persistence of TB, especially in the context of multidrug-resistant (MDR) and extensively drug-resistant (XDR) strains, has stimulated urgent interest in novel therapeutic approaches. Beyond traditional bactericidal drugs, antivirulence strategies targeting *M. tb* pathogenesis mechanisms offer a promising alternative. The ESX-1 T7SS is one such target. ESX-1 is indispensable for phagosomal rupture, cytosolic access, and host immune subversion, largely mediated through the effectors EsxA and EsxB. By compromising phagosomal integrity, ESX-1 triggers downstream pathways including type I IFN induction, inflammasome activation, and pro-inflammatory cascades that paradoxically benefit bacterial persistence (Table 1.1).

Table 1.1 Therapeutic strategies targeting the ESX-1 system in *Mycobacterium tuberculosis*.

Targets	Mechanism of action	Therapeutic implications	References
EccB1 (*Rv3869*)	Periplasmic scaffold, ATPase activity, and hexameric assembly support substrate translocation.	Targeting the ATPase activity of EccB1 could impair the energy-dependent translocation of ESX-1 virulence factors, as EccB1 has been structurally characterized as a periplasmic ATPase likely forming a hexameric channel and essential for the secretion of virulence proteins.	[16–18]
EccD1 (*Rv3877*)	Transmembrane channel, Ub-like domain, forms a pore for EsxA/B.	Inhibiting the transmembrane channel function of ESX-1 components (e.g., EccD1) may prevent the secretion of virulence factors through the ESX-1 system and thereby attenuate bacterial virulence	[1,23–27]
EccE1 (*Rv3882c*)	The membrane anchor, stabilizes the ESX-1 complex.	Disruption of ESX-1-associated membrane components destabilizes the secretion machinery and impairs export of effectors such as EsxA/B and EspA/C, as seen in multiple mechanistic studies of ESX-1 assembly and secretion.	[4,28]
EccA1 (*Rv3868*)	Cytosolic ATPase; substrate recognition + ATP hydrolysis.	Inhibiting EccA1's ATPase activity (and potentially its substrate interactions via TPR domains) could block ESX-1-mediated secretion and disrupt associated virulence pathways, including mycolic acid synthesis, impairing intracellular survival.	[29–31]
EspR	Regulates the espACD operon, controlled by PhoPR and MprAB.	EspR is a transcriptional regulator of the espACD operon and is itself secreted by ESX-1; disrupting its regulatory feedback may reduce virulence through modulation of ESX-1 activation, supporting anti-virulence strategies.	[32–34]
EspN + EspM	Regulates ESX-1 genes (whiB6, espE, eccA), potential lipid sensing.	ESX-1 activation is regulated via two-component systems (PhoPR, MprAB) and involves EspR/EspA; disrupting these regulatory pathways could block ESX-1 activation during infection, offering a strategic anti-virulence approach.	[37,38]
EsxA (ESAT-6)	Pore-forming effector enables phagosomal escape.	ESAT-6 forms pH-dependent pores in host membranes-blocking its oligomerization or membrane insertion could prevent phagosomal rupture and virulence.	[42–44]

Table 1.1 Therapeutic strategies targeting the ESX-1 system in *Mycobacterium tuberculosis*.—cont'd

Targets	Mechanism of action	Therapeutic implications	References
EspA	Required for EsxA/B secretion.	EspA is required for ESX-1 secretion and acts downstream of PhoPR and EspR regulation; inhibiting its interactions could selectively block ESX-1 without affecting bacterial viability.	[46]
EspB	Forms oligomers under acidic pH, binds host membranes.	Although specific studies on EspB's pH-dependent behavior are limited, ESX-1-mediated membrane disruption is largely contributed by ESAT-6 and associated components under acidic conditions.	[47–49]
PE35–PPE68 + EspG1	Substrate chaperoning and export complex.	PE/PPE proteins and EspG chaperones assist in substrate export via ESX-1; disrupting these interactions may prevent proper targeting and secretion of EsxA	[54,55]

Briefly listing the spanning structural components, effectors, regulatory proteins, and host-directed pathways, representing approaches aiming to impair virulence, enhance host defence, or modulate inflammation without exerting direct bactericidal pressure.

Recent discoveries demonstrate that pharmacological inhibition of ESX-1 can attenuate virulence without directly compromising bacterial viability, thus reducing the risk of resistance compared to classical antibiotics. A notable example is IMB-BZ, a small molecule that selectively blocks ESX-1-dependent secretion of EsxB by disrupting its interaction with EccCb1. This leads to decreased virulence in vitro and in animal models, without impairing bacterial growth [101]. Such antivirulence agents exploit the principle that neutralizing pathogenic mechanisms, rather than exerting bactericidal pressure, diminishes resistance selection while preserving therapeutic efficacy. The observation that ESX-1-deficient mutants are profoundly attenuated in vivo further supports the therapeutic relevance of this pathway. This section examines therapeutic opportunities organized into four major domains: targeting ESX-1 structural components, inhibiting ESX-1 effectors, enhancing host autophagy and repair pathways, and modulating inflammasome activity. Collectively, these strategies highlight the diverse avenues available for disrupting the host-pathogen interface.

6.1 Targeting structural components of ESX-1

The structural proteins of ESX-1 form the functional backbone of the secretion apparatus and represent attractive therapeutic targets. The non-redundant roles of EccB1, EccD1, EccE1, and EccA1 provide multiple points of intervention.

6.1.1 EccB1 (Rv3869)

This periplasmic anchor contributes both structural integrity and effector release regulation. Its hexameric ATPase activity powers translocation, while the flexible A2 domain fine-tunes effector secretion [16–18]. Inhibiting ATPase activity would paralyze the secretion machinery. Importantly, structural models suggest that disrupting its oligomerization could destabilize the entire ESX-1 complex, providing a high-leverage therapeutic node.

6.1.2 EccD1 (Rv3877)

As a polytopic transmembrane protein, EccD1 constitutes the central channel across the inner membrane. Its ubiquitin-like dimer interface and negatively charged groove mediate selective export of EsxA and EsxB [16,23–27]. Because it serves as the pore for secretion, inhibitors targeting its dimerization or charge-specific substrate recognition could block effector release entirely, akin to plugging the gateway of the ESX-1 system.

6.1.3 EccE1 (Rv3882c)

With 12 predicted transmembrane helices, EccE1 functions as a structural organizer, stabilizing EccB1 and EccD1. Knockouts reveal that its absence destabilizes the secretion system and prevents substrate export [23,28]. This makes EccE1 a precision target, intervening here collapses the system's architecture rather than just blocking one effector.

6.1.4 EccA1 (Rv3868)

This cytosolic AAA^+ ATPase mediates effector recognition and translocation via canonical Walker A/B motifs and an arginine finger essential for oligomerization [29–31]. Remarkably, EccA1 also regulates lipid metabolism, interacting with Pks13, KasA, and KasB, which drive mycolic acid biosynthesis. Thus, targeting EccA1 could simultaneously impair secretion and lipid metabolism, amplifying therapeutic impact. Together, these four components represent non-redundant choke points. Their conservation across mycobacterial strains enhances the likelihood of broad-spectrum efficacy, while their essentiality for secretion ensures that pharmacological disruption directly undermines virulence.

6.2 Targeting ESX-1 effectors

While structural proteins maintain the secretion apparatus, ESX-1 effectors directly mediate phagosomal rupture, immune evasion, and pathogenesis. Therapeutic strategies aimed at blocking their function or regulation could neutralize virulence.

6.2.1 EspR

A transcriptional regulator of the espACD operon, EspR integrates signals from the PhoPR and MprAB two-component systems [36]. Inhibiting EspR could downregulate ESX-1 effectors globally, attenuating virulence at the transcriptional level without applying bactericidal pressure.

6.2.2 EspN and EspM

Acting as a regulatory switch, EspN and EspM govern whiB6, espE, and eccA transcription. The sterol carrier protein 2 (Scp2) domain of EspN hints at lipid-sensing functionality [38]. Pharmacological interference here could impair infection-specific tuning of ESX-1 expression, selectively weakening virulence under host conditions.

6.2.3 EsxA

As the pore-forming effector, EsxA is the most direct target. Neutralizing antibodies or small molecules that block oligomerization, prevent conformational transitions at acidic pH, or inhibit membrane insertion could abolish phagosomal rupture [40–44]. Structural elucidation of EsxA helices provides a foundation for rational inhibitor design.

6.2.4 EspA

Required for EsxA/EsxB secretion [46], EspA dimerizes via disulfide bonds. Inhibitors targeting these bonds or EspA-machinery interactions could prevent effector export altogether.

6.2.5 EspB

Undergoes pH-dependent oligomerization into heptameric rings that interact with host phospholipids [47–49]. Blocking oligomerization or lipid binding would disrupt its role in phagosomal destabilization and mitochondrial interference.

6.2.6 PE35-PPE68 heterodimer

Required for stabilization of substrates and EspG1-dependent secretion [55], this complex can be disrupted by peptides or small molecules targeting heterodimer formation or chaperone binding. Targeting effectors directly is compelling because it interrupts virulence while sparing bacterial viability, thus reducing selective pressure for resistance. This approach also aligns with vaccine strategies seeking to generate neutralizing immunity against ESX-1 effectors.

6.3 Enhancing autophagy and phagosome repair

Host-directed therapies offer a complementary strategy by amplifying innate defense pathways that restrict ESX-1 activity. Two critical mechanisms are ESCRT-mediated membrane repair and selective autophagy (xenophagy). Pharmacological enhancement of ESCRT recruitment or autophagic flux could reinforce host defenses, reseal damaged membranes, and route bacilli toward lysosomal degradation. Combining such approaches with conventional antimicrobials could synergize by neutralizing the cytosolic survival advantage conferred by ESX-1.

GAL-8 is a prime example. As a molecular sensor of membrane rupture, GAL-8 recruits selective autophagy adaptors, including TAX1BP1, which then attract LC3-positive membranes [45]. Drugs that upregulate GAL-8 expression or stabilize GAL-8-TAX1BP1 interaction could potentiate xenophagic clearance. Moreover, autophagy-inducing agents, such as mTOR inhibitors (e.g., rapamycin derivatives), or small molecules enhancing lysosomal biogenesis could provide broad-spectrum reinforcement of this pathway. Importantly, these strategies do not act on bacteria directly, thereby avoiding resistance selection. Instead, they recalibrate the host environment to deny bacilli their cytosolic refuge.

6.4 Modulating inflammasome activity

The NLRP3 inflammasome is a double-edged sword in TB pathogenesis. On one hand, its activation drives caspase-1 cleavage, Gasdermin D pore formation, and IL-1β/IL-18 release, which are critical for bacterial containment [79,102]. On the other hand, uncontrolled activation fuels pyroptosis, necrosis, and tissue pathology, particularly in extrapulmonary sites like the central nervous system. This duality creates both opportunity and risk for therapy. MCC950, a selective NLRP3 inhibitor, has shown that inflammasome suppression can reduce pathological IL-1β and TNF-α secretion while limiting pyroptotic death in microglia infected with *M. tb* H37Ra [103]. Intriguingly, MCC950 enhanced bacterial clearance when combined with conventional TB drugs, indicating synergy rather than antagonism.

However, genetic and pharmacological disruption of inflammasome components in some models paradoxically increased bacterial persistence [104]. This suggests the need for context-dependent modulation: early activation may be beneficial for priming immunity, while chronic overactivation contributes to immunopathology. The therapeutic challenge is to calibrate inflammasome activity, attenuating excessive inflammation while preserving protective cytokine responses. This likely requires temporally tuned interventions or adjunctive regimens targeting specific patient populations, such as those with extrapulmonary TB where pathology outweighs benefit.

6.5 Challenges in targeting host-pathogen interactions

Despite the promise of antivirulence and host-directed strategies, several challenges remain.

1. **Specificity and Toxicity.** Many ESX-1 components share functional parallels with host trafficking proteins. Targeting them raises concerns of off-target toxicity and impaired cellular physiology [9,81].
2. **Immune Balance.** Enhancing autophagy or inflammasome activity carries the risk of overshooting into immunopathology. Conversely, suppressing inflammation may allow bacterial persistence [105,106].
3. **Heterogeneity of Responses.** Single-cell transcriptomics and spatial profiling reveal that *M. tb* exploits heterogeneity in macrophage subsets [106,107]. Therapies must therefore account for diverse host-pathogen interactions across tissues and disease states.
4. **Model Limitations.** Murine models incompletely recapitulate human TB, particularly in granuloma structure and immune regulation. Robust preclinical validation is essential before clinical translation.

Integrating multi-omics, structural biology, and high-resolution imaging will be critical for mapping ESX-1 dynamics and guiding rational drug design [106–108]. Precision-driven host-directed therapies that target pathogenic nodes without compromising essential host functions represent the next frontier in TB treatment. Targeting ESX-1 and its downstream host interactions offers an innovative path beyond conventional antibiotics. Whether by dismantling structural components, neutralizing effectors, amplifying autophagy and repair, or fine-tuning inflammasome activity, antivirulence and host-directed strategies hold significant potential. The goal is not merely to kill bacilli but to disarm their virulence while reinforcing host resilience, thereby reducing disease burden and mitigating resistance.

7. Conclusions and perspective

The pathogenic success of *M. tb* is intimately tied to its remarkable ability to manipulate host intracellular compartments, particularly the phagosome. At the center of this process lies the ESX-1 secretion system, encoded within the RD1 locus, which represents one of the most important virulence determinants in TB pathogenesis. ESX-1 is not a passive leak of bacterial products but a precisely orchestrated system in which conserved structural proteins (Ecc family members), regulatory factors, and a repertoire of effector proteins coordinate to promote controlled phagosomal rupture. Among its effectors, EsxA and EsxB are indispensable. Secreted as heterodimers, they dissociate under acidic conditions: EsxA inserts into the phagosomal membrane, forming pores that mediate cytosolic access, while EsxB stabilizes the complex during secretion. Accessory proteins, including EspA, EspB, EspC, and the PE35-PPE68 heterodimer, enhance ESX-1 function by stabilizing machinery, modulating host trafficking, and fine-tuning immune subversion. Together, these factors establish phagosomal rupture as a deliberate, regulated virulence mechanism, not a by-product of infection.

The consequences for the host are profound. ESX-1-driven membrane damage leads to the cytosolic exposure of bacterial DNA, activating the cGAS-STING-TBK1-IRF3 axis and inducing type I IFNs. Unlike in many viral infections, however, type I IFN in TB paradoxically promotes bacterial persistence by suppressing IL-1β and TNF-α production. Simultaneously, cytosolic pattern recognition receptors, including NLRs, detect membrane disruption and activate inflammasomes. This culminates in IL-1β and IL-18 release and, in some cases, Gasdermin D, driven pyroptosis. While these processes initiate antimicrobial responses, they also facilitate tissue damage and bacterial dissemination. Thus, ESX-1 ensures escape from the phagosome while manipulating host immunity to create a niche favorable for persistence.

The host, in turn, mobilizes counteractive defenses. GAL-8 acts as a sentinel of membrane integrity, detecting exposed glycans on ruptured phagosomes and recruiting TAX1BP1 to trigger selective autophagy. Cytosolic serpins, such as Serpina3g/Spi2A, neutralize cathepsin B released during lysosomal rupture, preserving cellular integrity. ESCRT-mediated membrane repair and phagosomal acidification further limit cytosolic access, while tightly regulated inflammasome activation seeks to balance protective immunity with the prevention of immunopathology. Yet, *M. tb* frequently evades or undermines these defenses through functional redundancy and adaptability, highlighting the complexity of this host-pathogen arms race.

From a therapeutic perspective, ESX-1 represents an exceptionally attractive antivirulence target. Pharmacological inhibition of ESX-1 can attenuate virulence without applying the same selective pressure for resistance as bactericidal antibiotics. A notable example is IMB-BZ, a small molecule that disrupts the CFP-10-EccCb1 interaction, thereby preventing secretion and attenuating virulence in vitro and in vivo [101]. Other strategies under investigation include small molecules against structural proteins such as EccA1 and EccD1, neutralizing antibodies or inhibitors targeting EsxA and EspD, and agents that enhance autophagic flux or ESCRT-mediated repair. Furthermore, inflammasome modulators such as MCC950 offer a means of curbing damaging hyperinflammation while preserving protective cytokine responses. These antivirulence and host-directed therapies are particularly relevant in the era of multidrug-resistant and extensively drug-resistant TB, where conventional treatments are increasingly inadequate.

Looking forward, several challenges remain. ESX-1 shares structural and functional parallels with host trafficking pathways, raising concerns about off-target toxicity. Host-directed therapies require temporal precision, for instance, augmenting autophagy or inflammasomes at one stage of infection may be beneficial but deleterious at another. Advances in single-cell RNA sequencing, spatial transcriptomics, and cryo-electron microscopy are providing unprecedented insight into ESX-1 structure, host-pathogen heterogeneity, and macrophage subset-specific responses. Integrating these technologies with functional genomics will be crucial for designing therapies that selectively disrupt pathogenic processes while safeguarding host physiology.

In conclusion, the ESX-1 system epitomizes the sophistication of *M. tb* virulence, functioning both as a molecular weapon for phagosomal rupture and as a manipulator of host immunity. The ongoing interplay between bacterial effectors and host defenses constitutes a molecular arms race that determines infection outcomes. High-resolution dissection of this dynamic not only advances our understanding of TB biology but also opens the door to novel therapeutic interventions. By combining antivirulence agents with host-directed therapies, the field moves closer to shifting the host-pathogen balance decisively in favor of durable infection control. Such integrative strategies may ultimately redefine TB treatment paradigms and provide a critical tool in the fight against one of humanity's most persistent pathogens.

Acknowledgments

Dr. Mohammad Shariq and Dr. Farhan Ahmed gratefully acknowledge the support of an internal seed grant and MURTHY fellowship that facilitated the initiation of laboratory research. Syeda Rushna and Onaiza Ansari sincerely acknowledges the Hamdard National Foundation for awarding a fellowship that supported the successful completion of the doctoral degree.

References

[1] Goletti D, Meintjes G, Andrade BB, Zumla A, Shan Lee S. Insights from the 2024 WHO global tuberculosis report - more comprehensive action, innovation, and investments required for achieving WHO end TB goals. Int J Infect Dis 2025;150:107325. https://doi.org/10.1016/j.ijid.2024.107325.

[2] Passos BBS, Araújo-Pereira M, Vinhaes CL, Amaral EP, Andrade BB. The role of ESAT-6 in tuberculosis immunopathology. Front Immunol 2024;15:1383098. https://doi.org/10.3389/fimmu.2024.1383098.

[3] Vergne I, Fratti RA, Hill PJ, Chua J, Belisle J, Deretic V. *Mycobacterium tuberculosis* phagosome maturation arrest: mycobacterial phosphatidylinositol analog phosphatidylinositol mannoside stimulates early endosomal fusion. Mol Biol Cell 2004;15(2):751–60. https://doi.org/10.1091/mbc.e03-05-0307.

[4] Queval CJ, Brosch R, Simeone R. The macrophage: a disputed fortress in the battle against *Mycobacterium tuberculosis*. Front Microbiol 2017;8:2284. https://doi.org/10.3389/fmicb.2017.02284.

[5] Sathkumara HD, Muruganandah V, Cooper MM, et al. Mucosal delivery of ESX-1-expressing BCG strains provides superior immunity against tuberculosis in murine type 2 diabetes. Proc Natl Acad Sci USA 2020; 117(34):20848–59. https://doi.org/10.1073/pnas.2003235117.

[6] Koiri D, Nandi M, Hameem PMA, et al. Real-time visualization reveals *Mycobacterium tuberculosis* ESAT-6 disrupts phagosome-like compartment via fibril-mediated vesiculation. Cell Rep 2025;44(3):115328. https://doi.org/10.1016/j.celrep.2025.115328.

[7] McLaughlin B, Chon JS, MacGurn JA, et al. A mycobacterium ESX-1-secreted virulence factor with unique requirements for export. PLoS Pathog 2007;3(8):e105. https://doi.org/10.1371/journal.ppat.0030105.

[8] Conrad WH, Osman MM, Shanahan JK, et al. Mycobacterial ESX-1 secretion system mediates host cell lysis through bacterium contact-dependent gross membrane disruptions. Proc Natl Acad Sci U S A 2017; 114(6):1371–6. https://doi.org/10.1073/pnas.1620133114.

[9] Wong KW. The role of ESX-1 in *Mycobacterium tuberculosis* pathogenesis. Microbiol Spectr 2017;5(3). https://doi.org/10.1128/microbiolspec.TBTB2-0001-2015.

[10] Malik AA, Shariq M, Sheikh JA, et al. Mechanisms of immune evasion by *Mycobacterium tuberculosis*: the impact of T7SS and cell wall lipids on host defenses. Crit Rev Biochem Mol Biol 2024;59(5):310–36. https://doi.org/10.1080/10409238.2024.2411264.

[11] Daugelat S, Kowall J, Mattow J, et al. The RD1 proteins of *Mycobacterium tuberculosis*: expression in Mycobacterium smegmatis and biochemical characterization. Microb Infect 2003;5(12):1082–95. https://doi.org/10.1016/s1286-4579(03)00205-3.

[12] Bunduc CM, Bitter W, Houben ENG. Structure and function of the mycobacterial type VII secretion systems. Annu Rev Microbiol 2020;74:315–35. https://doi.org/10.1146/annurev-micro-012420-081657.

[13] Bottai D, Frigui W, Clark S, et al. Increased protective efficacy of recombinant BCG strains expressing virulence-neutral proteins of the ESX-1 secretion system. Vaccine 2015;33(23):2710–8. https://doi.org/10.1016/j.vaccine.2015.03.083.

[14] Mustafa AS. Immunological characterization of proteins expressed by genes located in Mycobacterium tuberculosis-specific genomic regions encoding the ESAT6-like proteins. Vaccines 2021;9(1):27. https://doi.org/10.3390/vaccines9010027.

[15] Tiwari S, Casey R, Goulding CW, Hingley-Wilson S, Jacobs WR. Infect and inject: how *Mycobacterium tuberculosis* Exploits its major virulence-associated type VII secretion system, ESX-1. Microbiol Spectr 2019;7(3). https://doi.org/10.1128/microbiolspec.BAI-0024-2019.

[16] Wagner JM, Chan S, Evans TJ, et al. Structures of EccB1 and EccD1 from the core complex of the mycobacterial ESX-1 type VII secretion system. BMC Struct Biol 2016;16:5. https://doi.org/10.1186/s12900-016-0056-6.

[17] Zhang XL, Li DF, Fleming J, et al. Core component EccB1 of the *Mycobacterium tuberculosis* type VII secretion system is a periplasmic ATPase. FASEB J Off Publ Fed Am Soc Exp Biol 2015;29(12):4804–14. https://doi.org/10.1096/fj.15-270843.

[18] Xie XQ, Zhang XL, Qi C, et al. Crystallographic observation of the movement of the membrane-distal domain of the T7SS core component EccB1 from *Mycobacterium tuberculosis*. Acta Crystallogr Sect F Struct Biol Commun 2016;72(Pt 2):139–44. https://doi.org/10.1107/S2053230X16000212.

[19] Bandyopadhyay A, Saxena AK. Structural and ATPase activity analysis of nucleotide binding domain of Rv3870 enzyme of *M. tuberculosis* ESX-1 system. Int J Biol Macromol 2021;189:879–89. https://doi.org/10.1016/j.ijbiomac.2021.08.130.

[20] Feltcher ME, Sullivan JT, Braunstein M. Protein export systems of *Mycobacterium tuberculosis*: novel targets for drug development? Future Microbiol 2010;5(10):1581–97. https://doi.org/10.2217/fmb.10.112.

[21] Clemmensen HS, Knudsen NPH, Rasmussen EM, et al. An attenuated *Mycobacterium tuberculosis* clinical strain with a defect in ESX-1 secretion induces minimal host immune responses and pathology. Sci Rep 2017;7:46666. https://doi.org/10.1038/srep46666.

[22] Wang S, Zhou K, Yang X, et al. Structural insights into substrate recognition by the type VII secretion system. Protein Cell 2020;11(2):124–37. https://doi.org/10.1007/s13238-019-00671-z.

[23] Soler-Arnedo P, Sala C, Zhang M, Cole ST, Piton J. Polarly localized EccE1 is required for ESX-1 function and stabilization of ESX-1 membrane proteins in *Mycobacterium tuberculosis*. J Bacteriol 2020;202(5): e00662-19. https://doi.org/10.1128/JB.00662-19.

[24] Stinear TP, Seemann T, Harrison PF, et al. Insights from the complete genome sequence of Mycobacterium marinum on the evolution of *Mycobacterium tuberculosis*. Genome Res 2008;18(5):729–41. https://doi.org/10.1101/gr.075069.107.

[25] Zhang M, Adroub S, Ummels R, et al. Comprehensive pan-genome analysis of Mycobacterium marinum: insights into genomic diversity, evolution, and pathogenicity. Sci Rep 2024;14(1):27723. https://doi.org/10.1038/s41598-024-75228-0.

[26] Pym AS, Brodin P, Brosch R, Huerre M, Cole ST. Loss of RD1 contributed to the attenuation of the live tuberculosis vaccines Mycobacterium bovis BCG and Mycobacterium microti. Mol Microbiol 2002;46(3): 709–17. https://doi.org/10.1046/j.1365-2958.2002.03237.x.

[27] Majlessi L, Brodin P, Brosch R, et al. Influence of ESAT-6 secretion system 1 (RD1) of *Mycobacterium tuberculosis* on the interaction between mycobacteria and the host immune system. J Immunol 2005;174(6):3570–9. https://doi.org/10.4049/jimmunol.174.6.3570.

[28] Das C, Ghosh TS, Mande SS. Computational analysis of the ESX-1 region of *Mycobacterium tuberculosis*: insights into the mechanism of type VII secretion system. PLoS One 2011;6(11):e27980. https://doi.org/10.1371/journal.pone.0027980.

[29] Luthra A, Gaur A, Ramachandran R. Rv3868 (EccA1), an essential component of the *Mycobacterium tuberculosis* ESX-1 secretion system, is thermostable. Biochim Biophys Acta 2013;1834(6):1181–6. https://doi.org/10.1016/j.bbapap.2013.02.004.

[30] Wagner JM, Evans TJ, Korotkov KV. Crystal structure of the N-terminal domain of $EccA_1$ ATPase from the ESX-1 secretion system of *Mycobacterium tuberculosis*. Proteins 2014;82(1):159–63. https://doi.org/10.1002/prot.24351.

[31] Joshi SA, Ball DA, Sun MG, et al. EccA1, a component of the Mycobacterium marinum ESX-1 protein virulence factor secretion pathway, regulates mycolic acid lipid synthesis. Chem Biol 2012;19(3):372–80. https://doi.org/10.1016/j.chembiol.2012.01.008.

[32] Blasco B, Stenta M, Alonso-Sarduy L, et al. Atypical DNA recognition mechanism used by the EspR virulence regulator of *Mycobacterium tuberculosis*. Mol Microbiol 2011;82(1):251–64. https://doi.org/10.1111/j.1365-2958.2011.07813.x.

[33] Blasco B, Chen JM, Hartkoorn R, et al. Virulence regulator EspR of *Mycobacterium tuberculosis* is a nucleoid-associated protein. PLoS Pathog 2012;8(3):e1002621. https://doi.org/10.1371/journal.ppat.1002621.

[34] Rosenberg OS, Dovey C, Tempesta M, et al. EspR, a key regulator of *Mycobacterium tuberculosis* virulence, adopts a unique dimeric structure among helix-turn-helix proteins. Proc Natl Acad Sci USA 2011;108(33):13450–5. https://doi.org/10.1073/pnas.1110242108.

[35] Raghavan S, Manzanillo P, Chan K, Dovey C, Cox JS. Secreted transcription factor controls *Mycobacterium tuberculosis* virulence. Nature 2008;454(7205):717–21. https://doi.org/10.1038/nature07219.

[36] Cao G, Howard ST, Zhang P, et al. EspR, a regulator of the ESX-1 secretion system in *Mycobacterium tuberculosis*, is directly regulated by the two-component systems MprAB and PhoPR. Microbiol Read Engl 2015;161(Pt 3):477–89. https://doi.org/10.1099/mic.0.000023.

[37] Nicholson KR, Cronin RM, Menon AR, Jennisch MK, Tobin DM, Champion PA. In: The EspN transcription Factor is an Infection-dependent Regulator of the ESX-1 System in *M. marinum*; 2023. https://doi.org/10.1101/2023.02.15.528779.

[38] Nicholson KR, Cronin RM, Prest RJ, et al. The antagonistic transcription factors, EspM and EspN, regulate the ESX-1 secretion system in M. marinum. mBio 2024;15(4):e0335723. https://doi.org/10.1128/mbio.03357-23.

[39] Simeone R, Bottai D, Brosch R. ESX/type VII secretion systems and their role in host-pathogen interaction. Curr Opin Microbiol 2009;12(1):4–10. https://doi.org/10.1016/j.mib.2008.11.003.

[40] Bao Y, Wang L, Sun J. A small protein but with diverse roles: a review of esxa in mycobacterium-host interaction. Cells 2021;10(7):1645. https://doi.org/10.3390/cells10071645.

[41] Collars OA, Jones BS, Hu DD, et al. An N-acetyltransferase required for ESAT-6 N-terminal acetylation and virulence in Mycobacterium marinum. mBio 2023;14(5):e0098723. https://doi.org/10.1128/mbio.00987-23.

[42] de Jonge MI, Pehau-Arnaudet G, Fretz MM, et al. ESAT-6 from *Mycobacterium tuberculosis* dissociates from its putative chaperone CFP-10 under acidic conditions and exhibits membrane-lysing activity. J Bacteriol 2007;189(16):6028–34. https://doi.org/10.1128/JB.00469-07.

[43] Ma Y, Keil V, Sun J. Characterization of *Mycobacterium tuberculosis* EsxA membrane insertion: roles of N- and C-terminal flexible arms and central helix-turn-helix motif. J Biol Chem 2015;290(11):7314–22. https://doi.org/10.1074/jbc.M114.622076.

[44] Smith J, Manoranjan J, Pan M, et al. Evidence for pore formation in host cell membranes by ESX-1-secreted ESAT-6 and its role in Mycobacterium marinum escape from the vacuole. Infect Immun 2008; 76(12):5478–87. https://doi.org/10.1128/IAI.00614-08.

[45] Bell SL, Lopez KL, Cox JS, Patrick KL, Watson RO. Galectin-8 senses phagosomal damage and recruits selective autophagy adapter TAX1BP1 to control *Mycobacterium tuberculosis* infection in macrophages. mBio 2021;12(4):e0187120. https://doi.org/10.1128/mBio.01871-20.

[46] Garces A, Atmakuri K, Chase MR, et al. EspA acts as a critical mediator of ESX1-dependent virulence in *Mycobacterium tuberculosis* by affecting bacterial cell wall integrity. PLoS Pathog 2010;6(6):e1000957. https://doi.org/10.1371/journal.ppat.1000957.

[47] Korotkova N, Piton J, Wagner JM, et al. Structure of EspB, a secreted substrate of the ESX-1 secretion system of *Mycobacterium tuberculosis*. J Struct Biol 2015;191(2):236–44. https://doi.org/10.1016/j.jsb.2015.06.003.

[48] Gijsbers A, Vinciauskaite V, Siroy A, et al. Priming mycobacterial ESX-secreted protein B to form a channel-like structure. Curr Res Struct Biol 2021;3:153–64. https://doi.org/10.1016/j.crstbi.2021.06.001.

[49] Sengupta N, Padmanaban S, Dutta S. Cryo-EM reveals the membrane-binding phenomenon of EspB, a virulence factor of the mycobacterial type VII secretion system. J Biol Chem 2023;299(4):104589. https://doi.org/10.1016/j.jbc.2023.104589.

[50] Fortune SM, Jaeger A, Sarracino DA, et al. Mutually dependent secretion of proteins required for mycobacterial virulence. Proc Natl Acad Sci U S A 2005;102(30):10676–81. https://doi.org/10.1073/pnas.0504922102.

[51] Lou Y, Rybniker J, Sala C, Cole ST. EspC forms a filamentous structure in the cell envelope of *Mycobacterium tuberculosis* and impacts ESX-1 secretion. Mol Microbiol 2017;103(1):26–38. https://doi.org/10.1111/mmi.13575.

[52] Guo Q, Bi J, Li M, et al. ESX secretion-associated protein C from *Mycobacterium tuberculosis* induces macrophage activation through the toll-like receptor-4/mitogen-activated protein kinase signaling pathway. Front Cell Infect Microbiol 2019;9:158. https://doi.org/10.3389/fcimb.2019.00158.

[53] Brodin P, Majlessi L, Marsollier L, et al. Dissection of ESAT-6 system 1 of *Mycobacterium tuberculosis* and impact on immunogenicity and virulence. Infect Immun 2006;74(1):88–98. https://doi.org/10.1128/IAI.74.1.88-98.2006.

[54] Tiwari B, Soory A, Raghunand TR. An immunomodulatory role for the *Mycobacterium tuberculosis* region of difference 1 locus proteins PE35 (Rv3872) and PPE68 (Rv3873). FEBS J 2014;281(6):1556–70. https://doi.org/10.1111/febs.12723.

[55] Damen MPM, Meijers AS, Keizer EM, et al. The ESX-1 substrate PPE68 has a key function in ESX-1-mediated secretion in Mycobacterium marinum. mBio 2022;13(6):e0281922. https://doi.org/10.1128/mbio.02819-22.

[56] Mittal E, Skowyra ML, Uwase G, et al. *Mycobacterium tuberculosis* Type VII secretion system effectors differentially impact the ESCRT endomembrane damage response. mBio 2018;9(6):e01765-18. https://doi.org/10.1128/mBio.01765-18.

[57] Wiens KE, Ernst JD. The mechanism for type I interferon induction by *Mycobacterium tuberculosis* is bacterial strain-dependent. PLoS Pathog 2016;12(8):e1005809. https://doi.org/10.1371/journal.ppat.1005809.

[58] Watson RO, Bell SL, MacDuff DA, et al. The cytosolic sensor cGAS detects *Mycobacterium tuberculosis* DNA to induce type I interferons and activate autophagy. Cell Host Microbe 2015;17(6):811–9. https://doi.org/10.1016/j.chom.2015.05.004.

[59] Liu S, Guan L, Peng C, et al. *Mycobacterium tuberculosis* suppresses host DNA repair to boost its intracellular survival. Cell Host Microbe 2023;31(11):1820–1836.e10. https://doi.org/10.1016/j.chom.2023.09.010.

[60] Bobba S, Chauhan KS, Akter S, et al. A protective role for type I interferon signaling following infection with *Mycobacterium tuberculosis* carrying the rifampicin drug resistance-conferring RpoB mutation H445Y. PLoS Pathog 2024;20(4):e1012137. https://doi.org/10.1371/journal.ppat.1012137.

[61] Ji DX, Yamashiro LH, Chen KJ, et al. Type I interferon-driven susceptibility to *Mycobacterium tuberculosis* is mediated by IL-1Ra. Nat Microbiol 2019;4(12):2128–35. https://doi.org/10.1038/s41564-019-0578-3.

[62] Chai Q, Wang L, Liu CH, Ge B. New insights into the evasion of host innate immunity by *Mycobacterium tuberculosis*. Cell Mol Immunol 2020;17(9):901–13. https://doi.org/10.1038/s41423-020-0502-z.

[63] Theobald SJ, Müller TA, Lange D, Keck K, Rybniker J. The role of inflammasomes as central inflammatory hubs in *Mycobacterium tuberculosis* infection. Front Immunol 2024;15:1436676. https://doi.org/10.3389/fimmu.2024.1436676.

[64] Chai Q, Yu S, Zhong Y, et al. A bacterial phospholipid phosphatase inhibits host pyroptosis by hijacking ubiquitin. Science 2022;378(6616):eabq0132. https://doi.org/10.1126/science.abq0132.

[65] Qu Z, Zhou J, Zhou Y, et al. Mycobacterial EST12 activates a RACK1-NLRP3-gasdermin D pyroptosis-IL-1β immune pathway. Sci Adv 2020;6(43):eaba4733. https://doi.org/10.1126/sciadv.aba4733.

[66] Lovey A, Verma S, Kaipilyawar V, et al. Early alveolar macrophage response and IL-1R-dependent T cell priming determine transmissibility of *Mycobacterium tuberculosis* strains. Nat Commun 2022;13(1):884. https://doi.org/10.1038/s41467-022-28506-2.

[67] Theobald SJ, Gräb J, Fritsch M, et al. Gasdermin D mediates host cell death but not interleukin-1β secretion in Mycobacterium tuberculosis-infected macrophages. Cell Death Discov 2021;7(1):327. https://doi.org/10.1038/s41420-021-00716-5.

[68] Mehra A, Zahra A, Thompson V, et al. *Mycobacterium tuberculosis* type VII secreted effector EsxH targets host ESCRT to impair trafficking. PLoS Pathog 2013;9(10):e1003734. https://doi.org/10.1371/journal.ppat.1003734.

[69] Ahmad F, Rani A, Alam A, et al. Macrophage: a cell with many faces and functions in tuberculosis. Front Immunol 2022;13:747799. https://doi.org/10.3389/fimmu.2022.747799.

[70] Rabinovich GA, Toscano MA. Turning "sweet" on immunity: galectin-glycan interactions in immune tolerance and inflammation. Nat Rev Immunol 2009;9(5):338–52. https://doi.org/10.1038/nri2536.

[71] van Kooyk Y, Rabinovich GA. Protein-glycan interactions in the control of innate and adaptive immune responses. Nat Immunol 2008;9(6):593–601. https://doi.org/10.1038/ni.f.203.

[72] Vasta GR. Roles of galectins in infection. Nat Rev Microbiol 2009;7(6):424–38. https://doi.org/10.1038/nrmicro2146.

[73] Boyle KB, Randow F. The role of "eat-me" signals and autophagy cargo receptors in innate immunity. Curr Opin Microbiol 2013;16(3):339–48. https://doi.org/10.1016/j.mib.2013.03.010.

[74] Khaminets A, Behl C, Dikic I. Ubiquitin-dependent and independent signals in selective autophagy. Trends Cell Biol 2016;26(1):6–16. https://doi.org/10.1016/j.tcb.2015.08.010.

[75] Dupont N, Lacas-Gervais S, Bertout J, et al. Shigella phagocytic vacuolar membrane remnants participate in the cellular response to pathogen invasion and are regulated by autophagy. Cell Host Microbe 2009;6(2):137–49. https://doi.org/10.1016/j.chom.2009.07.005.

[76] Thurston TLM, Wandel MP, von Muhlinen N, Foeglein A, Randow F. Galectin 8 targets damaged vesicles for autophagy to defend cells against bacterial invasion. Nature 2012;482(7385):414–8. https://doi.org/10.1038/nature10744.

[77] Paz I, Sachse M, Dupont N, et al. Galectin-3, a marker for vacuole lysis by invasive pathogens. Cell Microbiol 2010;12(4):530–44. https://doi.org/10.1111/j.1462-5822.2009.01415.x.

[78] Feeley EM, Pilla-Moffett DM, Zwack EE, et al. Galectin-3 directs antimicrobial guanylate binding proteins to vacuoles furnished with bacterial secretion systems. Proc Natl Acad Sci U S A 2017;114(9):E1698–706. https://doi.org/10.1073/pnas.1615771114.

[79] Beckwith KS, Beckwith MS, Ullmann S, et al. Plasma membrane damage causes NLRP3 activation and pyroptosis during *Mycobacterium tuberculosis* infection. Nat Commun 2020;11(1):2270. https://doi.org/10.1038/s41467-020-16143-6.

[80] Augenstreich J, Arbues A, Simeone R, et al. ESX-1 and phthiocerol dimycocerosates of *Mycobacterium tuberculosis* act in concert to cause phagosomal rupture and host cell apoptosis. Cell Microbiol 2017;19(7). https://doi.org/10.1111/cmi.12726.

[81] Beatty WL, Rhoades ER, Hsu DK, Liu FT, Russell DG. Association of a macrophage galactoside-binding protein with Mycobacterium-containing phagosomes. Cell Microbiol 2002;4(3):167–76. https://doi.org/10.1046/j.1462-5822.2002.00183.x.

[82] Lerner TR, Queval CJ, Fearns A, Repnik U, Griffiths G, Gutierrez MG. Phthiocerol dimycocerosates promote access to the cytosol and intracellular burden of *Mycobacterium tuberculosis* in lymphatic endothelial cells. BMC Biol 2018;16(1):1. https://doi.org/10.1186/s12915-017-0471-6.

[83] Osman MM, Pagán AJ, Shanahan JK, Ramakrishnan L. Mycobacterium marinum phthiocerol dimycocerosates enhance macrophage phagosomal permeabilization and membrane damage. PLoS One 2020;15(7):e0233252. https://doi.org/10.1371/journal.pone.0233252.

[84] Jayaraman P, Sada-Ovalle I, Beladi S, et al. Tim3 binding to galectin-9 stimulates antimicrobial immunity. J Exp Med 2010;207(11):2343–54. https://doi.org/10.1084/jem.20100687.

[85] Sada-Ovalle I, Chávez Galán L, Torre-Bouscoulet L, et al. The Tim3-galectin 9 pathway induces antibacterial activity in human macrophages infected with *Mycobacterium tuberculosis*. J Immunol 2012;189(12):5896–902. https://doi.org/10.4049/jimmunol.1200990.

[86] Gutierrez MG, Master SS, Singh SB, Taylor GA, Colombo MI, Deretic V. Autophagy is a defense mechanism inhibiting BCG and *Mycobacterium tuberculosis* survival in infected macrophages. Cell 2004;119(6):753–66. https://doi.org/10.1016/j.cell.2004.11.038.

[87] Tufariello JM, Chapman JR, Kerantzas CA, et al. Separable roles for *Mycobacterium tuberculosis* ESX-3 effectors in iron acquisition and virulence. Proc Natl Acad Sci U S A 2016;113(3):E348–57. https://doi.org/10.1073/pnas.1523321113.

[88] López-Jiménez AT, Cardenal-Muñoz E, Leuba F, et al. The ESCRT and autophagy machineries cooperate to repair ESX-1-dependent damage at the Mycobacterium-containing vacuole but have opposite impact on containing the infection. PLoS Pathog 2018;14(12):e1007501. https://doi.org/10.1371/journal.ppat.1007501.

[89] Moreira-Teixeira L, Mayer-Barber K, Sher A, O'Garra A. Type I interferons in tuberculosis: foe and occasionally friend. J Exp Med 2018;215(5):1273–85. https://doi.org/10.1084/jem.20180325.

[90] Manzanillo PS, Shiloh MU, Portnoy DA, Cox JS. *Mycobacterium tuberculosis* activates the DNA-dependent cytosolic surveillance pathway within macrophages. Cell Host Microbe 2012;11(5):469–80. https://doi.org/10.1016/j.chom.2012.03.007.

[91] Stanley SA, Johndrow JE, Manzanillo P, Cox JS. The type I IFN response to infection with *Mycobacterium tuberculosis* requires ESX-1-mediated secretion and contributes to pathogenesis. J Immunol 2007;178(5):3143–52. https://doi.org/10.4049/jimmunol.178.5.3143.

[92] Hamerman JA, Hayashi F, Schroeder LA, et al. Serpin 2a is induced in activated macrophages and conjugates to a ubiquitin homolog. J Immunol 2002;168(5):2415–23. https://doi.org/10.4049/jimmunol.168.5.2415.

[93] Abdallah AM, Bestebroer J, Savage NDL, et al. Mycobacterial secretion systems ESX-1 and ESX-5 play distinct roles in host cell death and inflammasome activation. J Immunol 2011;187(9):4744–53. https://doi.org/10.4049/jimmunol.1101457.

[94] Lee J, Remold HG, Ieong MH, Kornfeld H. Macrophage apoptosis in response to high intracellular burden of *Mycobacterium tuberculosis* is mediated by a novel caspase-independent pathway. J Immunol 2006;176(7):4267–74. https://doi.org/10.4049/jimmunol.176.7.4267.

[95] Nobs E, Laschanzky K, Munke K, Movert E, Valfridsson C, Carlsson F. Cytosolic serpins act in a cytoprotective feedback loop that limits ESX-1-dependent death of Mycobacterium marinum-infected macrophages. mBio 2024;15(9):e0038424. https://doi.org/10.1128/mbio.00384-24.

[96] Liu N, Phillips T, Zhang M, et al. Serine protease inhibitor 2A is a protective factor for memory T cell development. Nat Immunol 2004;5(9):919–26. https://doi.org/10.1038/ni1107.
[97] Byrne SM, Aucher A, Alyahya S, et al. Cathepsin B controls the persistence of memory CD8+ T lymphocytes. J Immunol 2012;189(3):1133–43. https://doi.org/10.4049/jimmunol.1003406.
[98] Dev A, Byrne SM, Verma R, Ashton-Rickardt PG, Wojchowski DM. Erythropoietin-directed erythropoiesis depends on serpin inhibition of erythroblast lysosomal cathepsins. J Exp Med 2013;210(2):225–32. https://doi.org/10.1084/jem.20121762.
[99] Li L, Byrne SM, Rainville N, et al. Brief report: serpin Spi2A as a novel modulator of hematopoietic progenitor cell formation. Stem Cells Dayt Ohio 2014;32(9):2550–6. https://doi.org/10.1002/stem.1778.
[100] Simeone R, Sayes F, Song O, et al. Cytosolic access of *Mycobacterium tuberculosis*: critical impact of phagosomal acidification control and demonstration of occurrence in vivo. PLoS Pathog 2015;11(2):e1004650. https://doi.org/10.1371/journal.ppat.1004650.
[101] Jia P, Zhang Y, Xu J, et al. IMB-BZ as an inhibitor targeting ESX-1 secretion system to control mycobacterial infection. J Infect Dis 2022;225(4):608–16. https://doi.org/10.1093/infdis/jiab486.
[102] Xu J, Núñez G. The NLRP3 inflammasome: activation and regulation. Trends Biochem Sci 2023;48(4):331–44. https://doi.org/10.1016/j.tibs.2022.10.002.
[103] Xie Z, Hui H, Yao Q, et al. By Regulating the NLRP3 inflammasome can reduce the release of inflammatory factors in the co-culture model of tuberculosis H37Ra strain and rat microglia. Front Cell Infect Microbiol 2021;11:637769. https://doi.org/10.3389/fcimb.2021.637769.
[104] Subbarao S, Sanchez-Garrido J, Krishnan N, Shenoy AR, Robertson BD. Genetic and pharmacological inhibition of inflammasomes reduces the survival of *Mycobacterium tuberculosis* strains in macrophages. Sci Rep 2020;10(1):3709. https://doi.org/10.1038/s41598-020-60560-y.
[105] Wassermann R, Gulen MF, Sala C, et al. *Mycobacterium tuberculosis* differentially activates cGAS- and Inflammasome-dependent intracellular immune responses through ESX-1. Cell Host Microbe 2015;17(6):799–810. https://doi.org/10.1016/j.chom.2015.05.003.
[106] Bernard EM, Fearns A, Bussi C, et al. Tuberculosis infection of human iPSC-derived macrophages reveals complex membrane dynamics during xenophagy evasion. J Cell Sci 2020;134(5):jcs252973. https://doi.org/10.1242/jcs.252973.
[107] Zheng W, Borja M, Dorman LC, et al. Single-cell analysis reveals *Mycobacterium tuberculosis* ESX-1-mediated accumulation of permissive macrophages in infected mouse lungs. Sci Adv 2025;11(3):eadq8158. https://doi.org/10.1126/sciadv.adq8158.
[108] Thacker VV, Dhar N, Sharma K, Barrile R, Karalis K, McKinney JD. A lung-on-chip model of early *Mycobacterium tuberculosis* infection reveals an essential role for alveolar epithelial cells in controlling bacterial growth. eLife 2020;9:e59961. https://doi.org/10.7554/eLife.59961.

Further reading

[1] Abdallah AM, Gey van Pittius NC, Champion PAD, et al. Type VII secretion–mycobacteria show the way. Nat Rev Microbiol 2007;5(11):883–91. https://doi.org/10.1038/nrmicro1773.
[2] Wong KW, Jacobs WR. Critical role for NLRP3 in necrotic death triggered by *Mycobacterium tuberculosis*. Cell Microbiol 2011;13(9):1371–84. https://doi.org/10.1111/j.1462-5822.2011.01625.x.
[3] Stanley SA, Raghavan S, Hwang WW, Cox JS. Acute infection and macrophage subversion by *Mycobacterium tuberculosis* require a specialized secretion system. Proc Natl Acad Sci U S A 2003;100(22):13001–6. https://doi.org/10.1073/pnas.2235593100.
[4] Hsu T, Hingley-Wilson SM, Chen B, et al. The primary mechanism of attenuation of bacillus Calmette-Guerin is a loss of secreted lytic function required for invasion of lung interstitial tissue. Proc Natl Acad Sci U S A 2003;100(21):12420–5. https://doi.org/10.1073/pnas.1635213100.

[5] Volkman HE, Clay H, Beery D, Chang JCW, Sherman DR, Ramakrishnan L. Tuberculous granuloma formation is enhanced by a mycobacterium virulence determinant. PLoS Biol 2004;2(11):e367. https://doi.org/10.1371/journal.pbio.0020367.
[6] Carlsson F, Kim J, Dumitru C, et al. Host-detrimental role of Esx-1-mediated inflammasome activation in mycobacterial infection. PLoS Pathog 2010;6(5):e1000895. https://doi.org/10.1371/journal.ppat.1000895.
[7] Wolf AJ, Linas B, Trevejo-Nuñez GJ, et al. *Mycobacterium tuberculosis* infects dendritic cells with high frequency and impairs their function in vivo. J Immunol 2007;179(4):2509–19. https://doi.org/10.4049/jimmunol.179.4.2509.
[8] Lienard J, Munke K, Wulff L, et al. Intragranuloma accumulation and inflammatory differentiation of neutrophils underlie mycobacterial Esx-1-dependent immunopathology. mBio 2023;14(2):e0276422. https://doi.org/10.1128/mbio.02764-22.
[9] Cohen SB, Gern BH, Delahaye JL, et al. Alveolar macrophages provide an early *Mycobacterium tuberculosis* niche and initiate dissemination. Cell Host Microbe 2018;24(3):439–446.e4. https://doi.org/10.1016/j.chom.2018.08.001.
[10] Gao LY, Guo S, McLaughlin B, Morisaki H, Engel JN, Brown EJ. A mycobacterial virulence gene cluster extending RD1 is required for cytolysis, bacterial spreading and ESAT-6 secretion. Mol Microbiol 2004;53(6):1677–93. https://doi.org/10.1111/j.1365-2958.2004.04261.x.
[11] Koo IC, Wang C, Raghavan S, Morisaki JH, Cox JS, Brown EJ. ESX-1-dependent cytolysis in lysosome secretion and inflammasome activation during mycobacterial infection. Cell Microbiol 2008;10(9):1866–78. https://doi.org/10.1111/j.1462-5822.2008.01177.x.
[12] Kurenuma T, Kawamura I, Hara H, et al. The RD1 locus in the *Mycobacterium tuberculosis* genome contributes to activation of caspase-1 via induction of potassium ion efflux in infected macrophages. Infect Immun 2009;77(9):3992–4001. https://doi.org/10.1128/IAI.00015-09.
[13] Lienard J, Nobs E, Lovins V, Movert E, Valfridsson C, Carlsson F. The Mycobacterium marinum ESX-1 system mediates phagosomal permeabilization and type I interferon production via separable mechanisms. Proc Natl Acad Sci USA 2020;117(2):1160–6. https://doi.org/10.1073/pnas.1911646117.
[14] Watson RO, Manzanillo PS, Cox JS. Extracellular *M. tuberculosis* DNA targets bacteria for autophagy by activating the host DNA-sensing pathway. Cell 2012;150(4):803–15. https://doi.org/10.1016/j.cell.2012.06.040.
[15] Collins AC, Cai H, Li T, et al. Cyclic GMP-AMP synthase is an innate immune DNA sensor for *Mycobacterium tuberculosis*. Cell Host Microbe 2015;17(6):820–8. https://doi.org/10.1016/j.chom.2015.05.005.
[16] Movert E, Lienard J, Valfridsson C, Nordström T, Johansson-Lindbom B, Carlsson F. Streptococcal M protein promotes IL-10 production by cGAS-independent activation of the STING signaling pathway. PLoS Pathog 2018;14(3):e1006969. https://doi.org/10.1371/journal.ppat.1006969.
[17] Forsyth S, Horvath A, Coughlin P. A review and comparison of the murine alpha1-antitrypsin and alpha1-antichymotrypsin multigene clusters with the human clade A serpins. Genomics 2003;81(3):336–45. https://doi.org/10.1016/s0888-7543(02)00041-1
[18] de Mezer M, Rogaliński J, Przewoźny S, et al. SERPINA3: stimulator or inhibitor of pathological changes. Biomedicines 2023;11(1):156. https://doi.org/10.3390/biomedicines11010156.
[19] Morris EC, Dafforn TR, Forsyth SL, et al. Murine serpin 2A is a redox-sensitive intracellular protein. Biochem J 2003;371(Pt 1):165–73. https://doi.org/10.1042/BJ20021567.
[20] Horvath AJ, Forsyth SL, Coughlin PB. Expression patterns of murine antichymotrypsin-like genes reflect evolutionary divergence at the Serpina3 locus. J Mol Evol 2004;59(4):488–97. https://doi.org/10.1007/s00239-004-2640-9.
[21] Liu N, Raja SM, Zazzeroni F, et al. NF-kappaB protects from the lysosomal pathway of cell death. EMBO J 2003;22(19):5313–22. https://doi.org/10.1093/emboj/cdg510.
[22] Liu N, Wang Y, Ashton-Rickardt PG. Serine protease inhibitor 2A inhibits caspase-independent cell death. FEBS Lett 2004;569(1–3):49–53. https://doi.org/10.1016/j.febslet.2004.05.061.

[23] Kreuzaler PA, Staniszewska AD, Li W, et al. Stat3 controls lysosomal-mediated cell death in vivo. Nat Cell Biol 2011;13(3):303–9. https://doi.org/10.1038/ncb2171.
[24] Shamji MH, Temblay JN, Cheng W, et al. Antiapoptotic serine protease inhibitors contribute to survival of allergenic TH2 cells. J Allergy Clin Immunol 2018;142(2):569–581.e5. https://doi.org/10.1016/j.jaci.2017.07.055.
[25] Yoon MC, Solania A, Jiang Z, et al. Selective neutral pH inhibitor of Cathepsin B designed based on cleavage preferences at cytosolic and lysosomal pH conditions. ACS Chem Biol 2021;16(9):1628–43. https://doi.org/10.1021/acschembio.1c00138.

CHAPTER

2

Mycobacterial PE/PPE proteins: Key regulators of apoptosis and pathogenic outcomes

Gauri Shrivastava[1], Diksha Thakuri[1], Anwar Alam[1,2,3], Seyed E. Hasnain[1,4] and Nasreen Z. Ehtesham[1]

[1]*Department of Life Science, School of Bioscience and Technology, Sharda University, Greater Noida, Uttar Pradesh, India;* [2]*Department of Biotechnology, School of Bioscience and Technology, Sharda University, Greater Noida, Uttar Pradesh, India;* [3]*DST-FIST, Sharda University, Greater Noida, Uttar Pradesh, India;* [4]*Department of Biochemical Engineering and Biotechnology, IIT Delhi, New Delhi, India*

Highlights

1. *M.tb.* utilizes PE/PPE proteins to modulate the host immune responses and apoptotic pathways.
2. PE/PPE proteins interact with host cell receptors, influencing cytokine production and signaling cascades.
3. Certain PE/PPE proteins promote intracellular survival by inhibiting phagosome maturation and autophagy.
4. PE/PPE proteins regulate apoptosis through intrinsic, extrinsic, and ER stress induced pathways.
5. The ability of PE/PPE proteins to fine tune host cell fate facilitates *M.tb.* persistence.
6. The PE/PPE proteins contribute to immune evasion, granuloma formation, and bacterial dissemination.
7. Elucidating the molecular mechanisms of PE/PPE proteins may guide novel therapeutic strategies against tuberculosis.

Abbreviations

3D Three dimensional
AIM2 Absent in melanoma 2
Akt AKT serine/threonine kinase
Ams Alveolar macrophages
ASK1 Apoptosis signal regulating kinase 1
ATB Active tuberculosis
BAK BCL2 antagonist/killer
Bax BCL2 associated X protein
Bcl2 B cell lymphoma 2
CDC Centers for disease control and prevention
CFP-10 Culture filtrate protein 10 kDa
CHOP C/EBP homologous protein
CR Complement receptor

Navigating Tuberculosis. https://doi.org/10.1016/B978-0-443-27735-1.00006-X

CRISPRi Clustered regularly interspaced short palindromic repeats interference
DNA Deoxyribonucleic acid
EPTB Extrapulmonary tuberculosis
ER Endoplasmic reticulum
ERK1/2 Extracellular signal regulated kinase 1/2
ERK Extracellular signal regulated kinase
ESAT-6 Early secreted antigenic target 6 kDa
ESX ESAT 6 secretion system
FAS Fas cell surface death receptor
GBP-1 Guanylate-binding protein-1
ICAD Inhibitor of caspase activated DNase
ICAM-3 Intercellular adhesion molecule-3
IFNγ Interferon gamma
IFN Interferon
IGRA Interferon-gamma release assay
IL Interleukin
IRF3 Interferon regulatory factor 3
JNK Jun nterminal kinase
LAP LC3associated phagocytosis
LC3 Microtubule associated protein 1A/1B-light chain 3
LDL Low density lipoprotein
LTBI Latent tuberculosis infection
M.tb Mycobacterium tuberculosis
MAPK Mitogen activated protein kinase
MOMP Mitochondrial outer membrane permeabilization
MPTR Major polymorphic tandem repeat
NADPH Nicotinamide adenine dinucleotide phosphate hydrogen
NF-κB Nuclear factor kappa B
NLRP3 NLR family pyrin domain containing 3
PAF Platelet activating factor
PAMPs Pathogen associated molecular patterns
PARP Poly (ADP Ribose) polymerase
PE Proline glutamic acid
PGE2 Prostaglandin E2
PGRS Polymorphic GC rich sequences
PI3K Phosphatidylinositol 3 kinase
PPE Proline proline glutamic acid
PRRs Pattern recognition receptors
PS Phosphatidylserine
PTB Pulmonary tuberculosis
RNA Ribonucleic acid
ROS Reactive oxygen species
SOCS3 Suppressor of cytokine signaling 3
T7SS Type VII secretion system
TB Tuberculosis
TGF-β Transforming growth factor beta
Th1 T helper 1
Th2 T helper 2
TLR Toll like receptor

TNF Tumor necrosis factor
TNFR Tumor necrosis factor receptor
TNFRSF10A TNF receptor superfamily member 10a
TST Tuberculin skin test
UPR Unfolded protein response
UV Ultraviolet
WHO World Health Organization

1. Introduction

Tuberculosis (TB) is an infectious disease caused by the pathogenic bacterium *Mycobacterium tuberculosis* (*M.tb.*), an aerobic, non motile organism that predominantly targets the pulmonary system. *M.tb.* is a Gram positive, acid fast bacillus responsible for the onset of TB in humans, posing a significant public health challenge, particularly in developing regions. Initially, *M.tb.* is localized within the endosomes of infected macrophages. TB is responsible for more annual fatalities than many other infectious diseases [1]. It is regarded as one of the most adaptable human pathogens, employing unique strategies to persist within the host until conditions are conducive to its transmission [2]. According to the World Health Organization (WHO), TB affected approximately 10.6 million individuals globally in 2023, resulting in 1.3 million deaths attributed to the disease [3]. It is estimated that approximately 25% of the global population harbors latent TB infection (LTBI). Upon entering the alveoli of the lungs, *M.tb.* can disseminate into the bloodstream and spread throughout the body. Typically, between weeks 2 and 8, the immune system arrests bacterial replication and restricts further dissemination. At this juncture, the infection is classified as LTBI because the immune system has effectively contained the pathogen. Individuals with LTBI do not exhibit any signs or symptoms of disease and are non infectious (CDC, 2021). Nevertheless, they pose a potential risk for the future development of active tuberculosis (ATB) [4]. Upon entry, *M.tb.* frequently establishes a persistent latent infection that remains asymptomatic and generally non infectious, potentially enduring for extended periods. However, factors such as aging or compromised health can trigger reactivation, leading to active infection and clinical symptoms that can be transmitted to others [5]. ATB is characterized by clinical symptoms, radiological findings, and microbiological confirmation. The severity of the disease is associated with intensified symptoms, reduced likelihood of spontaneous recovery, compromised immune responses, and increased bacterial transmission. Factors such as pre existing health conditions or the use of immunosuppressive medications can elevate the risk of latent infections progressing to ATB [6,7]. TB primarily affects the lungs, referred to as pulmonary tuberculosis (PTB), but it can also impact other areas of the body. *M.tb.* may disseminate through the bloodstream or lymphatic system, resulting in a state of ATB, which is classified as extrapulmonary tuberculosis (EPTB). This condition may involve various organs, including bones, kidneys, brain, spine, and skin. Individuals in close contact with infected persons are at the highest risk of contracting TB.

PTB presents with nonspecific clinical symptoms that often develop gradually. Primary indicators include a chronic cough lasting at least 3 weeks, thoracic discomfort, hemoptysis or mucus expectoration, dyspnea, fever, nocturnal diaphoresis, progressive weight loss, generalized fatigue, and anorexia. PTB poses a significant risk to life if not addressed promptly. However, appropriate antibiotic treatment can lead to disease resolution, particularly when interventions occur in the early stages. In addition to its respiratory manifestations, pulmonary TB may cause systemic disturbances such as hyponatremia and glucose dysregulation, which frequently remain subclinical. Mycobacterial infection can extend beyond the lungs in cases of extrapulmonary dissemination, eliciting organ specific pathological manifestations, including febrile episodes, rigors, and localized discomfort, depending on the affected anatomical site [8,9].

Immunodiagnostic tools, including the tuberculin skin test (TST) and interferon gamma release assays (IGRAs), are employed to detect both latent and active *M.tb.* infections. While these methods are valuable, their predictive accuracy in determining the progression from LTBI to ATB remains limited [10–12].

Effective immune responses targeting and eliminating *M.tb.* are crucial for understanding the transition from LTBI to ATB. Upon inhalation, *M.tb.* specifically targets alveolar epithelial cells within the respiratory system, playing a critical role in the pathogenesis and dissemination of infection. Identifying LTBI and ATB is essential to impede the progression to ATB. Resistance to *M.tb.* infection relies on the coordinated action of innate and adaptive immune responses. Various myeloid cell populations respond to the bacterium by initiating inflammation, aiming to protect the host and potentially eradicate the infection. *M.tb.* is a highly adapted bacterium that employs sophisticated mechanisms to establish infection, evade immune responses, and persist within host tissues. Its pathogenicity is dependent on virulence factors such as lipid molecules, secretion systems, and PE/PPE proteins, which manipulate host cell processes, including autophagy and immune signaling [13].

2. Intracellular sabotage: The immune evasion strategies of *Mycobacterium tuberculosis* within alveolar macrophages

Upon inhalation of infectious aerosol droplets containing live *M.tb.* bacilli, the host's alveolar macrophages (AMs) engage in phagocytic uptake of the bacteria [14]. Macrophages play a crucial role in the initial defense against *M.tb.* infection, as they possess the ability to inhibit bacterial proliferation and serve as sites for bacterial multiplication. The entry of *M.tb.* into macrophages is facilitated by specific cell surface receptors, including integrin family complement receptors such as CR1 and CR3 [13]. As a slow growing facultative intracellular pathogen, *M.tb.* disrupts the normal maturation process of macrophages, thereby impairing their ability to eliminate infected cells [15]. Once the bacilli enter the alveoli, they are engulfed by alveolar macrophages and other phagocytic immune cells, where *M.tb.* initiates intracellular replication and establishes infection. The bacterium employs various virulence factors to bypass mucosal barriers, utilizing host macrophages to invade pulmonary tissue and other organs, ultimately promoting sustained granuloma formation. The progression of *M. tb.* infection represents a dynamic interaction between host defense mechanisms and bacterial immune evasion strategies. Key components of protective immunity, macrophages, granulomas, and $CD4^+$ T cell-derived interferon gamma (IFN-γ), constitute the primary defense against infection [2]. The bacterium effectively undermines these mechanisms to achieve varying clinical outcomes, ranging from LTBI to ATB. Upon entry into the host, *M.tb.* frequently establishes a latent infection, characterized by prolonged asymptomatic persistence and minimal transmissibility. These sophisticated strategies underscore the ability of *M.tb.* to persist within the host system and evade the immune system [5,16].

In vitro studies have demonstrated that viable bacilli disrupt phagosomal activity, blocking their maturation and subsequent fusion with the lysosomes. A subset of internalized bacilli utilize the ESAT 6 secretion system 1 (ESX-1) type VII secretion system (T7SS) to breach the phagosomal membrane, facilitating their escape into the cytosol of host cells [17,18]. By employing various immune evasion tactics, *M.tb.* effectively establish an infection within AMs, proliferating rapidly until the immune response begins to control the pathogen [19]. Consequently, *M.tb.* may undergo localized sterilization, resulting in the calcification of lesions and the formation of Ghon foci. Alternatively, huge caseation and tissue necrosis can occur, leading to pulmonary tissue impairment and liberation of *M.tb.* into the airways. Although macrophage activation has been widely studied in the context of *M.tb.* as a control, the intricate and multifaceted nature of these interactions remains an active area of research [20,21]. *M.tb* utilizes its ESX-1 T7SS to compromise phagosomal membrane integrity, facilitating its intracellular survival and pathogenicity. This system mediates the release of critical virulence factors; notably, the 6 kDa early secreted antigenic target (ESAT-6) and the 10 kDa culture filtrate protein (CFP-10) form a heterodimeric complex with a precise stoichiometric ratio of 1:1. These proteins play pivotal roles in host pathogen interactions, including membrane destabilization and host cell lysis, while simultaneously eliciting robust T cell mediated immune responses. Their multifaceted functionalities highlight their importance as significant determinants of *M.tb.* virulence [22].

3. Molecular arsenal of *Mycobacterium tuberculosis*: Dissecting protein and non-protein virulence determinants

Virulence factors are essential for *M.tb.* to establish infection, evade host defenses, and drive disease progression. These factors are categorized into two groups, encompassing a range of biomolecules such as glycolipids, lipids, nucleic acids, and glycans. Each of these biomolecules plays a crucial role in diagnostics, cellular persistence, interaction with host derived pathogenic elements, and regulation of virulence. The primary non-protein virulence determinants, predominantly located in lipid rich membranes, are vital for the survival and pathogenicity of *M.tb.* Notably, the cell wall of *M.tb.* consists of approximately 60% lipids. Additionally, the bacterium's surface expressed or secreted components facilitate interactions with host cells, significantly enhancing its pathogenic potential [23].

M.tb., an intracellular obligate pathogen, possesses specialized proteins within its proteome that are critical for its survival and pathogenicity. Alongside lipids, cell membranes incorporate various proteins that influence and modulate the host immune response [24]. Mycobacterial PE, PPE proteins, and lipoproteins employ diverse mechanisms to subvert host defenses, including inhibition of phagosome maturation, escape from phagosomes, alteration of cytokine production, modulation of autophagy, and induction of host apoptosis. The ESX secretion system is crucial for the pathogenicity of *M.tb.* by regulating the secretion of key proteins and their translocation from the cytoplasmic membrane to the bacterial cell wall. The ESX gene cluster encodes proteins such as EspB, EspA, EspC, and EspG, and secreted ESAT6 and CFP10, which constitute essential components of its virulence arsenal, along with PE-PPE family proteins (Table 2.1). Additional elements, including EccB, EccC, EccD, and MycP, are integral to the ESX system's function. Subtypes of ESX secretion systems, specifically ESX-1, ESX-3, and ESX-5, are localized within the bacterial inner membrane and facilitate the allocation of PE and PPE proteins, demonstrating their crucial role in mediating host pathogen interactions and enhancing bacterial survival [23,25].

Table 2.1 Virulence Factors of *Mycobacterium tuberculosis*.

S. No.	Non protein virulence factors	Function of the non protein virulence factors
1.	Phosphatidyl-Myo-Inositol Mannosides (PIMs) [23]	These glycolipids are notably abundant and act as biochemical precursors in the formation of LM and LAM, which are critical immunomodulatory components of the bacterial cell wall [26]
2.	Lipomannan (LM) [23]	LM potently triggers innate immune responses through its tetra acylated molecular configuration, which stimulates macrophages by engaging toll like receptor 2 (TLR2) and TLR4 signaling receptors [27]
3.	Lipoarabinomannan (LAM) [23,28]	LAM is a glycolipid made of LM attached to several arabinose units. Variations in the mannose rich cell wall of *Mycobacterium tuberculosis (M.tb)* influence its virulence and ability to adapt to the host [29]
4.	Phthiocerol Dimycocerosates (PDIM) [23,25]	PDIM helps *M.tb.* avoid detection by TLRs, which delays immune cell and adaptive response recruitment to infection sites. It is also essential for efficient phagocytosis and contributes to the rupture of phagosomes and mitochondrial membranes, aiding *M.tb*'s survival and spread within host cell [30,31]
5.	Lipids and Glycolipids [25]	Most lipids in the *M.tb* cell envelope have immunomodulatory effects that depend on several mechanisms: they can serve as ligands either activating or inhibiting of pattern recognition receptors (PRRs), conceal pathogen associated molecular patterns (PAMPs) on the bacterial surface to block PRR recognition, directly integrate into host cell membranes to disrupt their functions, and act as antigens that are presented by CD1 molecules to unconventional T cells [32] Glycolipids in the *M.tb* cell envelope are key regulators of the innate immune response during infection [33]
6.	Mycolic Acid [25,34]	*M.tb* evades host immune clearance by surrounding itself with a layer rich in mycolic acids, which are highly resistant to degradation by macrophages. This mycolic acid encapsulation not only helps the bacterium escape immune destruction but also shields it from the effects of antibiotics, contributing to its persistence and drug resistance [35]
7.	Trehalose Monomycolate (TMM), Trehalose Dimycolate (TDM) [25]; and Diacyl Trehalose [25,36]	This sugar based metabolite is vital in *M.tb.* It serves as a primary building block for cell wall glycolipids and functions as a carrier for mycolic acids during cell wall assembly [37]; Additionally, it provides *M.tb.* with carbon and energy, protects the bacterium from desiccation, freezing, and osmotic stress, and is crucial for its virulence [38]

Table 2.1 Virulence Factors of *Mycobacterium tuberculosis*.—cont'd

S. No.	Non protein virulence factors	Function of the non protein virulence factors
8.	Phenolic Glycolipids (PGL) [25,39]	Phenolic glycolipids (PGLs) suppress the host's innate immune response by reducing pro inflammatory cytokine production [40] Certain *M.tb* strains produce PGLs that counteract the immune stimulating effects of other non polar glycolipids in infected monocytes or macrophages, with their inhibitory activity showing concentration dependent suppression [41]
9.	Phospholipids [25,42]	Some *M.tb* proteins, like EspB, attach to host phospholipids such as phosphatidic acid and phosphatidylserine, which are key players in cell signaling. This binding enhances *M.tb's* virulence on its own, indicating that interacting with host phospholipids directly helps the bacterium modulate host defenses and successfully initiate infection [43]
10.	Glycans [25]	Peptidoglycan and arabinogalactan in the cell envelope are essential for *M.tb* infection and contribute significantly to the cell envelope's impermeability [44]; Glycans also influence the host and, as components of the bacterial capsule, are likely to engage with host cells before LAM and LM do [45]
11.	Nucleic acids [25,46]	Through its specialized ESAT-6 secretion system 1 (ESX-1) Type VII secretion system, *M.tb* can disrupt or break open the phagosomal membrane, allowing the bacteria to escape into the cytosol. Once in the cytosol, mycobacterial DNA is thought to be released, where it can interact with the host sensor cyclic GMP AMP synthase (cGAS). This interaction activates downstream immune responses, including type I interferon (IFN) production [47,48]

S. No.	Protein virulence factors	Function of the protein virulence factors
1.	Lipoprotein virulence factors [49]	Bacterial lipoproteins are a plentiful group of cell wall proteins anchored to membranes, playing diverse roles such as conferring antibiotic resistance, binding and transporting substrates, aiding in adherence, facilitating protein export and folding, participating in cell signaling, and contributing to sporulation [50].
2.	Phosphatase virulence factors [23]	SapM and PtpA are phosphatases involved in blocking phagosome maturation. SapM, is indispensable for *M.tb*'s ability to cause disease, as it interferes with phagosome maturation by hydrolyzing host phosphatidylinositol-3-phosphate (PI3P), thus preventing the fusion of phagosomes with lysosomes and promoting *M.tb* survival within host cells [51]. PtpA, on the other hand, is a small, secreted protein tyrosine phosphatase that dephosphorylates multiple host proteins to dampen the immune response, making it critical for both the survival and virulence of *M.tb* [52]

Continued

Table 2.1 Virulence Factors of *Mycobacterium tuberculosis*.—cont'd

S. No.	Protein virulence factors	Function of the protein virulence factors
3.	Kinase virulence factor [23]	*M.tb* secretes two serine/threonine kinases with distinct roles: protein kinase E (PknE), which helps the bacterium survive by inhibiting host cell apoptosis [53], and protein kinase G (PknG) which regulates autophagy [54], further contributing to the pathogen's ability to persist within host cells
4.	PE-PGRS family surface proteins [55]	PE-PGRS proteins are found in high concentrations within necrotic and caseous granulomas [56], where they can enhance inflammation by directly engaging Toll like receptors (TLRs) via their PGRS domain [57].
5.	Early secreted antigenic target 6 (ESAT-6) [58,59]	ESAT-6 triggers the disruption of the phagosomal membrane, enabling pathogens to escape into the cytosol, whereas mutants lacking ESAT-6 remain trapped within the phagosomes [60].
6.	Culture filtrate protein 10 (CFP-10) [61]	It seems to possess membrane lysing properties, which help microbes escape from the phagosome into the macrophage cytoplasm, and also promote host cell necrosis and the spread of bacteria to neighboring cells [62].
7.	MPT63 and MPT64 [63]	MPT63 and MPT64 are secreted, immunodominant antigens of *M.tb* that are abundantly produced during active tuberculosis (TB) [64]. MPT63 triggers macrophage cell death through a pH dependent conformational change [63]. In contrast, MPT64 stimulates the release of IFN-γ in both murine macrophages and TB patients, while also reducing apoptosis by upregulating TGF-β expression and dampening inflammation [65]
8.	Rv3655c, Rv3654c and Rv3364c putative proteins [66]; [67]	The Rv3654c and Rv3655c proteins, encoded within a seven gene operon, are transported into the cytoplasm of infected macrophages. These proteins inhibit apoptosis by targeting and disrupting the extrinsic apoptosis pathway [66] The Rv3364c protein from *M.tb* enters the host cell cytosol, where it binds to and inhibits the serine protease cathepsin G on the macrophage membrane. This inhibition disrupts cathepsin G's enzymatic activity, which in turn reduces the activation of caspase-1 and ultimately suppresses pyroptosis [23]
9.	Zmp1 [68]	Zmp1 is a zinc dependent peptidase that is crucial for the pathogenicity of *M.tb.* This enzyme has been shown to disrupt the maturation of the phagosome within macrophages, thereby aiding *M.tb.* survival. Zmp1 is believed to suppress inflammasome activation, which in turn inhibits the activation and secretion of caspase-1 and interleukin-1β (IL-1β) [69].

Table 2.1 Virulence Factors of *Mycobacterium tuberculosis*.—cont'd

S. No.	Protein virulence factors	Function of the protein virulence factors
10.	SecA2 [70]	Mycobacteria possess two SecA proteins, SecA1 and SecA2. SecA1 is vital for survival and functions as the primary SecA in the general Sec secretion pathway [71]. In contrast, certain proteins, such as SodA, have been identified as specifically relying on SecA2 for their export [72].
11.	Hydrolase porin CpnT [73]	CpnT supports the growth of *M.tb* within macrophages, likely by enhancing the bacterium's ability to acquire nutrients. Additionally, it helps the bacteria escape from the macrophage by releasing a toxin that triggers a form of cell death similar to necrosis [74]
12.	Hydrolase protease RipA [75]	RipA suppresses autophagy in RAW264.7 cells by activating the phosphatidyl inositol kinase AKT serine/threonine kinase mechanistic target of rapamycin kinase complex 1 (PI3K-AKT-mTORC1) signaling pathway through the TLR4 innate immune receptor. This protein targets the mitochondria, where it modulates mitochondrial enzyme levels, leading to metabolic reprogramming in macrophages. Additionally, RipA blocks caspase dependent apoptosis, thereby promoting the survival of the pathogen within host cells [75]
13.	Translocase secreted protein; NuoG [76]	NuoG carries out its antiapoptotic and virulence enhancing roles specifically through the enzymatic action of the nucleotide dihydrogen complex 1 (NDH-1) [76]

The table outlines major non protein and protein virulence factors that contribute to *M.tb.* pathogenesis. Non protein factors such as glycolipids, mycolic acids, and nucleic acids enable immune evasion, cell wall stability, and stress resistance. Protein factors, including lipoproteins, phosphatases, kinases, PE/PGRS proteins, and secretory antigens (ESAT-6, CFP-10, MPT63/64), disrupt phagosome maturation, modulate apoptosis, block autophagy, and suppress host immune responses. Together, these molecules orchestrate bacterial survival, persistence, and dissemination within the host.

4. PE/PPE proteins in *Mycobacterium tuberculosis*: Evolutionary expansion and virulence specialization

The PE and PPE protein families of *M.tb.* are distinctive to pathogenic mycobacterial species. These novel virulence determinants play essential roles in the pathogenicity of *M.tb.*, enabling the bacterium to evade host immune defenses through various sophisticated mechanisms. As immunogenic proteins, they significantly modulate the host immune response and apoptotic pathways, particularly by targeting mitochondrial functions, thereby facilitating intracellular survival and persistence of *M.tb.* inside the host [77]. PE/PPE proteins are characterized by conserved N-terminal motifs, specifically proline-glutamic acid (PE) or proline-proline-glutamic acid (PPE). In contrast, their C-terminal regions exhibit significant variability and frequently contain disordered domains that are expected to interact with host cellular processes. These proteins were initially discovered through comprehensive genomic sequencing of *M.tb.* [78]. The notable genomic allocation and expansion of PE/PPE proteins in pathogenic mycobacteria signifies their critical role in virulence [79]. For instance, the non pathogenic species of *Mycobacterium smegmatis* (*M. smegmatis*) possesses approximately 10 PE/PPE genes. In contrast, pathogenic species, such as *Mycobacterium bovis* (*M. bovis*) and *Mycobacterium marinum* (*M. marinum*) exhibit many more genes, with approximately 160 and 280 PE/PPE genes, respectively. Numerous PE/PPE proteins are connected to the genomic regions that encode the five ESX T7SSs in *M.tb.*, which are closely associated with its virulence [80,81]. Interestingly, the genome of M*ycobacterium leprae* (*M. leprae*) which has undergone significant reduction, contains only 12 potential PE/PPE genes, further highlighting the essential role of these proteins in the pathogenesis of mycobacteria [24,79].

4.1 Unraveling the PE/PPE mosaic: Structural diversity and functional complexity in *Mycobacterium tuberculosis* pathogenesis

The PE/PPE protein family in *M.tb.* exhibits a structurally distinctive organization characterized by specific sequence motifs and unique domain architectures. These proteins are classified based on their conserved N-terminal domains and heterogeneity observed in their C-terminal regions. While the N-terminal motifs PE and PPE are highly conserved, considerable variability in C-terminal extensions contributes to the formation of functionally diverse subclasses. PE proteins in *M.tb.* are defined by a conserved N-terminal PE domain (~110 amino acids), characterized by a Pro-Glu (PE) motif at residues 9–10, which is primarily involved in protein protein interactions, secretion, and localization to the bacterial cell surface. In contrast, their C-terminal regions exhibit variability, enabling their subclassification into distinct types [82]. PE only proteins contain a PE domain, without additional C-terminal extensions. They are considered to facilitate protein protein interactions and assist in the export of other PE/PPE proteins via T7SS, exemplified by PE3 (Rv0285) and PE10 (Rv0442c). The PE_PGRS subfamily includes proteins with an N-terminal PE domain and a C-terminal polymorphic GC rich sequences (PGRS) domain rich in glycine (G) and alanine (A) residues, which are often organized in GGX repeat motifs that vary significantly in sequence and length (100–500 residues), contributing to antigenic diversity and immune evasion. These are localized on the mycobacterial cell wall and modulate host pathogen interactions by acting as molecular decoys, binding to macrophage receptors, and influencing host cell responses, as seen in PE_PGRS proteins such as PE_PGRS17 (Rv0977) and PE_PGRS33 (Rv1818c). The PE_ major polymorphic tandem repeat (MPTR) subfamily consists of proteins with an N-terminal PE domain and C-terminal regions enriched by tandem repeat sequences distinct from PGRS motifs, where MPTR sequences involve conserved repeat motifs longer than PGRS repeats and are proposed to contribute to pathogen adaptation by altering host pathogen dynamics or are associated with bacterial stress responses, exemplified by PE_MPTR3 (Rv3477) [83].

PPE proteins in *M.tb.* are defined by a conserved N-terminal PPE domain (~180 amino acids) characterized by a PPE motif at positions seven to nine, with their C-terminal regions exhibiting significant variability, enabling subclassification into distinct types [19].

PPE only proteins, consisting solely of the conserved PPE domain, are hypothesized to function as scaffolding proteins in secretion systems, thereby facilitating the transport of effector molecules. Some proteins have been implicated in host immune modulation; however, their exact mechanisms remain unclear, as exemplified by PPE15 (Rv1039c) and PPE18 (Rv1196). The PPE MPTR subfamily contains a conserved PPE domain followed by long tandem repeat sequences, encoded in large genomic clusters and associated with virulence related secretion pathways; these proteins, often surface exposed, act as decoy antigens to evade host immunity and interact with macrophages to alter immune responses, promoting bacterial survival, as seen in PPE MPTR proteins such as PPE57 (Rv3425) and PPE68 (Rv3873) [84].

The PPE PPW and PPE SVP subfamilies, although less characterized, are distinguished by conserved C-terminal motifs, including proline rich sequences, which are proposed to contribute to host pathogen interactions through protein protein interactions with host immune receptors, aiding bacterial adhesion and invasion of host cells [19].

4.2 Immunoevasive allies: PE/PPE proteins as modulators of host signaling and intracellular survival in tuberculosis

The secretion of PE/PPE proteins into the extracellular milieu facilitates their direct interaction with host targets. Certain proteins have been demonstrated to engage with receptors such as TLR2/4 on macrophage surfaces, thereby initiating subsequent signaling pathways. The PE9-PE10 complex [81], PPE39 [85], and PE_PGRS5 [86] interact with TLR4 to activate downstream signaling pathways and modulate cytokine production. PPE18 is capable of promoting IL-10 secretion, potentially inducing a Th2 type immune response through its interaction with TLR2 [87] It has been shown to inhibit the production of proinflammatory cytokines mediated by nuclear factor kappa B (NF-κB) by upregulating the suppressor of cytokine signaling 3 (SOCS3) protein [88].

Generally, the interaction of PE/PPE proteins with cell surface receptors induces downstream signaling pathways such as NF-κB and mitogen activated protein kinase (MAPK) (including p38, Janus kinase (JNK), and extracellular regulated kinase (ERK). This activation influences cytokine production and elicits proinflammatory or antiinflammatory responses [89]. Once internalized by macrophages, *M.tb.* adapts to the intracellular environment, characterized by lower pH and reactive oxygen and nitrogen species, effectively establishing a niche. PPE60 [90] and PE13 [19] can enhance cellular resistance to low pH, surface stress, and antibiotic exposure, thereby promoting intracellular survival. Following acclimatization to a hostile environment, *M.tb.* persists within macrophages by inhibiting phagosomal acidification and preventing the fusion of phagosomes with lysosomes. Overexpression of PE-PGRS62 significantly impedes phagosome maturation. PE and PPE proteins are posited to play a role in host defense mechanisms that aim to restrict the survival of *M.tb.* or are closely linked to the bacterium's capacity to persist and proliferate within host cells. These interactions result in three distinct outcomes observed in host cellular environments, delineated as follows.

4.2.1 Cell death crossroads: Apoptosis, autophagy, and necrosis in the pathogenesis of tuberculosis

Apoptosis, a regulated form of programmed cell death, is orchestrated by host cells. While apoptosis induced by specific bacterial proteins may facilitate bacterial survival, it also plays an essential role in the eradication of intracellular bacteria and the activation of adaptive immunity [91,92]. Autophagy, characterized by the formation of autolysosomes, aids host cells in pathogen elimination. However, dysfunctional autolysosomes can degrade substantial portions of cellular proteins and organelles, resulting in autophagic cell death [92,93]. Autophagy, a cellular process involving self digestion, is essential for both innate and adaptive immunity, significantly contributing to the defense against bacterial infections and the clearance of intracellular pathogens [94]. Passive cell death, or necrosis, occurs in response to external stimuli and contributes to the spread and virulence of *M.tb.* by facilitating its release and dissemination [91,95].

5. Molecular camouflage: How *Mycobacterium tuberculosis* harnesses PE/PPE proteins and secretion systems to evade host defenses

The PE/PPE family of proteins in *M.tb.* is predominantly located within the bacterial cell wall, where they inhibit programmed cell death (apoptosis) in macrophages [96]. This mechanism facilitates bacterial survival and dissemination within lung macrophages, a process essential for evading host immune defenses and mediating interactions between pathogens and immune cells. Notably, clinical samples from granuloma tissues of patients with PTB exhibited significantly elevated levels of PE11, PE34, PE-PGRS14, PE-PGRS33, PE-PGRS57, and PE-PGRS62 compared to *M.tb.* cultured under laboratory conditions, with this upregulation demonstrating statistical significance [97]. The PE/PPE family proteins in *M.tb.* enhance bacterial survival in hostile host environments. PE-PGRS3 facilitates adhesion to type 2 pneumocytes via a distinctive arginine rich C-terminal domain. This interaction enables *M.tb.* to acquire host derived cardiolipin and phosphatidylinositol phosphate sources under conditions of phosphate deficiency. Such nutrient scavenging mechanisms are critical during TB pathogenesis, particularly in phosphate scarce niches, such as caseous granulomas and foamy macrophages, where *M.tb.* persists for extended periods. The ability of the protein to extract essential host lipids for bacterial metabolism demonstrates its role in the adaptation of *M.tb.* to nutrient stress during chronic infection. The capacity of the protein to extract vital materials for its growth and development from the host during key stages of TB pathogenesis is believed to contribute to the prolonged survival of *M.tb.* in caseous granulomas and foamy macrophages [98]. During *M.tb.* infection, PE/PPE family proteins influence several cell death pathways, including apoptosis and pyroptosis [99]. Mitochondria play a critical role in apoptosis, suggesting that proteins targeting host cell mitochondria may significantly modulate immune responses and apoptotic mechanisms [100]. Research spanning decades has focused on understanding how *M.tb.* resists and undermines various defense mechanisms of macrophages. As early as 1971, *M.tb.* was noted to impair the fusion of phagosomes and lysosomes, a critical step in the ability of macrophages to eliminate pathogens [101].

M.tb. produces a variety of lipids, including phosphatidylinositol Mannosides, lipoarabinomannan, diacyltrehaloses, polyacyltrehaloses, trehalose dimycolates, and sulfoglycolipid 1, which are implicated in modulating inflammatory signaling and inhibiting phagosome maturation [102] The bacterium employs several proteins to evade host immune defenses. Proteins such as NdkA, CpsA, and PPE2 impede the recruitment of nucleotide adenine dinucleotide phosphate hydrogen (NADPH) oxidase to phagosomes, a process essential for the generation of reactive oxygen species (ROS) necessary for pathogen elimination. Additionally, *M.tb.* catalase (KatG) and NuoG detoxify ROS, further safeguarding the bacteria. By inhibiting NADPH oxidase, *M.tb.* disrupts microtubule associated protein 1A/1B light chain 3 (LC3) associated phagocytosis, a pathway that typically targets pathogens for lysosomal degradation. This disruption also limits tumor necrosis factor (TNF) mediated apoptosis, facilitating the survival of *M.tb.* within host cells [92,103,104]. *M.tb.* encodes five distinct T7SSs, ESX-1 through ESX-5, which facilitate the export of proteins across the bacterial cell envelope. Notably, the ESX-1 and ESX-3 systems are particularly significant in interactions with macrophages and are essential for the bacterium's virulence [105]. By permeabilizing phagosomes, *M.tb.* gains access to host nutrients and transports effector proteins to the cytosol. Beyond influencing lysosomal trafficking, *M.tb.* effectors also suppress the activation of absent in melanoma 2 (AIM2) and NLR family pyrin domain containing 3 (NLRP3) inflammasomes and promote necrosis in host cells [13].

Phagosomal disruption initiates a complex interplay between host defense mechanisms and pathogen counterstrategies. Although macrophages possess intrinsic mechanisms to restore the integrity of the endolysosomal system, *M.tb.* employs the ESX-3 effector protein EsxH to impede these reparative processes, effectively subverting host cell recovery pathways [106]. Furthermore, members of the PE-PGRS protein family in *M.tb.* play a crucial role in inhibiting autophagy pathways. Additionally, within human lymphatic endothelial cells, cytosolic *M.tb.* aggregate into cord like structures, which confer resistance to selective autophagy, thereby promoting bacterial survival and persistence within the intracellular environment [107,108]. *M.tb.* employs multiple strategies to evade macrophage defenses. It neutralizes phagosomal pH using compounds such as 1-tuberculosinyladenosine and possesses mechanisms to resist acid mediated killing. Moreover, excessive inflammation triggered by type I interferons and TNF contributes to macrophage apoptosis, further aiding *M.tb.* survival. Overall, *M.tb.* employs a complex strategy to subvert macrophage antimicrobial responses and thrive in various intracellular environments [16].

6. Apoptosis at the crossroads of immunity: Host defense mechanisms and *Mycobacterium tuberculosis* evasion strategies

The immune system protects against infections by initiating rapid responses to microbial pathogens through the recognition of conserved molecular patterns unique to invading microorganisms. These molecular patterns, known as PAMPs, are detected by PRRs located on the surfaces of phagocytic immune cells, such as dendritic cells, macrophages, and neutrophils. Among the various PRRs, TLRs are the most extensively studied for their role in detecting microbial pathogens. The interaction between PRRs and PAMPs triggers conserved signaling cascades within the host defense system, ultimately regulating the expression of numerous immune response genes, primarily those involved in inflammatory processes [109]. PRRs are crucial for identifying dying cells. Programmed cell death, or apoptosis, is a tightly regulated cellular process executed through specific intracellular signaling pathways essential for proper development and maintenance of tissue homeostasis. During apoptosis, cells present distinct ligands commonly referred to as self engulfing signals, such as externalized phosphatidylserine (PS) [110] and intercellular adhesion molecule-3 (ICAM-3) [111]. Additionally, apoptotic cells display pre-existing molecules that undergo oxidative modifications, including oxidized low density lipoprotein (LDL) like structures [112]. Recognition of these apoptotic markers by the innate immune system activates macrophages, prompting the release of antiinflammatory mediators, such as TGF-β, prostaglandin E2 (PGE2), and platelet activating factor (PAF), thereby contributing to the resolution of inflammation and tissue homeostasis [110].

Apoptosis was initially described as a form of cell death characterized by morphological features, such as cell shrinkage, membrane blebbing, chromatin condensation, and apoptosome formation. Apoptosis occurs via caspases, which are cysteinyl aspartate directed proteases [113,114]. In mammalian cells, apoptosis is activated through the extrinsic and intrinsic pathways. The extrinsic pathway is mediated by cell surface receptors from the tumor necrosis factor receptor (TNFR) family, including TNFR1, Fas cell surface death receptor (FAS), and TNF receptor superfamily member 10a (TNFRSF10A) [115]. This activation leads to the initiation of caspases, particularly initiator caspases such as caspase-2, 8, 9, and 10, which subsequently activate executioner caspases such as caspase-3, 6, and 7. Conversely, the intrinsic pathway is initiated by various stressors, including withdrawal of growth factors, ultra violet (UV) radiation, DNA damage, and exposure to cytotoxic drugs [116]. Recent research indicates that *M.tb.* proteins, specifically PPE32 [117], and PE_PGRS5 [87], are associated with apoptosis in cells mediated by endoplasmic reticulum (ER) stress. In contrast, PE_PGRS62 [96], and PE_PGRS18 [118], reduce apoptosis and improve survival rates. Additionally, PE31 enhances guanylate binding protein-1 (GBP-1) expression and inhibits caspase-3 activation, thereby preventing macrophage apoptosis via the NF-κB pathway [119].

6.1 Pathogen-driven apoptosis: Decoding the multifaceted role of PE/PPE proteins in *Mycobacterium tuberculosis* infection

The PE/PPE protein family has garnered substantial research attention due to its role in modulating host cellular functions, particularly apoptosis. Apoptosis, a precisely regulated cellular mechanism responsible for the elimination of diseased or damaged cells, is crucial for maintaining immune homeostasis. Pathogens such as *M.tb.* exploit this process to their advantage, thereby influencing disease progression. PE/PPE proteins induce apoptosis in host cells, highlighting their essential role in the pathogenesis of TB.

PE/PPE proteins modulate host cell signaling cascades and initiate apoptotic processes. For instance, PE_PGRS33 facilitates apoptosis in macrophages by activating the extrinsic apoptotic pathway [120,121]. It engages TLR-2 [55] triggering the production of ROS, which subsequently activate caspase-8 and caspase-3. PE_PGRS33 stimulates the release of TNF and activates apoptosis signal regulating kinase 1 (ASK1). This activation initiates downstream signaling through MAPK pathways, specifically p38 and JNK, via its interaction with TLR-2, leading to the induction of apoptosis [122]. The mitochondrial or intrinsic apoptotic pathway is induced in response to cellular stress, ensuring the permeabilization of the mitochondrial outer membrane and the release of cytochrome C into the cytosol. Several PE/PPE proteins have been reported to modulate and disrupt this pathway. For example, PPE18 has been identified as a regulator of the phosphatidyl inositol 3 kinase (PI3K)/AKT serine/threonine kinase (Akt) signaling cascade, contributing to mitochondrial impairment and the induction of intrinsic apoptotic pathways [123]. PPE18 modulates host immune responses by interacting with TLR-2 on macrophages, activating the p38 MAPK pathway to induce IL-10, and suppressing proinflammatory cytokines such as IL-12p40. Unlike other mycobacterial TLR-2 ligands that drive TNF production, PPE18 skews the immune response toward an antiinflammatory state. The mechanism dictating whether TLR-2 signaling results in pro or antiinflammatory outcomes remains unclear. PPE18 also suppresses apoptotic processes by enhancing the expression of BCL2 apoptosis regulator (Bcl-2), an anti-apoptotic protein, while concurrently decreasing the expression of BCL2 associated X, apoptosis regulator (Bax), a pro apoptotic protein [124]. The PE25/PPE41 complex disrupts mitochondrial function, initiates the activation of caspase-9, and subsequently drives the progression of apoptosis [125]. In contrast, PPE57 regulates oxidative stress mechanisms, thereby exerting an indirect influence on apoptotic pathways [126]. PPE34 promotes ROS accumulation, leading to mitochondrial dysfunction and the subsequent induction of apoptotic processes [127]. The C-terminal domain of Rv1039c (PPE15) has been characterized as a molecular mimic of host pro apoptotic proteins, facilitating mitochondria dependent apoptosis in macrophages while simultaneously eliciting host immune responses [122].

The extrinsic apoptotic pathway is initiated through the activation of death receptors, such as Fas and TNF receptors, which subsequently lead to the activation of caspase-8. For instance, PPE2 suppresses Fas expression in macrophages, thereby inhibiting apoptotic processes and facilitating bacterial survival within the host [128]. Conversely, PPE26 enhances TNF mediated apoptotic pathways by increasing macrophage sensitivity to death receptor signaling [84]. The PE9-PE10 heterodimer represents a novel TLR-4 ligand that modulates host immune responses by inducing macrophage apoptosis. Unlike other PE/PPE proteins, which often inhibit apoptosis to promote intracellular persistence, PE9 and PE10 employ apoptosis as a strategy to eliminate macrophages and prevent immune mediated clearance of *M.tb.* by triggering TLR-4 signaling. This heterodimer activates interferon regulatory factor 3 (IRF 3) dependent pathways, leading to caspase activation and apoptotic cell death. The specificity of the PE9-PE10 heterodimer for TLR-4 provides insights into the functional diversity of the PE/PPE protein family [122].

6.2 Stress signals and cell death: PE/PPE-mediated modulation of the unfolded protein response in tuberculosis

PE/PPE proteins play a crucial role in regulating apoptosis through their involvement in endoplasmic reticulum (ER) stress and the unfolded protein response (UPR). ER stress activates the UPR signaling pathway, which can lead to apoptotic cell death under prolonged or severe conditions. Notably, PPE19 Ind.uces ER stress, triggering CHOP mediated apoptosis in infected cells. Furthermore, PE/PPE protein complexes are believed to influence UPR pathways, modulating the balance between cellular survival and apoptosis, thereby contributing to host pathogen interactions and the progression to TB [19]. PPE41 also induces ER stress, leading to the activation of caspase-12 and resulting in apoptosis [125]. Additionally, PE_PGRS29 has been implicated in the modulation of calcium signaling, a process closely associated with ER stress induced apoptotic pathways. These findings reveal the critical roles of PE/PPE family proteins in influencing ER stress responses and mediating apoptosis, thereby contributing to the pathophysiology of *M.tb.* infection [129].

6.3 Hijacking host fate: PE/PPE-mediated modulation of autophagy and apoptosis in TB pathogenesis

Autophagy and apoptosis are intricately linked cellular processes that are crucial in determining cell fate. *M.tb.* exploits this interaction to enhance its survival within the host. PPE37 has been identified as an inhibitor of apoptosis, thereby preventing apoptosis and creating a conducive environment for bacterial persistence [122]. Furthermore, PE/PPE proteins have been demonstrated to disrupt LC3-associated phagocytosis (LAP), effectively interfering with host immune responses. PPE60 impedes the activation of NF-kB, a key transcription factor that typically supports cell survival and mediates immune responses. This inhibition compromises the host's ability to effectively combat infection [90]. For instance, PE_PGRS47 modulates MAPK signaling pathways, leading to a reduction in cytokine production and the facilitation of apoptotic processes [130]. These findings reveal the critical role of PE/PPE proteins in subverting host defense mechanisms and modulating the autophagy apoptosis axis during infection.

PE13 modulates host cell apoptosis by orchestrating a complex interplay between the p38 MAPK, ERK, and NF-kB signaling pathways. PE13 triggers phosphorylation of p38 MAPK, which activates downstream ERK and NF-kB pathways. While ERK activation typically promotes cell survival, PE13 paradoxically drives pro apoptotic signals in immune cells. Additionally, PE13 Ind.uces a biphasic NF-kB response, with initial activation leading to the release of pro inflammatory cytokines, such as TNF-α and IL-1β, enhancing apoptosis in macrophages. The crosstalk between p38 MAPK, ERK, and NF-kB pathways governs the extent of apoptosis, with NF-kB establishing a feedback loop that sustains inflammation and ultimately shifts the balance toward apoptotic cell death. This targeted apoptosis facilitates the elimination of infected macrophages; however, if dysregulated, it can lead to necrosis like cell death that supports bacterial dissemination. PE13's ability to fine tune these pathways shows its crucial role in *M.tb.* pathogenesis and immune evasion [131]. PE_PGRS30 modulates the expression of cytokines, including TNF-α, IL-1β, and IL-6, thereby diminishing immune cell recruitment and apoptotic signaling. It interferes with NF-κB activation, consequently restricting the transcription of proapoptotic genes. Furthermore, PE_PGRS30 inhibits the MAPK pathways, specifically suppressing p38 and JNK phosphorylation, to enhance macrophage survival. By down-regulating proinflammatory cytokines, death receptors, and mitochondrial outer membrane permeabilization (MOMP) proteins such as BAX and BAK, PE_PGRS30 prevents apoptosis, thereby ensuring macrophage longevity as a replication niche for *M.tb.* The inhibition of apoptosis supports immune evasion, facilitates granuloma formation, and promotes bacterial persistence within the host [130].

PPE11, a protein of *M.tb.*, modulates host apoptosis through complex mechanisms involving inflammatory pathways, MAPK signaling, and mitochondrial regulation. PPE11 suppresses the release of proinflammatory cytokines, such as TNF-α, IL-1β, and IL-6, delaying apoptosis to create a survival niche for the pathogen. It impairs NF-κB activation, favoring antiapoptotic gene expression, down-regulating p38 and JNK MAPK pathways, sustaining ERK activation, and promoting cell survival. At the mitochondrial level, PPE11 alters the BAX/BAK balance, stabilizing the mitochondrial membrane and inhibiting cytochrome C secretion and caspase activation. By inhibiting apoptosis and maintaining mitochondrial integrity, PPE11 facilitates bacterial replication in macrophages Fig. 2.1.

This delay in apoptosis ultimately leads to necrosis like cell death, which aids bacterial dissemination. PPE11's ability to regulate inflammation and apoptosis demonstrates its role as a key virulence

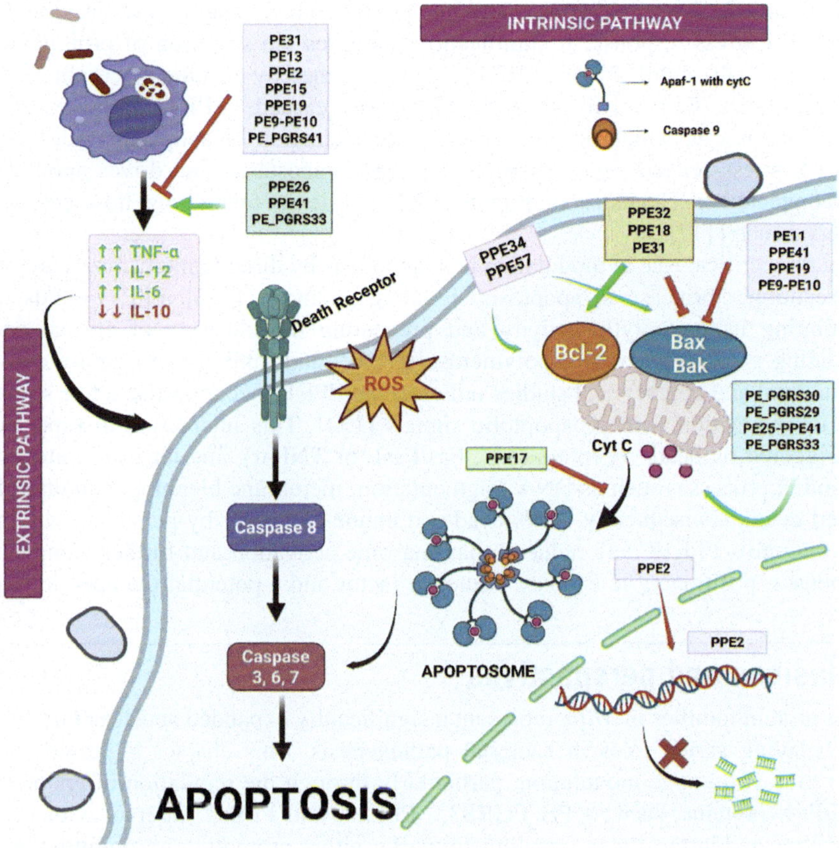

FIGURE 2.1 Regulation of Host Cell Apoptosis by PE/PPE Proteins of *Mycobacterium tuberculosis*.

This figure summarizes how PE and PPE family proteins of *M.tb.* influence host cell apoptosis through both extrinsic and intrinsic pathways. In the extrinsic route, proteins such as PE_PGRS33, PPE26, and the PE9-PE10 heterodimer promote apoptosis by enhancing death receptor signaling and caspase activation, whereas PPE2 reduces Fas expression and PPE11 limits NF-κB and MAPK signaling to suppress apoptosis. In the intrinsic pathway, proteins including PE25-PPE41, PPE34, PPE15, PPE19, and PPE41 disrupt mitochondrial stability, leading to cytochrome C release and caspase-9 activation, while PE_PGRS30, PPE11, and PE31 act to preserve mitochondrial integrity and inhibit apoptosis. Some proteins, such as PPE17, directly interfere with executioner caspases, further preventing cell death. Together, these findings highlight that PE/PPE proteins have diverse and sometimes opposing effects, enabling *Mtb.* to fine tune host cell death responses to balance immune evasion with long term survival.

factor in TB pathogenesis [130]. PPE11, PPE25, and PPE26 modulate host inflammatory and apoptotic signaling pathways to facilitate bacterial survival. By disrupting NF-κB activation, MAPK signaling cascades, and mitochondrial function, these proteins extend the lifespan of infected macrophages, providing a favorable intracellular environment for pathogen persistence [132]. PPE32 Ind.uces macrophage apoptosis by modulating cytokine production, activating the caspase cascade, and enhancing the ER stress response. It significantly increases the secretion of proinflammatory cytokines, such as IL-12p40 and IL-32, via ERK1/2 signaling, thereby priming macrophages for apoptosis while supporting bacterial survival. Through the caspase cascade, PPE32 promotes the cleavage of caspase-3 and caspase-9, which are essential for the execution of apoptosis. Additionally, PPE32 upregulates ER stress related genes, disrupts cellular homeostasis, and drives apoptotic pathways. These interconnected mechanisms highlight PPE32's role in facilitating host cell apoptosis and pathogen persistence [117].

PPE17 plays a critical role in modulating host apoptosis by directly interacting with caspase-3, the principal executioner protease in apoptosis signaling pathways [133]. PPE17 binds to caspase-3, thereby inhibiting its proteolytic activity and preventing the cleavage of critical apoptotic substrates, including poly (ADP-ribose) polymerase (PARP) and inhibitor of caspase activated DNase (ICAD). Structural and biochemical studies indicate that PPE17 occupies the active site of caspase-3, obstructing its interaction with proapoptotic signals [134]. This interaction disrupts both extrinsic apoptosis, triggered by death receptors (e.g., Fas/FasL or TNF-α), and intrinsic mitochondrial pathways, preventing processes such as DNA fragmentation, membrane blebbing, and other hallmarks of apoptotic cell death. Consequently, *M.tb.* evades immune clearance by persisting within the macrophages. Additionally, PPE17 may reduce inflammasome activation and further dampen host inflammatory responses, positioning it as a key virulence factor and a potential therapeutic target [135].

7. Conclusions and perspectives

The PE/PPE protein families of *M.tb.* represent a significantly expanded and structurally diverse class of proteins, playing varied roles in bacterial pathogenesis. This chapter highlights their essential contribution to host immune modulation, particularly through the regulation of apoptotic pathways. Specific PE/PPE proteins, such as PE_PGRS33, PPE18, and PPE37, interact with host cell membranes, receptors, and intracellular signaling networks, either promoting or inhibiting apoptosis. This dynamic modulation facilitates intracellular survival, enhances bacterial dissemination, and contributes to granuloma formation. Furthermore, these proteins are pivotal in immune evasion by impairing antigen presentation and attenuating proinflammatory responses, thereby supporting pathogen persistence within the host environment.

Elucidating the molecular mechanisms by which PE/PPE proteins modulate host cell apoptosis provides critical insights into *M.tb.* virulence strategies. Despite increasing evidence of their functional significance, key aspects remain unexplored, particularly concerning the precise molecular determinants and downstream signaling pathways affected by these proteins. Future research should aim to address these knowledge gaps, enabling the development of targeted therapeutics and immunomodulatory strategies to counteract *M.tb.* infections.

To comprehensively delineate the role of PE/PPE proteins in *M.tb.* pathogenesis and host interactions, several key research directions warrant further investigation. Structural characterization through advanced cryo electron microscopy and protein protein interaction studies will facilitate the resolution of PE/PPE protein domains and their engagement with host signaling pathways, potentially unveiling novel virulence determinants. Functional genomic approaches, including CRISPRi and transposon mutagenesis, can elucidate the essentiality of specific PE/PPE genes, particularly those implicated in apoptosis regulation. Comparative analyses of macrophages, dendritic cells, and epithelial cells are crucial for establishing the cell type specific effects of PE/PPE mediated apoptosis and determining how these interactions influence bacterial persistence and immune evasion across diverse tissue environments. In vivo models such as zebrafish and humanized mouse systems offer valuable platforms for assessing the contribution of individual PE/PPE proteins to granuloma formation, latency maintenance, and disease reactivation. Given their immunogenic properties and roles in immune modulation, certain PE/PPE proteins hold promise as antigenic candidates for subunit vaccine development or as molecular targets for host directed therapeutic strategies aimed at reestablishing apoptosis and immune homeostasis. Ultimately, PE/PPE proteins are not merely genomic expansions but function as critical modulators of *M.tb.* infection. Bridging the knowledge gap between their structural attributes and host specific activities will be crucial for advancing novel diagnostic and therapeutic interventions for TB control.

References

[1] Anishiya Chella Daisy ER, Vinothini K, Rajan M. Chapter 5—pH, thermo- and stimuli-responsive nanotherapy for tuberculosis. In: Rajan M, editor. A mechanistic approach to medicines for tuberculosis nanotherapy [Internet]. Academic Press; 2021. p. 119–48.

[2] Anes E, Pires D, Mandal M, Azevedo-Pereira JM. ESAT-6 a major virulence factor of *Mycobacterium tuberculosis*. Biomolecules June 9, 2023;13(6):968.

[3] Tuberculosis (TB). 2025 Apr 29. Available from: https://www.who.int/news-room/fact-sheets/detail/tuberculosis.

[4] Dirix V, Dauby N, Hites M, Watelet E, Van Praet A, Godefroid A, et al. Optimal detection of latent *Mycobacterium tuberculosis* infection by combined heparin-binding hemagglutinin (HBHA) and early secreted antigenic target 6 (ESAT-6) whole-blood interferon gamma release assays. J Clin Microbiol May 18, 2022;60(5):e0244321.

[5] Ehrt S, Schnappinger D, Rhee KY. Metabolic principles of persistence and pathogenicity in *Mycobacterium tuberculosis*. Nat Rev Microbiol August 2018;16(8):496–507.

[6] Khabibullina NF, Kutuzova DM, Burmistrova IA, Lyadova IV. The biological and clinical aspects of a latent tuberculosis infection. Trop Med Infect Dis March 8, 2022;7(3):48.

[7] Anton C, Machado FD, Ramirez JMA, Bernardi RM, Palominos PE, Brenol CV, et al. Latent tuberculosis infection in patients with rheumatic diseases. J Bras Pneumol Publicacao Soc Bras Pneumol E Tisilogia April 25, 2019;45(2).

[8] Luies L, du Preez I. The echo of pulmonary tuberculosis: mechanisms of clinical symptoms and other disease-induced systemic complications. Clin Microbiol Rev September 16, 2020;33(4).

[9] Loddenkemper R, Lipman M, Zumla A. Clinical aspects of adult tuberculosis. Cold Spring Harb Perspect Med February 6, 2015;6(1):a017848.

[10] Trajman A, Steffen RE, Menzies D. Interferon-gamma release assays versus tuberculin skin testing for the diagnosis of latent tuberculosis infection: an overview of the evidence. Pulm Med 2013;2013:601737.

[11] Sethi D, Goyal S. Miliary tuberculosis-related acute respiratory distress syndrome: early diagnosis can save life. Indian J Respir Care December 6, 2022;10(2):243–5.

[12] Mortazavi-Moghaddam SG, Pagheh AS, Ahmadpour E, Barac A, Ebrahimzadeh A. Tuberculous meningitis and miliary tuberculosis in Iran: a review. Asian Pac J Trop Med April 2022;15(4):143.

[13] Rastogi S, Ellinwood S, Augenstreich J, Mayer-Barber KD, Briken V. *Mycobacterium tuberculosis* inhibits the NLRP3 inflammasome activation via its phosphokinase PknF. PLoS Pathog July 2021;17(7):e1009712.

[14] Ahmad F, Rani A, Alam A, Zarin S, Pandey S, Singh H, et al. Macrophage: a cell with many faces and functions in tuberculosis. Front Immunol 2022;13:747799.

[15] Pandey S, Ehtesham NZ, Hasnain SE. Editorial: emerging concepts in *Mycobacterium tuberculosis* pathogenesis: host-pathogen interaction and stress adaption mechanisms. Front Cell Infect Microbiol 2023;13: 1148756.

[16] Kanabalan RD, Lee LJ, Lee TY, Chong PP, Hassan L, Ismail R, et al. Human tuberculosis and *Mycobacterium tuberculosis* complex: a review on genetic diversity, pathogenesis and omics approaches in host biomarkers discovery. Microbiol Res May 1, 2021;246:126674.

[17] Buter J, Cheng TY, Ghanem M, Grootemaat AE, Raman S, Feng X, et al. *Mycobacterium tuberculosis* releases an antacid that remodels phagosomes. Nat Chem Biol September 2019;15(9):889–99.

[18] Wong KW. The role of ESX-1 in *Mycobacterium tuberculosis* pathogenesis. Microbiol Spectr May 19, 2017;5(3). https://doi.org/10.1128/microbiolspec.tbtb2-0001–2015.

[19] Sharma T, Alam A, Ehtram A, Rani A, Grover S, Ehtesham NZ, et al. The *Mycobacterium tuberculosis* PE_PGRS protein family acts as an immunological decoy to subvert host immune response. Int J Mol Sci January 2022;23(1):525.

[20] Thakur M, Muniyappa K. Macrophage activation highlight an important role for NER proteins in the survival, latency and multiplication of *Mycobacterium tuberculosis*. Tuberc Edinb Scotl January 2023;138: 102284.

[21] Mata E, Tarancon R, Guerrero C, Moreo E, Moreau F, Uranga S, et al. Pulmonary BCG induces lung-resident macrophage activation and confers long-term protection against tuberculosis. Sci Immunol September 24, 2021;6(63):eabc2934.

[22] Malik AA, Shariq M, Sheikh JA, Jaiswal U, Fayaz H, Shrivastava G, et al. Mechanisms of immune evasion by *Mycobacterium tuberculosis*: the impact of T7SS and cell wall lipids on host defenses. Crit Rev Biochem Mol Biol October 2024;59(5):310–36.

[23] Ramon-Luing LA, Palacios Y, Ruiz A, Téllez-Navarrete NA, Chavez-Galan L. Virulence factors of *Mycobacterium tuberculosis* as modulators of cell death mechanisms. Pathog Basel Switz June 18, 2023;12 (6):839.

[24] Comparative genomics of metabolic pathways in Mycobacterium species: gene duplication, gene decay and lateral gene transfer | FEMS Microbiology Reviews | Oxford Academic. Available from: https://academic.oup.com/femsre/article/30/6/906/494109.

[25] Rahlwes KC, Dias BRS, Campos PC, Alvarez-Arguedas S, Shiloh MU. Pathogenicity and virulence of *Mycobacterium tuberculosis*. Virulence December 2023;14(1):2150449.

[26] Bansal-Mutalik R, Nikaido H. Mycobacterial outer membrane is a lipid bilayer and the inner membrane is unusually rich in diacyl phosphatidylinositol dimannosides. Proc Natl Acad Sci U S A April 1, 2014;111 (13):4958–63.

[27] Doz E, Rose S, Nigou J, Gilleron M, Puzo G, Erard F, et al. Acylation determines the toll-like receptor (TLR)-dependent positive versus TLR2-, mannose receptor-, and SIGNR1-independent negative regulation of pro-inflammatory cytokines by mycobacterial lipomannan. J Biol Chem September 7, 2007;282(36): 26014–25.

[28] Fukuda T, Matsumura T, Ato M, Hamasaki M, Nishiuchi Y, Murakami Y, et al. Critical roles for lipomannan and lipoarabinomannan in cell wall integrity of mycobacteria and pathogenesis of tuberculosis. mBio February 19, 2013;4(1).

[29] Torrelles JB, Schlesinger LS. Diversity in *Mycobacterium tuberculosis* mannosylated cell wall determinants impacts adaptation to the host. Tuberc Edinb Scotl March 2010;90(2):84–93.

[30] Rens C, Chao JD, Sexton DL, Tocheva EI, Av-Gay Y. Roles for phthiocerol dimycocerosate lipids in *Mycobacterium tuberculosis* pathogenesis. Microbiol Read Engl March 2021;167(3).

[31] Day TA, Mittler JE, Nixon MR, Thompson C, Miner MD, Hickey MJ, et al. *Mycobacterium tuberculosis* strains lacking surface lipid phthiocerol dimycocerosate are susceptible to killing by an early innate host response. Infect Immun December 2014;82(12):5214–22.

[32] Correia-Neves M, Nigou J, Mousavian Z, Sundling C, Källenius G. Immunological hyporesponsiveness in tuberculosis: the role of mycobacterial glycolipids. Front Immunol 2022;13.

[33] Ishikawa E, Mori D, Yamasaki S. Recognition of mycobacterial lipids by immune receptors. Trends Immunol January 2017;38(1):66–76.

[34] Wang H, Liu D, Zhou X. Effect of mycolic acids on host immunity and lipid metabolism. Int J Mol Sci December 28, 2023;25(1):396.

[35] Kumar N, Srivastava R, Mongre RK, Mishra CB, Kumar A, Khatoon R, et al. Identifying the novel inhibitors against the mycolic acid biosynthesis pathway target "mtFabH" of *Mycobacterium tuberculosis*. Front Microbiol May 6, 2022;13.

[36] Scalavino V, Liso M, Cavalcanti E, Gigante I, Lippolis A, Mastronardi M, et al. miR-369-3p modulates inducible nitric oxide synthase and is involved in regulation of chronic inflammatory response. Sci Rep September 29, 2020;10(1):15942.

[37] Ryll R, Kumazawa Y, Yano I. Immunological properties of trehalose dimycolate (cord factor) and other mycolic acid-containing glycolipids-a review. Microbiol Immunol 2001;45(12):801–11.

[38] Shi L, Zhang H, Qiu Y, Wang Q, Wu X, Wang H, et al. Biochemical characterization and ligand-binding properties of trehalose-6-phosphate phosphatase from *Mycobacterium tuberculosis*. Acta Biochim Biophys Sin October 2013;45(10):837–44.

[39] Cambier CJ, O'Leary SM, O'Sullivan MP, Keane J, Ramakrishnan L. Phenolic glycolipid facilitates mycobacterial escape from microbicidal tissue-resident macrophages. Immunity September 19, 2017;47 (3):552–65.

[40] Oldenburg R, Demangel C. Pathogenic and immunosuppressive properties of mycobacterial phenolic glycolipids. Biochimie October 2017;141:3–8.

[41] Reed MB, Domenech P, Manca C, Su H, Barczak AK, Kreiswirth BN, et al. A glycolipid of hypervirulent tuberculosis strains that inhibits the innate immune response. Nature September 2, 2004;431(7004):84–7.

[42] Ruhl CR, Pasko BL, Khan HS, Kindt LM, Stamm CE, Franco LH, et al. *Mycobacterium tuberculosis* sulfolipid-1 activates nociceptive neurons and induces cough. Cell April 16, 2020;181(2):293–305.e11.

[43] Chen JM, Zhang M, Rybniker J, Boy-Röttger S, Dhar N, Pojer F, et al. *Mycobacterium tuberculosis* EspB binds phospholipids and mediates EsxA-independent virulence. Mol Microbiol September 2013;89(6): 1154–66.

[44] Dulberger CL, Rubin EJ, Boutte CC. The mycobacterial cell envelope—a moving target. Nat Rev Microbiol January 2020;18(1):47–59.

[45] Glatman-Freedman A, Casadevall A, Dai Z, Jacobs WR, Li A, Morris SL, et al. Antigenic evidence of prevalence and diversity of *Mycobacterium tuberculosis* arabinomannan. J Clin Microbiol July 2004;42(7): 3225–31.

[46] Zhang L, Jiang X, Pfau D, Ling Y, Nathan CF. Type I interferon signaling mediates Mycobacterium tuberculosis–induced macrophage death. J Exp Med October 30, 2020;218(2).

[47] Wassermann R, Gulen MF, Sala C, Perin SG, Lou Y, Rybniker J, et al. *Mycobacterium tuberculosis* differentially activates cGAS- and inflammasome-dependent intracellular immune responses through ESX-1. Cell Host Microbe June 10, 2015;17(6):799–810.

[48] Watson RO, Bell SL, MacDuff DA, Kimmey JM, Diner EJ, Olivas J, et al. The cytosolic sensor cGAS detects *Mycobacterium tuberculosis* DNA to induce type I interferons and activate autophagy. Cell Host Microbe June 10, 2015;17(6):811–9.

[49] Mycobacterium tuberculosis lipoprotein LprG binds lipoarabinomannan and determines its cell envelope localization to control phagolysosomal fusion | PLoS Pathog. 2025 Apr 30. Available from: https://journals.plos.org/plospathogens/article?id=10.1371/journal.ppat.1004471.

[50] Sutcliffe IC, Russell RR. Lipoproteins of gram-positive bacteria. J Bacteriol March 1995;177(5):1123–8.
[51] Puri RV, Reddy PV, Tyagi AK. Secreted acid phosphatase (SapM) of *Mycobacterium tuberculosis* is indispensable for arresting phagosomal maturation and growth of the pathogen in Guinea pig tissues. PLoS One 2013;8(7):e70514.
[52] Bach H, Papavinasasundaram KG, Wong D, Hmama Z, Av-Gay Y. *Mycobacterium tuberculosis* virulence is mediated by PtpA dephosphorylation of human vacuolar protein sorting 33B. Cell Host Microbe May 15, 2008;3(5):316–22.
[53] Kumar D, Narayanan S. pknE, a serine/threonine kinase of *Mycobacterium tuberculosis* modulates multiple apoptotic paradigms. Infect Genet Evol J Mol Epidemiol Evol Genet Infect Dis June 2012;12(4):737–47.
[54] Ge P, Lei Z, Yu Y, Lu Z, Qiang L, Chai Q, et al. Tuberculosis PknG manipulates host autophagy flux to promote pathogen intracellular survival. Autophagy March 2022;18(3):576–94.
[55] Kramarska E, Squeglia F, De Maio F, Delogu G, Berisio R. PE_PGRS33, an important virulence factor of *Mycobacterium tuberculosis* and potential target of host humoral immune response. Cells January 15, 2021; 10(1):161.
[56] Kruh NA, Troudt J, Izzo A, Prenni J, Dobos KM. Portrait of a pathogen: the *Mycobacterium tuberculosis* proteome in vivo. PLoS One November 11, 2010;5(11):e13938.
[57] Camassa S, Palucci I, Iantomasi R, Cubeddu T, Minerva M, De Maio F, et al. Impact of pe_pgrs33 gene polymorphisms on *Mycobacterium tuberculosis* infection and pathogenesis. Front Cell Infect Microbiol 2017;7:137.
[58] Naqvi N, Ahuja Y, Zarin S, Alam A, Ali W, Shariq M, et al. BCG's role in strengthening immune responses: implications for tuberculosis and comorbid diseases. Infect Genet Evol J Mol Epidemiol Evol Genet Infect Dis January 2025;127:105703.
[59] Mustafa AS. Chemical and biological characterization of Mycobacterium tuberculosis-specific ESAT6-like proteins and their potentials in the prevention of tuberculosis and asthma. Med Princ Pract September 13, 2023;32(4–5):217–24.
[60] Simeone R, Bobard A, Lippmann J, Bitter W, Majlessi L, Brosch R, et al. Phagosomal rupture by *Mycobacterium tuberculosis* results in toxicity and host cell death. PLoS Pathog February 2, 2012;8(2): e1002507.
[61] Valizadeh A, Imani Fooladi AA, Sedighian H, Mahboobi M, Gholami Parizad E, Behzadi E, et al. Evaluating the performance of PPE44, HSPX, ESAT-6 and CFP-10 factors in tuberculosis subunit vaccines. Curr Microbiol July 19, 2022;79(9):260.
[62] de Jonge MI, Pehau-Arnaudet G, Fretz MM, Romain F, Bottai D, Brodin P, et al. ESAT-6 from *Mycobacterium tuberculosis* dissociates from its putative chaperone CFP-10 under acidic conditions and exhibits membrane-lysing activity. J Bacteriol August 2007;189(16):6028–34.
[63] Sannigrahi A, Nandi I, Chall S, Jawed JJ, Halder A, Majumdar S, et al. Conformational switch driven membrane pore formation by Mycobacterium secretory protein MPT63 induces macrophage cell death. ACS Chem Biol July 19, 2019;14(7):1601–10.
[64] Mustafa T, Leversen NA, Sviland L, Wiker HG. Differential in vivo expression of mycobacterial antigens in Mycobacterium tuberculosisinfected lungs and lymph node tissues. BMC Infect Dis October 3, 2014;14(1): 535.
[65] Fan Q, Lu M, Xia ZY, Bao L. *Mycobacterium tuberculosis* MPT64 stimulates the activation of murine macrophage modulated by IFN-γ. Eur Rev Med Pharmacol Sci December 2013;17(24):3296–305.
[66] Danelishvili L, Yamazaki Y, Selker J, Bermudez LE. Secreted *Mycobacterium tuberculosis* Rv3654c and Rv3655c proteins participate in the suppression of macrophage apoptosis. PLoS One May 4, 2010;5(5): e10474.
[67] Danelishvili L, Everman JL, McNamara MJ, Bermudez LE. Inhibition of the plasma-membrane-associated serine protease cathepsin G by *Mycobacterium tuberculosis* Rv3364c suppresses caspase-1 and pyroptosis in macrophages. Front Microbiol 2011;2:281.

[68] Kurane T, Matsunaga T, Ida T, Sawada K, Nishimura A, Fukui M, et al. GRIM-19 is a target of mycobacterial Zn2+ metalloprotease 1 and indispensable for NLRP3 inflammasome activation. FASEB J Off Publ Fed Am Soc Exp Biol January 2022;36(1):e22096.

[69] Dak M, Šlachtová V, Šebela M, Bazgier V, Berka K, Smiejkowska N, et al. Novel heterocyclic hydroxamates as inhibitors of the mycobacterial zinc metalloprotease Zmp1 to probe its mechanism of function. Eur J Med Chem December 15, 2022;244:114831.

[70] The two distinct types of SecA2-dependent export systems—PubMed. 2025 Apr 30. Available from: https://pubmed.ncbi.nlm.nih.gov/31215505/.

[71] Silencing essential protein secretion in mycobacterium smegmatis by using tetracycline repressors. J. Bacteriol. Available from: https://journals.asm.org/doi/full/10.1128/jb.00216-07.

[72] Molecular microbiology | Microbiol J | Wiley Online Library. 2025 May 9. Available from: https://onlinelibrary.wiley.com/doi/full/10.1046/j.1365-2958.2003.03438.x.

[73] Sun J, Siroy A, Lokareddy RK, Speer A, Doornbos KS, Cingolani G, et al. The tuberculosis necrotizing toxin kills macrophages by hydrolyzing NAD. Nat Struct Mol Biol September 2015;22(9):672–8.

[74] Danilchanka O, Sun J, Pavlenok M, Maueröder C, Speer A, Siroy A, et al. An outer membrane channel protein of *Mycobacterium tuberculosis* with exotoxin activity. Proc Natl Acad Sci U S A May 6, 2014;111 (18):6750–5.

[75] Shariq M, Quadir N, Sharma N, Singh J, Sheikh JA, Khubaib M, et al. *Mycobacterium tuberculosis* RipA dampens TLR4-mediated host protective response using a multi-pronged approach involving autophagy, apoptosis, metabolic repurposing, and immune modulation. Front Immunol 2021;12:636644.

[76] Velmurugan K, Chen B, Miller JL, Azogue S, Gurses S, Hsu T, et al. *Mycobacterium tuberculosis* nuoG is a virulence gene that inhibits apoptosis of infected host cells. PLoS Pathog July 2007;3(7):e110.

[77] Priyanka SS, Sharma M. Role of PE/PPE proteins of *Mycobacterium tuberculosis* in triad of host mitochondria, oxidative stress and cell death. Microb Pathog August 1, 2024;193:106757.

[78] Koleske BN, Jacobs WR, Bishai WR. The *Mycobacterium tuberculosis* genome at 25 years: lessons and lingering questions. J Clin Investig 2023;133(19):e173156.

[79] Kapopoulou A, Lew JM, Cole ST. The MycoBrowser portal: a comprehensive and manually annotated resource for mycobacterial genomes. Tuberc Edinb Scotl January 2011;91(1):8–13.

[80] Wu J, Ru H wei, Xiang Z hao, Jiang J, chen WY, Zhang L, et al. WhiB4 regulates the PE/PPE gene family and is essential for virulence of Mycobacterium marinum. Sci Rep June 7, 2017;7(1):3007.

[81] Gey van Pittius NC, Sampson SL, Lee H, Kim Y, van Helden PD, Warren RM. Evolution and expansion of the *Mycobacterium tuberculosis* PE and PPE multigene families and their association with the duplication of the ESAT-6 (esx) gene cluster regions. BMC Evol Biol November 15, 2006;6:95.

[82] D'Souza C, Kishore U, Tsolaki AG. The PE-PPE family of *Mycobacterium tuberculosis*: proteins in disguise. Immunobiology March 2023;228(2):152321.

[83] Ehtram A, Shariq M, Ali S, Quadir N, Sheikh JA, Ahmad F, et al. Teleological cooption of *Mycobacterium tuberculosis* PE/PPE proteins as porins: role in molecular immigration and emigration. Int J Med Microbiol April 1, 2021;311(3):151495.

[84] Ehtram A, Shariq M, Quadir N, Jamal S, Pichipalli M, Zarin S, et al. Deciphering the functional roles of PE18 and PPE26 proteins in modulating *Mycobacterium tuberculosis* pathogenesis and immune response. Front Immunol January 30, 2025;16:1517822.

[85] Tiwari B, Ramakrishnan UM, Raghunand TR. The *Mycobacterium tuberculosis* protein pair PE9 (Rv1088)-PE10 (Rv1089) forms heterodimers and induces macrophage apoptosis through Toll-like receptor 4. Cell Microbiol November 2015;17(11):1653–69.

[86] Choi HH, Kwon KW, Han SJ, Kang SM, Choi E, Kim A, et al. PPE39 of the *Mycobacterium tuberculosis* strain Beijing/K induces Th1-cell polarization through dendritic cell maturation. J Cell Sci September 5, 2019;132(17).

[87] Grover S, Sharma T, Singh Y, Kohli S, Manjunath P, Singh A, et al. The PGRS domain of *Mycobacterium tuberculosis* PE_PGRS protein Rv0297 is involved in endoplasmic reticulum stress-mediated apoptosis through toll-like receptor 4. mBio June 19, 2018;9(3):e01017–8.
[88] Nair S, Pandey AD, Mukhopadhyay S. The PPE18 protein of *Mycobacterium tuberculosis* inhibits NF-κB/rel-mediated proinflammatory cytokine production by upregulating and phosphorylating suppressor of cytokine signaling 3 protein. J Immunol Baltim Md 1950 May 1, 2011;186(9):5413–24.
[89] Yang G, Luo T, Sun C, Yuan J, Peng X, Zhang C, et al. PPE27 in Mycobacterium smegmatis enhances mycobacterial survival and manipulates cytokine secretion in mouse macrophages. J Interferon Cytokine Res Off J Int Soc Interferon Cytokine Res September 2017;37(9):421–31.
[90] Gong Z, Kuang Z, Li H, Li C, Ali MK, Huang F, et al. Regulation of host cell pyroptosis and cytokines production by *Mycobacterium tuberculosis* effector PPE60 requires LUBAC mediated NF-κB signaling. Cell Immunol January 2019;335:41–50.
[91] Fink SL, Cookson BT. Apoptosis, pyroptosis, and necrosis: mechanistic description of dead and dying eukaryotic cells. Infect Immun April 2005;73(4):1907–16.
[92] Srinivasan L, Ahlbrand S, Briken V. Interaction of *Mycobacterium tuberculosis* with host cell death pathways. Cold Spring Harb Perspect Med June 26, 2014;4(8):a022459.
[93] Mariño G, Niso-Santano M, Baehrecke EH, Kroemer G. Self-consumption: the interplay of autophagy and apoptosis. Nat Rev Mol Cell Biol February 2014;15(2):81–94.
[94] Shariq M, Quadir N, Alam A, Zarin S, Sheikh JA, Sharma N, et al. The exploitation of host autophagy and ubiquitin machinery by *Mycobacterium tuberculosis* in shaping immune responses and host defense during infection. Autophagy 2023;19(1):3–23.
[95] Behar SM, Divangahi M, Remold HG. Evasion of innate immunity by *Mycobacterium tuberculosis*: is death an exit strategy? Nat Rev Microbiol September 2010;8(9):668–74.
[96] Long Q, Xiang X, Yin Q, Li S, Yang W, Sun H, et al. PE_PGRS62 promotes the survival of Mycobacterium smegmatis within macrophages via disrupting ER stress-mediated apoptosis. J Cell Physiol November 2019;234(11):19774–84.
[97] Sampson SL. Mycobacterial PE/PPE proteins at the host-pathogen interface. Clin Dev Immunol 2011;2011: 497203.
[98] De Maio F, Salustri A, Battah B, Palucci I, Marchionni F, Bellesi S, et al. PE_PGRS3 ensures provision of the vital phospholipids cardiolipin and phosphatidylinositols by promoting the interaction between *M. tuberculosis* and host cells. Virulence 2021;12(1):868–84.
[99] Shariq M, Quadir N, Sheikh JA, Singh AK, Bishai WR, Ehtesham NZ, et al. Post translational modifications in tuberculosis: ubiquitination paradox. Autophagy March 2021;17(3):814–7.
[100] Vringer E, Tait SWG. Mitochondria and cell death-associated inflammation. Cell Death Differ February 2023;30(2):304–12.
[101] Armstrong JA, Hart PDA. Response of cultured macrophages to *Mycobacterium tuberculosis*, with observations on fusion of lysosomes with phagosomes. J Exp Med 1971;134(3):713–40.
[102] Augenstreich J, Briken V. Host cell targets of released lipid and secreted protein effectors of *Mycobacterium tuberculosis*. Front Cell Infect Microbiol 2020;10.
[103] Köster S, Upadhyay S, Chandra P, Papavinasasundaram K, Yang G, Hassan A, et al. *Mycobacterium tuberculosis* is protected from NADPH oxidase and LC3-associated phagocytosis by the LCP protein CpsA. Proc Natl Acad Sci U S A October 10, 2017;114(41):E8711–20.
[104] Miller JL, Velmurugan K, Cowan MJ, Briken V. The type I NADH dehydrogenase of *Mycobacterium tuberculosis* counters phagosomal NOX2 activity to inhibit TNF-alpha-mediated host cell apoptosis. PLoS Pathog April 22, 2010;6(4):e1000864.
[105] Ehtram A, Shariq M, Ali S, Ehtesham NZ, Hasnain SE. ESX5 associated PE proteins of *Mycobacterium tuberculosis* drives Th1 type polarization: an immunomodulatory paradigm. J Immunol May 1, 2020;204 (1_Suppl.).

[106] Mittal E, Skowyra ML, Uwase G, Tinaztepe E, Mehra A, Köster S, et al. *Mycobacterium tuberculosis* type VII secretion system effectors differentially impact the ESCRT endomembrane damage response. mBio November 27, 2018;9(6).

[107] Strong EJ, Jurcic Smith KL, Saini NK, Ng TW, Porcelli SA, Lee S. Identification of autophagy-inhibiting factors of *Mycobacterium tuberculosis* by high-throughput loss-of-function screening. Infect Immun November 16, 2020;88(12).

[108] Lerner TR, Queval CJ, Lai RP, Russell MR, Fearns A, Greenwood DJ, et al. *Mycobacterium tuberculosis* cords within lymphatic endothelial cells to evade host immunity. JCI Insight May 21, 2020;5(10):e136937.

[109] Akira S, Uematsu S, Takeuchi O. Pathogen recognition and innate immunity. Cell February 24, 2006;124 (4):783–801.

[110] Fadok VA, Voelker DR, Campbell PA, Cohen JJ, Bratton DL, Henson PM. Exposure of phosphatidylserine on the surface of apoptotic lymphocytes triggers specific recognition and removal by macrophages. J Immunol Baltim Md 1950 1992 Apr 1;148(7):2207–16.

[111] Moffatt OD, Devitt A, Bell ED, Simmons DL, Gregory CD. Macrophage recognition of ICAM-3 on apoptotic Leukocytes1. J Immunol June 1, 1999;162(11):6800–10.

[112] Chang MK, Bergmark C, Laurila A, Hörkkö S, Han KH, Friedman P, et al. Monoclonal antibodies against oxidized low-density lipoprotein bind to apoptotic cells and inhibit their phagocytosis by elicited macrophages: evidence that oxidation-specific epitopes mediate macrophage recognition. Proc Natl Acad Sci U S A May 25, 1999;96(11):6353–8.

[113] Thornberry NA, Lazebnik Y. Caspases: enemies within. Science August 28, 1998;281(5381):1312–6.

[114] Kroemer G, El-Deiry WS, Golstein P, Peter ME, Vaux D, Vandenabeele P, et al. Classification of cell death: recommendations of the nomenclature committee on cell death. Cell Death Differ November 2005;12 (Suppl. 2):1463–7.

[115] Krammer PH. CD95's deadly mission in the immune system. Nature October 12, 2000;407(6805):789–95.

[116] Brenner D, Mak TW. Mitochondrial cell death effectors. Curr Opin Cell Biol December 2009;21(6):871–7.

[117] Deng W, Yang W, Zeng J, Abdalla AE, Xie J. *Mycobacterium tuberculosis* PPE32 promotes cytokines production and host cell apoptosis through caspase cascade accompanying with enhanced ER stress response. Oncotarget October 11, 2016;7(41):67347–59.

[118] Yang W, Deng W, Zeng J, Ren S, Ali MK, Gu Y, et al. *Mycobacterium tuberculosis* PE_PGRS18 enhances the intracellular survival of M. smegmatis via altering host macrophage cytokine profiling and attenuating the cell apoptosis. Apoptosis Int J Program Cell Death April 2017;22(4):502–9.

[119] Ali MK, Zhen G, Nzungize L, Stojkoska A, Duan X, Li C, et al. *Mycobacterium tuberculosis* PE31 (Rv3477) attenuates host cell apoptosis and promotes recombinant M. Smegmatis intracellular survival via up-regulating GTPase guanylate binding protein-1. Front Cell Infect Microbiol 2020;10:40.

[120] Wei L, Liu L, Meng Z, Qi K, Gao X, Feng J, et al. Recognition of *Mycobacterium tuberculosis* by macrophage Toll-like receptor and its role in autophagy. Inflamm Res Off J Eur Histamine Res Soc Al May 2024;73(5):753–70.

[121] Advancing the understanding of the PE/PPE protein families through PPE_MPTR protein expression and analysis of pe/ppe single nucleotide variants in differentially drug sensitive Mycobacterium tuberculosis. 2025 Apr 30. Available from: https://scholar.sun.ac.za/items/af813245-d052-472e-8712-d31cd799c45c.

[122] Priyanka, Sharma S, Varma-Basil M, Sharma M. C-terminal region of Rv1039c (PPE15) protein of *Mycobacterium tuberculosis* targets host mitochondria to induce macrophage apoptosis. Apoptosis Int J Program Cell Death October 2024;29(9–10):1757–79.

[123] Rv3539 (PPE63) of mycobacterium tuberculosis promotes survival of mycobacterium smegmatis in human macrophages cell line via cell wall modulation of bacteria and altering host's immune response—PubMed. 2025 Apr 30. Available from: https://pubmed.ncbi.nlm.nih.gov/37401981/.

[124] Guo F, Wei J, Song Y, Li B, Qian Z, Wang X, et al. Immunological effects of the PE/PPE family proteins of *Mycobacterium tuberculosis* and related vaccines. Front Immunol September 27, 2023;14.

[125] Veerapandian R, Gadad SS, Jagannath C, Dhandayuthapani S. Live attenuated vaccines against tuberculosis: targeting the disruption of genes encoding the secretory proteins of mycobacteria. Vaccines May 12, 2024;12(5):530.
[126] Xu Y, Yang E, Huang Q, Ni W, Kong C, Liu G, et al. PPE57 induces activation of macrophages and drives Th1-type immune responses through TLR2. J Mol Med Berl Ger June 2015;93(6):645–62.
[127] Patni K, Agarwal P, Kumar A, Meena LS. Computational evaluation of anticipated PE_PGRS39 protein involvement in host-pathogen interplay and its integration into vaccine development. 3 Biotech April 2021; 11(4):204.
[128] Bisht MK, Pal R, Dahiya P, Naz S, Sanyal P, Nandicoori VK, et al. The PPE2 protein of *Mycobacterium tuberculosis* is secreted during infection and facilitates mycobacterial survival inside the host. Tuberc Edinb Scotl December 2023;143:102421.
[129] Reverse vaccinology approach to design a multi-epitope vaccine construct based on the Mycobacterium tuberculosis biomarker PE_PGRS17—PubMed. 2025 Apr 30. Available from: https://pubmed.ncbi.nlm.nih.gov/35554858/.
[130] Li W, Yan Z, Zhang N, Zhang Z, Xiang X. Novel role of PE_PGRS47 in the alteration of mycobacterial cell wall integrity and drug resistance. Arch Microbiol April 6, 2023;205(5):174.
[131] Anand PK, Kaur J. Rv3539 (PPE63) of Mycobacterium tuberculosis promotes survival of Mycobacterium smegmatis in human macrophages cell line via cell wall modulation of bacteria and altering host's immune response. Curr Microbiol July 4, 2023;80(8):267.
[132] Mi Y, Bao L, Gu D, Luo T, Sun C, Yang G. *Mycobacterium tuberculosis* PPE25 and PPE26 proteins expressed in Mycobacterium smegmatis modulate cytokine secretion in mouse macrophages and enhance mycobacterial survival. Res Microbiol April 2017;168(3):234–43.
[133] Najafi A, Tafaghodi M, Sankian M, Amini Y, Ghazvini K. Cloning, expression, and refolding of PPE17 protein of Mycobacterium tuberculosis as a promising vaccine candidate. Iran J Basic Med Sci 2019;44(1): 53–9.
[134] Bartizal TJ. Advancing the understanding of the PE/PPE protein families through PPE_MPTR protein expression and analysis of pe/ppe single nucleotide variants in differentially drug sensitive Mycobacterium tuberculosis. Stellenbosch: Stellenbosch University; 2022. http://hdl.handle.net/10019.1/126325.
[135] Udgata A, Qureshi R, Mukhopadhyay S. Transduction of functionally contrasting signals by two mycobacterial PPE proteins downstream of TLR2 receptors. J Immunol Baltim Md 1950 September 1, 2016;197 (5):1776–87.

CHAPTER

3

Unraveling *Mycobacterium tuberculosis* pathogenesis: Insights into PE/PPE proteins from a molecular viewpoint

Shivangi Prandiyal*, [1], Dilpreet Kaur*, [1], Waseem Ali[1], Nasreen Z. Ehtesham[1], Seyed E. Hasnain[1, 2] and Anwar Alam[1, 3, 4]

[1]*Department of Life Science, School of Bioscience and Technology, Sharda University, Greater Noida, Uttar Pradesh, India;* [2]*Department of Biochemical Engineering and Biotechnology, IIT Delhi, New Delhi, India;* [3]*Department of Biotechnology, School of Bioscience and Technology, Sharda University, Greater Noida, Uttar Pradesh, India;* [4]*DST-FIST, Sharda University, Greater Noida, Uttar Pradesh, India*

Highlights

1. PE/PPE proteins are unique to Mycobacterium and play crucial roles in pathogenesis.
2. They are involved in cell wall integrity, antigenic variation, and immune evasion.
3. PE/PPE proteins interact with host receptors and modulate immune signaling pathways.
4. Their expression is regulated by environmental triggers and stress response mechanisms.
5. Structural diversity and intrinsic disorder contribute to their functional adaptability.
6. Genomic variations in PE/PPE genes impact virulence and immunogenicity.
7. Understanding PE/PPE proteins is crucial for developing novel TB diagnostics and therapies.

Abbreviations

3D Three dimensional
AhpC Alkyl hydroperoxide reductase C
AIM2 Absent in melanoma 2
Akt AKT serine/threonine kinase
Ams Alveolar macrophages
ASK1 Apoptosis signal regulating kinase 1
ATB Active tuberculosis
BAK BCL2 antagonist/killer
Bax BCL2 associated X protein
Bcl-2 B cell lymphoma 2
cAMP Cyclic adenosine monophosphate

*Equal contributing authors.

Navigating Tuberculosis. https://doi.org/10.1016/B978-0-443-27735-1.00005-8

CDC Centers for disease control and prevention
CFP-10 Culture filtrate protein 10 kDa
CHOP C/EBP homologous protein
COX2 Cyclooxygenase-2
CR Complement receptor
CRISPRi Clustered regularly interspaced short palindromic repeats interference
DNA Deoxyribonucleic Acid
E3 Ubiquitin protein ligase
EPTB Extrapulmonary tuberculosis
ER Endoplasmic reticulum
ERK1/2 Extracellular signal regulated kinase 1/2
ERK Extracellular signal regulated kinase
ESAT-6 Early secreted antigenic target 6 kDa
EspR ESX-1 secretion regulator
ESX ESAT-6 secretion system
FAS Fas Cell surface death receptor
GBP-1 Guanylate binding protein-1
HA Hemagglutinin
HIV Human immunodeficiency virus
ICAD Inhibitor of caspase activated DNase
ICAM-3 Intercellular adhesion molecule-3
IdeR Iron dependent regulator
IFN-γ Interferon gamma
IFN Interferon
IGRA Interferon gamma release assay
IL Interleukin
iNOS Inducible nitric oxide synthase
IRF-3 Interferon regulatory factor 3
JNK Jun N-Terminal kinase
KatG Catalase peroxidase
LAP LC3-associated Phagocytosis
LC3 Microtubule associated Protein 1A/1B-Light Chain 3
LDL; Low density lipoprotein
Lsr2 Nucleoid associated transcriptional inhibitor
LTBI Latent tuberculosis infection
MA Mycolic acid
MAPK Mitogen activated protein kinase
MOMP Mitochondrial outer membrane permeabilization
MPTR Major polymorphic tandem repeat
Mtb Mycobacterium tuberculosis
MTBC Mycobacterium tuberculosis complex
MyD88 MYD88 Innate immune signal transduction adaptor
NADPH Nicotinamide adenine dinucleotide phosphate hydrogen
NF-κB Nuclear factor kappa B
NLRP3 NLR family pyrin pomain containing 3
ORF Open reading frame
PAF Platelet activating factor
PAMPs Pathogen associated molecular patterns
PARP Poly (ADP Ribose) polymerase

PCR Polymerase chain reaction
PE Proline glutamic acid
PGE2 Prostaglandin E2
PGRS Polymorphic GC rich sequences
PhoPR Phosphate specific regulatory system
PI3K Phosphatidylinositol 3-Kinase
PPE Proline proline glutamic acid
PRRs Pattern recognition receptors
PS Phosphatidylserine
PTB Pulmonary tuberculosis
RFLP Restriction fragment length polymorphism
RNA-Seq RNA sequencing
RNA Ribonucleic acid
ROS Reactive oxygen species
SCID Severe Combined Immunodeficiency
SH3 SRC homology 3
SigB Sigma factor B
SigD Sigma factor D
SigF Sigma factor F
SigG Sigma factor G
SigL Sigma factor L
SigM Sigma factor M
Smurf1 SMAD specific E3 ubiquitin protein ligase 1
SNP Single nucleotide polymorphism
SOCS3 Suppressor of cytokine signaling 3
STB Smooth tubercle bacilli
T7SS Type VII secretion system
TB Tuberculosis
TCS Two component system
TGF-β Transforming growth factor Beta
Th1 T-Helper 1
Th2 T-Helper 2
TLR Toll-Like Receptor
TNF Tumor necrosis factor
TNFR Tumor necrosis factor receptor
TNFRSF10A TNF receptor superfamily member 10a
TST Tuberculin skin test
UPR Unfolded protein response
UV Ultraviolet
VEGF Vascular endothelial growth factor
WHO World health organization

1. Introduction

Mycobacterium tuberculosis (*Mtb*) is a highly efficient pathogen primarily affecting the lungs and causing the characteristic symptoms of pulmonary tuberculosis (TB). It is the etiological agent of TB, a contagious infectious disease. TB poses a significant global health threat due to its ability to infect multiple organs [1]. Understanding the pathogenesis and transmission of TB is crucial for developing effective treatments. Despite being identified more than a 100 years ago, *Mtb* continues to be a key health concern, causing substantial disease burden and fatality. According to the latest WHO Global Tuberculosis Report, an estimated 400,000 new cases of rifampicin-resistant (RIF-resistant) TB occurred worldwide in 2023, revealing the persistent burden of drug resistant TB and the critical need for more effective therapies The WHO Global TB Report 2024 estimated that about 10.8 million people developed active tuberculosis (ATB) in 2023, with an incidence rate of 134 cases per 100,000 population. Despite global efforts, TB remains one of the deadliest infectious diseases, surpassing coronavirus disease (COVID-19) to become the leading cause of mortality caused by an infectious agent [2,3]. Approximately 1.25 million people died from TB in 2023, a slight drop from 1.32 million in 2022. Furthermore, about 25% of the world population is predicted to have latent TB infection (LTBI), reflecting prior exposure to *Mtb* without active disease. TB continues to impose a significant public health and socioeconomic burden worldwide, necessitating intensified strategies for its prevention, diagnosis, and treatment.

Mtb is recognized as an obligate pathogen, necessitating a living host for survival. Its life cycle commences upon entry into the respiratory system and lungs. Anatomically modern humans have coexisted with the *Mtb* complex (MTBC) throughout their evolutionary history, facilitating its global dissemination over the past 70,000 years. As an obligate intracellular organism, *Mtb* initiates infection by penetrating the respiratory system and establishing itself in the lungs [4,5]. This phase, termed the primary infection, marks the first interaction between *Mtb* and the host's respiratory system. Research suggests that *Mtb*'s interaction with its human host is influenced more by immunological modulation mechanisms than by direct competition, diverging from the conventional evolutionary arms race model [4]. Following contact with various host cell receptors, including NOD-like receptors (NLRs), C-type lectin receptors (CLRs), dendritic cell receptors (DCs), mannose receptors (MRs), and toll-like receptors (TLRs), *Mtb* enters an intracellular replication phase. Subsequently, alveolar macrophages and dendritic cells internalize *Mtb*.

To eliminate the infection, *Mtb*-infected cells migrate to the lymph nodes, and release cytokines, including Tumor Necrosis Factor-alpha (TNF-α) and interferon-gamma (IFN-γ), along with proteolytic enzymes. However, this immune response facilitates the dissemination of *Mtb* within the lung parenchyma rather than its effective eradication [4,5]. The activation of macrophages enhances inflammatory responses that are essential for both host defence and disease progression by recruiting additional innate immune cells. Neutrophils, which exhibit greater phagocytic activity than macrophages, produce an intensified oxidative burst through reactive oxygen species (ROS) [6]. Lymphocytes at the infection site initiate a series of cell mediated immune responses that further mobilize immune cells to restrict bacterial growth [7,8]. Nevertheless, an intrinsic delay in the T-cell response during early infection provides an opportunity for *Mtb* to establish a persistent infection [7]. Complete bacterial clearance might be achieved at this stage if the host immune system remains robust [9]. However, in most cases, the immune response is insufficient to eliminate the infection, leading to the accumulation of monocytes near infected macrophages. This process promotes the formation of solid granulomas, a hallmark pathological feature of TB. By infecting alveolar epithelial cells and inducing cellular necrosis, *Mtb* can potentially breach the mucosal barrier, exacerbating the disease course [10,11]. This reveals the significance of localized immune responses in preventing the spread of TB [12]. Granulomas are formed to contain the spread of microbes, yet paradoxically, they can also serve as niches that shield *Mtb* from immune recognition and clearance. This results in LTBI, a clinically defined condition where the bacteria remain dormant within the host [9,13]. Effective LTBI screening, diagnosis, and treatment are crucial to reducing TB incidence and advancing toward global TB eradication.

In the initial stages of granuloma formation, vascular endothelial growth factor (VEGF) signaling induces extensive vascularization, resulting in well perfused structures characterized by prominent lymphocytic cuffs. As granulomas develop, macrophages differentiate into distinct subtypes, leading to a stratified structure. This structure consists of an outer fibrous layer encircled by aggregated lymphocytes, which in turn surround a macrophage-rich core essential for bacterial containment and immunological regulation. At this juncture, granulomas are asymptomatic and non-contagious [14,15]. Previous studies indicate that *Mtb* utilizes mycolic acid (MA) to facilitate the transformation of macrophages into foam cells, thereby contributing to granuloma formation. The core of the granuloma may eventually disintegrate, resulting in a caseous granuloma replete with foam cells and necrotic immune cells. As TB progresses, the center of this granuloma may rupture, permitting bacterial escape and initiating the transition from LTBI to active infection [16]. Upon reactivation of LTBI, symptoms manifest, and the bacteria can disseminate to new hosts, perpetuating the infection. *Mtb* is also capable of spreading via the bloodstream [17], and the lymphatic system [18], leading to *extra pulmonary* TB. This form of TB can affect nearly every organ, including the lymph nodes, joints, genitourinary system, pleura, and bones [19].

2. The enigma of the PE/PPE gene family

The PE and PPE multigene families, unique to the Mycobacterium genus, exhibit considerable variability across species that correlates with differences in growth rate and pathogenicity (Fig. 3.1). It is confirmed that saprophytic, fast growing, and nonpathogenic species such as *M. smegmatis* and *M. abscessus* hold a relatively limited number of pe/ppe genes and are phylogenetically closer to the genus's last common ancestor. In contrast, these gene families are substantially expanded in pathogenic, slow growing species within the MTBC [20,21]. The initial identification of the PE/PPE protein families in 1998 [22], instigated extensive research into their biological functions in *Mtb*. Their prevalence within the MTBC genome suggests a significant role in TB pathogenesis; however, despite extensive experimental investigation, the precise biological functions of these proteins remain incompletely defined.

Previous analyses focused on the evolutionary history and genomic structure of pe/ppe genes within the MTBC. Utilizing in silico genomic analyses, comparative genomics, and DNA hybridization assays, the researchers have reconstructed phylogenetic relationships and classified these gene families into five distinct subfamilies. The ancestral members, associated with the esx -1 gene cluster, belong to subfamily I and appear pivotal to *Mtb* biology. Among them, the ancestral members associated with the esx-1 gene cluster (subfamily I) appear to be central to the biology and pathogenicity of *Mtb* [20].

Their identification occurred before the complete genome sequence of the *Mtb* reference strain H37Rv was made available in 1998, which marked a crucial shift in understanding the genomic complexity of this pathogen. The availability of the full genome facilitated the recognition of a large number of genes belonging to the pe/ppe family, many of which encode surface expressed or secreted proteins potentially involved in immune interactions and pathogenesis [23].

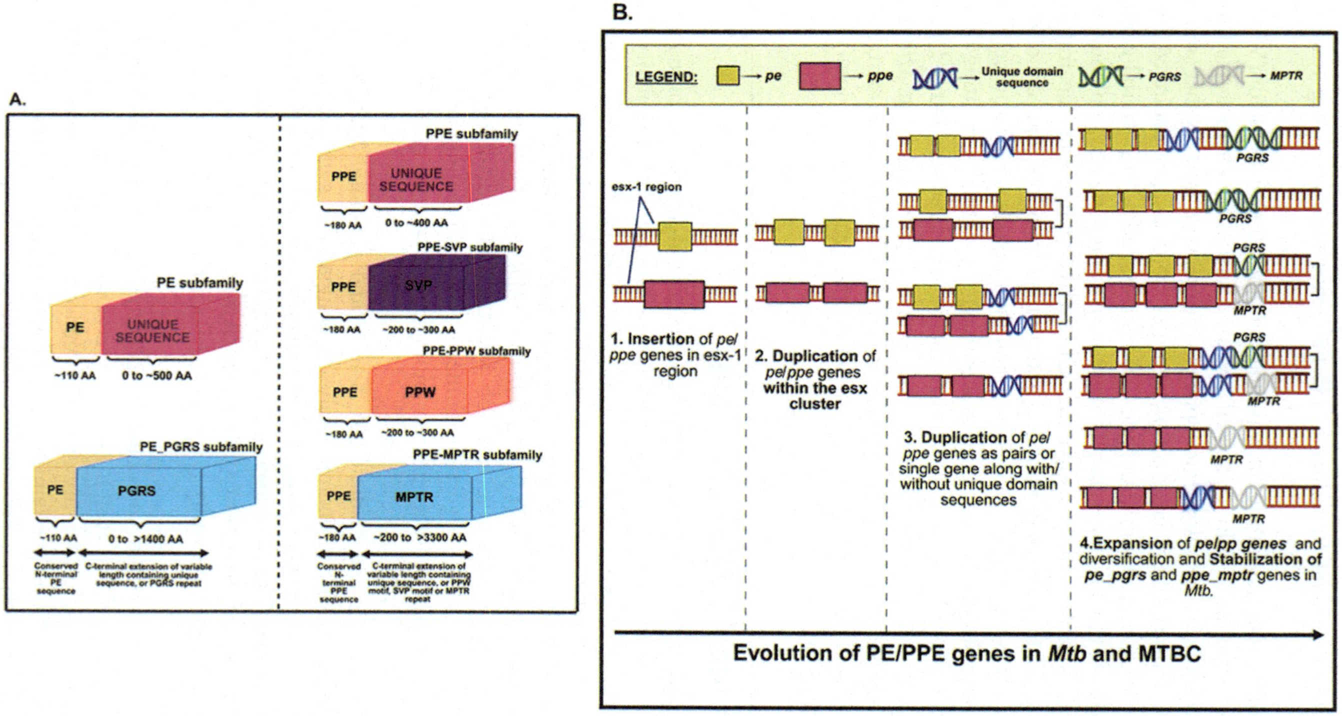

FIGURE 3.1 PE/PPE gene structure.

(A) Schematic representation of gene architectures within the PE and PPE gene families, depicting conserved N-terminal domains, motif organization, and distinctions among subfamilies present in each family. (B) Comparative analysis of the evolutionary trajectories of PE/PPE genes in *Mtb* and the *Mycobacterium tuberculosis* complex (MTBC).

3. Structure and function of PE/PPE, PGRS, and MPTR gene family: Unraveling the genetic enigma

The polymorphic G-C rich sequences (PGRS) in *Mtb* were first analyzed by Poulet and Cole (1995), who identified them as tandem repeats of the core sequence CGGCGGCAA distributed at 26 chromosomal sites, most likely originating from triplet expansion mutations (Fig. 3.1). Their repetitive nature significantly contributes to genomic variation in *Mtb*, and the association of several PGRS copies with conserved open reading frames (ORFs) suggests a potential regulatory function [24]. Genome annotation of *Mtb* was particularly challenging due to the abundance of repetitive sequences linked to the pe, ppe, pe_pgrs, and pe_mptr gene families. These families are marked by high sequence conservation in their N-terminal domains but exhibit substantial variability in their C-terminal regions, particularly among the PE_PGRS subgroup, making accurate gene prediction and functional annotation difficult [22]. Despite extensive efforts, the precise biological roles of a majority of these proteins remain poorly characterized due to their sequence redundancy and the technical difficulties in studying such highly similar gene products.

According to Cole and colleagues, nearly 10% of the *Mtb* genome is dedicated to genes encoding PE and PPE proteins, which together account for around 7%–8% of the bacterium's total coding capacity [22]. The names of these families are derived from their conserved N-terminal motifs: PE proteins begin with a Pro-Glu sequence, while PPE proteins contain a Pro-Pro-Glu sequence, forming the basis for their classification [20]. Variations in the C-terminal regions of these proteins have enabled their further subdivision into distinct structural and functional subfamilies. One such subgroup, the PE_PGRS family, is characterized by glycine-rich C-terminal domains composed of repetitive tripeptide motifs such as Gly-Gly-Ala and Gly-Gly-Asn. The number and arrangement of these motifs differ among members, reflecting polymorphic features that may underlie antigenic variability and immune evasion [25].

Functionally, PE and PPE proteins often exist as adjacent operons in the genome and are believed to be co-transcribed as bicistronic mRNA units, forming stable heterodimeric complexes essential for their secretion and surface localization [22,26]. This organization is thought to allow coordinated expression of interacting protein partners, possibly facilitating their assembly and function at host pathogen interfaces. The pe_pgrs genes, however, are well distributed throughout the genome and typically transcribed independently. This difference in genomic arrangement may reflect functional diversification within the PE/PPE superfamily, with PE_PGRS proteins potentially fulfilling broader or more independent roles in pathogenesis [27].

Additionally, a subgroup of PE proteins referred to as "PE-unique" has been identified. These lack pairing with PPE genes and are characterized by novel C-terminal domains not found in other family members [26]. The functional significance of these unique domains remains under investigation, but their presence hints at specialized roles beyond canonical PE/PPE interactions. Similarly, PPE proteins are characterized by a well conserved N-terminal domain which may serve as a recognition or secretion signal, akin to the PE domain, suggesting a shared mechanism for subcellular trafficking [28,29]. Several PPE proteins are co-transcribed with their cognate PE partners, but others are found as standalone genes, hinting at diverse regulatory and functional paradigms.

A particular subset of PPE proteins includes a C-terminal MPTR domain, consisting of Asn-X-Gly-X-Gly-Asn-X-Gly motifs arranged in repetitive units [20]. This repetitive region is thought to influence protein structure, antigen presentation, and immune recognition. The presence of MPTR domains has been implicated in modulating the host immune responses and may play a role in immune evasion by limiting T-cell recognition or altering antigen processing pathways. Collectively, the structural heterogeneity, genomic organization, and immunological properties of the PE, PPE, and PE_PGRS families highlight their evolutionary refinement as key mediators of host pathogen interaction and suggest their contribution to the success of *Mtb* [30].

Among pe subfamilies, the pe_pgrs group is the largest and most diverse, representing more than 60% of all pe genes in *Mtb*. However, functional analyses indicate that only 51 members of this subfamily remain transcriptionally active or structurally intact across the *Mtb* complex, suggesting that the remainder may represent pseudogenes or lineage specific remnants. Members of this group share several defining features: (i) high redundancy with overlapping sequence identity, (ii) intrinsically disordered glycine-rich regions, and (iii) strong antigenic potential driven by repetitive amino acid motifs. Collectively, these characteristics are thought to contribute to immune evasion by presenting a shifting antigenic profile to host immune cells, thereby limiting the effectiveness of memory-based immunity [31].

4. Evolution and diversification of the PE/PPE gene family

The expansion of pe_pgrs genes in *Mtb*, despite the bacterium's overall trend toward reductive genome evolution, strongly suggests that these proteins provide a selective advantage, particularly for chronic infection and long-term intracellular persistence [22,26]. Their enrichment in pathogenic mycobacteria such as *Mtb*, *M. bovis*, and *M. marinum*, and relative absence in environmental nonpathogens, supports the hypothesis that PE_PGRS proteins have co-evolved with host immune pressures to enhance pathogen adaptability and virulence. These proteins may interfere with host signaling, promote granuloma formation, or alter antigen processing to delay or suppress immune activation [32].

Interestingly, the distribution of pe/ppe genes is not exclusive to pathogenic mycobacteria. For example, *M. smegmatis*, a nonpathogenic and fast growing species, contains a limited set of pe/ppe genes, most of which are clustered within the esx-6 type VII secretion system locus [20]. This furthers the belief that the fundamental pe/ppe gene architecture predates the evolution of pathogenicity and that subsequent expansion and diversification arose as adaptive responses to host-specific pressures. Thus, the pe/ppe family likely originated from a small ancestral gene pool and underwent lineage-specific amplification to support intracellular survival, immune modulation, and colonization of host tissues.

Overall, the structure function landscape of the PE/PPE protein family illustrates a remarkable example of pathogen-driven genomic innovation. These proteins, under their repetitive sequences, immunogenic motifs, and adaptive transcriptional profiles, contribute significantly to *Mtb*'s ability to persist in hostile intracellular environments. Their evolutionary trajectory reflects a fine balance between retaining essential functional domains and introducing sufficient variability to evade immune detection and adapt to diverse host conditions [20]. This complexity receals the importance of these proteins not only as virulence factors but also as potential candidates for vaccine targets and therapeutic interventions.

Within this context, the PE_PGRS family represents a lineage specific subset of the *Mtb* genome. Originating from GC-rich polymorphic repetitive sequences (PGRS), first described for their role in strain differentiation and molecular typing [31]. These genes have undergone remarkable diversification. Further insights into the role of PE/PPE proteins in virulence were provided by Simeone and colleagues, who demonstrated that ESX/type VII secretion systems are critical determinants of *Mtb* pathogenicity [33–35]. Consistent with these findings, slow-growing pathogenic mycobacteria within the MTBC harbor a significantly expanded repertoire of pe/ppe genes compared with their fast-growing, non-pathogenic counterparts. This expansion is especially marked in subfamily V, which includes the highly polymorphic pe_pgrs and ppe_mptr genes, hypothesized to enhance genetic variability and immune evasion. The duplication of the esx-1 cluster and the subsequent emergence of the esx-5 locus are thought to have driven this expansion, with proliferation of pe_pgrs and ppe_mptr genes coinciding with these genomic duplications [20].

Genomic studies on smooth tubercle bacilli (STB) have provided important clues about the origin and evolution of pe/ppe genes and their role in the pathogenesis of *Mtb*. Namouchi and his colleagues showed that pe_pgrs genes in STB undergo frequent recombination both within and between genes, creating extensive genetic diversity. In addition, certain genes were found to be under positive selection, suggesting that these changes improved the ability of STB to adapt to host environments [36]. Karboul et al. further reported an ancient duplication event that produced PE_PGRS17 and PE_PGRS18. Later genetic changes introduced a highly variable region (12/40) within PE_PGRS7, which persisted through the evolution of the *M. africanum-M. bovis* lineage. This region was later converted to resemble pe_pgrs18 through gene conversion. Interestingly, this polymorphism is still present in some MTBC strains but absent in others, indicating a correlation between its acquisition and the rise of highly successful MTBC lineages, thus, implicating PE/PGRS proteins in shaping host pathogen interactions [37].

Further evidence of diversification was provided by McEvoy et al., who described rapid molecular changes in the hypervariable ppe38 region. These findings highlight the importance of homologous recombination as a driving force behind the expansion and diversification of pe/ppe gene families. Such recombination driven genetic changes likely form the basis for functional diversity and help MTBC strains adapt to the human host [38,39].

5. Polymorphic nature of the PE/PPE gene diversity, gene expression and mutation

PE/PPE proteins are characterized by a three dimensional structure comprising conserved N-terminal domains and variable C-terminal regions. Structural analyses have revealed the presence of beta-helical and alpha-helical conformations, which contribute to their functional versatility. The crystal structure of these proteins indicates their potential role in protein-protein interactions and antigen presentation. Furthermore, the crystal structure suggests that these proteins frequently adopt lipid-binding or host-interacting configurations, which are critical for bacterial survival [40]. The structural features of PE domains include α-helical, coiled structures that facilitate protein-protein interactions, while PPE domains possess a β-sheet-rich structure, often involved in surface interactions. Molecular modeling studies, including homology modeling and docking, have been employed to predict the binding interactions of PE/PPE proteins with host receptors [41].

Comparative genomic approaches have shed light on their role in immune evasion. A large scale study of 216 *Mtb* strains from seven lineages identified seven pe/ppe genes under diversifying selection, suggesting adaptive changes in immune-related regions. Although overall antigenic diversity appeared low, most mutations were located in predicted T cell epitopes. In fact, 91% of nonsynonymous mutations occurred within these epitopes. When tested in TB patients in the Gambia, 72% of individuals showed distinct immune responses to ancestral versus mutant epitopes, highlighting the biological impact of these sequence changes [42].

Classification studies have further clarified the organization of the PE protein family. Based on domain structure and phylogeny, Gey van Pittius and colleagues divided them into three subgroups: PE-only, PE-unique, and PE_PGRS proteins [20]. PE-only proteins are generally very small, often less than 100 amino acids, and usually pair with PPE proteins to form heterodimers. These complexes may act as structural or regulatory elements at the host -pathogen interface. In contrast, PE_PGRS proteins carry large repetitive regions that may contribute to antigenic variation and immune modulation. Together, this classification highlights the structural and functional diversity of the PE/PPE family and its significance in *Mtb* adaptability [24].

6. Regulation of PE/PPE proteins

In *Mtb*, the regulation of PE and PPE protein expression is orchestrated by a multilayered network involving sigma factors, transcriptional regulators, and two component systems. These regulatory pathways respond dynamically to environmental cues, particularly those encountered within the host, to ensure appropriate expression of virulence-associated genes, including those belonging to the pe/ppe family [43].

Among the sigma factors, several play critical roles in transcriptional modulation of pe/ppe genes. SigM, SigB, SigG, SigF, SigL, and SigD have all been implicated in this complex regulatory framework [44,45]. SigL, for instance, has been demonstrated to regulate polyketide synthase genes, as well as proteins associated with secretion and the cell membrane. The regulatory role of SigL has been shown to extend to genes encoding PE/PPE proteins, many of which contribute to immune modulation and pathogenesis. Hahn et al. established that SigL is not only crucial for stress adaptation, but also essential for full virulence of the bacillus, supporting its role in regulating genes central to infection and survival within the host [44].

SigF is another sigma factor that contributes to pe/ppe regulation. It has been reported to control several genes linked to cell wall biosynthesis, including those directly involved in the synthesis and export of PE/PPE proteins. SigF also directly regulates phoY1, a gene encoding a transcriptional regulator linked to phosphate metabolism and intracellular survival [46]. The regulation of phoY1 further demonstrates the integrative role of SigF in coordinating metabolic adaptation with antigenic modulation. In addition, both SigG and SigM are induced during DNA damage and oxidative stress, indicating their importance in upregulating pe/ppe genes during genotoxic and host imposed challenges [45].

SigD is yet another sigma factor that modulates the expression of genes involved in virulence, lipid metabolism, and the bacterial stress response. Its role in pe/ppe gene regulation is part of a broader transcriptional strategy to adapt the pathogen's outer membrane and secretory pathways during infection. Raman et al. demonstrated that SigD governs genes required during late stages of infection, many of which overlap with those of the pe/ppe family [47].

SigB, however, appears to play a central and conserved role within this network. It is upregulated in response to environmental stress and has been implicated in the regulation of recently evolved pe/ppe genes that are unique to pathogenic mycobacteria. These genes are thought to encode immunologically relevant proteins that assist the bacterium in evading host responses. Manganelli et al. reported that SigB is a major regulator under nutrient limitation and oxidative stress [48]. This is further supported by studies showing that SigB drives the expression of pe/ppe genes under stress and stationary phase conditions, contributing to *Mtb* survival in latent or persistent infection phases [45,49]. These sigma factors do not work in isolation; Lee et al. emphasized that SigB and SigF work in tandem, sometimes coregulating overlapping sets of pe/ppe and stress response genes, thus ensuring transcriptional robustness under varying conditions [45].

In addition to sigma factors, several transcriptional regulators independently modulate pe/ppe gene expression. Ahmed et al. reported that many PE/PPE proteins are under the control of specialized regulatory proteins, which respond to intracellular cues such as metal ion concentrations and oxidative stress [50]. One of the most studied among these is IdeR, an iron responsive dual function transcriptional regulator. According to Gold et al., IdeR can act as both an activator and repressor, and controls the expression of approximately 11 pe and ppe genes in addition to genes involved in Fe acquisition and storage [51]. This positions IdeR as a critical regulatory nexus linking nutrient availability with virulence factor expression.

The two component system PhoPR adds another regulatory layer by modulating the expression of lipid biosynthesis genes and pe/ppe family members. Solans et al. demonstrated that the PhoP dependent noncoding RNA Mcr7 regulates the twin arginine translocation (TAT) system, which is involved in the secretion of numerous proteins, including PE/PPE components [52]. This coupling of transcriptional regulation with secretion pathways ensures not only the proper expression but also the timely delivery of PE/PPE proteins to host interfaces.

Further complexity is added by transcriptional regulators such as Rv0485, which modulate specific operons within the pe/ppe family. Goldstone et al. reported that Rv0485 specifically controls the pe13-ppe18 operon, and disruption of this regulator leads to decreased expression of these genes and attenuated virulence in mice models [53]. This supports the hypothesis that precise regulation of individual pe/ppe loci is essential for *Mtb* pathogenicity and immune modulation.

Together, these findings depict a highly evolved regulatory circuit that allows *Mtb* to respond to host-derived signals by fine tuning the expression of immunomodulatory PE/PPE proteins. Sigma factors such as SigL [44], SigB, SigG [45], SigF [46,47], and SigD form the primary response mechanisms, while global transcriptional regulators like IdeR [51], Rv0485 [54], and the PhoPR system provide specificity and environmental responsiveness. The integration of transcriptional signals with post transcriptional mechanisms like noncoding RNAs further enriches the regulation, ensuring optimal antigen presentation and immune evasion. Such sophisticated control over pe/ppe expression enables *Mtb* to persist, replicate, and disseminate within the host while avoiding immune detection, thus unveiling its potential as a target for therapeutic intervention [55].

7. PE/PPE gene recombination and implications in virulence

Polymorphic variation within the pe/ppe family facilitates the pathogen's evasion of immune detection, while its structural characteristics further promote mycobacterial adaptation within the host. The pe/ppe genes are subject to frequent genetic recombination, and variability in their C-terminal domains influences immune recognition. This antigenic polymorphism poses challenges for vaccine development, suggesting that strategies targeting conserved regions may enhance vaccine efficacy [38]. The ppe18 gene in *Mtb* encodes a cell membrane associated protein involved in immune interactions. It is reported that genetic alterations in ppe18, such as singlenucleotide polymorphisms (SNPs), insertions, and deletions, were identified in 20.9% of clinical strains. Many of these alterations occurred in T-cell epitope regions, potentially affecting immunological recognition. Regional genetic variability is indicated by the higher frequency of ppe18 mutations in strains from Arkansas (26.2%) compared to Turkey (14.6%). The efficacy of the *Mtb*72F vaccine is questioned due to the prevalence of mutations in Genetic Group 1 strains, which are common in regions with high TB burdens. These findings implicate the significance of targeting broadly conserved vaccine antigens to enhance TB protection, highlighting the role of ppe18 in antigenic variation and immune evasion [56].

In 2008, a study conducted a comparative analysis with the reference H37Rv strain, examining genetic polymorphism in pe_pgrs16 and pe_pgrs26 genes of *Mtb* across 200 clinical isolates. The pe_pgrs gene family is associated with *Mtb* persistence within the host, evasion of the immune system, and antigenic variation. The study concluded that variations in pe_pgrs16 were present in 51% of the isolates, while variations in PE_PGRS26 were observed in 50%. Both genes exhibited SNPs, inframe insertions, and frameshifts. Among the 200 strains, 72 distinct PE_PGRS protein sequence patterns were identified. Notably, pe_pgrs26 exhibited a larger frequency of deletions, whereas pe_pgrs16 demonstrated a greater incidence of frameshift mutations (40%) compared to pe_pgrs26. The study posits that genetic variability in these genes influences bacterial persistence and immunological recognition [29]. Furthermore, Delogu and colleagues proposed that despite similarities, slight sequence variations can confer functional roles, enabling the expression of specific protein subsets based on the bacterial milieu [23]. The genomic arrangement of pe and ppe genes into operonic structures suggests a tightly regulated coexpression system, allowing synchronous translation and interaction of these protein pairs [39]. Structural analyses, particularly of the rv2430c and rv2431c genes, have revealed that their protein products share a common helical structure critical for dimer formation and are likely involved in protein protein interactions required for virulence factor trafficking or immune modulation [40]. This heterodimerization potentially facilitates bacterial survival by mediating specific host interactions, promoting immune modulation, or aiding in immune evasion. The discovery of such conserved architectures in otherwise variable proteins emphasizes the functional convergence of these molecules, despite their sequence diversity.

8. Transcriptional plasticity: Differential expression of PE/PPE genes in *Mtb*

Voskuil and colleagues conducted a comprehensive analysis of pe-ppe gene expression levels utilizing DNA microarrays across 15 distinct environmental conditions. The study examined 169 pe/ppe genes and identified differential expression in 128 of them. However, no consistent expression patterns were observed within the pe/ppe families or their subsets. For example, pe3 was uniquely upregulated in activated macrophages, indicating that certain genes are specifically expressed in particular environments. Notably, in 10 out of the 15 conditions examined, PE11 exhibited differential regulation. This suggests that *Mtb* may evade host immune detection by promoting antigenic variation through the dynamic synthesis of unique PE/PPE proteins in response to environmental stimuli such as iron stress [57], nutritional starvation, cAMP stress, etc. [58].

In vivo studies conducted by Rachman and his colleagues (2006) explored the differential expression of PE and PPE proteins. They found that two PE, three PPEMPTR, and four PEPGRS proteins were overexpressed in lung granulomas of TB patients. Similarly, Ramakrishnan et al. (2000) identified that two pe_pgrs genes from *M. marinum* are exclusively expressed in macrophages. Further analysis of ppe_mptr gene expression during mouse infection, as reported by Soldini and colleagues, indicated that some genes were constitutively expressed, while others were suppressed or activated at various stages of infection [59].

9. PE/PPE-mediated cell signaling events

PE/PPE proteins represent a unique family of proteins in *Mtb* that significantly influence the interaction between the bacterium and host macrophages. These proteins are involved in multiple stages of infection, including adhesion, immune modulation, intracellular survival, and dissemination. During the initial infection, PE/PPE proteins facilitate *Mtb* adhesion to host macrophages by interacting with receptors such as TLR2 and TLR4 [60]. Proteins like PPE18, PPE26, and PPE57 modulate immune signaling pathways, suppressing proinflammatory responses and aiding bacterial survival [61]. Additionally, PPE38 and PPE29 enhance phagocytosis, ensuring that *Mtb* is taken up by macrophages. However, once inside, the bacterium employs various PE/PPE proteins to inhibit phagolysosome maturation. Proteins such as PE_PGRS62, PE_PGRS30, and PPE25 prevent lysosomal fusion, allowing *Mtb* to persist in a protective intracellular niche [62].

Mtb employs various strategies to counteract host defences, notably through the regulation of autophagy and apoptosis via PE/PPE proteins. Specifically, PE_PGRS41, PE_PGRS47, and PE_PGRS29 are implicated in the modulation of autophagy, whereas PPE32 and PE_PGRS55 facilitate apoptosis, thereby aiding in bacterial dissemination. In contrast, proteins such as PE_PGRS18 and PE31 inhibit apoptosis, thereby extending bacterial survival within macrophages. Additionally, these proteins contribute to inflammation and tissue damage. Proteins like PE_PGRS33, PPE27, and PPE25/PPE41 enhance inflammatory responses, which lead to the recruitment of immune cells and potential bacterial spread. *Mtb* further exploits efferocytosis to transition from dying macrophages to other immune cells, including neutrophils, dendritic cells, and eosinophils. Collectively, PE/PPE proteins are integral to *Mtb* pathogenesis by manipulating host immune responses, ensuring intracellular persistence, and facilitating bacterial dissemination [63]. A comprehensive understanding of these interactions may inform the development of targeted therapies against TB (Fig. 3.2) [64].

10. Disrupting the defenders: PE/PPE proteins of *Mtb* as master regulators of host immune signaling

PE and PPE proteins have the capacity to modulate host cell signaling by interacting with essential immune molecules, including nuclear factor kappa B (NF-κB), signal transducer and activator of transcription (STAT) pathways, and inflammasome activation. Their involvement in undermining host immune surveillance mechanisms renders them essential in *Mtb* pathogenesis. PE/PPE proteins manipulate host signaling cascades to facilitate mycobacterial survival, notably through the inhibition

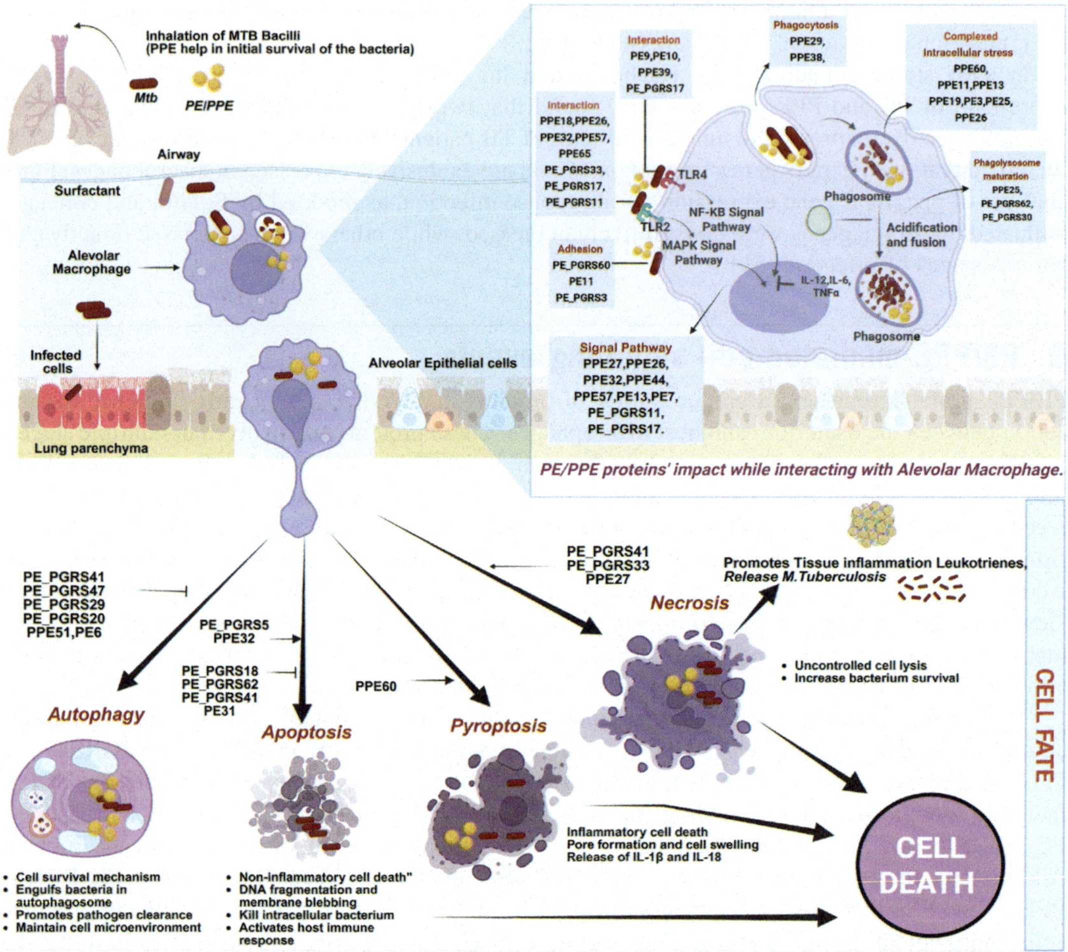

FIGURE 3.2 Role of PE/PPE proteins in modulating host cell fate during *Mtb* infection.

Upon airway entry, *Mtb* bacilli are phagocytosed by alveolar macrophages, initiating complex immune responses. The figure illustrates how representative PE/PPE proteins interact with host Toll-like receptors (TLR2, TLR4), modulate intracellular processes including phagocytosis, phagosome maturation, and stress adaptation, and alter immune signal pathways. Specific PE/PPE members (e.g., PPE25, PE_PGRS52, PE_PGRS30) inhibit phagosome-lysosome fusion, supporting bacterial persistence, while others regulate key macrophage cell fate decisions. Downstream effects include autophagy (enhancing clearance and homeostasis), apoptosis (inducing programmed death and adaptive immunity), pyroptosis (triggering inflammatory responses via IL-1β/IL-18), and necrosis (leading to tissue damage and bacterial dissemination). Collectively, these interactions and manipulations of host pathways by PE/PPE proteins determine cell fate, immune evasion, and disease progression in tuberculosis.

of apoptosis. Certain PE/PPE proteins suppress proapoptotic pathways in infected macrophages. PE/PPE proteins modulate NF-κB and map kinase (MAPK) pathways, with a reduction in proinflammatory cytokine secretion aiding immune evasion. A critical immune response mechanism to *Mtb* infection is granuloma formation. Granuloma development prompts the release of proinflammatory cytokines and chemokines that attract other immune cells to the infection site [65]. Granulomas are classified into two types: active and inactive. Active granulomas are characterized by high levels of inflammation and contain numerous infected macrophages and replicating mycobacteria, whereas inactive granulomas exhibit low inflammation and a reduced number of infected macrophages compared to active granulomas [66]. Zhangli Peng et al. hypothesized that PPE36 plays additional roles in promoting *Mtb* survival and infection, particularly through immune modulation. The study investigates how the host immune response is modulated by the *Mtb* protein PPE36. In virulent *Mtb* strains, the cell wall associated protein PPE36 is enriched. According to the study, PPE36 facilitates *Mtb* in evading immunological responses by promoting the E3 ligase SMAD-specific E3 ubiquitin protein ligase 1 (Smurf1) to degrade MYD88, an innate immune signal transduction adaptor (MyD88), a crucial adaptor protein in immune signaling (Fig. 3.3). This degradation enables *Mtb* to persist in host macrophages and mice by suppressing the inflammatory NF-κB and MAPK pathways [67].

10.1 PE/PPE as regulators of cell physiology

A key member of the PE/PPE family, PE_PGRS30, has been implicated in promoting *Mtb* virulence by inducing apoptosis in host macrophages. Its functional significance is further supported by studies involving *M. marinum*, where its homolog, mag24, appears to influence the course of chronic infection [68]. Additional experimental evidence using *M. smegmatis* expressing recombinant PE_PGRS30 demonstrated that this protein prolongs the bacterial lag phase in culture medium and modifies colony morphology, suggesting its role in growth regulation. Moreover, deletion of the rv1651c gene, which encodes PE_PGRS30, significantly attenuates the virulence of *Mtb* H37Rv, confirming that this protein is essential for full pathogenic potential [68]. Several PE/PPE proteins suppress phagolysosome fusion, a vital step in innate defence.

Within macrophages, *Mtb* adapts to environmental stress (low pH, ROS, RNS) using PE/PPE proteins. PPE60 [69], and PE13 improve bacterial endurance under such stressors. PPE11 supports survival in hostile intracellular settings, maintaining elevated bacterial loads and causing severe tissue damage [70]. PPE36 downregulates key inflammatory cytokines (TNF-α, IL-6, IL-1β), thereby curbing immune activation [67]. By obstructing phagosome-lysosome fusion, *Mtb* ensures long-term macrophage survival. This function is impaired in pe_pgrs30 and pe_pgrs47 deletion mutants [71,72], and overexpression of PE_PGRS62 significantly hinders phagosome maturation [73]. PPE25 expression surges during phagocytosis, and its mutant strain cannot inhibit lysosomal fusion or survive intracellularly [74].

10.2 The role of PE/PPE proteins in *Mtb* resistance to host-derived stress

Mtb employs various mechanisms to counteract oxidative stress, including the expression of genes that neutralize free radicals. Additionally, pathogenic PE/PPE proteins enhance resistance to oxidative stress, thereby facilitating persistent infection [64]. PE6 evokes an inflammatory response with the host macrophage and is shown to enhance the intracellular survival [75]. Srivastava and colleagues

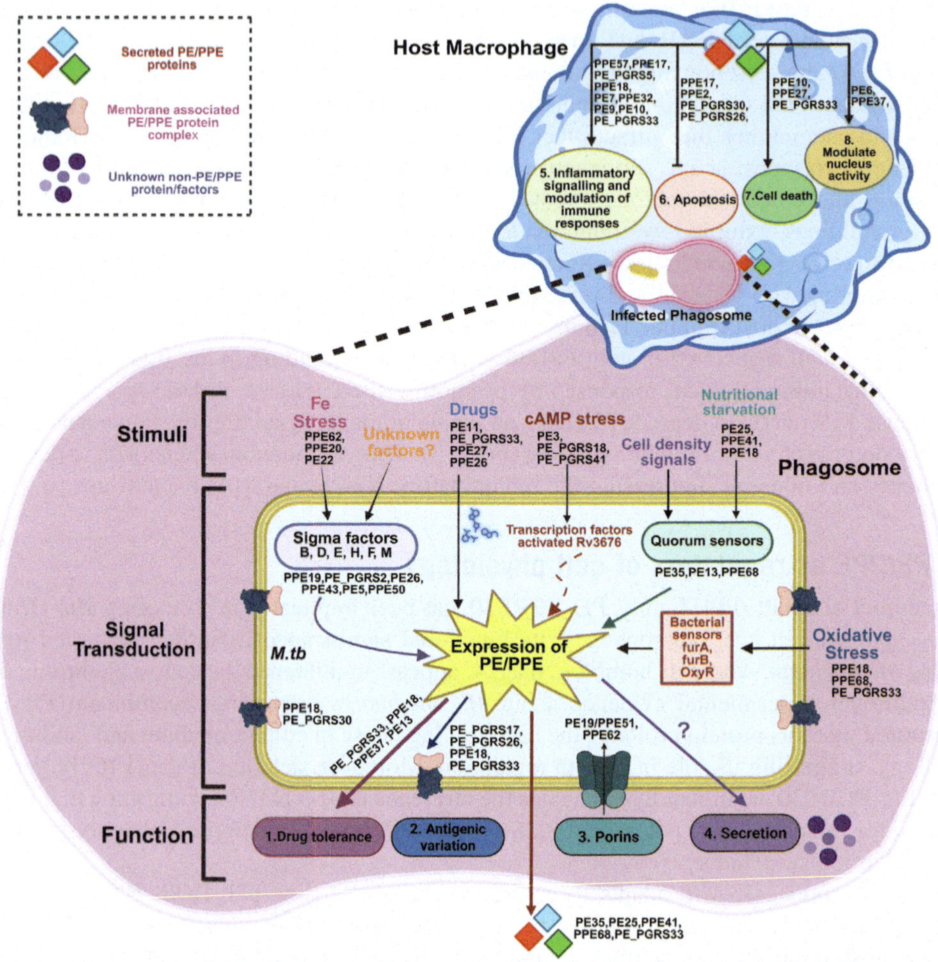

FIGURE 3.3 Schematic representation of PE/PPE protein functions within *Mtb* and the host macrophage.

PE/PPE proteins, which may be surface-localized or secreted, are expressed in response to various cellular stresses and regulatory factors. Within the bacterium, these proteins contribute to drug tolerance (via overexpression under drug pressure), antigenic variation (through their variable C-termini), and solute transport (by acting as porins across the outer membrane). The secretion of PE/PPE proteins via the ESX system is a widely discussed function, though its precise role remains to be fully defined. In the host macrophage, secreted PE/PPE proteins interact with host proteins and signaling pathways, modulating immune responses, inhibiting or inducing apoptosis, promoting cell death, and influencing nuclear activities. Collectively, these multifaceted roles reveal the significance of PE/PPE proteins in mycobacterial survival, immune modulation, and pathogenesis.

demonstrated that the overexpression of PE15 and PE5 in *M. smegmatis* leads to a reduction in inducible nitric oxide synthase (iNOS) levels, thereby enhancing bacterial viability within macrophages. Furthermore, PE_PGRS62 and PPE2 inhibit nitric oxide production, while PE_PGRS11 and PPE34 activate cyclooxygenase-2 (COX2), thereby augmenting oxidative stress resistance. Notably, the SRC Homology 3 (SH3) domain of PPE2 interacts with the host's p67phox subunit, disrupting the assembly of NADPH oxidase and consequently reducing ROS levels, which promotes pathogen survival [76].

Beyond oxidative stress tolerance, these proteins play an important role in modulating host cell death processes. They can induce both apoptosis and necrosis, contributing to the persistence and dissemination of *Mtb* [77]. Moreover, several PE/PPE proteins inhibit autophagy, thereby enhancing bacterial survival post infection. These proteins are predominantly expressed during the chronic infection stage, allowing *Mtb* to propagate silently within the host. However, additional research is required to elucidate the precise timing and regulatory mechanisms governing their expression [69,78]. Modifications in the fatty acid composition of the mycobacterial cell wall, facilitated by PE11, have been shown to enhance its permeability, potentially influencing bacterial interactions with the host environment [79]. Similarly, the overexpression of PE19 increases the permeability of the *Mtb* envelope, which may constitute an adaptive response to stress induced by the host [80]. Notably, several members of the PE_PGRS family are specifically expressed in granulomatous lesions, suggesting their involvement in host pathogen interactions [81]. PPE11 enhances bacterial resistance to host immune responses by providing protection against lysozymes, reactive nitrogen intermediates, and acidic conditions, in addition to inducing membrane alterations. Moreover, PPE60 and PE13 contribute to bacterial viability by augmenting cellular stress tolerance. Thus, the variable production of PE-PPE proteins by *Mtb* likely represents an adaptive response to specific environmental challenges encountered during infection. Several pe/ppe genes are implicated in *Mtb* persistence and are highly expressed in various stress states [58]. Another recent study demonstrates differential expression of 128 out of 169 genes in response to stressors such as respiratory inhibition (KCN), protonophore-induced uncoupling (CCCP), oxidative stress (H_2O_2), lipid metabolism perturbation (palmitic acid treatment), nutrient starvation, antibiotic exposure (ethambutol), and heat shock (45°C). These findings suggest the dynamic regulation of PE and PPE proteins, indicating their critical role in bacterial adaptability and persistence in adverse environments. To counteract oxidative stress, a significant threat posed by the host immune system, *Mtb* has developed several defensive mechanisms. Enzymes such as alkyl hydroperoxide reductase C (AhpC) and catalase-peroxidase (KatG) are essential for detoxifying reactive oxygen species. Additionally, by inhibiting free radical activity, pathogenic PE and PPE proteins extend the pathogen's survival within host cells, thereby promoting bacterial persistence.

10.3 PE/PPE role in pathogenicity and immune evasion

Despite high variability in sequence, pgrs family genes may be utilized in conjunction with other genetic markers to track *Mtb* strains, even if it is not an ideal standalone marker for epidemiology. While many researchers have previously employed PCR to explore pgrs as a target for strain differentiation, IS6110-restriction fragment length polymorphism (RFLP), a more established typing method, demonstrated greater discriminatory power. Understanding pgrs linked ORFs may provide insights into *Mtb* virulence, persistence, and immune evasion. Future research should focus on the functional characterization of PGRS, its potential role in treatment resistance, and its integration with other molecular markers to improve TB strain typing. [82]. Furthermore, the pe_pgrs33 gene encodes a surface-expressed protein that interacts with host immune cells, contributing to cytotoxic effects on macrophages and facilitating *Mtb*'s evasion of immune clearance. The PE domain of the protein induces a cellular immune response, while the PGRS domain elicits a humoral immune response. The host immune system is evaded *by Mtb* through the action of PE/PPE proteins, which interact with the host's macrophages and dendritic cells. The change in cytokine production affects the immune cell functioning, leading to the inhibition caused by T cells [83]. Consequently, the lack of an effective immune response allows *Mtb* to invade, survive, and proliferate, resulting in chronic infection or latency. Genomic research on antigenic variation in *Mtb* directs that most of the T cell epitopes are largely conserved across *Mtb* strains [84]. However, some variation exists within the PE/PPE protein family, particularly in regions not recognized by T cells. While certain segments of the ESX protein family exhibit variability, the dominant epitopes remain constant [42]. This constancy enhances the potential for transmission within the host. Following this, antimicrobial resistance facilitates the evolution and adaptation of the pathogen, which is then countered by the host immune system [71]. PE/PPE proteins contribute to antigenic variation by altering epitope presentation to host immune cells, thereby enabling *Mtb* to evade antibody mediated clearance and adaptive immune recognition. In terms of evading innate immunity, PE/PPE proteins inhibit phagolysosome fusion, and modulation of cytokine production reduces immune activation. Unlike many other infections, *Mtb* does not commonly utilize antigenic variation to evade immune detection. Instead, it modifies the inflammatory response, delays T cell responses, and adjusts host immune signaling. In comparison, other pathogens such as the influenza virus frequently alter the hemagglutinin (HA) protein to evade immune recognition, human immune deficiency virus (HIV) achieves rapid immune escape through a high mutation rate, and *Streptococcus pneumoniae* employs capsular polysaccharide variation to avoid antibody detection [42].

10.4 PE/PPE role in cell wall modulation

PE/PPE proteins are crucial for maintaining cell wall integrity and facilitating biogenesis. These proteins interact with mycolic acids, contributing to the assembly of the outer layer of the cell wall, thereby influencing virulence and host pathogen interactions. Certain PPE proteins engage with lipid components of the cell wall, affecting cell wall permeability and antibiotic resistance. Mutations in pe/ppegenes impact bacterial virulence, with the deletion of specific pe/ppe resulting in attenuated virulence [26]. The transport of PPE proteins in *Mtb* is dependent on the ESX-5 secretion system, which is important for virulence and cell wall integrity. It has been shown that disruption of esx-5 results in the loss of PPE41 secretion, rendering *Mtb* more susceptible to detergents and antibiotics due to a compromised cell wall [85].

Surface-exposed PE/PPE proteins, including PE_PGRS33 and Rv3097c, are likely involved in evading host immune detection mechanisms. PE_PGRS33, in particular, has been observed to associate loosely with the outer membrane, which may facilitate its release into the extracellular environment and potentially modulate immune responses [61]. According to the study, PE11 (Rv1169c), a PE/PPE family protein found in *Mtb*, is essential for virulence, stress tolerance, and cell wall remodeling. Knock in Expression of PE11 in *M. smegmatis* resulted in increased glycolipid content, enhanced antibiotic resistance, and improved survival under stressors such as sodium dodecyl sulfate

(SDS), acidic pH, and oxidative stress. PE11 modifies cell wall integrity and lipid metabolism to facilitate immune evasion. In mouse models, PE11 expression exacerbated lung injury, bacterial burdens, and cytokine modulation (IL-10, IL-4, and TNF-α), all of which contributed to virulence and persistence [86].

Modifications in the fatty acid composition of the mycobacterial cell wall, facilitated by PE11, have been shown to enhance its permeability, potentially influencing bacterial interactions with the host environment [79]. Similarly, the over expression of PE19 increases the permeability of the *Mtb* envelope, which may constitute an adaptive response to stress induced by the host [80]. Notably, several members of the PE_PGRS family are specifically expressed in granulomatous lesions, suggesting their involvement in host-pathogen interactions [81].

11. Clinical implications of PE/PPE gene and diversity: PE/PPE as targets for TB therapy

The PE/PPE protein families play crucial roles in the virulence and survival of *Mtb*, making them attractive targets for therapeutic intervention. Mutants deficient in ppe25 pe19 or eccD5 exhibited severe attenuation and failed to survive in Severe combined immunodeficiency (SCID) mice or macrophages. *Mtb* strains lacking ESX-5 could potentially be utilized as live-attenuated vaccine candidates. Targeting ESX-5 may enhance TB treatment by increasing *Mtb* susceptibility to drugs and reducing its ability to evade host immunity [85]. These findings suggest that ESX-5 may serve as a viable therapeutic target, as it is essential for *Mtb* pathogenicity.

12. Conclusions and perspectives

The PE/PPE protein family of *Mtb* presents a molecular enigma, characterized by its structural diversity and functional significance in pathogenesis. Phylogenetic analyses indicate that these genes have coevolved with esx loci, which encode secretion systems responsible for exporting PE/PPE proteins. The genomic organization of these proteins is intricate, with several pairs of pe/ppe genes colocalized and transcribed as operonic clusters, suggesting coordinated functional roles. Many of these proteins form heterodimeric complexes that localize to the cell surface or are secreted, demonstrating their potential roles in host-pathogen interactions and immune modulation. Despite significant advancements in biological and molecular characterization, structural elucidation of PE/PPE proteins remain limited. The presently available crystal structures reveal a high degree of intrinsic disorder, likely enhancing their functional adaptability and facilitating protein-protein interactions and moonlighting functions. This structural plasticity provides an evolutionary advantage, potentially contributing to immune evasion and antigenic variation. The scarcity of atomic-level structural data hinders our understanding of their precise molecular mechanisms, necessitating further research into their three-dimensional architecture and structure-function relationships.

Genomic variation among pe/ppe genes introduces additional complexity. Nucleotide sequence variations among homologous proteins can lead to significant physicochemical changes, impacting their virulence and immunogenicity. Their ability to manipulate host immune responses suggests a crucial role in modulating both innate and adaptive immunity, allowing *Mtb* to persist within the host. Regulation of pe/ppe gene expression is finely tuned, influenced by environmental stimuli such as host immune pressure, nutrient availability, and stress conditions. This dynamic regulation is essential for the bacterium's ability to transition between different stages of infection, from dormancy to active disease. Furthermore, studies have identified several PE/PPE proteins as key players in host cell death pathways, reinforcing their importance in immune evasion and bacterial survival. A deeper understanding of these proteins at the molecular, structural, and immunological levels holds immense promise for TB research. Their roles in immune modulation make them attractive candidates for novel vaccine design, diagnostic markers, and drug development. However, to fully harness their potential, extensive efforts are needed to elucidate their structures, interaction networks, and precise functions in pathogenesis. Future studies should integrate structural biology, functional genomics, and immunological profiling to uncover new therapeutic strategies aimed at curbing TB and potentially other infectious diseases where these proteins may play a role. Collectively, these insights highlight PE/PPE proteins as both a challenge and an opportunity—while their variability and immune modulatory functions complicate our understanding of Mtb pathogenesis, they also present exploitable vulnerabilities that could be harnessed to advance vaccines, diagnostics, and host-directed therapies in the fight against tuberculosis.

References

[1] Pai M, Behr MA, Dowdy D, Dheda K, Divangahi M, Boehme CC, et al. Tuberculosis. Nat Rev Dis Primers 2016;2. https://doi.org/10.1038/NRDP.2016.76.

[2] Sheikh JA, Malik AA, Quadir N, Ehtesham NZ, Hasnain SE. Learning from COVID-19 to tackle TB pandemic: from despair to hope. Lancet Reg Health Southeast Asia 2022;2. https://doi.org/10.1016/j.lansea.2022.05.004.

[3] Malik AA, Sheikh JA, Ehtesham NZ, Hira S, Hasnain SE. Can Mycobacterium tuberculosis infection lead to cancer? Call for a paradigm shift in understanding TB and cancer. Int J Med Microbiol 2022;312. https://doi.org/10.1016/j.ijmm.2022.151558.

[4] Gagneux S. Host–pathogen coevolution in human tuberculosis. Phil Trans Biol Sci 2012;367:850. https://doi.org/10.1098/RSTB.2011.0316.

[5] Leung AN. Pulmonary tuberculosis: the essentials. Radiology 1999;210:307–22. https://doi.org/10.1148/RADIOLOGY.210.2.R99JA34307.

[6] Hilda JN, Das S, Tripathy SP, Hanna LE. Role of neutrophils in tuberculosis: a bird's eye view. Innate Immun 2019;26:240. https://doi.org/10.1177/1753425919881176.

[7] Winslow GM, Cooper A, Reiley W, Chatterjee M, Woodland DL. Early T-cell responses in tuberculosis immunity. Immunol Rev 2008;225. https://doi.org/10.1111/j.1600-065X.2008.00693.x.

[8] Cooper AM. Cell-mediated immune responses in tuberculosis. Annu Rev Immunol 2009;27:393–422. https://doi.org/10.1146/ANNUREV.IMMUNOL.021908.132703.

[9] Schluger NW. The pathogenesis of tuberculosis: the first one hundred (and twenty-three) years. Am J Respir Cell Mol Biol 2005;32:251–6. https://doi.org/10.1165/RCMB.F293.

[10] Bermudez LE, Goodman J. Mycobacterium tuberculosis invades and replicates within type II alveolar cells. Infect Immun 1996;64:1400–6. https://doi.org/10.1128/IAI.64.4.1400-1406.1996.

[11] Danelishvili L, McGarvey J, Li YJ, Bermudez LE. Mycobacterium tuberculosis infection causes different levels of apoptosis and necrosis in human macrophages and alveolar epithelial cells. Cell Microbiol 2003;5:649–60. https://doi.org/10.1046/J.1462-5822.2003.00312.X.

[12] Torrelles JB, Schlesinger LS. Integrating lung physiology, immunology, and tuberculosis. Trends Microbiol 2017;25:688–97. https://doi.org/10.1016/j.tim.2017.03.007.

[13] Luies L, Du Preez I. The echo of pulmonary tuberculosis: mechanisms of clinical symptoms and other disease-induced systemic complications. Clin Microbiol Rev 2020;33:1–19. https://doi.org/10.1128/CMR.00036-20.
[14] Chee CBE, Reves R, Zhang Y, Belknap R. Latent tuberculosis infection: opportunities and challenges. Respirology 2018;23:893–900. https://doi.org/10.1111/RESP.13346.
[15] Cohen A, Mathiasen VD, Schön T, Wejse C. The global prevalence of latent tuberculosis: a systematic review and meta-analysis. Eur Respir J 2019;54. https://doi.org/10.1183/13993003.00655-2019.
[16] Grosset J. Mycobacterium tuberculosis in the extracellular compartment: an underestimated adversary. Antimicrob Agents Chemother 2003;47:833–6. https://doi.org/10.1128/AAC.47.3.833-836.2003.
[17] Smith DT, Abernathy RS, Smith GB, Bondurant S. The apical localization of reinfection pulmonary tuberculosis. I. The stream flow theory. Am Rev Tuberc 1954;70:547–56. https://doi.org/10.1164/ART.1954.70.4.547.
[18] Lerner TR, Wilkinson RJ, Gutierrez MG. Mycobacterium tuberculosis cords within lymphatic endothelial cells to evade host immunity. JCI Insight 2020;5:e136937. https://doi.org/10.1172/jci.insight.136937.
[19] Heye T, Stoijkovic M, Kauczor HU, Junghanss T, Hosch W. Extrapulmonale Tuberkulose: Die radiologische Bildgebung eines fast vergessenen Verwandlungskünstlers. RoFo Fortschritte Auf Dem Gebiet Der Rontgenstrahlen Und Der Bildgebenden Verfahren 2011;183:1019–29. https://doi.org/10.1055/S-0031-1273429.
[20] Gey Van Pittius NC, Sampson SL, Lee H, Kim Y, Van Helden PD, Warren RM. Evolution and expansion of the Mycobacterium tuberculosis PE and PPE multigene families and their association with the duplication of the ESAT-6 (esx) gene cluster regions. BMC Evol Biol 2006;6. https://doi.org/10.1186/1471-2148-6-95.
[21] Sassi M, Drancourt M. Genome analysis reveals three genomospecies in Mycobacterium abscessus. BMC Genom 2014;15:1–10. https://doi.org/10.1186/1471-2164-15-359/FIGURES/2.
[22] Cole ST, Brosch R, Parkhill J, Garnier T, Churcher C, Harris D, et al. Deciphering the biology of mycobacterium tuberculosis from the complete genome sequence. Nature 1998;393:537–44. https://doi.org/10.1038/31159;KWRD=SCIENCE.
[23] Delogu G, Bigi F, Hasnain SE, Cataldi A. Enigmatic proteins from the surface: the Erp, PE, and PPE protein families. Mycobact Cell Envelope 2014:133–51. https://doi.org/10.1128/9781555815783.CH8.
[24] Akhter Y, Ehebauer MT, Mukhopadhyay S, Hasnain SE. The PE/PPE multigene family codes for virulence factors and is a possible source of mycobacterial antigenic variation: perhaps more? Biochimie 2012;94:110–6. https://doi.org/10.1016/J.BIOCHI.2011.09.026.
[25] Sharma T, Alam A, Ehtram A, Rani A, Grover S, Ehtesham NZ, et al. The Mycobacterium tuberculosis PE_PGRS protein family acts as an immunological decoy to subvert host immune response. Int J Mol Sci 2022;23. https://doi.org/10.3390/IJMS23010525.
[26] Dheenadhayalan V, Delogu G, Sanguinetti M, Fadda G, Brennan MJ. Variable expression patterns of Mycobacterium tuberculosis PE_PGRS genes: evidence that PE_PGRS16 and PE_PGRS26 are inversely regulated in vivo. J Bacteriol 2006;188:3721. https://doi.org/10.1128/JB.188.10.3721-3725.2006.
[27] Grover S, Sharma T, Singh Y, Kohli S, Manjunath P, Singh A, et al. The PGRS domain of Mycobacterium tuberculosis PE_PGRS protein Rv0297 is involved in Endoplasmic reticulum stress-mediated apoptosis through toll-like receptor 4. mBio 2018;9. https://doi.org/10.1128/MBIO.01017-18/SUPPL_FILE/MBO003183943ST4.DOCX.
[28] Daleke MH, Cascioferro A, De Punder K, Ummels R, Abdallah AM, Van Der Wel N, et al. Conserved Pro-Glu (PE) and Pro-Pro-Glu (PPE) protein domains target LipY lipases of pathogenic mycobacteria to the cell surface via the ESX-5 pathway. J Biol Chem 2011;286:19024–34. https://doi.org/10.1074/jbc.M110.204966.
[29] Talarico S, Zhang L, Marrs CF, Foxman B, Cave MD, Brennan MJ, et al. Mycobacterium tuberculosis PE_PGRS16 and PE_PGRS26 genetic polymorphism among clinical isolates. Tuberculosis 2008;88:283–94. https://doi.org/10.1016/J.TUBE.2008.01.001.
[30] Chakhaiyar P, Nagalakshmi Y, Aruna B, Murthy KJR, Katoch VM, Hasnain SE. Regions of high antigenicity within the hypothetical PPE major polymorphic tandem repeat open-reading frame, Rv2608, show a

differential humoral response and a low T cell response in various categories of patients with tuberculosis. JID (J Infect Dis) 2004;190:1237–44. https://doi.org/10.1086/423938/2/M_190-7-1237-FIG001.GIF.

[31] Poulet S, Cole ST. Characterization of the highly abundant polymorphic GC-rich-repetitive sequence (PGRS) present in Mycobacterium tuberculosis. Arch Microbiol 1995;163:87–95. https://doi.org/10.1007/BF00381781.

[32] Kohli S, Singh Y, Sharma K, Mittal A, Ehtesham NZ, Hasnain SE. Comparative genomic and proteomic analyses of PE/PPE multigene family of Mycobacterium tuberculosis H37Rv and H37Ra reveal novel and interesting differences with implications in virulence. Nucleic Acids Res 2012;40:7113–22. https://doi.org/10.1093/NAR/GKS465.

[33] Simeone R, Bottai D, Brosch R. ESX/type VII secretion systems and their role in host-pathogen interaction. Curr Opin Microbiol 2009;12:4–10. https://doi.org/10.1016/j.mib.2008.11.003.

[34] Serafini G, Pompili M, Elena Seretti M, Stefani H, Palermo M, Coryell W, et al. The role of inflammatory cytokines in suicidal behavior: a systematic review. Eur Neuropsychopharmacol 2013;23:1672–86. https://doi.org/10.1016/j.euroneuro.2013.06.002.

[35] Tufariello JAM, Chapman JR, Kerantzas CA, Wong KW, Vilchèze C, Jones CM, et al. Separable roles for Mycobacterium tuberculosis ESX-3 effectors in iron acquisition and virulence. Proc Natl Acad Sci U S A 2016;113:E348–57. https://doi.org/10.1073/PNAS.1523321113.

[36] Namouchi A, Karboul A, Fabre M, Gutierrez MC, Mardassi H. Evolution of smooth tubercle bacilli PE and PE_PGRS genes: evidence for a prominent role of recombination and Imprint of positive selection. PLoS One 2013;8:e64718. https://doi.org/10.1371/JOURNAL.PONE.0064718.

[37] Karboul A, Mazza A, Gey Van Pittius NC, Ho JL, Brousseau R, Mardassi H. Frequent homologous recombination events in Mycobacterium tuberculosis PE/PPE multigene families: potential role in antigenic variability. J Bacteriol 2008;190:7838–46. https://doi.org/10.1128/JB.00827-08.

[38] McEvoy CR, Van Helden PD, Warren RM, Van Pittius NCG. Evidence for a rapid rate of molecular evolution at the hypervariable and immunogenic Mycobacterium tuberculosis PPE38 gene region. BMC Evol Biol 2009;9. https://doi.org/10.1186/1471-2148-9-237.

[39] Supply P, Marceau M, Mangenot S, Roche D, Rouanet C, Khanna V, et al. Genomic analysis of smooth tubercle bacilli provides insights into ancestry and pathoadaptation of Mycobacterium tuberculosis. Nat Genet 2013 45:2 2013;45:172–9. https://doi.org/10.1038/ng.2517.

[40] Manjunath P, Ahmad J, Samal J, Rani A, Sheikh JA, Zarin S, et al. Expression of a unique M. tuberculosis DNA MTase Rv1509 in M. smegmatis alters the gene expression pattern and enhances virulence. Front Microbiol 2024;15:1344857. https://doi.org/10.3389/FMICB.2024.1344857/BIBTEX.

[41] Strong M, Sawaya MR, Wang S, Phillips M, Cascio D, Eisenberg D. Toward the structural genomics of complexes: crystal structure of a PE/PPE protein complex from Mycobacterium tuberculosis. Proc Natl Acad Sci U S A 2006;103:8060–5. https://doi.org/10.1073/PNAS.0602606103.

[42] Ernst JD. Antigenic variation and immune escape in the MTBC. Adv Exp Med Biol 2017;1019:171. https://doi.org/10.1007/978-3-319-64371-7_9.

[43] Zarin S, Shariq M, Rastogi N, Ahuja Y, Manjunath P, Alam A, et al. Rv2231c, a unique histidinol phosphate aminotransferase from Mycobacterium tuberculosis, supports virulence by inhibiting host-directed defense. Cell Mol Life Sci 2024;81:1–17. https://doi.org/10.1007/S00018-024-05200-8/FIGURES/8.

[44] Hahn MY, Raman S, Anaya M, Husson RN. The Mycobacterium tuberculosis extracytoplasmic-function sigma factor SigL regulates polyketide synthases and secreted or membrane proteins and is required for virulence. J Bacteriol 2005;187:7062–71. https://doi.org/10.1128/JB.187.20.7062-7071.2005.

[45] Lee J-H, Karakousis PC, Bishai WR. Roles of SigB and SigF in the Mycobacterium tuberculosis sigma factor network. J Bacteriol 2008;190:699–707. https://doi.org/10.1128/JB.01273-07.

[46] Williams EP, Lee JH, Bishai WR, Colantuoni C, Karakousis PC. Mycobacterium tuberculosis SigF regulates genes encoding cell wall-associated proteins and directly regulates the transcriptional regulatory gene phoY1. J Bacteriol 2007;189:4234. https://doi.org/10.1128/JB.00201-07.

[47] Raman S, Hazra R, Dascher CC, Husson RN. Transcription regulation by the Mycobacterium tuberculosis alternative sigma factor SigD and its role in virulence. J Bacteriol 2004;186:6605–16. https://doi.org/10.1128/JB.186.19.6605-6616.2004.

[48] Manganelli R, Dubnau E, Tyagi S, Kramer FR, Smith I. Differential expression of 10 sigma factor genes in Mycobacterium tuberculosis. Mol Microbiol 1999;31:715–24. https://doi.org/10.1046/J.1365-2958.1999.01212.X.

[49] Calamita H, Ko C, Tyagi S, Yoshimatsu T, Morrison NE, Bishai WR. The Mycobacterium tuberculosis SigD sigma factor controls the expression of ribosome-associated gene products in stationary phase and is required for full virulence. Cell Microbiol 2005;7:233–44. https://doi.org/10.1111/J.1462-5822.2004.00454.X.

[50] Ahmed A, Das A, Mukhopadhyay S. Immunoregulatory functions and expression patterns of PE/PPE family members: roles in pathogenicity and impact on anti-tuberculosis vaccine and drug design. IUBMB Life 2015;67:414–27. https://doi.org/10.1002/IUB.1387;WGROUP:STRING:PUBLICATION.

[51] Gold B, Marcela RG, Marras SAE, Pentecost M, Smith I. The Mycobacterium tuberculosis ideR is a dual functional regulator that controls transcription of genes involved in iron acquisition, iron storage and survival in macrophages. Mol Microbiol 2001;42:851–65. https://doi.org/10.1046/J.1365-2958.2001.02684.X.

[52] Solans L, Gonzalo-Asensio J, Sala C, Benjak A, Uplekar S, Rougemont J, et al. The PhoP-dependent ncRNA Mcr7 modulates the TAT secretion system in Mycobacterium tuberculosis. PLoS Pathog 2014;10:e1004183. https://doi.org/10.1371/JOURNAL.PPAT.1004183.

[53] Goldstone RM, Goonesekera SD, Bloom BR, Sampson SL. The transcriptional regulator Rv0485 modulates the expression of a pe and ppe gene pair and is required for Mycobacterium tuberculosis virulence. Infect Immun 2009;77:4654–67. https://doi.org/10.1128/IAI.01495-08.

[54] Klemperer EM, Hughes JR. Commentary on Wu et al. (2017): do very brief reduction interventions increase quitting among smokers not ready to quit? Addiction 2017;112:2041–2. https://doi.org/10.1111/ADD.13970.

[55] Mendoza-Espinosa P, García-González V, Moreno A, Castillo R, Mas-Oliva J. Disorder-to-order conformational transitions in protein structure and its relationship to disease. Mol Cell Biochem 2009;330(1):105–20. https://doi.org/10.1007/S11010-009-0105-6.

[56] Hebert AM, Talarico S, Yang D, Durmaz R, Marrs CF, Zhang L, et al. DNA polymorphisms in the pepA and PPE18 genes among clinical strains of Mycobacterium tuberculosis: implications for vaccine efficacy. Infect Immun 2007;75:5798–805. https://doi.org/10.1128/IAI.00335-07/ASSET/BAD328D3-362F-4526-A802-3512AE753CC1/ASSETS/GRAPHIC/ZII0120770210001.JPEG.

[57] Ahmad J, Farhana A, Pancsa R, Arora SK, Srinivasan A, Tyagi AK, et al. Contrasting function of structured N-terminal and unstructured C-terminal segments of mycobacterium tuberculosis PPE37 protein. mBio 2018;9. https://doi.org/10.1128/MBIO.01712-17/SUPPL_FILE/MBO006173677ST4.DOCX.

[58] Voskuil MI, Schnappinger D, Rutherford R, Liu Y, Schoolnik GK. Regulation of the Mycobacterium tuberculosis PE/PPE genes. Tuberculosis 2004;84:256–62. https://doi.org/10.1016/j.tube.2003.12.014.

[59] Soldini S, Palucci I, Zumbo A, Sali M, Ria F, Manganelli R, et al. PPE-MPTR genes are differentially expressed by Mycobacterium tuberculosis in vivo. Tuberculosis 2011;91:563–8. https://doi.org/10.1016/j.tube.2011.08.002.

[60] Nair S, Ramaswamy PA, Ghosh S, Joshi DC, Pathak N, Siddiqui I, et al. The PPE18 of Mycobacterium tuberculosis interacts with TLR2 and activates IL-10 induction in macrophage. J Immunol 2009;183:6269–81. https://doi.org/10.4049/JIMMUNOL.0901367.

[61] Ehtram A, Shariq M, Quadir N, Jamal S, Pichipalli M, Zarin S, et al. Deciphering the functional roles of PE18 and PPE26 proteins in modulating Mycobacterium tuberculosis pathogenesis and immune response. Front Immunol 2025;16:1517822. https://doi.org/10.3389/FIMMU.2025.1517822/BIBTEX.

[62] Malik AA, Shariq M, Sheikh JA, Jaiswal U, Fayaz H, Shrivastava G, et al. Mechanisms of immune evasion by Mycobacterium tuberculosis: the impact of T7SS and cell wall lipids on host defenses. Crit Rev Biochem Mol Biol 2024;59. https://doi.org/10.1080/10409238.2024.2411264.

[63] Qian J, Chen R, Wang H, Zhang X. Role of the PE/PPE family in host–pathogen interactions and prospects for anti-tuberculosis vaccine and diagnostic tool design. Front Cell Infect Microbiol 2020;10:594288. https://doi.org/10.3389/FCIMB.2020.594288/BIBTEX.
[64] Mohareer K, Tundup S, Hasnain SE. Transcriptional regulation of mycobacterium tuberculosis pe/PPE genes: a molecular switch to virulence? J Mol Microbiol Biotechnol 2011;21:97–109. https://doi.org/10.1159/000329489.
[65] Talreja J, Peng C, Nguyen T-M, Draghici S, Samavati L. Discovery of novel transketolase epitopes and the development of IgG-based tuberculosis serodiagnostics. Microbiol Spectr 2023;11. https://doi.org/10.1128/SPECTRUM.03377-22.
[66] Kolloli A, Singh P, Subbian S. Granulomatous response to Mycobacterium tuberculosis infection. In: Understanding the host immune response against Mycobacterium tuberculosis infection. Springer; 2018. p. 41–66. https://doi.org/10.1007/978-3-319-97367-8_3.
[67] Peng Z, Yue Y, Xiong S. Mycobacterial PPE36 modulates host inflammation by promoting E3 ligase Smurf1-mediated MyD88 degradation. Front Immunol 2022;13. https://doi.org/10.3389/FIMMU.2022.690667.
[68] Iantomasi R, Sali M, Cascioferro A, Palucci I, Zumbo A, Soldini S, et al. PE_PGRS30 is required for the full virulence of Mycobacterium tuberculosis. Cell Microbiol 2012;14:356–67. https://doi.org/10.1111/J.1462-5822.2011.01721.X.
[69] Gong Z, Kuang Z, Li H, Li C, Ali MK, Huang F, et al. Regulation of host cell pyroptosis and cytokines production by Mycobacterium tuberculosis effector PPE60 requires LUBAC mediated NF-κB signaling. Cell Immunol 2019;335:41–50. https://doi.org/10.1016/J.CELLIMM.2018.10.009.
[70] Peng X, Luo T, Zhai X, Zhang C, Suo J, Ma P, et al. PPE11 of Mycobacterium tuberculosis can alter host inflammatory response and trigger cell death. Microb Pathog 2019;126:45–55. https://doi.org/10.1016/j.micpath.2018.10.031.
[71] Ilesanmi A, Odeniran OM, Tatsipie L, Osam Duodu E, Ankrah PK. The role of proline-proline-glutamic acid (PPE) proteins in Mycobacterium tuberculosis virulence: mechanistic insights and therapeutic implications. Cureus 2024;16. https://doi.org/10.7759/CUREUS.51955.
[72] Saini V, Miller SA, Fisher WW. Multiple schedules in practical application: research trends and implications for future investigation. J Appl Behav Anal 2016;49:421–44. https://doi.org/10.1002/JABA.300.
[73] Huang T, Zhang J, Xu ZP, Hu L Le, Chen L, Shao JL, et al. Deciphering the effects of gene deletion on yeast longevity using network and machine learning approaches. Biochimie 2012;94:1017–25. https://doi.org/10.1016/j.biochi.2011.12.024.
[74] Jha AP, Stanley EA, Kiyonaga A, Wong L, Gelfand L. Examining the protective effects of mindfulness training on working memory capacity and affective experience. Emotion 2010;10:54–64. https://doi.org/10.1037/A0018438.
[75] Sharma N, Shariq M, Quadir N, Singh J, Sheikh JA, Hasnain SE, et al. Mycobacterium tuberculosis protein PE6 (Rv0335c), a novel TLR4 agonist, evokes an inflammatory response and modulates the cell death pathways in macrophages to enhance intracellular survival. Front Immunol 2021;12:696491. https://doi.org/10.3389/FIMMU.2021.696491/BIBTEX.
[76] Srivastava R, Duan R, Ahn SH. Multiple roles of CTDK-I throughout the cell. Cell Mol Life Sci 2019;76:2789–97. https://doi.org/10.1007/s00018-019-03118-0.
[77] Tundup S, Mohareer K, Hasnain SE. Mycobacterium tuberculosis PE25/PPE41 protein complex induces necrosis in macrophages: role in virulence and disease reactivation? FEBS Open Bio 2014;4:822–8. https://doi.org/10.1016/J.FOB.2014.09.001;WGROUP:STRING:PUBLICATION.
[78] Tiwari U, Yadav P, Nigam D. Study on phytochemical screening and antibacterial potential of methanolic flower and leaf extracts of Hibiscus rosa sinensis. Int J Innov Appl Res 2015;3(6):9–14.
[79] Deng W, Zeng J, Xiang X, Li P, Xie J. PE11 (Rv1169c) selectively alters fatty acid components of mycobacterium smegmatis and host cell interleukin-6 level accompanied with cell death. Front Microbiol 2015;6:141377. https://doi.org/10.3389/FMICB.2015.00613/BIBTEX.

[80] Ramakrishnan P, Aagesen AM, McKinney JD, Tischler AD. Mycobacterium tuberculosis resists stress by regulating PE19 expression. Infect Immun 2016;84:735–46. https://doi.org/10.1128/IAI.00942-15.

[81] Ramakrishnan L, Federspiel NA, Falkow S. Granuloma-specific expression of Mycobacterium virulence proteins from the glycine-rich PE-PGRS family. Science 2000;288:1436–9. https://doi.org/10.1126/SCIENCE.288.5470.1436.

[82] Poulet S, Cole ST. Repeated DNA sequences in mycobacteria. Arch Microbiol 1995;163:79–86. https://doi.org/10.1007/BF00381780/METRICS.

[83] Ehtram A, Shariq M, Ali S, Quadir N, Sheikh JA, Ahmad F, et al. Teleological cooption of Mycobacterium tuberculosis PE/PPE proteins as porins: role in molecular immigration and emigration. Int J Med Microbiol 2021;311. https://doi.org/10.1016/j.ijmm.2021.151495.

[84] Tundup S, Pathak N, Ramanadham M, Mukhopadhyay S, Murthy KJR, Ehtesham NZ, et al. The co-operonic PE25/PPE41 protein complex of Mycobacterium tuberculosis elicits increased humoral and cell mediated immune response. PLoS One 2008;3:e3586. https://doi.org/10.1371/JOURNAL.PONE.0003586.

[85] Bottai D, di Luca M, Majlessi L, Frigui W, Simeone R, Sayes F, et al. Disruption of the ESX-5 system of Mycobacterium tuberculosis causes loss of PPE protein secretion, reduction of cell wall integrity and strong attenuation. Mol Microbiol 2012;83:1195–209. https://doi.org/10.1111/J.1365-2958.2012.08001.X.

[86] Singh P, Rao RN, Reddy JRC, Prasad R, Kotturu SK, Ghosh S, et al. PE11, a PE/PPE family protein of Mycobacterium tuberculosis is involved in cell wall remodeling and virulence. Sci Rep 2016;6. https://doi.org/10.1038/SREP21624.

CHAPTER

4

Host-pathogen dynamics in *Mycobacterium tuberculosis*: The impact of PE/PPE proteins on innate and adaptive immunity

Sajal Dey[1,2], Vandana Maurya[1,3], Niteen Pathak[1] and Sangita Mukhopadhyay[1]

[1]*BRIC-Centre for DNA Fingerprinting and Diagnostics (CDFD), Hyderabad, Telangana, India;* [2]*Graduate Studies, Manipal Academy of Higher Education, Manipal, Karnataka, India;* [3]*Graduate Studies, Regional Center for Biotechnology, Faridabad, Haryana, India*

Abbreviation

AB Antibody
BCG Bacillus Calmette–Guérin
DC Dendritic cells
DC-SIGN Dendritic cell-specific intercellular adhesion molecule -3- grabbing nonintegrin
ERK Extracellular signal regulated protein kinase
ESAT-6 Early secreted antigenic target 6
ESX-1 ESAT-6 secretion system-1
IFN Interferon
IL Interleukin
MHC Major histocompatibility complex
Mtb *Mycobacterium tuberculosis*
NF-κB Nuclear factor kappa-B
NO Nitric oxide
P-L Phagosome-lysosome
P38 MAPK P38 Mitogen activated protein kinases
PE Proline-Glutamic acid
PPE Proline-proline-glutamic acid
RNI Reactive nitrogen intermediates
ROS Reactive oxygen species
TB Tuberculosis
Th T-helper
TLR Toll-like receptors
TNF-α, Tumor necrosis factor alpha

Navigating Tuberculosis. https://doi.org/10.1016/B978-0-443-27735-1.00009-5

1. Introduction

Mycobacterium tuberculosis (*Mtb*), the causative agent for tuberculosis (TB), is estimated to infect roughly a quarter of the world's population. Slow growth, complex cell wall architecture, and various other devious survival strategies adopted by the bacilli help them to survive successfully inside the host. It also modulates the host metabolic and immune machineries to create a favorable growth environment. Despite Bacillus Calmette–Guérin (BCG) vaccination and Directly Observed Treatment, Short Course (DOTS) therapy, TB persists in the population of developing countries. Additionally, the emergence of multidrug-resistant (MDR) and extensively drug-resistant (XDR) strains has amplified the treatment failure. According to the WHO Global Tuberculosis Report (2024), 10.8 million new TB cases were registered in 2023 worldwide, with a reversal in a downward trend [1]. The net reduction in the TB incidence rate from 2015 to 2023 globally was 8.3%, which is far from the WHO End TB Strategy milestone of a 50% reduction by 2025. Eight countries in 2023 accounted for more than two-thirds of global TB cases. These countries are India (26%), Indonesia (10%), China (6.8%), the Philippines (6.8%), Pakistan (6.3%), Nigeria (4.6%), Bangladesh (3.5%), and the Democratic Republic of the Congo (3.1%).

Genome-based *Mtb* studies were started with the *Mtb* H37Rv genome sequencing in 1998 by Cole et al. [2]. It contains 4,411,529 base pairs, ~4000 genes with GC content more than 60%, which is unique to mycobacteria. This also led to the finding of a distinct kind of genes with conserved proline and glutamic acid residue in the 7th–10th position of the protein. Based on these characteristics, they are divided into two gene families, PE, Proline-Glutamic acid, with 99 members, and PPE, Proline-proline-glutamic acid, with 69 members. These gene families are only found in the mycobacteria [2]. Despite approximately 10% of the *Mtb* genome belonging to the PE/PPE family, their functional relevance was largely unknown [3]. From fast-growing saprophytic species like *M. smegmatis* to slow-growing pathogenic species like *M. leprae,* the *pe/ppe* gene family went through evolutionary expansion [4]. Their unique location inside the esx secretory system genomic locus, expansion from nonpathogenic to pathogenic species, and redundancy have made them an attractive set of genes to understand their role in the context of host-pathogen interaction. It has been shown that eight of the sigma factors, namely SigB, SigD, SigE, SigF, SigG, SigH, SigL, and SigM, are involved in the regulation of *pe/ppe* gene expression. Mainly, SigB, along with SigE and SigH, form a network and manage to regulate the transcription of *Mtb pe/ppe* genes [5]. Here, we have briefly summarized the classification, evolution, and variation of the PE/PPE family of proteins and their roles in the modulation of host innate and adaptive immune responses. Also, their contributions as potential vaccine candidates are discussed.

2. Features of PE and PPE proteins and their classification

PE and PPE family proteins are unique to *Mycobacterium* sp. and found to be present in pathogenic as well as saprophytic mycobacterial species [6,7]. However, their number is less in nonpathogenic mycobacteria but expanded more in the pathogenic species within the *Mtb* complex (MTBC), *M. leprae*, *M. marinum*, *M. ulcerans,* and *M. avium* possess the highest number of *pe/ppe* genes [4]. In the *Mtb* H37Rv strain, the genome encodes approximately 168 proteins from the unique mycobacterial PE and PPE gene families, of which 99 belong to the PE and 69 to the PPE family. These gene families account for approximately 10% of the total coding capacity of the *Mtb* genome [2,8]. PE and PPE proteins are characterized by the presence of conserved Proline-Glutamic acid (PE) and Proline-Proline-Glutamic acid (PPE) motifs within the N-terminal sequence of ~110 and ~180 amino acids, respectively [2]. In contrast, the C-terminal region is found to be highly variable.

Both families can be further subdivided based on the sequence variability in the C-terminal region. The PE family can be classified into two main subfamilies: the PE_PGRS subfamily, containing polymorphic GC-rich sequences, and the PE subfamily, which does not contain any distinctive features [9]. The PE_PGRS subfamily consists of 65 members with a guanine and cytosine (GC) content (~80%) and presence of Gly-Gly-Ala/Gly-Gly-Asn repeats, which can vary in sequence and size [10]. These proteins are also comparatively larger in size, with some members reaching up to 1400 amino acids in length. The presence of extended C-terminal regions and repeat domains renders them more susceptible to recombination and mutations, resulting in significantly high variability. The members of the PE_PGRS family are scattered throughout the genome and are generally not cotranscribed with other genes [11] (Fig. 4.1).

The PPE family is further divided into subfamilies based on the presence of specific motifs in the C-terminal region. These include PPE-PPW (ProX-X-Pro-X-X-Trp motif), PPE-SVP (Gly-X-X-Ser-Val-Pro-X-X-Trp motif), and the evolutionarily conserved PPE-MPTR (major polymorphic tandem repeat, containing many repeats of Asn-X-Gly-X-Gly-X-Asn-X-Gly signature motif). Interestingly, many *pe* genes are located adjacent to *ppe* genes, and these *pe/ppe* couplets are mostly coexpressed

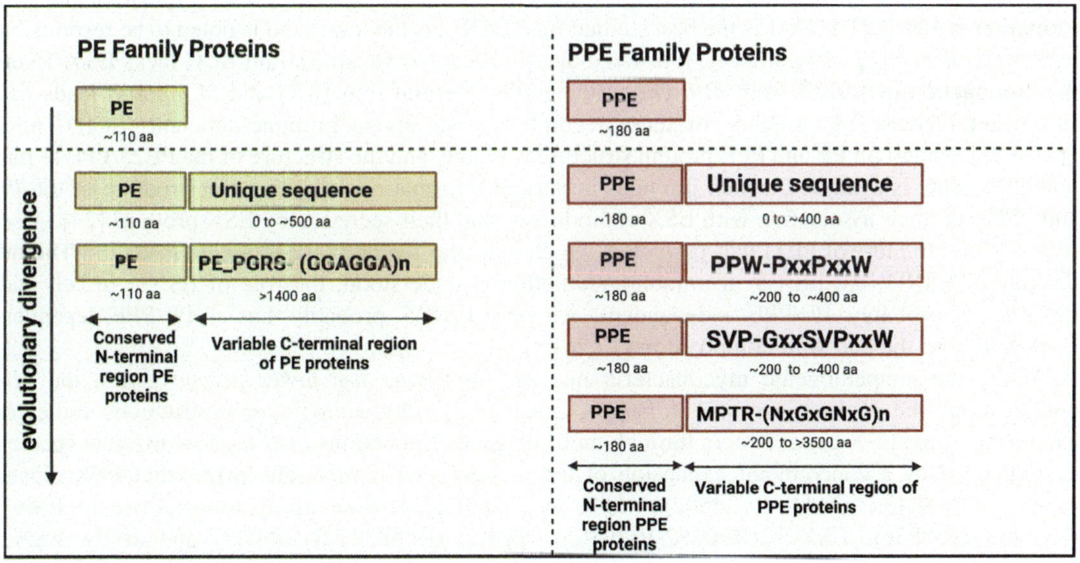

FIGURE 4.1 Diagrammatic representation of *pe* and *ppe* gene families.

PE and PPE family proteins possess a relatively conserved N-terminus of ~110 and ~180 amino acids (aa), respectively. These proteins were further divided into subgroups based on their C-terminal variable.

[10,11] (Fig. 4.1). The functional diversity in PE/PPE proteins is largely attributed to their highly variable C-terminal regions characterized by the presence of the conserved YxxxD/E motif and WxG motif, respectively. These motifs are critical for their recognition and secretion through the Type VII secretion system (T7SS) [12].

3. The *pe/ppe* genomic distribution and evolution

The *pe/ppe* family of genes is exclusively present in mycobacterial species and is absent from their immediate relatives, such as *Nocardia farcinica* [13]. The origin of *pe/ppe* genes is still not known, but there are speculations that the earliest form of *pe* and *ppe* genes were in close association with the ESX or T7SS [14]. Most PE/PPE proteins are found as cooperonic pairs. PE members are placed at the 5′ end of the operon, followed by PPE members, and then early secretory proteins (ESX) family members. These operons are generally part of the ESX secretion system. The ESX system is involved in PE/PPE protein export [15]. The *Mtb* genome possesses five types of the ESX secretion system (also known as T7SS) gene clusters, which are named after its first known secreted substrate ESAT-6 [Rv3875, EsxA] [16,17]. The *pe/ppe* genes have evolved and duplicated in association with the duplication of the five *esx* gene clusters [6,18] (Fig. 4.2). It is observed that ESX-1 encodes secreted proteins EsxA and EsxB, which are flanked directly upstream by *pe35* and *ppe68* [19]. While ESX-1 substrates are involved in membrane lysis and phagosomal escape, ESX-3, which contains *pe5* and *ppe4,* is engaged in iron hemostasis and zinc acquisition [16,17,20]. ESX-5, which contains two PE (PE18 and PE19) and three PPE (PPE25, PPE26, and PPE27) proteins, is mainly involved in immunomodulation [15,20] and nutrient uptake [21]. ESX-4 with no PE/PPE proteins is the most ancestral among these secretion systems and is shown to be involved in heme-iron acquisition and pore formation in *Mtb* [22]. ESX-1 is the best studied (and ESX-2 is the least) and is noted to be responsible for *Mtb* virulence [12]. Importantly, it is noted that the *M. bovis* vaccine strain BCG lacks ESX-1, and the introduction of ESX-1 from *Mtb via* an integrating cosmid into BCG and *M. microti* leads to a significant increase in virulence of these recombinant strains in immune-deficient (SCID) mice [12,19,23]. Report on PE and PPE protein structure is scarce, and the structure of the PE25/PPE41 pair bound to EspG is available, which provides important information regarding the structure of the PE and PPE pair, their association with ESX components, and their secretion by ESX proteins [24]. Also, recently, the structure of PE11 was predicted, which indicates the presence of an α/β fold [25]. Though the role of ESX-1 and ESX-5 in immunomodulation is understood, the role of ESX-1 in cell wall integrity is probably PE/PPE independent, whereas ESX-5 probably has a PE/PPE-dependent contribution to the cell wall composition [17,26].

While the nonpathogenic mycobacteria like *M. smegmatis* has fewer *pe/ppe* genes, the *Mtb* complex has the highest number of *pe/ppe* genes [4,7]. Phylogenetic reconstructions indicated expansion of the ESX gene clusters through multiple gene duplications over time in different species, accompanied by a concomitant expansion of the *pe/ppe* gene family. The primordial ESX cluster region is ESX-4 (Rv3444c-Rv3450c), which was duplicated sequentially to give rise to ESX-1 (Rv3866-Rv3883c), ESX-3 (Rv0282-Rv0292), ESX-2 (Rv3884c-Rv3895c), and lastly ESX-5 (Rv1782-Rv1798). Interestingly, the initial ESX cluster region, ESX-4, does not have any *pe/ppe* genes, and therefore, it is believed that these genes must have first integrated into ESX-1 and then expanded during the subsequent gene duplication events [6,27]. The *pe/ppe* genes are incorporated into

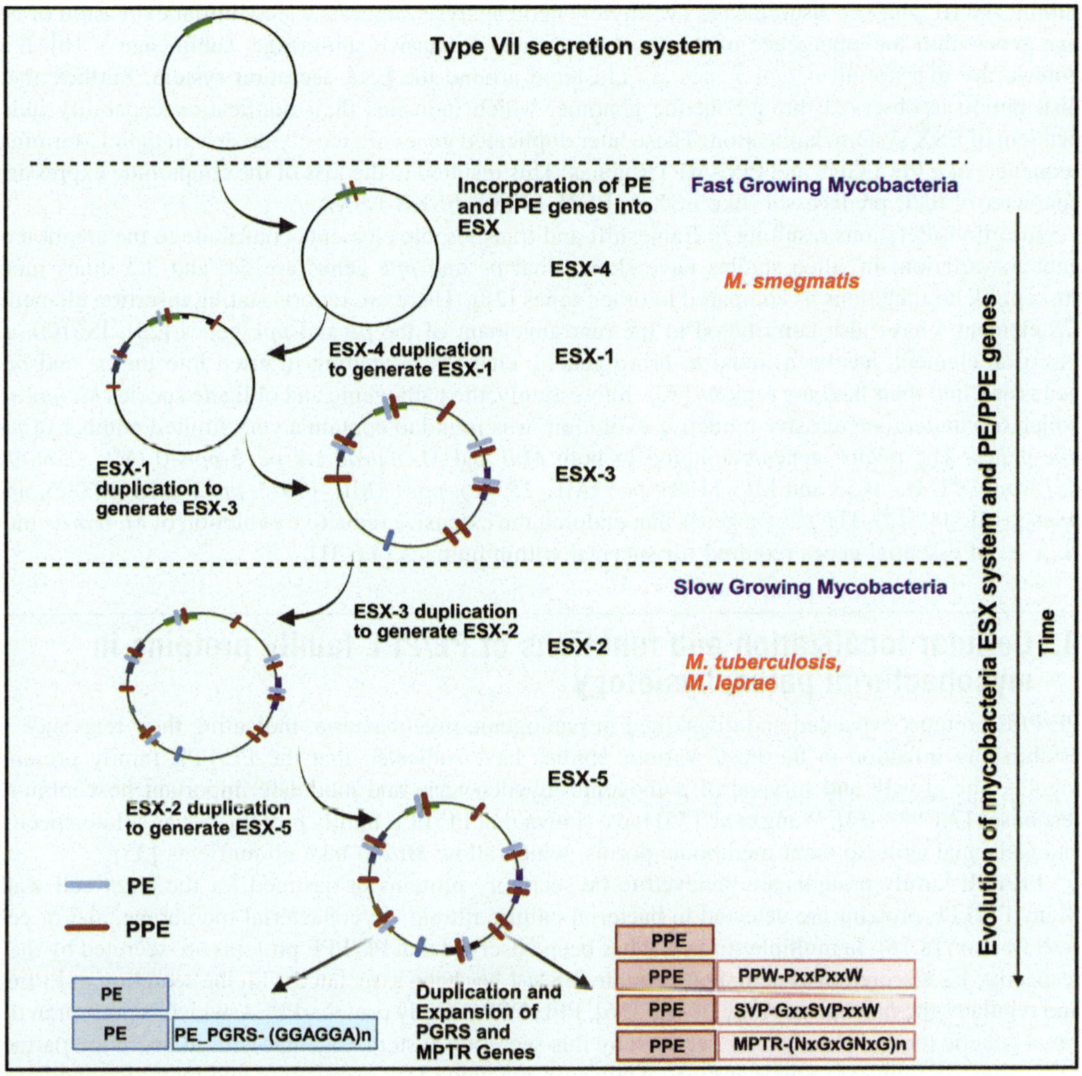

FIGURE 4.2 Flow diagram predicting the evolution and expansion of *pe* and *ppe* genes.

During the course of evolution, *pe* and *ppe* were first inserted into the *esx* gene cluster. Duplication of *the esx* cluster led to duplication of the *pe* and *ppe* genes. Spontaneous homologous recombination, random mutation, insertion/deletion, and *IS* elements contributed to the expansion and evolution of *pe* and *ppe* genes into different subgroups.

the ESX cluster and amplified and are divided into different sublineages to assign them to their respective ESX system (Fig. 4.2). SublineageI constitute the *pe/ppe* genes inserted into and/or associated with the *esx-1* gene cluster. A similar relationship is hypothesized for *esx-3*/sublineage II, *esx-2*/

sublineage III, and *esx-5*/sublineage IV. Phylogenetic analysis suggests a slow initial expansion of *pe/ppe* genes until the emergence of the most recent and expansive sublineage, sublineage V [6]. It is noteworthy that not all *pe/ppe* genes are clustered around the ESX secretion system. Further, their distribution is observed throughout the genome, which indicates their duplication capability independent of ESX system duplication. These later duplicated genes are mostly diverse in their C-terminal sequence, like PE_PGRS and PPE-MPTR genes. This resulted in the loss of the cooperonic expression character of their predecessor, like ESX-1, PE35-PPE68-ESXB-ESXA.

Insertions/deletions resulting in frameshift and transposable elements contribute to the creation of genetic variation. In silico studies have shown that *pe* and *ppe* genes are 3.0 and 3.3 times more susceptible to mutations as compared to other genes [28]. There are reports stating insertion elements (*IS* elements) have also contributed to the rearrangement of the *pe* and *ppe* genes [29]. IS6100, an insertion element, has been found to bring genetic changes by getting inserted into the *pe* and *ppe* genes and into their flanking regions [30]. Interestingly, the pathogenic and obligate species *M. leprae,* which has undergone massive reductive evolution, was found to contain a very limited number of *pe/ppe* genes. The *pe/ppe* genes belonging to both *Mtb* and *M. leprae* are *pe15-ppe20* (ML_0538-9), *pe13-ppe18* (ML_1053 and ML_1182), *pe5* (ML_2534c), *ppe1* (ML_1991), *ppe2* (ML_1828c), and *ppe68* (ML_0051c). The *pe/ppe* genes that endured the extensive reductive evolution of *M. leprae* may be a set of essential genes required for survival within humans [4,6,31].

4. Cellular localization and functions of PE/PPE family proteins in mycobacterial pathophysiology

PE/PPE proteins expanded and diversified in pathogenic mycobacteria, indicating their relevance in establishing infection in the host. Various studies have indicated that the PE/PPE family proteins regulate the growth and survival of pathogenic mycobacteria and modulate important host immune responses [7,12,32–34]. Wang et al. [35] have shown that PE/PPE family proteins act as solute-specific channels analogous to outer membrane porins, which allow *Mtb* to take up nutrients [35].

PE/PPE family proteins are believed to be secretory proteins or destined for the outer cell wall. Many PE/PPE proteins are detected in bacterial culture filtrate, mycobacterial membrane, and/or cell wall fraction [8,16]. In multiple studies, it has been observed that PE/PPE proteins are secreted by their respective ESX secretion system. For example, ESX-1 has been associated with the secretion of PPE68 and regulates the pathophysiology of *Mtb* [36]. PPE-PPW family protein, PPE4, which is present in the *esx-3* genetic locus, is likely to be secreted by this secretion system, together with its secretion partner PE5. These proteins are important for mycobactin-mediated iron acquisition and metal ion-mediated gene expression. Also, it is found that the PPE20 and its cognate partner PE15 are ESX-3 substrates [37]. Two other PPE-PPW proteins, PPE36 and PPE37, also utilize this system and maintain iron homeostasis through functions in heme-iron acquisition [38,39]. The ESX-5 secretion system is mainly responsible for the secretion of PE18, PE19, PPE25, PPE26, and PPE27 proteins, as well as the sublineage IV and V PE/PPE proteins [21,40,41]. In *M. marinum* and *Mtb*, the N-terminal PE domain of the well-defined PE_PGRS33 was found to be important for protein localization to the cell wall [26,42]. Also, it was shown that the N-terminal domain enables secretion of PE_PGRS30 while the C-terminal domain confers lipase activity [43]. When PE_PGRS30 was expressed in nonpathogenic fast-growing *M. smegmatis*, colony size decreased with alteration in growth kinetics [44]. Thus, the role of

PE_PGRS30 is thought to be involved in *Mtb* latency. Association of several PE proteins with the mycobacterial cell wall and its lipase or esterase enzymatic activity suggests their involvement in dictating the biochemical composition of the cell wall and colony morphology of the bacilli [25,45]. Expression of *Mtb* PE11 protein in *M. smegmatis* alters colony morphology and changes the lipid profile of the cell envelope with greater amounts of glycolipids and polar lipids, indicating its role in modulating cell wall architecture [46]. Again, its role in favoring mycobacterial replication in vivo is reported [47]. Sultana et al. [48] have shown that PE16 of *Mtb* has an esterase activity [48]. Further, in silico studies by the group reveal that certain PE and PPE proteins like PE1, PE2, PE3, PE4, PE16, PPE28, PPE42, and PPE63 have a hydrolase activity domain [48]. Structural integrity was compromised in the *ppe38* mutant of *M. marinum* due to alteration of the cell surface, and as a result, biofilm formation was interrupted [49]. PPE proteins also have roles in maintaining the cell wall permeability, as was shown by disrupting some of the *ppe* genes by transposon mutagenesis, which leads to increased resistance against ampicillin [50]. Deletion of the *ppe27-pe19 locus* from *Mtb* increases its sensitivity to cell wall stress, confirming the role of this locus in strengthening the bacterial cell wall to withstand host-induced stress [51]. Direct interaction of PE and PPE proteins with the host depends on their surface exposure or secretion into the extracellular milieu. Some of the proteins of the PE and PPE families are cell wall localized and therefore can directly interact with host cell surface molecules. PPE17 and PPE18 are present on the *Mtb* cell wall and interact with toll-like receptor 2 (TLR2) and influence cytokine signaling [52,53]. Detection of PPE41 in late endosomes [54] and PPE2 in the macrophage nucleus [55], suggesting that some family members are secretory in nature. High-throughput proteomics data have also confirmed the localization of PE and PPE proteins on the *Mtb* cell wall and beyond.

5. Modulation of host immune responses by PE/PPE family proteins

To protect against any nonself agent, the human body develops a well-orchestrated immune response to eliminate or neutralize the foreign agent. Innate immune cells are components of the immune system that act as a first responder to the job. The TLRs are one of the pattern recognition receptors (PRRs) on the surface of immune cells that recognize mycobacteria and initiate intracellular signaling. This intracellular signaling instructs the immune cells to induce effector molecules to restrict infection. Innate response is accompanied by the initiation of adaptive immune responses. The adaptive immune response is very much dependent on the signals provided by the innate cells. Thus, a coordinated response from both the innate and adaptive arms are needed for optimum functioning of the immune system.

Mtb spreads from an infected person to a healthy individual through bacilli containing air droplets, which directly enter the lungs [56]. Inside the lung, it is phagocytosed by resident alveolar macrophages. Macrophages are one of the important cells of the innate immune system that induce protective responses like secretion of cytokines like TNF-α, IL-12, IL-1β, and production of free radicals like nitric oxide (NO) and reactive oxygen species (ROS), which are critical for anti-*Mtb* immunity [57]. *Mtb* utilizes PE/PPE family proteins to suppress various innate as well as adaptive immune responses for its successful replication and establishment of infection inside the hosts [34,58] (Fig. 4.3).

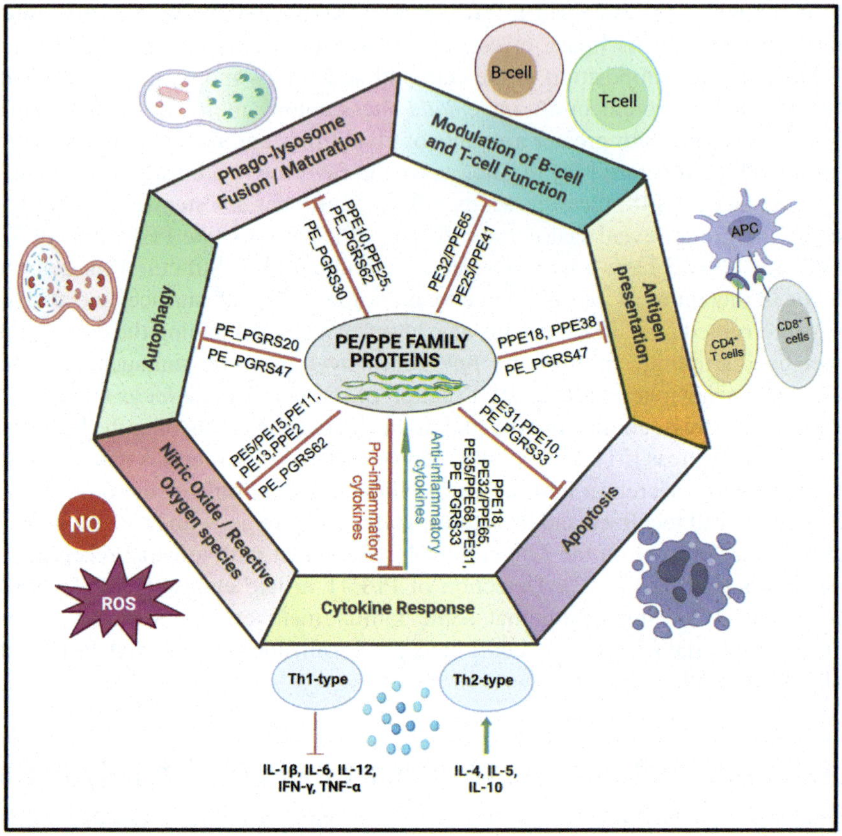

FIGURE 4.3 Important innate and adaptive host immune responses are modulated by the PE/PPE family proteins of *Mycobacterium tuberculosis*.

NO, nitric oxide; *ROS*, reactive oxygen species.
Various *Mtb* PE/PPE family proteins modulate crucial host defence functions, such as antigen presentation, cell death, cytokine induction, phagosome maturation, free radical production, and B-cell and T-cell functions, to favor its growth and persistence inside host cells and establish a successful infection.

5.1 Modulation of innate immune responses by PE/PPE family proteins

5.1.1 Phagosome-lysosome fusion

Following phagocytosis, mycobacteria are trapped in the phagosome. Fusion of phagosome with lysosome is a critical and highly regulated process where the phagosome undergoes a series of stages of vesicular protein recruitment followed by fusion with lysosome, leading to acidification of the phagolysosome that is responsible for killing the bacilli. An important survival strategy of *Mtb* is to arrest phagosome-lysosome (P-L) fusion and phago-lysosomal function/acidification so that the bacilli can safely persist and multiply inside macrophages [57] (Fig. 4.3). It could be observed that *Mtb* has employed PE/PPE family proteins to counter this crucial step. For example, PE_PGRS62 arrests phagosome maturation in J774A macrophage by preventing the recruitment of Rab7 and LAMP1 in phagosomes [59], as well as PPE54 prevents phagosome maturation [60]. PPE10 is shown to favor mycobacterial survival by interfering with phagolysosomal function, as the *ppe10* deleted BCG strain failed to prevent phagolysosome acidification in J774 macrophages [61]. Another study indicates that PE_PGRS47 inhibits autophagy, antigen processing, and MHC class II-mediated antigen presentation, which could be due to its inhibitory effect on P-L fusion/function [62]. BCG with four *pe_pgrs* mutants (*pe_pgrs5, pe_pgrs28, pe_pgrs44,* and *pe_pgrs59*) and three *ppe_mptr* mutants (*ppe_mptr10, ppe_mptr16,* and *ppe_mptr21*) were found to be localized more predominantly in acidified phagosomes as compared to the wild-type bacteria, suggesting their roles in preventing phagosome maturation [61,63]. MAV-2928 (a homolog of PPE25) of *M. avium* was shown to prevent the fusion of phagosome with lysosomes inside macrophages [64]. PE_PGRS30 and its homologue in *M. marinum*, MAG24-1, have also been found to interfere with phagosome maturation in THP-1 cells via reduced recruitment of LAMP1 [43].

5.1.2 Cytokine responses

Activated macrophages secrete various cytokines like IL-1β, IL-10, IL-12, and TNF-α during *Mtb* infection, and *Mtb* adopts strategies to regulate the release of these cytokines (Fig. 4.3). While TNF-α and IL-12 are mainly involved in inducing a T-helper 1 (Th1)-type immune response, which is responsible for cell-mediated immunity and Mtb clearance [57,65–68], IL-10 induction from macrophages leads to the development of a Th2-type immune response, which is known to favor *Mtb* persistence [57,69–73]. Thus, the pathogen adopts various mechanisms to specifically upregulate IL-10/Th2-type response and turn down IL-1β/IL-12/TNF-α cytokines [8,17] (Fig. 4.3).

Though TNF-α contributes to defense against *Mtb* [74] but there is also increasing evidence indicating that excessive levels of TNF-α contribute to *Mtb* disease pathology [75,76]. In addition, IL-6 plays a key role in the inflammatory response and is considered to act as a proinflammatory cytokine [77]. It could be noticed that PPE38 increases IL-6 and TNF-α response downstream of TLR2 in THP-1 macrophages, in addition to several other inflammatory factors [49], and PPE26 and PPE57 increase these proinflammatory cytokines in RAW 264.7 macrophages via the MAPK-NF-κB signaling pathway [78,79]. Also, PE6 increased levels of IL-2, IL-6, and TNF-α in splenocytes of mice [80]. In contrast, PE11 protein induces antiinflammatory cytokines, which probably help the bacterium in persistence and progression of disease in a mouse model of infection [46]. In another study, it could be observed that PPE13 activates secretion of IL-1β and NLRP3 inflammasome by inducing cleavage of caspase 1 [81]. Interestingly, PPE17, a cell wall-associated protein of *Mtb*, specifically interacts with the 16–20 leucine-rich repeat (LRR) domain of TLR2 in macrophages and induces predominantly a proinflammatory type immune response via PKC-IRAK3-p38MAPK signaling axis by TLR1/2 heterodimerization [53]. Whereas, PPE18, another cell wall localized protein, which is expressed in *Mtb* 14.8-fold higher than in *M. bovis* [82], binds to the LRR 11–15 domain of TLR2 and facilitates TLR2/TLR2 homodimerization and induces an antiinflammatory-type immune response [52,53]. PPE18 induces high IL-10 through activation of p38MAPK [52,53]. It has also been shown that PPE18 simultaneously inhibits IL-12 and TNF-α proinflammatory cytokines involving the SOCS3-Rel signaling pathway [83]. This increase in Th2-type response with a simultaneous decrease in Th1-type response by PPE18 is probably responsible for the increased virulence of *Mtb*, as mice infected with *ppe18* knock-out *Mtb* had a lesser bacterial burden with reduced organ pathology and increased survival rate as compared to mice infected with wild-type *Mtb* [84]. PPE18 strongly skewed the macrophage response towards the M2-phenotype [85] and biased the antiPPD T-cell response towards the Th2-type [52]. Interestingly, disruption of the transcriptional regulator Rv0485 of the cooperonic pair of PE13 and PPE18 caused reduced secretion of proinflammatory cytokines and attenuation of *Mtb* virulence in mice [86]. All these studies indicate that PPE18 probably acts as an important *Mtb* virulence factor. Similar to PPE18, PPE37 also reduces the level of proinflammatory cytokines like TNF-α and IL-6, which is due to reduced levels of phosphorylated p65 subunit of NF-κB, p38MAPK, and ERK1/2 [85,87].Similarly, the *Mtb* proteins, PE_PGRS26 and PE_PGRS33, also induce IL-10 during infection with *M. smegmatis* expressing these proteins and favor bacterial survival [88]. These studies indicate that all of these factors probably contribute to the better survival of the bacilli in infected mice. *Mtb* PE32 and PPE65 are cooperonic proteins transcribed and translated together. Recombinant PE32/PPE65 protein cotreatment reduces TNF-α and IL-6 expression, whereas upregulation of IL-10 was observed [89]. Also, the RD1 located PE/PPE pair consisting of PE35 and PPE68 increased IL-10 and monocyte chemoattractant protein (MCP)-1 but decreased IL-12 levels in THP-1 macrophages [90]. It is interesting to note that MCP-1 knock-out mice are defective in granuloma formation and are resistant to *Mtb* infection [91]. Thus, PE35/PPE68 might have a pleiotropic role in modulating immune functions to favor mycobacterial survival. Further, PE_PGRS62 downregulates the expression of IL-1β and IL-6 and inhibits the inflammatory response [92]. Whereas PE_PGRS33 expression resulted in upregulation of IL-10 and reduction of IL-12 levels [93]. Similarly, PE_PGRS41 and PE_PGRS61 generate an antiinflammatory response [94,95]. Also, PE5/PE15 upregulates antiinflammatory cytokines but downregulates proinflammatory cytokine production, targeting the MAPK signaling pathway and providing a survival advantage to the bacilli inside macrophages [96]. Other PE proteins like PE8, PE12, and PE17 act as inflammation suppressors to favor bacterial persistence inside the host [97–100]. Another study by Deng et al. [101] hints that PPE32 (Rv1808) can bind to TLR2 and induce IL-10 as well as TNF-α and IL-6 cytokines, involving the MAPK-NF-κB pathway [101]. Also, PE31 protein increases IL-10 but decreases the induction of IL-6 and IL-12 cytokines, targeting the NF-κB signaling pathway and favoring mycobacterial survival [102]. All these studies indicate that though some proteins activate proinflammatory cytokines and are involved in disease pathogenesis, *Mtb* probably predominantly upregulates the IL-10/Th2-type response while downregulating the protective IL-12/TNF-α proinflammatory responses to favor bacterial survival. In addition to macrophages, PE/PPE family proteins also target dendritic cells (DCs) for manipulating immune functions. For example, matured DCs secrete IL-10 in response to PPE34 (Rv1917c) and promote Th2-type T-cell response and secretion of the nonprotective IL-4, IL-5, and IL-10 but not IFN-γ from $CD4^+$ T-cells that involves cyclooxygenase-2 (COX-2) [103]. In contrast, PE27 augments the production of IL-1β, IL-6, IL-12p70, and TNF-α in DCs targeting the MAPKs-NF-κB signaling pathways. PE27-treated DCs preferentially activate naïve $CD4^+$ T-cells to secrete IFN-γ and activate T-bet but not GATA-3, creating a Th1 cell bias [104]. Thus, PE/PPE family proteins induce both pro- and antiinflammatory immune responses in DCs.

5.1.3 Reactive oxygen species and reactive nitrogen intermediates

A macrophage infected with *Mtb* gets activated predominantly by the pathogen-associated molecular patterns (PAMPs) and also by IFN-γ, which is secreted by T-cells. These activated macrophages produce reactive oxygen species (ROS) using nicotinamide adenine dinucleotide phosphate (NADPH) oxidase and reactive nitrogen intermediates (RNIs) through inducible NO synthase (iNOS). Superoxide ions (O^{2-}) are converted to H_2O_2 by superoxide dismutase, which finally generates toxic hydroxyl radicals (OH·). Similarly, RNIs, NO, and other chemical intermediates, i.e., NO^-, NO_2^-, $\bullet NO_2$, N_2O_3, N_2O_4, are produced by iNOS, an enzyme that synthesizes NO from L-arginine in response to endogenous cytokines like IFN-γ and TNF-α, or pathogen-derived molecules [105]. In addition to activated macrophages, polymorphonuclear leukocytes also generate ROS and RNIs in response to *Mtb* infection [57]. These ROS and RNIs act as bactericidal agents by reacting with a wide range of molecules, including nucleic acids, proteins, lipids, and carbohydrates. *Mtb* possesses weapons to counter such stresses. For example, the thick cell wall of *Mtb* containing lipoarabinomannan (LAM), cyclopropanated mycolic acids, and phenolic glycolipid I (PGL-1) acts as a scavenger of free oxygen radicals [68,106]. Also, *Mtb* produces ROS scavenging enzymes like catalase (KatG) and superoxide dismutase (SodA and SodC) for neutralization of ROS [106–109]. The other enzymes produced by *Mtb,* like peroxidase and peroxynitrite reductase complex of AhpC, AhpD, SucB (DlaT), and Lpd, are also involved in deactivation of host-generated ROS and RNIs [106,109].

PE/PPE family proteins are involved in protecting the bacterium from ROS and RNIs (Fig. 4.3). For example, *Mtb* PPE2 protein suppresses NO production in murine macrophages by downregulating the *inos* gene transcription, resulting in increased survival of the bacilli inside macrophages [55,84,110]. *Mtb* PPE2 is a secretory protein that has a nuclear localization signal (NLS), which enables it to translocate into the nucleus. It then binds to DNA via leucine zipper motif at the GATA-1 binding site of the *inos* promoter, leading to its reduced expression [55]. Interestingly, the NLS part of PPE2 contributes to *Mtb* virulence [110]. Also, PPE2 inhibits ROS generation in macrophages via its SRC homology 3 (SH3)-like domain [111]. The Trp236 residue present in the SH3-like domain was responsible for its interaction with the $p67^{phox}$ component of the NADPH oxidase complex, thus preventing the assembly of the complex, leading to reduced ROS production [111]. Also, PE_PGRS62, a protein expected to alter the cell wall-related component, when expressed in *M. smegmatis*, has curtailed *inos* expression in infected macrophages without changing the transcriptional level by some unknown mechanism [59]. Also, PE5/PE15 downregulates the *inos* gene expression and helps the bacilli to survive better [96]. Importantly, PE15 is expressed six fold more in *Mtb* as compared to *M. bovis* [82]. PE11 (LipX or Rv1169c), an esterase, which is upregulated upon oxidative stress in macrophages as well as in patients with active TB, changes the cell wall architecture of the bacilli to resist free radicals and various environmental and antibiotic stress [46,109]. When PE11 is heterologously expressed in *M. smegmatis*, the protein enhances the survival of the bacilli in murine macrophages as well as in a mouse model of infection [46]. PE13 is a cell wall-associated protein that has a role in cell wall integrity and increases the survival of the bacilli against various stressors like low pH, H_2O_2, SDS, and diamides [112]. *Mtb* PE_PGRS5 (Rv0297) protein causes disruption of intracellular Ca^{2+} homeostasis and increased production of NO/ROS. The PGRS domain of Rv0297 is responsible for the protein to localize into the ER, and the protein is involved in the regulation of NO/ROS [113]. Also, the hypoxia-induced gene PE_PGRS11, having phosphoglycerate mutase activity, could induce resistance against oxidative stress by increasing the expression of Bcl2 and COX-2 [114].

5.1.4 Modulation of immune cell survival pathways

After infection, innate immune cells can undergo different cell death pathways like apoptosis, necrosis, autophagy, and pyroptosis to prevent intracellular growth of the bacilli. However, *Mtb does so* through various mechanisms, either suppressing or inducing cell death depending on the stages of growth (Fig. 4.3). By inhibiting apoptosis, *Mtb* gains an environment to replicate and survive inside macrophages, and many studies have indicated that the virulent mycobacteria inhibit apoptosis in infected macrophages [115,116]. In contrast, by activating necrosis, it acquires the advantage to disseminate and form a granuloma [117,118]. It could be observed that purified PPE32 promotes endoplasmic reticulum (ER) stress-related gene expression, inducing apoptosis in a caspase-3- and caspase-9-dependent manner in THP-1 macrophages [119], but heterologously expressed PE_PGRS41 in *M. smegmatis* showed reduced cleavage of caspase-3 and caspase-9, thereby inhibiting apoptosis [94]. In contrast, *M. smegmatis* expressing PE9/PE10 induces apoptosis in THP-1 by interacting with TLR and inducing IRF-3 signaling, which leads to upregulation of proapoptotic genes like bax, bim, and bid, as well as cleavage of caspase-3 [120]. PE_PGRS41 and PE_PGRS62 inhibit host cell apoptosis and help in the persistence of the bacilli [94,121]. PE_PGRS33 induces necrosis in macrophages, as evident by the increased levels of lactate dehydrogenase and nucleosomes in the supernatant of murine bone marrow macrophage culture infected with recombinant *M. smegmatis* [93]. However, Basu et al. (2007) reported that PE_PGRS33 induces apoptosis by TLR2-dependent release of TNF-α in macrophages [122]. It could be noted that deletions of the PGRS domain of PE_PGRS33 resulted in the failure of apoptosis [122]. Importantly, in clinical isolates, the PGRS domain of PE_PGRS33 is highly variable [123]. Another study reveals that PE_PGRS33 localizes to the host cell mitochondria preceding apoptosis [124]. Thus, PE_PGRS33 can influence macrophage cell death in many ways, which probably depends on its sequence, expression, and stage of infection. In an interesting study, it was noted that the PE_PGRS11 protein triggers TLR2-dependent expression of COX-2 and Bcl2 and provides resistance against oxidative stress [114]. Further, *Mtb* deletion mutants of *pe_pgrs20* and *pe_pgrs47* failed to prevent autophagy of infected macrophages, causing reduced intracellular survival of the bacilli [125]. Importantly, macrophages transfected to express PE_PGRS20 or PE_PGRS47 revealed that these proteins could inhibit autophagy initiation directly by interacting with Ras-related protein Rab1A [125]. Chai et al. (2019) showed that PE_PGRS29, a cell surface protein that contains an eukaryotic-like ubiquitin-associated domain, binds to ubiquitin and triggers xenophagy of murine bone marrow-derived macrophages [126]. PPE31 contributes to macrophage apoptosis [127], whereas PPE10 inhibits apoptosis and is involved in the establishment of infection by downregulating the expression of caspases and proinflammatory cytokines, targeting the LUBAC-HOIP-NF-κB signaling cascades [128]. In *M. smegmatis*, PE9 and PE10 were found to form a heterodimer and induce apoptosis in THP-1 cells through TLR4 binding and upregulation of proapoptotic genes like bax, bim, and bid [120]. The C-terminal domain of PE6 is shown to resemble the eukaryotic apoptotic Bcl2 protein and causes mitochondrial perturbations and induction of apoptosis [129]. Interestingly, PE8 and PE12 inhibit late-stage macrophage apoptosis, thus promoting bacterial survival [97,130]. Also, PE31 inhibits apoptosis in THP-1 macrophages, targeting the NF-κB-caspase-3 signaling pathway [102]. In contrast, PE17 and PE13 are involved in modulating macrophage signaling cascades and are critically important in the induction of apoptosis [98,112].

PE25/PPE41, a cooperonic protein complex, was found to induce necrosis in murine macrophages and pathogen dissemination [131]. PE_PGRS5 promotes ER stress-mediated apoptosis involving caspase-3 and -9 activation, helping in the cell-to-cell spread of *Mtb* [113]. Also, PPE68 protein is involved in host cell necrosis and bacterial escape from macrophages [88,132]. It is reported that PE_PGRS26 is involved in cell death and cell-to-cell spread, and PE_PGRS17 promotes the death of host cells [133]. PPE25 increases proinflammatory cytokine production targeting the p38-MAPK-ERK-NF-κB signaling and induces host cell necrosis [134]. An interesting study by Li et al. [100] reveals that infection of peritoneal macrophages with *M. smegmatis* expressing PE17 causes necrosis, inducing severe damage to macrophages [100].

Pyroptosis is a form of cell death with high inflammatory consequences. It is associated with plasma-membrane rupture, as well as the release of inflammatory factors and bacilli. In *Mtb* infection condition, it has been shown that *Mtb* protein can suppress pyroptosis in the initial stage of infection. PE/PPE family proteins have been recently shown to have contrasting functions assigned to their N- and C-terminal domains. For example, full-length PPE37 is cleaved into N- and C-terminal domains under low iron conditions [135]. The purified N-terminal domain of PPE37 promotes the proliferation and differentiation of monocytic THP-1 cells into CD11c, dendritic cell-specific intercellular adhesion molecule 3-grabbing nonintegrin (DC-SIGN) positive semimature dendritic cells. In contrast, the C-terminal domain of PPE37 induces apoptosis in a caspase 3-dependent manner in transfected THP-1 macrophages [135]. Contrasting results observed in different PE/PPE family proteins indicate that the temporal expression of these PE/PPE proteins during the course of infection influences the execution of different cell death pathways to benefit the pathogen.

5.1.5 Other immune functions

Pioneering studies on *M. marinum* PE_PGRS genes showed that these genes are involved in granuloma formation and improved survival of the bacilli inside macrophages [136]. The PE_PGRS33 mutant BCG has shown limited entry into macrophages, signifying its role in promoting uptake of bacilli inside macrophages during infection [137]. The PE_PGRS17 (Rv0978c) and PE_PGRS11 (Rv0754) proteins bind to TLR2 and induce activation and maturation of human DCs, and enhance their ability to stimulate T-cell activation and differentiation [138]. In contrast, PPE34 (Rv1917c) interacts with TLR2 and stimulates the maturation of human DCs.

It has been shown that inactivation of PE_PGRS33 in *Mtb* leads to reduced bacterial load in J774 murine macrophages [139]. Similar observations were made when PE_PGRS33 was expressed in nonpathogenic *M. smegmatis.* [88]. Also, studies suggest that inactivation of PPE46 could result in reduced *Mtb* growth in vivo [140]. Further, Cadieux et al. (2011) reported that the PGRS region of PE_PGRS33 is required for its localization into mitochondria, but the PE domain was crucial for the induction of cell death [124]. A related study has shown that inactivation of PE_PGRS33 leads to reduced infection in macrophages [141]. Also, its expression in the nonpathogenic fast-growing *M. smegmatis* revealed similar results in vitro [122]. PE_PGRS3 [142], PE35/PPE68 [90], and PE_PGRS29 [126] play roles in the establishment of mycobacterial infection. *Mtb* PE_PGRS protein Rv1818c, which is also present in BCG, is a cell wall-associated protein that helps in host-pathogen interaction and infectivity [9]. *Mtb* PE8 and PPE15 can induce inflammasome activation. It is observed that *Mtb* lacking ESX-5a is not efficient in activating the inflammasome compared to wild-type *Mtb* [143].

5.2 Role of PE/PPE family proteins in manipulating adaptive immune responses

Adaptive immune response is characterized by its precise, antigen-specific, long-term memory response. It mainly refers to the B-cell and T-cell mediated immunity, also called humoral and cell-mediated immunity, respectively. The adaptive immune response banks on the innate immune cells for antigen presentation and cytokine response for its activation and differentiation [144,145]. In the antigen presentation process, the antigenic peptide fragment is loaded onto the Major Histocompatibility Complex (MHC), which is expressed on the cell surface of antigen presenting cells (APCs). These MHC-peptide complexes are recognized by subsets of T-cells depending upon the type of MHC molecules presenting the antigens/peptides [146]. It is observed that *Mtb* adopts strategies to influence the antigen processing, presentation, and cytokine response in the APC-T-cell junction, thus strongly influencing the adaptive immune response [57] (Fig. 4.3). It is to be noted that healthy immunity depends on a delicate balance between the Th1-type and Th2-type T-cell responses. The Th1-type response promotes cellular immunity, which is crucial for killing intracellular pathogens like *Mtb*, while the Th2-type T-cell response facilitates bacterial survival inside the host [57,65,70,73,147–149]. Importantly, progression of clinical TB is found to be associated with increased Th2-type T-cell response, which is shifted to Th1-type after treatment with anti -TB drugs [57,65,148,150,151].

5.2.1 Manipulation of the antigen presentation function

Antigen presentation is a process by which APCs like macrophages, DCs, and B-cells load degraded pathogenic peptides onto MHC class I or class II molecules and present them on the cell surface. The MHC-peptide complexes are subsequently recognized by $CD8^+$ or $CD4^+$ T-cells to generate pathogen-specific effector and memory T-cells [152]. Thus, *Mtb* employs strategies to interfere with these responses to suppress or delay T-cell responses, and PE/PPE proteins are responsible for inhibiting antigen presentation by interfering with various stages of antigen processing/presentation pathways.

Phago-lysosomal acidification is essential in the degradation and processing of peptides required for MHC class II-restricted antigen presentation [152]. *Mtb* cell wall-associated protein PPE18 (Rv1196) was able to inhibit antigen presentation in macrophages by preventing phagosome acidification and antigen degradation, without affecting costimulatory signaling [153]. Interestingly, PE_PGRS47-deleted *Mtb* showed increased acidification and lysosomal fusion of phagosomes during infection, which ultimately resulted in higher MHC class II-mediated antigen presentation [62]. Therefore, *Mtb* uses PE/PPE family proteins in order to inhibit antigen presentation, dampening overall adaptive immune cell responses, which contributes to better persistence of the bacilli in the host. However, some PE/PPE family proteins like PE27, PE_PGRS11, and PE_PGRS17 activate maturation of dendritic cells and costimulatory signalling and stimulate host immune responses [104,138,154]. Bansal et al. [103] have shown that 2 cell wall-associated/secretory PE_PGRS proteins, PE_PGRS17 (Rv0978c) and PE_PGRS11 (Rv0754), could bind to TLR2 and induce maturation and activation of human DCs through MAPK-NF-κB signaling pathways and enhance the ability of DCs to stimulate $CD4^+$ T-cells [138]. Also, PE25/PPE41 dimer is involved in the maturation of mouse DCs, [155]. Recombinant PE27 treatment inducing DC maturation by up-regulating CD80, CD86, MHC class I, and MHC class II expression on the DC surface [104]. PPE38 downregulates class I antigen presentation in peritoneal macrophages, thereby reducing/delaying the protective $CD8^+$ T-cell activity [156]. PPE60 activates maturation of DC and various costimulatory molecules like CD80 and CD86, as well as MHC class II, and promotes antigen processing by DCs in a TLR2-dependent manner [157].

5.2.2 Modulation of B-cell and T-cell functions

B-cells are essential components of the adaptive immune system. Naive B-cells exit from the bone marrow and populate the bloodstream and peripheral secondary lymphoid organs. After infection, naïve B-cells undergo a differentiation process to form memory B-cells or plasma cells [158]. Plasma cells secrete antibodies specific to pathogens. These antibody molecules clear invading pathogens by direct neutralization, by activating the complement system, or by interacting with other immune cells [159]. Memory B-cells are long-lived cells and can be reactivated to differentiate into plasma cells to produce antibody following secondary infection, and therefore provide a long-term, ready-made reservoir for antibody-mediated protection against infections [160]. This property of memory B-cells to recall previous antigen signature and mount an antibody (Ab) response has been widely utilized to assess the vaccination-mediated immunity [161]. PE and PPE family members are a potentially rich source of B-cell epitopes, and they are mostly surface-exposed or secreted, and a number of them are likely to generate humoral immune responses. The highly repetitive domains of both PE and PPE proteins are immunodominant. TB patient's sera analysis showed predominant antibody responses against the N-terminal domain of the PPE17 protein [162]. Importantly, PPE17 could discriminate active TB and latent TB positive cases from the QFT (QuantiFERON-TB)-negative subjects and LTBI (latent tuberculosis infection) subjects showed higher Ab responses to PPE17 as compared to ESAT-6, CFP-10, and PPD, which are currently used as latent TB diagnostic antigens [163]. Also, the N-terminal part of PE_PGRS33 manipulates macrophage effector functions, whereas a full-length construct promotes B-cell response [141]. Another protein, PE_PGRS62, showed a higher antibody (IgG) response in patients with active and latent TB than in treated TB patients [164]. The PE32/PPE65 cooperonic protein suppresses IFN-γ in splenocytes and reduces IgG2a but increases IgG1, which is an indication of higher Th2-type responses [89]. There was also a reduction in IFN-γ and IL-2 positive $CD4^+$ and $CD8^+$ T-cell populations [89]. Ahmad et al. (2018) reveal that N- and C-terminal domains of PPE37 perform different functions during infection conditions [135]. Under iron-depletion state inside macrophages, N- and C-terminal domains of PPE37 are cleaved, the N-terminal part resides in the cytoplasm, whereas the C-terminal fragment localizes to the nucleus. The N-terminal part was found to promote THP-1 cells to differentiate into CD11c and DC-SIGN positive semimature dendritic cells with upregulation of IL-10 [135]. It also lowers IL-12 and TNF-α secretion to provide a conducive condition for Th2-type response. The C-terminal domain, in contrast, induces apoptosis in macrophages for the dissemination of the bacilli to the neighboring cells [135]. It is known that PE25 and PPE41 are coded by a single operon. In sera obtained from TB patients, it was observed that the humoral response induced in TB patients against the PE25/PPE41 complex was higher than that against PPE41 and PE25 alone [165]. PPE-MPTR42 elicits a humoral response primarily against domains rich in glycine and asparagine [166]. Both recombinantly purified PPE44 and BCG infection showed PPE44-specific IgG in mouse sera [167].

More than 1 week after *Mtb* infection, DCs from the lung migrate to the lymph nodes carrying *Mtb* antigen and stimulate T-cells for inducing effector responses. It has been observed in multiple instances that the presence of a high IL-10/IL-12 ratio and TGF-β positively correlates with high regulatory T (Treg) cells and higher bacterial burden. Thus, T-cell receptor-specific antigens have always been the center of attraction for the selection of peptide-based vaccine candidates [168,169]. Anti-mycobacterial immunity is mediated by a Th1-type T-cell response [170]. Th1-type T-cells that produces IFN-γ, IL-2, and TNF-α/β, which activate macrophages and cell-mediated immunity [171,172]. Interestingly, *Mtb* reduces the IL-12/IL-10 ratio for the establishment of a nonprotective Th2-type response in the host [171]. The recombinant *Mtb* PPE44 protein reduces the IFN-γ response in splenocytes, and PPE44 increases IgG1 antibody and reduces IgG2a in mice due to a skewed T-cell response towards the Th2-type [167]. It has been shown that *Mtb* PPE18 protein generates a strong antiinflammatory response by elevating IL-10/IL-12 ratio, skewing the T-cell response towards the Th2-type [52]. PPE18 is shown to favor bacterial survival, which could be through influencing the Th1/Th2 balance [84]. Interestingly, PPE18 increases expression and phosphorylation of SOCS3 [52], and active pulmonary TB individuals were also shown to have higher levels of SOCS3 protein in their lungs [150], which supports the hypothesis that SOCS3 signaling is probably involved in the regulation of Th1/Th2 balance, favoring bacterial survival. PPE60 activates Th1/Th17 responses via the TLR2–dependent maturation of dendritic cells [157]. PPE57 causes macrophage activation and drives Th1-type immune responses through TLR2 [78]. PE27 also polarizes T-cell response towards the Th1-type and improves dendritic cell maturation, leading to upregulation of CD80, CD86, MHC class I/II expression, and induction of TNF-α, IL-1β, IL-6, and IL-12 cytokines [104]. Other proteins like PE18, PE31, and PPE26, though, induce a robust $CD4^+$ and $CD8^+$ T-cell response, and PE18 especially generates memory T-cells in lungs; however, these three proteins fail to confer protective immunity against *Mtb* [173]. It could be observed that PE proteins like PE5 and PE15 suppress Th1-type T-cell response and promote the Th2-type immune response by decreasing the key Th1-type cytokine (IL-12) and increasing the Th2-type cytokine (IL-10), favoring bacterial survival [96]. This manipulation aligns with observations in TB patients, where a Th2 dominance and elevated levels of a Th1 suppressor protein (SOCS3) are often seen [150]. While PE11's role is less clear, studies suggest a complex interplay as mice infected with *M. smegmatis* expressing PE11 showed both increased proinflammatory and Th2-type cytokine expression, potentially contributing to higher mortality of mice [46].

6. PE/PPE family proteins as vaccine candidates

For more than a century, the only vaccine used around the globe has been BCG to prevent tuberculosis. It can prevent extrapulmonary tuberculosis effectively in children but fails to prevent pulmonary tuberculosis in adults. Further, BCG vaccine efficacy varies in demographically distinct populations, and the emergence of MDR and XDR incidences demands the design of novel TB vaccines.

Multiple studies have shown PE/PPE proteins as potent immunogens, and incorporating PE and PPE proteins in vaccine design is a rational approach. In various proof-of-concept studies, PE and PPE are being used as subunit vaccines or used in combination with BCG to augment its efficacy. PE/PPE family proteins have also been used in the development of novel subunit vaccine candidates. Human T-cell receptor sequencing shows the enrichment of T-cell receptors that recognize PE13 (Rv1195) [174]. This shows the potential of PE/PPE family proteins as vaccine candidates for effective TB elimination. A noteworthy development is the M72/AS01E vaccine. M72/AS01E is composed of a recombinant fusion protein derived from *Mtb* antigens, MTB39A (PPE18) and MTB32A (PepA), in combination with an AS01E adjuvant [175]. M72 showed a good safety profile with no weight loss or abnormal inflammatory indicators in cynomolgus monkeys, and M72 in combination with BCG revealed higher antituberculosis cytokines, long-term survival, and reversal of tuberculosis progression during challenge of the monkeys with *Mtb* [176]. A phase IIb trial is initiated to test the efficacy of the vaccine against pulmonary TB in adults with latent TB infections [177]. The results demonstrated a reduction in the incidence of TB and an overall vaccine efficacy of 54%. Individuals who received the vaccine showed higher levels of M72/AS01E-specific antibodies and $CD4^+$ T-cells throughout a 3-year period [178]. Although this is promising, it is still unsatisfactory.

In another study, the PE domain of PE_PGRS33 was added to the MPT65 antigen of *M. bovis* BCG as an N-terminal fusion protein. The recombinant *M. bovis* BCG strain was then able to express the MPT65 antigen on its surface, resulting in enhanced protection against *Mtb* infection in mice, compared to the parental *M. bovis* BCG strain [179]. Using Peripheral blood mononuclear cells

(PBMCs) from PPD-positive individuals, it could be possible to detect at least 10 distinct T-cell epitopes in PPE18 (mtb39A) [180]. Interestingly, mice immunized with *ppe18* DNA showed protection with lower bacterial burden upon challenge with *Mtb,* which indicates that PPE18 can be an important component of a subunit vaccine for *Mtb* [180]. Also, vaccination of mice with ACP (a peptide-based vaccine containing three Th1-immunodominant peptides from Ag85B) [12–26], CFP21 [12–26] and PPE18 [149–163]) showed increased number of IFN-γ positive $CD4^+$ T-cells, higher levels of IFN-γ cytokine and antibodies, however, it failed to improve the protective efficacy of BCG in iumanized or wild-type mice [181]. Another recombinant vector vaccine (INP-0288-1196-0125) was developed by constructing a "mosaic" ice nucleation protein (INP) on the surface of *Escherichia coli,* which was then used to display the Rv0288, Rv1196, and Rv0125 proteins on its surface. Although the vaccine could induce strong B-cell and $CD4^+$ T-cell responses in mice, there is a concern about the efficacy of this vaccine due to the high variability of the *ppe18* gene among *Mtb* complex (MTBC) [182,183].

PE/PPE proteins have also played an important role in the development of attenuated/whole-cell vaccine candidates. For example, the preclinical *Mtb Δppe25 pe19* vaccine candidate is attenuated via a partial deletion of the esx-5 genetic locus, involving three *ppe* genes (*ppe25, ppe26,* and *ppe27*) and two *pe* genes (*pe18* and *pe19*) [41]. The *Mtb Δppe25-pe19* strain is able to elicit a robust $CD4^+$ T-cell response compared to the BCG vaccine and is strongly protective against *Mtb* infection in mice [184]. Importantly, the attenuated *Mtb* vaccine candidate, *Mtb Δppe25-pe19,* is able to induce a Th1-type effector subset and stimulate the induction of innate immune response and represents a potential antituberculosis vaccine candidate [184,185]. The other vaccine ID93: GLA-SE (engineered by combining Rv1813, Rv3620, Rv3629, and PPE42) showed antiTB efficacy in controlling TB in terms of reduced pathology, fewer bacilli, and prevention of death of animals after challenge with virulent *Mtb* [186]. In a phase IIa clinical trial, in patients with previously treated tuberculosis, ID93:GLA-SE was found to be safe and induce robust and durable humoral and $CD4^+$ T-cell responses, thus can be considered for therapeutic vaccination for improving tuberculosis treatment outcomes [187]. Further, a liposomal formulation of GLA with QS-21 (GLA-LSQ) combined with ID93 revealed better protection over drug treatment alone in terms of survival of mice, increased Th1-type T-cell response, and enhanced resident memory T-cells in the lung [188].

Vaccination with *Mtb* PPE14 protein increases protection against *Mtb* infection both in mice and in the guinea pig model [189]. Xu et al. [79] showed that the PPE57 protein can be considered to work as a probable vaccine antigen against *Mtb* [79]. Also, it has been noticed that immunization of mice with recombinant PE3 protein induced a strong protective immune response against challenge with live mycobacteria [190]. Similarly, PPE15 was shown to induce significant protection when administered alone and as a boost to BCG vaccination in mice [191]. There are studies that show improved efficacy of polyprotein subunit candidates when fused with PE and PPE proteins. The fusion of MPT65 with the PE domain of PE_PGRS33 (cell wall localized protein) results in the localization of MPT65 antigen on the surface of BCG and enhances the efficiency of the BCG vaccine against *Mtb* infection in mice [179]. ID93, another vaccine candidate, contains PPE42 as one of the components of the 93 kDa polyprotein complex (Rv3619-Rv1813-Rv3620-Rv2608/PPE42). This polyprotein subunit vaccine, along with adjuvant, has been shown to provide protection against drug-resistant *Mtb* in mice and guinea pigs and therefore can be considered as a candidate molecule for improving the efficacy of the existing BCG vaccines [186]. Further, Chen et al. [155] indicated that the PE25/PPE41 protein complex could induce DC maturation and may serve as a potential antiTB vaccine [155]. Again, Hanif et al. [192] revealed that the PE35 recombinant DNA vaccine specifically generates Th1-type effector T-cell subsets and therefore could become a useful antiTB vaccine [192]. Amoudy et al. [193] reported that immunization of mice with *recombinant M. vaccae* expressing PE35 and CFP-10 could be a potential antiTB vaccine as it induces effective cellular and humoral immune responses [193]. Importantly, a multistage fusion protein vaccine consisting of PE28 (Rv3020c) and HspX targets the early and late phases of *Mtb* growth [194]. Certain PPE proteins like PPE25, PPE26, and PPE27 may be good antiTB vaccine candidates as they are potent inducers of cell-mediated immunity [15]. In another study, mice immunized with a subunit vaccine consisting of ESAT-6 and consensus $CD4^+$ T-cell epitopes of PE-PPE protein (N-terminus) had lower bacterial load when challenged with the highly pathogenic *Mtb* strain HN878 [195]. Other promising vaccines trialed in mice include the fusion protein Ag85B-Rv3425 [196], a triple-antigen fusion molecule consisting of EsxN (Rv1793), PPE42 (Rv2608), and the latency-associated Rv2628 PPE42-Rv1793-Rv2628 [197]. The other vaccine candidate, tetrafu56 (consisting of EspC, TB10.4, HspX, PPE57, CFP21, and Rv1352), appears to be a potential candidate for developing an effective multistage vaccine to prevent *Mtb* infection [198]. Thus, PE/PPE family proteins hold significant potential in TB vaccine research, highlighting the need for identifying the most suitable candidate genes for vaccine development.

7. Conclusion

The PE/PPE family proteins of *Mtb* have been studied with immense attention since their discovery, and research is yet to be continued to unfold their true biological roles. Whilst *pe/ppe* genes have been found in both pathogenic and saprophytic *Mycobacteria* sp., comparative genomic analysis has shown that *pe/ppe* genes are predominantly present within the pathogenic and slow-growing species of *Mycobacteria* sp. The highest number of *pe/ppe* genes is found in the *Mtb* complex. This indicates a favorable evolutionary selection for these genes specifically within the pathogenic species of *Mycobacteria* sp. This also indicates that the expansion of the PE/PPE family within the pathogenic species of *Mycobacteria* sp. is involved in the adaptation toward a pathogenic and intracellular lifestyle [18]. *Mtb* expresses a distinct set of genes that enable it to evade the protective immune responses of the host, and the expression profiles of *pe/ppe* genes inside different host milieus are a challenge to understand the pathophysiological role of these proteins in human TB [8,57]. Further, MDR and XDR are posing new challenges in the field of TB vaccination and treatment. Genome-wide expansion of PE and PPE proteins only in the pathogenic species is a strong indicator of their roles in *Mtb* pathogenesis. These proteins are either cell wall-associated or secretory. Various studies have shown the effect of these proteins in the modulation of immune cell responses. Therefore, extensive studies are required in TB to comprehend the role of PE and PPE proteins in host immunomodulation and the mechanisms of *Mtb* virulence, pathogenesis, and latency. Elucidation of the molecular mechanism by which PE/PPE family proteins exert their function can help to develop therapeutic interventions to control this deadly pathogen.

Acknowledgments

The authors gratefully acknowledge the financial support by the Science and Engineering Research Board (SERB), Department of Science and Technology (DST), Government of India (JCB/2021/000035), Council of Scientific and Industrial Research (CSIR), Govt. of India (37WS(0020)/2023–24/EMR-II/ASPIRE), Department of Biotechnology (DBT), Govt of India (BT/PR51149/MED/29/1660/2023), Indian Council of Medical Research (ICMR), Govt. of India (2021–10087/GTGE/ADHOC-BMS and IIRPSG-2024-01–01453) and a core grant from CDFD by DBT. SD is supported by the fellowship from Council of Scientific and Industrial Research (CSIR), Govt. of India and VM is supported by DST-INSPIRE fellowship.

References

[1] Global tuberculosis report. Geneva: World Health Organisation; 2024. https://www.who.int/teams/global-programme-on-tuberculosis-and-lung-health/tb-reports/global-tuberculosis-report-2024.

[2] Cole ST, Brosch R, Parkhill J, Garnier T, Churcher C, Harris D, et al. Deciphering the biology of *Mycobacterium tuberculosis* from the complete genome sequence. Nature 1998;393(6685):537–44. https://doi.org/10.1038/31159.

[3] Copin R, Coscolla M, Seiffert SN, Bothamley G, Sutherland J, Mbayo G, et al. Sequence diversity in the *pe_pgrs* genesof *Mycobacterium tuberculosis* is independent of human T cell recognition. mBio 2014;5(1): e00960-13. https://doi.org/10.1128/mBio.00960-13.

[4] McGuire AM, Weiner B, Park ST, Wapinski I, Raman S, Dolganov G, et al. Comparative analysis of Mycobacterium and related Actinomycetes yields insight into the evolution of *Mycobacterium tuberculosis* pathogenesis. BMC Genom 2012;13:120. https://doi.org/10.1186/1471-2164-13-120.

[5] Mehra S, Kaushal D. Functional genomics reveals extended roles of the *Mycobacterium tuberculosis* stress response factor σ^H. J Bacteriol 2009;191(12):3965–80. https://doi.org/10.1128/jb.00064-09.

[6] Gey van Pittius NC, Sampson SL, Lee H, Kim Y, van Helden PD, Warren RM. Evolution and expansion of the *Mycobacterium tuberculosis* PE and PPE multigene families and their association with the duplication of the ESAT-6 (esx) gene cluster regions. BMC Evol Biol 2006;6:95. https://doi.org/10.1186/1471-2148-6-95.

[7] Akhter Y, Ehebauer MT, Mukhopadhyay S, Hasnain SE. The PE/PPE multigene family codes for virulence factors and is a possible source of mycobacterial antigenic variation: perhaps more? Biochimie 2012;94(1): 110–6. https://doi.org/10.1016/j.biochi.2011.09.026.

[8] Mukhopadhyay S, Balaji KN. The PE and PPE proteins of *Mycobacterium tuberculosis*. Tuberculosis 2011; 91(5):441–7. https://doi.org/10.1016/j.tube.2011.04.004.

[9] Brennan MJ, Delogu G. The PE multigene family: a "molecular mantra" for mycobacteria. Trends Microbiol 2002;10(5):246–9. https://doi.org/10.1016/s0966-842x(02)02335-1.

[10] Brennan MJ. The enigmatic PE/PPE multigene family of mycobacteria and tuberculosis vaccination. Infect Immun 2017;85(6). https://doi.org/10.1128/iai.00969-16.

[11] Strong M, Sawaya MR, Wang S, Phillips M, Cascio D, Eisenberg D. Toward the structural genomics of complexes: crystal structure of a PE/PPE protein complex from *Mycobacterium tuberculosis*. Proc Natl Acad Sci U S A 2006;103(21):8060–5. https://doi.org/10.1073/pnas.0602606103.

[12] Ehtram A, Shariq M, Ali S, Quadir N, Sheikh JA, Ahmad F, et al. Teleological cooption of *Mycobacterium tuberculosis* PE/PPE proteins as porins: role in molecular immigration and emigration. Int J Med Microbiol 2021;311(3):151495. https://doi.org/10.1016/j.ijmm.2021.151495.

[13] Ishikawa J, Yamashita A, Mikami Y, Hoshino Y, Kurita H, Hotta K, et al. The complete genomic sequence of *Nocardia farcinica* IFM 10152. Proc Natl Acad Sci U S A 2004;101(41):14925–30. https://doi.org/10.1073/pnas.0406410101.

[14] Abdallah AM, Gey van Pittius NC, Champion PA, Cox J, Luirink J, Vandenbroucke-Grauls CM, et al. Type VII secretion—mycobacteria show the way. Nat Rev Microbiol 2007;5(11):883–91. https://doi.org/10.1038/nrmicro1773.

[15] Qian J, Chen R, Wang H, Zhang X. Role of the PE/PPE family in host-pathogen interactions and prospects for anti-tuberculosis vaccine and diagnostic tool design. Front Cell Infect Microbiol 2020;10:594288. https://doi.org/10.3389/fcimb.2020.594288.

[16] Pal R, Bisht MK, Mukhopadhyay S. Secretory proteins of *Mycobacterium tuberculosis* and their roles in modulation of host immune responses: focus on therapeutic targets. FEBS J 2022;289(14):4146–71. https://doi.org/10.1111/febs.16369.

[17] Ahmed A, Das A, Mukhopadhyay S. Immunoregulatory functions and expression patterns of PE/PPE family members: roles in pathogenicity and impact on anti-tuberculosis vaccine and drug design. IUBMB Life 2015;67(6):414–27. https://doi.org/10.1002/iub.1387.

[18] Fishbein S, van Wyk N, Warren RM, Sampson SL. Phylogeny to function: PE/PPE protein evolution and impact on *Mycobacterium tuberculosis* pathogenicity. Mol Microbiol 2015;96(5):901–16. https://doi.org/10.1111/mmi.12981.

[19] Majlessi L, Brodin P, Brosch R, Rojas MJ, Khun H, Huerre M, et al. Influence of ESAT-6 secretion system 1 (RD1) of *Mycobacterium tuberculosis* on the interaction between mycobacteria and the host immune system. J Immunol 2005;174(6):3570–9. https://doi.org/10.4049/jimmunol.174.6.3570.

[20] Roy S, Ghatak D, Das P, BoseDasgupta S. ESX secretion system: the gatekeepers of mycobacterial survivability and pathogenesis. Eur J Microbiol Immunol (Bp) 2020;10(4):202–9. https://doi.org/10.1556/1886.2020.00028.

[21] Ates LS, Ummels R, Commandeur S, van de Weerd R, Sparrius M, Weerdenburg E, et al. Essential role of the ESX-5 secretion system in outer membrane permeability of pathogenic mycobacteria. PLoS Genet 2015;11(5):e1005190. https://doi.org/10.1371/journal.pgen.1005190.

[22] Sankey N, Merrick H, Singh P, Rogers J, Reddi A, Hartson SD, et al. Role of the *Mycobacterium tuberculosis* ESX-4 secretion system in heme Iron utilization and pore formation by PPE proteins. mSphere 2023;8(2):e0057322. https://doi.org/10.1128/msphere.00573-22.

[23] Pym AS, Brodin P, Brosch R, Huerre M, Cole ST. Loss of RD1 contributed to the attenuation of the live tuberculosis vaccines *Mycobacterium bovis* BCG and *Mycobacterium microti*. Mol Microbiol 2002;46(3):709–17. https://doi.org/10.1046/j.1365-2958.2002.03237.x.

[24] Ekiert DC, Cox JS. Structure of a PE-PPE-EspG complex from *Mycobacterium tuberculosis* reveals molecular specificity of ESX protein secretion. Proc Natl Acad Sci U S A 2014;111(41):14758–63. https://doi.org/10.1073/pnas.1409345111.

[25] Dahiya P, Banerjee A, Saha A, Nandicoori VK, Ghosh S, Mukhopadhyay S. Structure-function relationship of PE11 esterase of *Mycobacterium tuberculosis* with respect to its role in virulence. Biochem Biophys Res Commun 2024;739:150927. https://doi.org/10.1016/j.bbrc.2024.150927.

[26] Daleke MH, Cascioferro A, de Punder K, Ummels R, Abdallah AM, van der Wel N, et al. Conserved Pro-Glu (PE) and Pro-Pro-Glu (PPE) protein domains target LipY lipases of pathogenic mycobacteria to the cell surface via the ESX-5 pathway. J Biol Chem 2011;286(21):19024–34. https://doi.org/10.1074/jbc.M110.204966.

[27] Gey Van Pittius NC, Gamieldien J, Hide W, Brown GD, Siezen RJ, Beyers AD. The ESAT-6 gene cluster of *Mycobacterium tuberculosis* and other high G+C Gram-positive bacteria. Genome Biol 2001;2(10). https://doi.org/10.1186/gb-2001-2-10-research0044.

[28] McEvoy CR, Cloete R, Muller B, Schurch AC, van Helden PD, Gagneux S, et al. Comparative analysis of *Mycobacterium tuberculosispe and ppe* genes reveals high sequence variation and an apparent absence of selective constraints. PLoS One 2012;7(4):e30593. https://doi.org/10.1371/journal.pone.0030593.

[29] Sampson SL, Warren RM, Richardson M, van der Spuy GD, van Helden PD. Disruption of coding regions by IS6110 insertion in *Mycobacterium tuberculosis*. Tuber Lung Dis 1999;79(6):349–59. https://doi.org/10.1054/tuld.1999.0218.

[30] Perez-Lago L, Herranz M, Lirola MM, Group I-T, Bouza E, Garcia de Viedma D. Characterization of microevolution events in *Mycobacterium tuberculosis* strains involved in recent transmission clusters. J Clin Microbiol 2011;49(11):3771–6. https://doi.org/10.1128/JCM.01285-11.

[31] Cole ST, Eiglmeier K, Parkhill J, James KD, Thomson NR, Wheeler PR, et al. Massive gene decay in the leprosy bacillus. Nature 2001;409(6823):1007–11. https://doi.org/10.1038/35059006.

[32] Kohli S, Singh Y, Sharma K, Mittal A, Ehtesham NZ, Hasnain SE. Comparative genomic and proteomic analyses of PE/PPE multigene family of *Mycobacterium tuberculosis* H_{37}Rv and H_{37}Ra reveal novel and interesting differences with implications in virulence. Nucleic Acids Res 2012;40(15):7113–22. https://doi.org/10.1093/nar/gks465.

[33] De Maio F, Berisio R, Manganelli R, Delogu G. PE_PGRS proteins of *Mycobacterium tuberculosis*: a specialized molecular task force at the forefront of host-pathogen interaction. Virulence 2020;11(1): 898–915. https://doi.org/10.1080/21505594.2020.1785815.

[34] Yu X, Feng J, Huang L, Gao H, Liu J, Bai S, et al. Molecular basis underlying host immunity subversion by *Mycobacterium tuberculosis* PE/PPE family molecules. DNA Cell Biol 2019;38(11):1178–87. https://doi.org/10.1089/dna.2019.4852.

[35] Wang Q, Boshoff HIM, Harrison JR, Ray PC, Green SR, Wyatt PG, et al. PE/PPE proteins mediate nutrient transport across the outer membrane of *Mycobacterium tuberculosis*. Science 2020;367(6482):1147–51. https://doi.org/10.1126/science.aav5912.

[36] Damen MPM, Phan TH, Ummels R, Rubio-Canalejas A, Bitter W, Houben ENG. Modification of a PE/PPE substrate pair reroutes an Esx substrate pair from the mycobacterial ESX-1 type VII secretion system to the ESX-5 system. J Biol Chem 2020;295(18):5960–9. https://doi.org/10.1074/jbc.RA119.011682.

[37] Tufariello JM, Chapman JR, Kerantzas CA, Wong KW, Vilcheze C, Jones CM, et al. Separable roles for *Mycobacterium tuberculosis* ESX-3 effectors in iron acquisition and virulence. Proc Natl Acad Sci U S A 2016;113(3):E348–57. https://doi.org/10.1073/pnas.1523321113.

[38] Mitra A, Speer A, Lin K, Ehrt S, Niederweis M. PPE Surface proteins are required for heme utilization by *Mycobacterium tuberculosis*. mBio 2017;8(1). https://doi.org/10.1128/mBio.01720-16.

[39] Tullius MV, Nava S, Horwitz MA. PPE37 is essential for *Mycobacterium tuberculosis* Heme-Iron Acquisition (HIA), and a defective PPE37 in *Mycobacterium bovis* BCG prevents HIA. Infect Immun 2019;87(2). https://doi.org/10.1128/IAI.00540-18.

[40] Abdallah AM, Verboom T, Weerdenburg EM, Gey van Pittius NC, Mahasha PW, Jimenez C, et al. PPE and PE_PGRS proteins of *Mycobacterium marinum* are transported via the type VII secretion system ESX-5. Mol Microbiol 2009;73(3):329–40. https://doi.org/10.1111/j.1365-2958.2009.06783.x.

[41] Bottai D, Di Luca M, Majlessi L, Frigui W, Simeone R, Sayes F, et al. Disruption of the ESX-5 system of *Mycobacterium tuberculosis* causes loss of PPE protein secretion, reduction of cell wall integrity and strong attenuation. Mol Microbiol 2012;83(6):1195–209. https://doi.org/10.1111/j.1365-2958.2012.08001.x.

[42] Cascioferro A, Daleke MH, Ventura M, Dona V, Delogu G, Palu G, et al. Functional dissection of the PE domain responsible for translocation of PE_PGRS33 across the mycobacterial cell wall. PLoS One 2011;6 (11):e27713. https://doi.org/10.1371/journal.pone.0027713.

[43] Iantomasi R, Sali M, Cascioferro A, Palucci I, Zumbo A, Soldini S, et al. PE_PGRS30 is required for the full virulence of *Mycobacterium tuberculosis*. Cell Microbiol 2012;14(3):356–67. https://doi.org/10.1111/j.1462-5822.2011.01721.x.

[44] Chatrath S, Gupta VK, Dixit A, Garg LC. The Rv1651c-encoded PE-PGRS30 protein expressed in *Mycobacterium smegmatis* exhibits polar localization and modulates its growth profile. FEMS Microbiol Lett 2011;322(2):194–9. https://doi.org/10.1111/j.1574-6968.2011.02354.x.

[45] Dahiya P, Bisht MK, Mukhopadhyay S. Role of PE family of proteins in mycobacterial virulence: potential on anti-TB vaccine and drug design. Int Rev Immunol 2025:1–16. https://doi.org/10.1080/08830185.2025.2455161.

[46] Singh P, Rao RN, Reddy JR, Prasad RB, Kotturu SK, Ghosh S, et al. PE11, a PE/PPE family protein of *Mycobacterium tuberculosis* is involved in cell wall remodeling and virulence. Sci Rep 2016;6:21624. https://doi.org/10.1038/srep21624.
[47] Rastogi S, Singh AK, Pant G, Mitra K, Sashidhara KV, Krishnan MY. Down-regulation of PE11, a cell wall associated esterase, enhances the biofilm growth of *Mycobacterium tuberculosis* and reduces cell wall virulence lipid levels. Microbiology (Read) 2017;163(1):52–61. https://doi.org/10.1099/mic.0.000417.
[48] Sultana R, Vemula MH, Banerjee S, Guruprasad L. The PE16 (Rv1430) of *Mycobacterium tuberculosis* is an esterase belonging to serine hydrolase superfamily of proteins. PLoS One 2013;8(2):e55320. https://doi.org/10.1371/journal.pone.0055320.
[49] Dong D, Wang D, Li M, Wang H, Yu J, Wang C, et al. PPE38 modulates the innate immune response and is required for *Mycobacterium marinum* virulence. Infect Immun 2012;80(1):43–54. https://doi.org/10.1128/IAI.05249-11.
[50] Danilchanka O, Mailaender C, Niederweis M. Identification of a novel multidrug efflux pump of *Mycobacterium tuberculosis*. Antimicrob Agents Chemother 2008;52(7):2503–11. https://doi.org/10.1128/AAC.00298-08.
[51] Ramakrishnan P, Aagesen AM, McKinney JD, Tischler AD. *Mycobacterium tuberculosis* resists stress by regulating PE19 expression. Infect Immun 2015;84(3):735–46. https://doi.org/10.1128/IAI.00942-15.
[52] Nair S, Ramaswamy PA, Ghosh S, Joshi DC, Pathak N, Siddiqui I, et al. The PPE18 of *Mycobacterium tuberculosis* interacts with TLR2 and activates IL-10 induction in macrophage. J Immunol 2009;183(10):6269–81. https://doi.org/10.4049/jimmunol.0901367.
[53] Udgata A, Qureshi R, Mukhopadhyay S. Transduction of functionally contrasting signals by twomycobacterialPPE proteins downstream of TLR2 receptors. J Immunol 2016;197(5):1776–87. https://doi.org/10.4049/jimmunol.1501816.
[54] Abdallah AM, Verboom T, Hannes F, Safi M, Strong M, Eisenberg D, et al. A specific secretion system mediates PPE41 transport in pathogenic *mycobacteria*. Mol Microbiol 2006;62(3):667–79. https://doi.org/10.1111/j.1365-2958.2006.05409.x.
[55] Bhat KH, Srivastava S, Kotturu SK, Ghosh S, Mukhopadhyay S. The PPE2 protein of *Mycobacterium tuberculosis* translocates to host nucleus and inhibits nitric oxide production. Sci Rep 2017;7:39706. https://doi.org/10.1038/srep39706.
[56] Kaufmann SH. How can immunology contribute to the control of tuberculosis? Nat Rev Immunol 2001;1(1):20–30. https://doi.org/10.1038/35095558.
[57] Hussain Bhat K, Mukhopadhyay S. Macrophage takeover and the host-bacilli interplay during tuberculosis. Future Microbiol 2015;10(5):853–72. https://doi.org/10.2217/fmb.15.11.
[58] Medha SS, Sharma M. Proline-Glutamate/Proline-Proline-Glutamate (PE/PPE) proteins of *Mycobacterium tuberculosis*: the multifaceted immune-modulators. Acta Trop 2021;222:106035. https://doi.org/10.1016/j.actatropica.2021.106035.
[59] Thi EP, Hong CJ, Sanghera G, Reiner NE. Identification of the *Mycobacterium tuberculosis* protein PE-PGRS62 as a novel effector that functions to block phagosome maturation and inhibit iNOS expression. Cell Microbiol 2013;15(5):795–808. https://doi.org/10.1111/cmi.12073.
[60] Brodin P, Poquet Y, Levillain F, Peguillet I, Larrouy-Maumus G, Gilleron M, et al. High content phenotypic cell-based visual screen identifies *Mycobacterium tuberculosis* acyltrehalose-containing glycolipids involved in phagosome remodeling. PLoS Pathog 2010;6(9):e1001100. https://doi.org/10.1371/journal.ppat.1001100.
[61] Stewart GR, Patel J, Robertson BD, Rae A, Young DB. Mycobacterial mutants with defective control of phagosomal acidification. PLoS Pathog 2005;1(3):269–78. https://doi.org/10.1371/journal.ppat.0010033.
[62] Saini NK, Baena A, Ng TW, Venkataswamy MM, Kennedy SC, Kunnath-Velayudhan S, et al. Suppression of autophagy and antigen presentation by *Mycobacterium tuberculosis* PE_PGRS47. Nat Microbiol 2016;1(9):16133. https://doi.org/10.1038/nmicrobiol.2016.133.

[63] Li Y, Miltner E, Wu M, Petrofsky M, Bermudez LE. A *Mycobacterium avium* PPE gene is associated with the ability of the bacterium to grow in macrophages and virulence in mice. Cell Microbiol 2005;7(4): 539–48. https://doi.org/10.1111/j.1462-5822.2004.00484.x.
[64] Jha SS, Danelishvili L, Wagner D, Maser J, Li YJ, Moric I, et al. Virulence-related *Mycobacterium avium* subsp *hominissuis* MAV_2928 gene is associated with vacuole remodeling in macrophages. BMC Microbiol 2010;10:100. https://doi.org/10.1186/1471-2180-10-100.
[65] Orme IM, Cooper AM. Cytokine/chemokine cascades in immunity to tuberculosis. Immunol Today 1999; 20(7):307–12. https://doi.org/10.1016/s0167-5699(98)01438-8.
[66] Roach DR, Bean AG, Demangel C, France MP, Briscoe H, Britton WJ. TNF regulates chemokine induction essential for cell recruitment, granuloma formation, and clearance of mycobacterial infection. J Immunol 2002;168(9):4620–7. https://doi.org/10.4049/jimmunol.168.9.4620.
[67] Robinson CM, Nau GJ. Interleukin-12 and interleukin-27 regulate macrophage control of *Mycobacterium tuberculosis*. J Infect Dis 2008;198(3):359–66. https://doi.org/10.1086/589774.
[68] Flynn JL, Chan J. Immunology of tuberculosis. Annu Rev Immunol 2001;19:93–129. https://doi.org/10.1146/annurev.immunol.19.1.93.
[69] Murray PJ, Young RA. Increased antimycobacterial immunity in interleukin-10-deficient mice. Infect Immun 1999;67(6):3087–95. https://doi.org/10.1128/IAI.67.6.3087-3095.1999.
[70] Redpath S, Ghazal P, Gascoigne NR. Hijacking and exploitation of IL-10 by intracellular pathogens. Trends Microbiol 2001;9(2):86–92. https://doi.org/10.1016/s0966-842x(00)01919-3.
[71] Feng CG, Kullberg MC, Jankovic D, Cheever AW, Caspar P, Coffman RL, et al. Transgenic mice expressing human interleukin-10 in the antigen-presenting cell compartment show increased susceptibility to infection with *Mycobacterium avium* associated with decreased macrophage effector function and apoptosis. Infect Immun 2002;70(12):6672–9. https://doi.org/10.1128/IAI.70.12.6672-6679.2002.
[72] Beamer GL, Flaherty DK, Assogba BD, Stromberg P, Gonzalez-Juarrero M, de Waal Malefyt R, et al. Interleukin-10 promotes *Mycobacterium tuberculosis* disease progression in CBA/J mice. J Immunol 2008; 181(8):5545–50. https://doi.org/10.4049/jimmunol.181.8.5545.
[73] Redford PS, Boonstra A, Read S, Pitt J, Graham C, Stavropoulos E, et al. Enhanced protection to *Mycobacterium tuberculosis* infection in IL-10-deficient mice is accompanied by early and enhanced Th1 responses in the lung. Eur J Immunol 2010;40(8):2200–10. https://doi.org/10.1002/eji.201040433.
[74] Flynn JL, Goldstein MM, Chan J, Triebold KJ, Pfeffer K, Lowenstein CJ, et al. Tumor necrosis factor-α is required in the protective immune response against *Mycobacterium tuberculosis* in mice. Immunity 1995;2 (6):561–72. https://doi.org/10.1016/1074-7613(95)90001-2.
[75] Bekker LG, Moreira AL, Bergtold A, Freeman S, Ryffel B, Kaplan G. Immunopathologic effects of tumor necrosis factor alpha in murine mycobacterial infection are dose dependent. Infect Immun 2000;68(12): 6954–61. https://doi.org/10.1128/IAI.68.12.6954-6961.2000.
[76] Blackmore TK, Manning L, Taylor WJ, Wallis RS. Therapeutic use of infliximab in tuberculosis to control severe paradoxical reaction of the brain and lymph nodes. Clin Infect Dis 2008;47(10):e83–5. https://doi.org/10.1086/592695.
[77] Saunders BM, Frank AA, Orme IM, Cooper AM. Interleukin-6 induces early gamma interferon production in the infected lung but is not required for generation of specific immunity to *Mycobacterium tuberculosis* infection. Infect Immun 2000;68(6):3322–6. https://doi.org/10.1128/IAI.68.6.3322-3326.2000.
[78] Su H, Kong C, Zhu L, Huang Q, Luo L, Wang H, et al. PPE26 induces TLR2-dependent activation of macrophages and drives Th1-type T-cell immunity by triggering the cross-talk of multiple pathways involved in the host response. Oncotarget 2015;6(36):38517–37. https://doi.org/10.18632/oncotarget.5956.
[79] Xu Y, Yang E, Huang Q, Ni W, Kong C, Liu G, et al. PPE57 induces activation of macrophages and drives Th1-type immune responses through TLR2. J Mol Med (Berl) 2015;93(6):645–62. https://doi.org/10.1007/s00109-014-1243-1.

[80] Sharma N, Shariq M, Quadir N, Singh J, Sheikh JA, Hasnain SE, et al. *Mycobacterium tuberculosis* protein PE6 (Rv0335c), a novel TLR4 agonist, evokes an inflammatory response and modulates the cell death pathways in macrophages to enhance intracellular survival. Front Immunol 2021;12:696491. https://doi.org/10.3389/fimmu.2021.696491.

[81] Yang Y, Xu P, He P, Shi F, Tang Y, Guan C, et al. Mycobacterial PPE13 activates inflammasome by interacting with the NATCH and LRR domains of NLRP3. FASEB J 2020;34(9):12820–33. https://doi.org/10.1096/fj.202000200RR.

[82] Rehren G, Walters S, Fontan P, Smith I, Zarraga AM. Differential gene expression between *Mycobacterium bovis* and *Mycobacterium tuberculosis*. Tuberculosis 2007;87(4):347–59. https://doi.org/10.1016/j.tube.2007.02.004.

[83] Nair S, Pandey AD, Mukhopadhyay S. The PPE18 protein of *Mycobacterium tuberculosis* inhibits NF-κB/rel-mediated proinflammatory cytokine production by upregulating and phosphorylating suppressor of cytokine signaling 3 protein. J Immunol 2011;186(9):5413–24. https://doi.org/10.4049/jimmunol.1000773.

[84] Bhat KH, Ahmed A, Kumar S, Sharma P, Mukhopadhyay S. Role of PPE18 protein in intracellular survival and pathogenicity of *Mycobacterium tuberculosis* in mice. PLoS One 2012;7(12):e52601. https://doi.org/10.1371/journal.pone.0052601.

[85] Ahmed A, Dolasia K, Mukhopadhyay S. *Mycobacterium tuberculosis* PPE18 protein reduces inflammation and increases survival in animal model of sepsis. J Immunol 2018;200(10):3587–98. https://doi.org/10.4049/jimmunol.1602065.

[86] Goldstone RM, Goonesekera SD, Bloom BR, Sampson SL. The transcriptional regulator Rv0485 modulates the expression of a *pe and ppe* gene pair and is required for *Mycobacterium tuberculosis* virulence. Infect Immun 2009;77(10):4654–67. https://doi.org/10.1128/IAI.01495-08.

[87] Daim S, Kawamura I, Tsuchiya K, Hara H, Kurenuma T, Shen Y, et al. Expression of the *Mycobacterium tuberculosis* PPE37 protein in *Mycobacterium smegmatis* induces low tumour necrosis factor alpha and interleukin 6 production in murine macrophages. J Med Microbiol 2011;60(Pt 5):582–91. https://doi.org/10.1099/jmm.0.026047-0.

[88] Singh PP, Parra M, Cadieux N, Brennan MJ. A comparative study of host response to three *Mycobacterium tuberculosis* PE_PGRS proteins. Microbiology 2008;154(Pt 11):3469–79. https://doi.org/10.1099/mic.0.2008/019968-0.

[89] Khubaib M, Sheikh JA, Pandey S, Srikanth B, Bhuwan M, Khan N, et al. *Mycobacterium tuberculosis* co-operonic PE32/PPE65 proteins alter host immune responses by hampering Th1 response. Front Microbiol 2016;7:719. https://doi.org/10.3389/fmicb.2016.00719.

[90] Tiwari B, Soory A, Raghunand TR. An immunomodulatory role for the *Mycobacterium tuberculosis* region of difference 1 locus proteins PE35 (Rv3872) and PPE68 (Rv3873). FEBS J 2014;281(6):1556–70. https://doi.org/10.1111/febs.12723.

[91] Lu B, Rutledge BJ, Gu L, Fiorillo J, Lukacs NW, Kunkel SL, et al. Abnormalities in monocyte recruitment and cytokine expression in monocyte chemoattractant protein 1-deficient mice. J Exp Med 1998;187(4):601–8. https://doi.org/10.1084/jem.187.4.601.

[92] Huang Y, Wang Y, Bai Y, Wang ZG, Yang L, Zhao D. Expression of PE_PGRS 62 protein in *Mycobacterium smegmatis* decrease mRNA expression of proinflammatory cytokines IL-1β, IL-6 in macrophages. Mol Cell Biochem 2010;340(1–2):223–9. https://doi.org/10.1007/s11010-010-0421-x.

[93] Dheenadhayalan V, Delogu G, Brennan MJ. Expression of the PE_PGRS 33 protein in *Mycobacterium smegmatis* triggers necrosis in macrophages and enhanced mycobacterial survival. Microbes Infect 2006;8(1):262–72. https://doi.org/10.1016/j.micinf.2005.06.021.

[94] Deng W, Long Q, Zeng J, Li P, Yang W, Chen X, et al. *Mycobacterium tuberculosis* PE_PGRS41 enhances the intracellular survival of *M. smegmatis* within macrophages via blocking innate immunity and inhibition of host defense. Sci Rep 2017;7:46716. https://doi.org/10.1038/srep46716.

[95] Yeruva VC, Kulkarni A, Khandelwal R, Sharma Y, Raghunand TR. The PE_PGRS proteins of *Mycobacterium tuberculosis* are Ca^{2+} binding mediators of host-pathogen interaction. Biochemistry 2016;55 (33):4675–87. https://doi.org/10.1021/acs.biochem.6b00289.

[96] Tiwari BM, Kannan N, Vemu L, Raghunand TR. The *Mycobacterium tuberculosis* PE proteins Rv0285 and Rv1386 modulate innate immunity and mediate bacillary survival in macrophages. PLoS One 2012;7(12): e51686. https://doi.org/10.1371/journal.pone.0051686.

[97] Xu T, Wang C, Li M, Yuan M, Wei J, Li B, et al. *Mycobacterium tuberculosis* PE8 (Rv1040c) promotes the intracellular survival of recombinant *Mycobacterium* by regulating host inflammatory cytokines and inhibiting cell late apoptosis. DNA Cell Biol 2023;42(5):254–64. https://doi.org/10.1089/dna.2022.0316.

[98] Abo-Kadoum MA, Assad M, Ali MK, Uae M, Nzaou SAE, Gong Z, et al. *Mycobacterium tuberculosis* PE17 (Rv1646) promotes host cell apoptosis via host chromatin remodeling mediated by reduced H3K9me3 occupancy. Microb Pathog 2021;159:105147. https://doi.org/10.1016/j.micpath.2021.105147.

[99] Zhang J, Cui Y, Zang X, Feng T, Chen F, Wang H, et al. PE12 interaction with TLR4 promotes intracellular survival of *Mycobacterium tuberculosis* by suppressing inflammatory response. Int J Biol Macromol 2023; 253(Pt 8):127547. https://doi.org/10.1016/j.ijbiomac.2023.127547.

[100] Li Z, Liu H, Li H, Dang G, Cui Z, Song N, et al. PE17 protein from *Mycobacterium tuberculosis* enhances *Mycobacterium smegmatis* survival in macrophages and pathogenicity in mice. Microb Pathog 2019;126: 63–73. https://doi.org/10.1016/j.micpath.2018.10.030.

[101] Deng W, Li W, Zeng J, Zhao Q, Li C, Zhao Y, et al. *Mycobacterium tuberculosis* PPE family protein Rv1808 manipulates cytokines profile via co-activation of MAPK and NF-κB signaling pathways. Cell Physiol Biochem 2014;33(2):273–88. https://doi.org/10.1159/000356668.

[102] Ali MK, Zhen G, Nzungize L, Stojkoska A, Duan X, Li C, et al. *Mycobacterium tuberculosis* PE31 (Rv3477) attenuates host cell apoptosis and promotes recombinant *M. smegmatis* intracellular survival via up-regulating GTPase guanylate binding protein-1. Front Cell Infect Microbiol 2020;10:40. https://doi.org/10.3389/fcimb.2020.00040.

[103] Bansal K, Sinha AY, Ghorpade DS, Togarsimalemath SK, Patil SA, Kaveri SV, et al. Src homology 3-interacting domain of Rv1917c of *Mycobacterium tuberculosis* induces selective maturation of human dendritic cells by regulating PI3K-MAPK-NF-κB signaling and drives Th2 immune responses. J Biol Chem 2010;285(47):36511–22. https://doi.org/10.1074/jbc.M110.158055.

[104] Kim WS, Kim JS, Cha SB, Kim SJ, Kim H, Kwon KW, et al. *Mycobacterium tuberculosis* PE27 activates dendritic cells and contributes to Th1-polarized memory immune responses during *in vivo* infection. Immunobiology 2016;221(3):440–53. https://doi.org/10.1016/j.imbio.2015.11.006.

[105] Ehrt S, Schnappinger D. Mycobacterial survival strategies in the phagosome: defence against host stresses. Cell Microbiol 2009;11(8):1170–8. https://doi.org/10.1111/j.1462-5822.2009.01335.x.

[106] Voskuil MI, Bartek IL, Visconti K, Schoolnik GK. The response of *Mycobacterium tuberculosis* to reactive oxygen and nitrogen species. Front Microbiol 2011;2:105. https://doi.org/10.3389/fmicb.2011.00105.

[107] Ng VH, Cox JS, Sousa AO, MacMicking JD, McKinney JD. Role of KatG catalase-peroxidase in mycobacterial pathogenesis: countering the phagocyte oxidative burst. Mol Microbiol 2004;52(5):1291–302. https://doi.org/10.1111/j.1365-2958.2004.04078.x.

[108] Piddington DL, Fang FC, Laessig T, Cooper AM, Orme IM, Buchmeier NA. Cu, Zn superoxide dismutase of *Mycobacterium tuberculosis* contributes to survival in activated macrophages that are generating an oxidative burst. Infect Immun 2001;69(8):4980–7. https://doi.org/10.1128/IAI.69.8.4980-4987.2001.

[109] Manca C, Paul S, Barry 3rd CE, Freedman VH, Kaplan G. *Mycobacterium tuberculosis* catalase and peroxidase activities and resistance to oxidative killing in human monocytes in vitro. Infect Immun 1999;67 (1):74–9. https://doi.org/10.1128/IAI.67.1.74-79.1999.

[110] Bisht MK, Pal R, Dahiya P, Naz S, Sanyal P, Nandicoori VK, et al. The PPE2 protein of *Mycobacterium tuberculosis* is secreted during infection and facilitates mycobacterial survival inside the host. Tuberculosis 2023;143:102421. https://doi.org/10.1016/j.tube.2023.102421.

[111] Srivastava S, Battu MB, Khan MZ, Nandicoori VK, Mukhopadhyay S. *Mycobacterium tuberculosis* PPE2 protein interacts with p67^{phox} and inhibits reactive oxygen species production. J Immunol 2019;203(5): 1218–29. https://doi.org/10.4049/jimmunol.1801143.

[112] Li H, Li Q, Yu Z, Zhou M, Xie J. *Mycobacterium tuberculosis* PE13 (Rv1195) manipulates the host cell fate via p38-ERK-NF-κB axis and apoptosis. Apoptosis 2016;21(7):795–808. https://doi.org/10.1007/s10495-016-1249-y.

[113] Grover S, Sharma T, Singh Y, Kohli S, P M, Singh A, et al. The PGRS domain of *Mycobacterium tuberculosis* PE_PGRS protein Rv0297 is involved in endoplasmic reticulum stress-mediated apoptosis through toll-like receptor 4. mBio 2018;9(3). https://doi.org/10.1128/mBio.01017-18.

[114] Chaturvedi R, Bansal K, Narayana Y, Kapoor N, Sukumar N, Togarsimalemath SK, et al. The multi-functional PE_PGRS11 protein from *Mycobacterium tuberculosis* plays a role in regulating resistance to oxidative stress. J Biol Chem 2010;285(40):30389–403. https://doi.org/10.1074/jbc.M110.135251.

[115] Keane J, Remold HG, Kornfeld H. Virulent *Mycobacterium tuberculosis* strains evade apoptosis of infected alveolar macrophages. J Immunol 2000;164(4):2016–20. https://doi.org/10.4049/jimmunol.164.4.2016.

[116] Velmurugan K, Chen B, Miller JL, Azogue S, Gurses S, Hsu T, et al. *Mycobacterium tuberculosis* nuoG is a virulence gene that inhibits apoptosis of infected host cells. PLoS Pathog 2007;3(7):e110. https://doi.org/10.1371/journal.ppat.0030110.

[117] Dobos KM, Spotts EA, Quinn FD, King CH. Necrosis of lung epithelial cells during infection with *Mycobacterium tuberculosis* is preceded by cell permeation. Infect Immun 2000;68(11):6300–10. https://doi.org/10.1128/IAI.68.11.6300-6310.2000.

[118] Saunders BM, Britton WJ. Life and death in the granuloma: immunopathology of tuberculosis. Immunol Cell Biol 2007;85(2):103–11. https://doi.org/10.1038/sj.icb.7100027.

[119] Deng W, Yang W, Zeng J, Abdalla AE, Xie J. *Mycobacterium tuberculosis* PPE32 promotes cytokines production and host cell apoptosis through caspase cascade accompanying with enhanced ER stress response. Oncotarget 2016;7(41):67347–59. https://doi.org/10.18632/oncotarget.12030.

[120] Tiwari B, Ramakrishnan UM, Raghunand TR. The *Mycobacterium tuberculosis* protein pair PE9 (Rv1088)-PE10 (Rv1089) forms heterodimers and induces macrophage apoptosis through Toll-like receptor 4. Cell Microbiol 2015;17(11):1653–69. https://doi.org/10.1111/cmi.12462.

[121] Long Q, Xiang X, Yin Q, Li S, Yang W, Sun H, et al. PE_PGRS62 promotes the survival of *Mycobacterium smegmatis* within macrophages via disrupting ER stress-mediated apoptosis. J Cell Physiol 2019;234(11): 19774–84. https://doi.org/10.1002/jcp.28577.

[122] Basu S, Pathak SK, Banerjee A, Pathak S, Bhattacharyya A, Yang Z, et al. Execution of macrophage apoptosis by PE_PGRS33 of *Mycobacterium tuberculosis* is mediated by Toll-like receptor 2-dependent release of TNF-α. J Biol Chem 2007;282(2):1039–50. https://doi.org/10.1074/jbc.M604379200.

[123] Talarico S, Cave MD, Marrs CF, Foxman B, Zhang L, Yang Z. Variation of the *Mycobacterium tuberculosis* PE_PGRS 33 gene among clinical isolates. J Clin Microbiol 2005;43(10):4954–60. https://doi.org/10.1128/JCM.43.10.4954-4960.2005.

[124] Cadieux N, Parra M, Cohen H, Maric D, Morris SL, Brennan MJ. Induction of cell death after localization to the host cell mitochondria by the *Mycobacterium tuberculosis* PE_PGRS33 protein. Microbiology (Read) 2011;157(Pt 3):793–804. https://doi.org/10.1099/mic.0.041996-0.

[125] Strong EJ, Ng TW, Porcelli SA, Lee S. *Mycobacterium tuberculosis* PE_PGRS20 and PE_PGRS47 proteins inhibit autophagy by interaction with Rab1A. mSphere 2021;6(4):e0054921. https://doi.org/10.1128/mSphere.00549-21.

[126] Chai Q, Wang X, Qiang L, Zhang Y, Ge P, Lu Z, et al. A *Mycobacterium tuberculosis* surface protein recruits ubiquitin to trigger host xenophagy. Nat Commun 2019;10(1):1973. https://doi.org/10.1038/s41467-019-09955-8.

[127] Feng S, Hong Z, Zhang G, Li J, Tian GB, Zhou H, et al. *Mycobacterium* PPE31 contributes to host cell death. Front Cell Infect Microbiol 2021;11:629836. https://doi.org/10.3389/fcimb.2021.629836.

[128] Asaad M, Kaisar Ali M, Abo-Kadoum MA, Lambert N, Gong Z, Wang H, et al. *Mycobacterium tuberculosis* PPE10 (Rv0442c) alters host cell apoptosis and cytokine profile via linear ubiquitin chain assembly complex HOIP-NF-κB signaling axis. Int Immunopharmacol 2021;94:107363. https://doi.org/10.1016/j.intimp.2020.107363.

[129] Medha P, Bhatt P, Sharma S, Sharma M. Role of C-terminal domain of *Mycobacterium tuberculosis* PE6 (Rv0335c) protein in host mitochondrial stress and macrophage apoptosis. Apoptosis 2023;28(1–2):136–65. https://doi.org/10.1007/s10495-022-01778-1.

[130] Xu A, Yang K, Xu J, Meng X, Liang Q, Huang B, et al. Mycobacterial PE12 protein promotes bacterial survival through inhibiting cell apoptosis. Vet Microbiol 2024;288:109922. https://doi.org/10.1016/j.vetmic.2023.109922.

[131] Tundup S, Mohareer K, Hasnain SE. *Mycobacterium tuberculosis* PE25/PPE41 protein complex induces necrosis in macrophages: role in virulence and disease reactivation? FEBS Open Bio 2014;4:822–8. https://doi.org/10.1016/j.fob.2014.09.001.

[132] Danelishvili L, Everman J, Bermudez LE. *Mycobacterium tuberculosis* PPE68 and Rv2626c genes contribute to the host cell necrosis and bacterial escape from macrophages. Virulence 2016;7(1):23–32. https://doi.org/10.1080/21505594.2015.1102832.

[133] Chen T, Zhao Q, Li W, Xie J. *Mycobacterium tuberculosis* PE_PGRS17 promotes the death of host cell and cytokines secretion via Erk kinase accompanying with enhanced survival of recombinant *Mycobacterium smegmatis*. J Interferon Cytokine Res 2013;33(8):452–8. https://doi.org/10.1089/jir.2012.0083.

[134] Mi Y, Bao L, Gu D, Luo T, Sun C, Yang G. *Mycobacterium tuberculosis* PPE25 and PPE26 proteins expressed in *Mycobacterium smegmatis* modulate cytokine secretion in mouse macrophages and enhance mycobacterial survival. Res Microbiol 2017;168(3):234–43. https://doi.org/10.1016/j.resmic.2016.06.004.

[135] Ahmad J, Farhana A, Pancsa R, Arora SK, Srinivasan A, Tyagi AK, et al. Contrasting function of structured N-terminal and unstructured C-terminal segments of *Mycobacterium tuberculosis* PPE37 protein. mBio 2018;9(1). https://doi.org/10.1128/mBio.01712-17.

[136] Ramakrishnan L, Federspiel NA, Falkow S. Granuloma-specific expression of *Mycobacterium* virulence proteins from the glycine-rich PE-PGRS family. Science 2000;288(5470):1436–9. https://doi.org/10.1126/science.288.5470.1436.

[137] Brennan MJ, Delogu G, Chen Y, Bardarov S, Kriakov J, Alavi M, et al. Evidence that mycobacterial PE_PGRS proteins are cell surface constituents that influence interactions with other cells. Infect Immun 2001;69(12):7326–33. https://doi.org/10.1128/IAI.69.12.7326-7333.2001.

[138] Bansal K, Elluru SR, Narayana Y, Chaturvedi R, Patil SA, Kaveri SV, et al. PE_PGRS antigens of *Mycobacterium tuberculosis* induce maturation and activation of human dendritic cells. J Immunol 2010;184(7):3495–504. https://doi.org/10.4049/jimmunol.0903299.

[139] Palucci I, Camassa S, Cascioferro A, Sali M, Anoosheh S, Zumbo A, et al. PE_PGRS33 contributes to *Mycobacterium tuberculosis* entry in macrophages through interaction with TLR2. PLoS One 2016;11(3):e0150800. https://doi.org/10.1371/journal.pone.0150800.

[140] Camacho LR, Ensergueix D, Perez E, Gicquel B, Guilhot C. Identification of a virulence gene cluster of *Mycobacterium tuberculosis* by signature-tagged transposon mutagenesis. Mol Microbiol 1999;34(2):257–67. https://doi.org/10.1046/j.1365-2958.1999.01593.x.

[141] Delogu G, Brennan MJ. Comparative immune response to PE and PE_PGRS antigens of *Mycobacterium tuberculosis*. Infect Immun 2001;69(9):5606–11. https://doi.org/10.1128/IAI.69.9.5606-5611.2001.

[142] De Maio F, Battah B, Palmieri V, Petrone L, Corrente F, Salustri A, et al. PE_PGRS3 of *Mycobacterium tuberculosis* is specifically expressed at low phosphate concentration, and its arginine-rich C-terminal domain mediates adhesion and persistence in host tissues when expressed in *Mycobacterium smegmatis*. Cell Microbiol 2018;20(12):e12952. https://doi.org/10.1111/cmi.12952.

[143] Shah S, Cannon JR, Fenselau C, Briken V. A duplicated ESAT-6 region of ESX-5 is involved in protein export and virulence of mycobacteria. Infect Immun 2015;83(11):4349–61. https://doi.org/10.1128/IAI.00827-15.

[144] Pfeiffer C, Stein J, Southwood S, Ketelaar H, Sette A, Bottomly K. Altered peptide ligands can control CD4 T lymphocyte differentiation in vivo. J Exp Med 1995;181(4):1569–74. https://doi.org/10.1084/jem.181.4.1569.

[145] D'Andrea A, Aste-Amezaga M, Valiante NM, Ma X, Kubin M, Trinchieri G. Interleukin 10 (IL-10) inhibits human lymphocyte INF-γ production by suppressing natural killer cell stimulatory factor/IL-12 synthesis in accessory cells. J Exp Med 1993;178(3):1041–8. https://doi.org/10.1084/jem.178.3.1041.

[146] Underhill DM, Bassetti M, Rudensky A, Aderem A. Dynamic interactions of macrophages with T cells during antigen presentation. J Exp Med 1999;190(12):1909–14. https://doi.org/10.1084/jem.190.12.1909.

[147] Rook G. Th1- or Th2-cell commitment during infectious disease: an oversimplification? Trends Immunol 2001;22(9):481–2. https://doi.org/10.1016/s1471-4906(01)01980-9.

[148] Lienhardt C, Azzurri A, Amedei A, Fielding K, Sillah J, Sow OY, et al. Active tuberculosis in Africa is associated with reduced Th1 and increased Th2 activity *in vivo*. Eur J Immunol 2002;32(6):1605–13. https://doi.org/10.1002/1521-4141(200206)32:6<1605::AID-IMMU1605>3.0.CO;2-1613.

[149] Abebe F. Synergy between Th1 and Th2 responses during *Mycobacterium tuberculosis* infection: a review of current understanding. Int Rev Immunol 2019;38(4):172–9. https://doi.org/10.1080/08830185.2019.1632842.

[150] Ashenafi S, Aderaye G, Bekele A, Zewdie M, Aseffa G, Hoang AT, et al. Progression of clinical tuberculosis is associated with a Th2 immune response signature in combination with elevated levels of SOCS3. Clin Immunol 2014;151(2):84–99. https://doi.org/10.1016/j.clim.2014.01.010.

[151] Ramalingam G, Khan JM, Jasmine S, Saminathan G, Manickan E, Rajagopal P, et al. Role of Th1/Th2 cytokine balance in predicting treatment outcomes and disease severity in tuberculosis. J King Saud Univ Sci 2024;36(11):103538. https://doi.org/10.1016/j.jksus.2024.103538.

[152] Harding CV, Boom WH. Regulation of antigen presentation by *Mycobacterium tuberculosis*: a role for Toll-like receptors. Nat Rev Microbiol 2010;8(4):296–307. https://doi.org/10.1038/nrmicro2321.

[153] Dolasia K, Nazar F, Mukhopadhyay S. *Mycobacterium tuberculosis* PPE18 protein inhibits MHC class II antigen presentation and B cell response in mice. Eur J Immunol 2021;51(3):603–19. https://doi.org/10.1002/eji.201848071.

[154] Meena LS. An overview to understand the role of PE_PGRS family proteins in *Mycobacterium tuberculosis* H37Rv and their potential as new drug targets. Biotechnol Appl Biochem 2015;62(2):145–53. https://doi.org/10.1002/bab.1266.

[155] Chen W, Bao Y, Chen X, Burton J, Gong X, Gu D, et al. *Mycobacterium tuberculosis* PE25/PPE41 protein complex induces activation and maturation of dendritic cells and drives Th2-biased immune responses. Med Microbiol Immunol 2016;205(2):119–31. https://doi.org/10.1007/s00430-015-0434-x.

[156] Meng L, Tong J, Wang H, Tao C, Wang Q, Niu C, et al. PPE38 protein of *Mycobacterium tuberculosis* inhibits macrophage MHC class I expression and dampens $CD8^+$ T cell responses. Front Cell Infect Microbiol 2017;7:68. https://doi.org/10.3389/fcimb.2017.00068.

[157] Su H, Zhang Z, Liu Z, Peng B, Kong C, Wang H, et al. *Mycobacterium tuberculosis* PPE60 antigen drives Th1/Th17 responses viaToll-like receptor 2-dependent maturation of dendritic cells. J Biol Chem 2018;293(26):10287–302. https://doi.org/10.1074/jbc.RA118.001696.

[158] Nothelfer K, Sansonetti PJ, Phalipon A. Pathogen manipulation of B cells: the best defence is a good offence. Nat Rev Microbiol 2015;13(3):173–84. https://doi.org/10.1038/nrmicro3415.

[159] LeBien TW, Tedder TF. B lymphocytes: how they develop and function. Blood 2008;112(5):1570–80. https://doi.org/10.1182/blood-2008-02-078071.

[160] Seifert M, Küppers R. Human memory B cells. Leukemia 2016;30(12):2283–92. https://doi.org/10.1038/leu.2016.226.

[161] Seifert M, Kuppers R. Molecular footprints of a germinal center derivation of human IgM^{+}(IgD^{+})CD27^{+} B cells and the dynamics of memory B cell generation. J Exp Med 2009;206(12):2659–69. https://doi.org/10.1084/jem.20091087.

[162] Abraham PR, Pathak N, Pradhan G, Sumanlatha G, Mukhopadhyay S. The N-terminal domain of *Mycobacterium tuberculosis* PPE17 (Rv1168c) protein plays a dominant role in inducing antibody responses in active TB patients. PLoS One 2017;12(6):e0179965. https://doi.org/10.1371/journal.pone.0179965.

[163] Abraham PR, Devalraju KP, Jha V, Valluri VL, Mukhopadhyay S. PPE17 (Rv1168c) protein of *Mycobacterium tuberculosis* detects individuals with latent TB infection. PLoS One 2018;13(11):e0207787. https://doi.org/10.1371/journal.pone.0207787.

[164] Koh KW, Soh SE, Seah GT. Strong antibody responses to *Mycobacterium tuberculosis* PE-PGRS62 protein are associated with latent and active tuberculosis. Infect Immun 2009;77(8):3337–43. https://doi.org/10.1128/IAI.01175-08.

[165] Tundup S, Pathak N, Ramanadham M, Mukhopadhyay S, Murthy KJ, Ehtesham NZ, et al. The co-operonic PE25/PPE41 protein complex of *Mycobacterium tuberculosis* elicits increased humoral and cell mediated immune response. PLoS One 2008;3(10):e3586. https://doi.org/10.1371/journal.pone.0003586.

[166] Chakhaiyar P, Nagalakshmi Y, Aruna B, Murthy KJ, Katoch VM, Hasnain SE. Regions of high antigenicity within the hypothetical PPE major polymorphic tandem repeat open-reading frame, Rv2608, show a differential humoral response and a low T cell response in various categories of patients with tuberculosis. J Infect Dis 2004;190(7):1237–44. https://doi.org/10.1086/423938.

[167] Bonanni D, Rindi L, Lari N, Garzelli C. Immunogenicity of mycobacterial PPE44 (Rv2770c) in *Mycobacterium bovis* BCG-infected mice. J Med Microbiol 2005;54(Pt 5):443–8. https://doi.org/10.1099/jmm.0.45960-0.

[168] Flynn JL, Goldstein MM, Triebold KJ, Koller B, Bloom BR. Major histocompatibility complex class I-restricted T cells are required for resistance to *Mycobacterium tuberculosis* infection. Proc Natl Acad Sci U S A 1992;89(24):12013–7. https://doi.org/10.1073/pnas.89.24.12013.

[169] Caruso AM, Serbina N, Klein E, Triebold K, Bloom BR, Flynn JL. Mice deficient in CD4 T cells have only transiently diminished levels of IFN-γ, yet succumb to tuberculosis. J Immunol 1999;162(9):5407–16.

[170] Cooper AM, Magram J, Ferrante J, Orme IM. Interleukin 12 (IL-12) is crucial to the development of protective immunity in mice intravenously infected with *Mycobacterium tuberculosis*. J Exp Med 1997;186(1):39–45. https://doi.org/10.1084/jem.186.1.39.

[171] Romagnani S. Th1/Th2 cells. Inflamm Bowel Dis 1999;5(4):285–94. https://doi.org/10.1097/00054725-199911000-00009.

[172] Sun L, Su Y, Jiao A, Wang X, Zhang B. T cells in health and disease. Signal Transduct Target Ther 2023;8(1):235. https://doi.org/10.1038/s41392-023-01471-y.

[173] García-Bengoa M, Vergara EJ, Tran AC, Bossi L, Cooper AM, Pearl JE, et al. Immunogenicity of PE18, PE31, and PPE26 proteins from *Mycobacterium tuberculosis* in humans and mice. Front Immunol 2023;14:1307429. https://doi.org/10.3389/fimmu.2023.1307429.

[174] Musvosvi M, Huang H, Wang C, Xia Q, Rozot V, Krishnan A, et al. T cell receptor repertoires associated with control and disease progression following *Mycobacterium tuberculosis* infection. Nat Med 2023;29(1):258–69. https://doi.org/10.1038/s41591-022-02110-9.

[175] Skeiky YA, Alderson MR, Ovendale PJ, Guderian JA, Brandt L, Dillon DC, et al. Differential immune responses and protective efficacy induced by components of a tuberculosis polyprotein vaccine, Mtb72F, delivered as naked DNA or recombinant protein. J Immunol 2004;172(12):7618–28. https://doi.org/10.4049/jimmunol.172.12.7618.

[176] Reed SG, Coler RN, Dalemans W, Tan EV, DeLa Cruz EC, Basaraba RJ, et al. Defined tuberculosis vaccine, Mtb72F/AS02A, evidence of protection in cynomolgus monkeys. Proc Natl Acad Sci U S A 2009;106(7):2301–6. https://doi.org/10.1073/pnas.0712077106.

[177] Van Der Meeren O, Hatherill M, Nduba V, Wilkinson RJ, Muyoyeta M, Van Brakel E, et al. Phase 2b placebo-controlled trial of M72/AS01$_E$ vaccine to prevent tuberculosis. N Engl J Med 2018;379(17):1621–34. https://doi.org/10.1056/NEJMoa1803484.

[178] Tait DR, Hatherill M, Van Der Meeren O, Ginsberg AM, Van Brakel E, Salaun B, et al. Final analysis of a Trial of M72/AS01$_E$ vaccine to prevent tuberculosis. N Engl J Med 2019;381(25):2429–39. https://doi.org/10.1056/NEJMoa1909953.

[179] Sali M, Di Sante G, Cascioferro A, Zumbo A, Nicolo C, Dona V, et al. Surface expression of MPT64 as a fusion with the PE domain of PE_PGRS33 enhances *Mycobacterium bovis* BCG protective activity against *Mycobacterium tuberculosis* in mice. Infect Immun 2010;78(12):5202–13. https://doi.org/10.1128/IAI.00267-10.

[180] Dillon DC, Alderson MR, Day CH, Lewinsohn DM, Coler R, Bement T, et al. Molecular characterization and human T-cell responses to a member of a novel *Mycobacterium tuberculosis* mtb39 gene family. Infect Immun 1999;67(6):2941–50. https://doi.org/10.1128/IAI.67.6.2941-2950.1999.

[181] Gong W, Liang Y, Mi J, Xue Y, Wang J, Wang L, et al. A peptide-based vaccine ACP derived from antigens of *Mycobacterium tuberculosis* induced Th1 response but failed to enhance the protective efficacy of BCG in mice. Indian J Tuberc 2022;69(4):482–95. https://doi.org/10.1016/j.ijtb.2021.08.016.

[182] Hakim JMC, Yang Z. Predicted structural variability of *Mycobacterium tuberculosis* PPE18 protein with immunological implications among clinical strains. Front Microbiol 2020;11:595312. https://doi.org/10.3389/fmicb.2020.595312.

[183] Guo F, Wei J, Song Y, Li B, Qian Z, Wang X, et al. Immunological effects of the PE/PPE family proteins of *Mycobacterium tuberculosis* and related vaccines. Front Immunol 2023;14:1255920. https://doi.org/10.3389/fimmu.2023.1255920.

[184] Sayes F, Sun L, Di Luca M, Simeone R, Degaiffier N, Fiette L, et al. Strong immunogenicity and cross-reactivity of *Mycobacterium tuberculosis* ESX-5 type VII secretion: encoded PE-PPE proteins predicts vaccine potential. Cell Host Microbe 2012;11(4):352–63. https://doi.org/10.1016/j.chom.2012.03.003.

[185] Sayes F, Pawlik A, Frigui W, Groschel MI, Crommelynck S, Fayolle C, et al. CD4$^+$ T cells recognizing PE/PPE antigens directly or via cross reactivity are protective against pulmonary *Mycobacterium tuberculosis* infection. PLoS Pathog 2016;12(7):e1005770. https://doi.org/10.1371/journal.ppat.1005770.

[186] Bertholet S, Ireton GC, Ordway DJ, Windish HP, Pine SO, Kahn M, et al. A defined tuberculosis vaccine candidate boosts BCG and protects against multidrug-resistant *Mycobacterium tuberculosis*. Sci Transl Med 2010;2(53):53ra74. https://doi.org/10.1126/scitranslmed.3001094.

[187] Day TA, Penn-Nicholson A, Luabeya AKK, Fiore-Gartland A, Du Plessis N, Loxton AG, et al. Safety and immunogenicity of the adjunct therapeutic vaccine ID93 + GLA-SE in adults who have completed treatment for tuberculosis: a randomised, double-blind, placebo-controlled, phase 2a trial. Lancet Respir Med 2021;9(4):373–86. https://doi.org/10.1016/S2213-2600(20)30319-2.

[188] Baldwin SL, Reese VA, Larsen SE, Pecor T, Brown BP, Granger B, et al. Therapeutic efficacy against *Mycobacterium tuberculosis* using ID93 and liposomal adjuvant formulations. Front Microbiol 2022;13:935444. https://doi.org/10.3389/fmicb.2022.935444.

[189] Skeiky YA, Alderson MR, Ovendale PJ, Lobet Y, Dalemans W, Orme IM, et al. Protection of mice and guinea pigs against tuberculosis induced by immunization with a single *Mycobacterium tuberculosis* recombinant antigen, MTB41. Vaccine 2005;23(30):3937–45. https://doi.org/10.1016/j.vaccine.2005.03.003.

[190] Singh SK, Kumari R, Singh DK, Tiwari S, Singh PK, Sharma S, et al. Putative roles of a proline-glutamic acid-rich protein (PE3) in intracellular survival and as a candidate for subunit vaccine against *Mycobacterium tuberculosis*. Med Microbiol Immunol 2013;202(5):365–77. https://doi.org/10.1007/s00430-013-0299-9.

[191] Stylianou E, Harrington-Kandt R, Beglov J, Bull N, Pinpathomrat N, Swarbrick GM, et al. Identification and evaluation of novel protective antigens for the development of a candidate tuberculosis subunit vaccine. Infect Immun 2018;86(7). https://doi.org/10.1128/iai.00014-18.

[192] Hanif SN, Al-Attiyah R, Mustafa AS. Cellular immune responses in mice induced by *M. tuberculosis* PE35-DNA vaccine construct. Scand J Immunol 2011;74(6):554–60. https://doi.org/10.1111/j.1365-3083.2011.02604.x.

[193] Amoudy HA, Ebrahimi BH, Mustafa AS. Immune responses against *Mycobacterium tuberculosis*-specific proteins PE35 and CFP10 in mice immunized with recombinant *Mycobacterium vaccae*. Saudi Med J 2014; 35(4):350–9.

[194] Khademi F, Yousefi-Avarvand A, Derakhshan M, Meshkat Z, Tafaghodi M, Ghazvini K, et al. *Mycobacterium tuberculosis* HspX/EsxS fusion protein: gene cloning, protein expression, and purification in *Escherichia coli*. Rep Biochem Mol Biol 2017;6(1):15–21.

[195] Choi SY, Kwon KW, Kim H, Choi HH, Shin SJ. Vaccine potential of ESAT-6 protein fused with consensus $CD4^+$ T-cell epitopes of PE/PPE proteins against highly pathogenic *Mycobacterium tuberculosis* strain HN878. Biochem Biophys Res Commun 2018;503(4):2195–201. https://doi.org/10.1016/j.bbrc.2018.06.017.

[196] Huang H, Wang F, Yang E, Wang H, Gao P, Shen H. Assessment of recombinant plasmid expressing fusion antigen Ag85B-Rv3425 in management of acute tuberculosis infection in mice. Exp Ther Med 2018;15(3): 3034–9. https://doi.org/10.3892/etm.2018.5785.

[197] Sulman S, Savidge BO, Alqaseer K, Das MK, Nezam Abadi N, Pearl JE, et al. Balance between protection and pathogenic response to aerosol challenge with *Mycobacterium tuberculosis* (Mtb) in mice vaccinated with TriFu64, a fusion consisting of three Mtb antigens. Vaccines (Basel) 2021;9(5). https://doi.org/10.3390/vaccines9050519.

[198] Arif S, Akhter M, Khaliq A, Akhtar MW. Fusion peptide constructs from antigens of *M. tuberculosis* producing high T-cell mediated immune response. PLoS One 2022;17(9):e0271126. https://doi.org/10.1371/journal.pone.0271126.

SECTION

II

Host defense mechanisms and cellular responses

CHAPTER

Mechanisms of phagosomal damage detection and response to *Mycobacterium tuberculosis*: Repair, autophagy, and biogenesis

5

Anam Mursaleen[1,a], Onaiza Ansari[1,a], Syeda Rushna[1], Mohd Shariq[2], Farhan Ahmed[2] and Javaid Ahmad Sheikh[1]

[1]*Department of Biotechnology, School of Chemical and Life Sciences, Jamia Hamdard, New Delhi, India;* [2]*GITAM School of Life Sciences, Gandhi Institute of Technology and Management (GITAM) University, Hyderabad Campus, Hyderabad, Telangana, India*

Highlights

1. Phagosome maturation is a tightly regulated process involving Rab GTPases, phosphoinositide's, and SNARE proteins.
2. M.tb employs effector proteins and lipids to modulate phagosomal maturation and destabilize membrane integrity.
3. Host cells detect phagosomal damage through galectins, calcium signaling, ATG8 E3-like ligases, and cGAS-STING pathway.
4. ESCRT-III, autophagy, vesicular trafficking, and calcium influx-driven responses are key mechanisms for phagosomal damage repair.
5. Host molecules like VPS18, CISH, CORO1A, ATG7, and ATG14 play crucial roles in reversing phagosomal damage.
6. Ubiquitination and TRIM proteins are involved in sensing and repairing phagosomal membrane damage.
7. The interplay between M.tb and host phagosomal environment is a dynamic battlefield of survival and defense strategies.

1. Introduction

Phagosome maturation is a meticulously regulated, multi-step process characterized by the intricate interplay of various cellular determinants that facilitate the degradation of internalized pathogens. An essential feature of this process is the progressive acidification of the phagosomal lumen, primarily

[a]Equal contributions.

Navigating Tuberculosis. https://doi.org/10.1016/B978-0-443-27735-1.00016-2

driven by vacuolar H^+-ATPase (VoH^+-ATPase), which actively transports protons into the phagosome [1]. Furthermore, phagosomal maturation is a well-coordinated process governed by the sequential recruitment of Ras-associated binding (Rab) GTPases, which function as critical regulators of membrane trafficking. These GTPases act as "regulatory switches," transitioning between the GDP-bound (inactive) and GTP-bound (active) states, thereby orchestrating the vesicular transport, membrane fusion, and intracellular compartmentalization essential for phagosomal development [2]. Rab5, a key regulator of early phagosomal maturation, mediates fusion with early endosomes and facilitates the recruitment of downstream effector proteins crucial for vesicular trafficking. As the maturation process progresses, Rab5 is sequentially replaced by Rab7, which promotes the fusion of phagosomes with late endosomal and lysosomal compartments, thereby enhancing their degradative potential. This transition from Rab5 to Rab7 serves as a critical regulatory checkpoint in phagosome maturation, ensuring the spatiotemporal coordination of membrane trafficking and degradative processes [3]. In addition to Rab proteins, other regulators such as phosphoinositides (phospholipid signatures), SNARE (Soluble N-ethylmaleimide-sensitive factor Attachment protein REceptor) proteins, and tethering factors contribute to the orchestration of this process. Phosphoinositides, in particular, are vital in defining membrane identity and recruiting specific effector proteins [4,5]. The composition of phosphoinositides on the phagosomal membrane changes during maturation, with phosphatidylinositol 3-phosphate PI3P being predominant in early phagosomes and phosphatidylinositol 3,5-bisphosphate PI(3,5)P2 increasing in later stages. SNARE proteins are essential for membrane fusion events during phagosome maturation. These proteins form complexes that bring membranes into proximity, facilitating fusion between phagosomes and other vesicles. Tethering factors, such as the homotypic fusion and protein sorting (HOPS) complex, work in concert with SNAREs to ensure the specificity and efficiency of fusion events [5].

The evolutionary pressure exerted by host defense mechanisms has prompted the development of sophisticated evasion strategies among various intracellular bacterial pathogens. For instance, *Mtb* secretes the effector protein SapM, which selectively dephosphorylates host proteins crucial for phagosome maturation, thereby inhibiting its fusion with lysosomes [6]. Similarly, *Legionella pneumophila* (*L. pneumophila*) releases effector proteins such as RalF, which activate the host Arf1 GTPase to facilitate the formation of the *Legionella*-containing vacuole (LCV). The LCV selectively acquires vesicles from the endoplasmic reticulum (ER) and mitochondria, creating an environment conducive to bacterial replication while evading the degradative mechanisms of mature phagolysosomes [7]. Concurrently, various intracellular bacterial pathogens, such as *Listeria monocytogenes* (*L. monocytogenes*) and Shigella flexneri (*S. flexneri*), employ specialized tactics to escape from the phagosome entirely, allowing them to replicate freely in the host cell cytosol. *L. monocytogenes* utilizes the pore-forming toxin listeriolysin O (LLO) and two phospholipases, PlcA and PlcB, to compromise the integrity of the phagosomal membrane, facilitating its escape into the host cytosol [8]. Similarly, *S. flexneri* employs a type III secretion system (T3SS) to deliver effector proteins, such as IpaB and IpaC, which induce perforation of the phagosomal membrane, enabling the cytosolic escape of the bacteria [9]. The evolutionary arms race between hosts and pathogens has significantly influenced the development of phagosomal damage detection and cellular defense mechanisms. Hosts have evolved a diverse array of pattern recognition receptors (PRRs), including toll-like receptors (TLRs), which detect various pathogen-associated molecular patterns (PAMPs) and danger-associated molecular patterns (DAMPs). Additionally, cytosolic sensors, such as cyclic GMP-AMP synthase (cGAS), play a critical role in detecting bacterial DNA from intracellular pathogens like *Mtb* and L. *monocytogenes*

[5]. Furthermore, host cells have developed repair mechanisms to maintain phagosomal integrity and limit pathogen escape into the cytosol. For example, the Endosomal Sorting Complex Required for Transport (ESCRT-III) is a key molecule in repairing membrane damage in *Mtb*-containing phagosomes [10], thereby preventing bacterial translocation into the cytosol. Moreover, mechanisms such as VPS18-mediated membrane repair and ATG7/ATG14-mediated recapture of escaped pathogens actively counteract *Mtb*'s disruption, restoring phagosomal function [11,12].

This chapter examines the complex interactions between *Mtb* and the host phagosomal environment, a critical arena where survival mechanisms and defense strategies are continually evolving. *Mtb* has developed a repertoire of proteins and lipids to influence phagosomal maturation, compromise membrane integrity, and manipulate host antimicrobial processes to ensure its persistence within macrophages. In response, host cells employ a multifaceted defense strategy that includes damage recognition, membrane repair, autophagic clearance, and restoration of phagosomal function. These dynamic host-pathogen interactions are critical not only for understanding tuberculosis (TB) pathogenesis but also for the development of host-directed therapies (HDTs) aimed at enhancing bacterial clearance and immune resilience.

2. Evasion tactics of *Mycobacterium tuberculosis* in host macrophages

The capacity of the *Mtb* to infiltrate host macrophages and reside within pathogen-favorable phagosomal vacuoles, termed mycobacteria containing vacuoles (MCVs), results in the inhibition of phagolysosome maturation. This process is fundamental to the infection, dormancy, reactivation, and dissemination of TB [13]. The phagocytosis of mycobacteria is an actin-dependent process mediated by a diverse array of receptors, including dectin-1, complement receptor 3 (CR3), TLRs, the mannose receptor, dendritic cell-specific intercellular adhesion molecule-3-grabbing nonintegrin (DC-SIGN), Fc receptors, scavenger receptors, and CD1 [14,15]. Following initial entry, *Mtb* resides within the phagolysosome for up to 4 days [16]. This is achieved by maintaining phagosomal acidification at a pH range of 6.3–6.5, facilitated by impaired retention of the VoH + -ATPase complex on the phagosome surface, preventing its tethering to MCV. Additionally, various mycobacterial components, such as lipoarabinomannan (LAM), trehalose dimycolate (TDM), and glycolipids containing acyl trehalose, impede the maturation and acidification process of the phagosome [17]. Consequently, *Mtb* hinder phagosomal maturation by disrupting membrane trafficking mechanisms, targeting host phosphatidylinositols, and interfering with calcium (Ca^{2+}) fluxes, Rab GTPases, and their associated effectors [13,18]. This interference inhibits the fusion of phagosomes with lysosomes and efficient acidification, thereby enabling *Mtb* to persist in immature phagocytic compartments Table 5.1 [19].

2.1 Phagosomal dynamics: Battleground of tuberculosis

The membrane of the MCVs is tightly apposed to the residing bacilli [18,31], resulting in minimal luminal space. A defining characteristic of the mycobacterial vacuole is its sustained retention of Rab5, an early endosomal marker that is typically subjected to elimination during phagosomal maturation, which is also associated with iron acquisition [32]. *M.tb*-mediated manipulation of Ca^{2+} signaling is associated with salient characteristics of the MCVs that are: (a) partial lumen acidification due to reduced levels of active VoH+-ATPase, as well as NADPH oxidase [18,32,33] and (b) inadequate

Table 5.1 Sensing rupture: phagosomal damage detection in *Mtb* infection.

Sensor	Mechanism of detection	Mechanism of*Mycobacterium tuberculosis (Mtb)*-mediated damage detection	References
Galectins (Gal-3, 8, 9)	Bind exposed β-galactoside glycans on damaged phagosomal membranes. Recruit autophagy adaptors and inhibit mechanistic target of rapamycin kinase (mTOR) to initiate autophagy.	*M.tb*'s ESAT-6 secretion system 1 (ESX-1) secretion system disrupts phagosomal membranes, exposing glycans. Gal-8 recruits Tax1 binding protein 1 (TAX1BP1) to target bacteria for xenophagy via ubiquitin (Ub)-microtubule associated protein 1 light chain 3 (MAP1LC3/LC3) interactions.	[20,21]
Ca^{2+} signaling ALG2 alpha-1,3/1,6-mannosyltransferase/ALG-2 interacting protein X (ALG-2/ALIX)	Detect uncontrolled Ca^{2+} leakage from damaged phagosomes. ALG-2 binds Ca^{2+} and recruits ALIX to activate endosomal sorting complex required for transport III (ESCRT-III) for membrane repair.	*Mtb*'s ESX-1 causes phagosomal rupture, triggering Ca^{2+} efflux. ALG-2/ALIX complex recruits ESCRT-III to repair damage or signal for autophagy.	[22]
Autophagy-related 8 (ATG8) E3-like ligases	Sense proton (H^+) or calcium (Ca^{2+}) leakage: Vacuolar-ATPase-autophagy related 16 like 1 (V-ATPase-ATG16L1) complex responds to H^+ leakage. - tectonin beta-propeller repeat containing 1-ATG5-ATG12 (TECPR1-ATG5-ATG12) responds to Ca^{2+} leakage.	*M.tb*-induced phagosomal damage via ESX-1 disrupts ion gradients. ATG8 ligases recruit ESCRT-III for repair or tag damaged compartments for autophagy.	[23]
Cyclic GMP-AMP synthase-cGAS- stimulator of interferon response cGAMP interactor (STING) pathway	Detects cytosolic DNA via cGAS, which synthesizes 2′3′-cGAMP to activate STING and TANK binding kinase 1-interferon regulatory factor 3 (TBK1-IRF3) signaling for type I interferon (IFN-I) production.	ESX-1-mediated phagosomal rupture releases *Mtb* DNA into the cytosol. cGAS-STING activation drives IFN-I responses, which *Mtb* exploits for immune evasion.	[24,25]
NLRs (nucleotide binding oligomerization domain containing 1/nucleotide binding oligomerization domain containing 2/NLR family pyrin domain containing 3) (NOD1/NOD2/NLRP3))	NOD1/NOD2: Detect bacterial peptidoglycan (PG) fragments. NLRP3: Senses K^+ efflux, reactive oxygen species (ROS), or lysosomal damage to activate inflammasome.	*Mtb*'s ESX-1 causes phagosomal damage, triggering K^+ efflux and NLRP3 inflammasome activation. NLRP3 drives interleukin 1 beta (IL-1β) release and pyroptosis, aiding bacterial spread.	[26,27]

Table 5.1 Sensing rupture: phagosomal damage detection in *Mtb* infection.—cont'd

Sensor	Mechanism of detection	Mechanism of*Mycobacterium tuberculosis (Mtb)*-mediated damage detection	References
Polyubiquitination (ring finger protein 115 (RNF115))	Interferon gamma (IFN-γ) enhances RNF115 recruitment to phagosomes, where it suppresses maturation via K48/K63 ubiquitination.	RNF115 ubiquitinates phagosomal proteins in IFN-γ-activated macrophages, delaying acidification and promoting *Mtb* survival. Depletion accelerates lysosomal degradation.	[28]
Tripartite motif containing (TRIM) proteins (TRIM16/32)	TRIM16: Binds Gal-3 on damaged phagosomes, promoting ubiquitination and autophagy. TRIM32: Mediates K63 ubiquitination of *Mtb*, recruiting autophagy-related (ATG) protein and adaptors/receptors.	TRIM16-Gal-3 complex targets ESX-1-damaged phagosomes. TRIM32 deficiency impairs *Mtb* ubiquitination and autophagic clearance, enhancing bacterial growth.	[29,30]

levels of active hydrolases within the lysosome, for instance, Cathepsin D [18,33]. MCVs also exhibit anomalies in the trafficking of plasma membrane markers, including major histocompatibility (MHC) molecules [18,34]. The capacity of mycobacteria to access transferrin-bound iron [18,35], the transferrin receptor enrichment on the mycobacterial phagosome [18,36], and the accessibility to glycosphingolipids indicates that the mycobacterial phagosome is a dynamic vacuole rather than a stable vacuole, despite an impediment in the access to the components of the lysosome. Additionally, MCV membrane exhibits a marked reduction in PI3P levels due to the release of mycobacterial mannose-capped LAM, a lipid that integrates into the phagosome membrane [37]. Thus, these hallmark features of mycobacterial phagosome create a conducive environment, safeguarding the TB pathogen and ensuring its successful escape to the cytosol. This section will elucidate the various stages of phagosome formation, starting with the formation of the phagosomal membrane, progressing through different maturation stages, and culminating in fusion with the lysosome, as illustrated in Fig. 5.1.

Upon internalization into macrophages or other phagocytes, *Mtb* encounters extracellular receptors that trigger a cascade of events within the phagocytes, facilitated by the activation of actin molecules. Consequently, the actin cytoskeleton undergoes reorganization, and the cell membrane extends outward to envelop the mycobacteria. The culmination of these events is the formation of a specialized compartment known as the phagocytic cup [5,38,39]. Upon closure of the phagocytic cup, it develops into a phagosome that subsequently undergoes maturation, acidification, and lysosomal fusion [38,39]. Phagosomal acidification is initiated with the immediate closure of the phagocytic cup [40]. The acidification is predominantly facilitated via VoH+-ATPase pump recruitment thereby regulates phagosomal pH [40,38]. This process induces a substantial reduction in pH within the phagosome, from approximately 6.5 to about 4.5 [40,39].

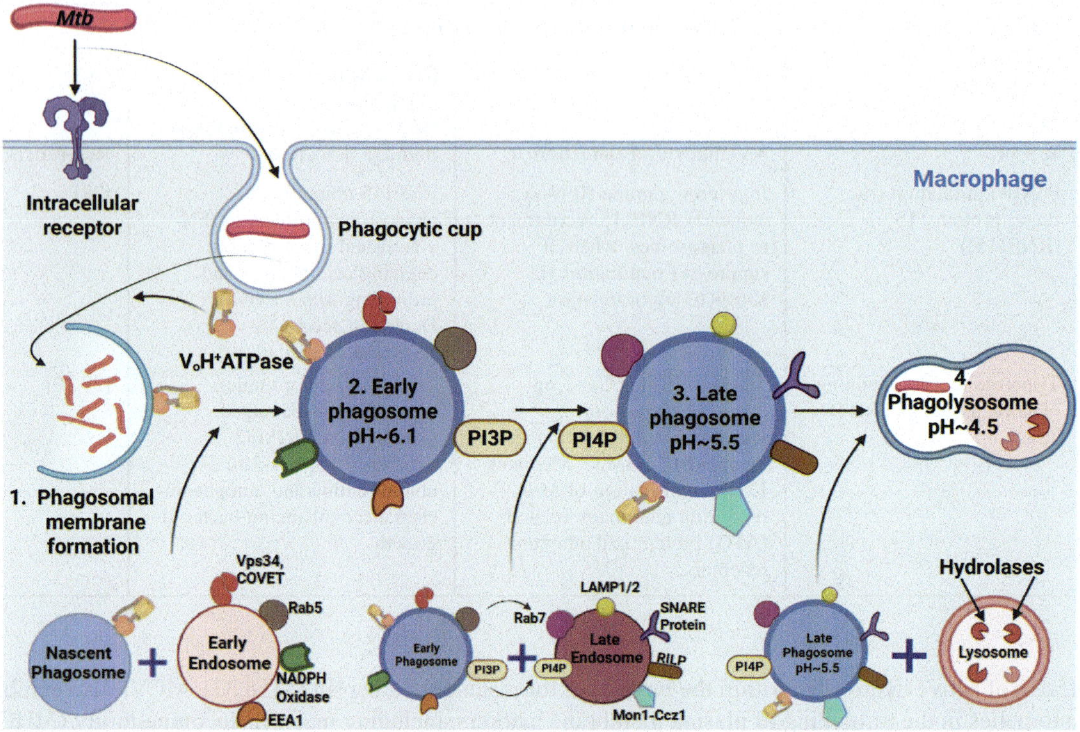

FIGURE 5.1 Molecular mechanism of phagosome formation and maturation in macrophages.

Upon the uptake of *Mycobacterium tuberculosis* (*M.tb*) by macrophages, an initial phagosome is formed, which undergoes a meticulously regulated maturation process crucial for pathogen elimination. Early phagosomes, characterized by a pH of approximately 6.1, acquire membrane components from early endosomes, such as Ras-related protein Rab-5A (Rab5), endosome antigen 1 (EEA1), and nucleotide adenine dinucleotide phosphate hydrogen (NADPH) oxidase, and generate phosphatidylinositol 3-phosphate (PI3P). The vacuolar-type H+-ATPase (V-ATPase) initiates the acidification of the compartment. The transition to a late phagosome, with a pH of approximately 5.5, involves the recruitment of Ras-related protein Rab-7A (Rab7), an increase in phosphatidylinositol 4-phosphate (PI4P), and interaction with late endosomes. Ultimately, fusion with lysosomes results in the formation of phagolysosomes, which have a pH of approximately 4.5 and are enriched with hydrolytic enzymes (hydrolases) necessary for microbial degradation. The V-ATPase plays a critical role throughout this process by facilitating progressive acidification.

ATP hydrolysis drives the transport of 3H+ into the phagosome [41] that is facilitated by ion-exchanging transporters such as CIC-7 and CIC-3 (Cl– Channels), or through cation export channels such as TPC2 (Two Pore Channel 2) and TRPML1 (transient receptor potential mucolipin 1) [41–43]. PIKfyve (phosphoinositide kinase, FYVE-type zinc finger containing)-mediated transition of PI3P to PI(3,5)P2 modulates TRMPL1, TCP-2, and potentially V-H+-ATPase itself. Several pathways facilitate the efflux of H+ from phagosomes, encompassing antiporters (NHE and CHE) and symporters (Natural resistance-associated macrophage protein 1 (NRAMP1) for monovalent and divalent cations/H+, H+-selective channels (Hv1), and amino acid-H+ cotransporters belonging to the SLC (Solute Carrier) family [41]. Phagosomal acidification is further accompanied by the NADPH oxidase complex (NOX) [5,41]. The by-products of the NOX complex utilize H+ to produce microbicidal free hydroxyl radicals, H2O2 (Hydrogen peroxide), and HOCl (Hypochlorous acid) that are catalyzed by phagosomal myeloperoxidase (MPO) [41,44,45]. Moreover, low pH conditions enable the transition of the active state of zymogens (nucleases, lipases, and proteases) into their degradative state [41]. For instance, the transport of pro-cathepsin B to the nascent phagosome then undergoes autoactivation and cleavage, indicating a progression to a more acidic environment within the phagosome [46]. Hence, modulation in the phagosomal microenvironment correlates with the prevention of phagosomal acidification that sustains the intraphagosomal survival of the intracellular pathogens.

2.2 The Rab code: Driving phagosome maturation

Phagosomes maturation is predominantly regulated by the Rab family of small GTPases, particularly Rab5 and Rab7 [47]. Phagosomal maturation involves the transformation of early phagosomes into late phagosomes, which subsequently mature into phagolysosomes [3], as illustrated in Fig. 5.1. Early endosomes, functioning as sorting centers, are characterized by a slightly acidic environment with a pH of approximately 6.1 and limited hydrolase activity (3). These endosomes express maturation markers, including early endosomal antigen 1 (EEA1), Rab5, SNAREs, and phosphoinositide kinases [3,40]. Upon fusion with the early-stage endosomal compartment, the early-stage phagosome acquires active Rab5 (GTP-bound) along with EEA 1 [5,40]. Notably, early endosomes also contain GTPase exchange factors (GEFs) such as Rabex-5, Rabaptin-5, and Rab22a [3,48]. Rab22a-mediated acquisition of Rabex-5 on the early-stage phagosome results in Rab5 activation, as Rabex-5 facilitates the exchange of GDP for GTP [49]. Rabaptin-5 aids in recruiting Class III phosphoinositide 3-kinase vacuolar protein-sorting 34 (Vps34) and enhances the catalytic function of Rabex-5. This interaction establishes a positive feedback loop that amplifies Rab5 activation [50]. Vps34 contributes to the accumulation of PI3P on the membrane of the nascent phagosome and is involved in various processes during the early stages of phagosome development [51]. Synergistically, PI-3-P and active Rab5 direct EEA1 and class C core vacuole/endosome tethering (CORVET) complexes to the phagosomal membrane [52,53]. The fusion between the nascent phagosome and early-stage endosome is mediated by EEA1, which functions to tether the target membrane (3). This fusion is further facilitated by the interaction between EEA1 and SNARE proteins Syntaxin6 and Syntaxin13 [54]. The CORVET complex, comprises six subunits, including four core components (Vps11, Vps16, Vps18, and Vps33) and two specific units (Vps3 and Vps8). Notably, Vps33 facilitates the attachment of soluble SNAREs to the phagosomal membrane, thereby promoting membrane fusion [55].

2.3 HOPS and SNAREs: Orchestrating phagosome-lysosome fusion

The maturation of the phagosome involves a transition from a nascent phagosome enriched with PI3P to a more acidic late phagosome through interactions with PI4P-abundant late-stage endosomes [3]. The synthesis of PI4P, catalyzed by phosphatidylinositol 4-kinase 2a (PI4K2A), is an essential step for the activation of Rab7 and functions directly as a precursor for the production of phosphatidylinositol 4,5-bisphosphate (PI4,5P2) on phagolysosomes [51]. The Rab conversion process is critically influenced by the MON1-CCZ1 complex, which acts as an intermediary, facilitating the transition from

Rab5 to Rab7 on the phagosomal membrane [56,57]. Initially, MON1 displaces Rabex5 on the phagosomal membrane, thereby terminating the self-perpetuating loop that activates Rab5 [3,58]. Subsequently, MON1 forms a dimeric complex with CCZ1, which serves as a Rab7 guanine nucleotide exchange factor (GEF). This MON1-CCZ1 dimer engages in a sequential interaction with Rab5 in its GTP-bound state, resulting in the dissociation of Rab7 from the GDP dissociation inhibitor (GDI). Consequently, this enables the recruitment of numerous Rab7 molecules to the phagosomal membrane [3]. Furthermore, the transition from Rab5 to Rab7 depends on a six-subunit tethering complex known as homotypic fusion and vacuole sorting (HOPS), which functions to activate Rab7 [59]. The HOPS complex mediate the Rab5 to Rab7 transition, thereby facilitating the maturation of the phagosome [54]. Moreover, SNARE proteins, such as vesicle-associated membrane protein 7 and 8 (VAMP7, VAMP8), in conjunction with syntaxin7 and syntaxin8, facilitate the proximal positioning of late phagosomes and lysosomes to enable their fusion [60]. The matured phagosome (late-stage) incorporates lysosome-associated membrane proteins 1 and 2 (LAMP1, LAMP2) along with various hydrolytic enzymes which are essential for late phagosome-lysosome fusion [61]. Ultimately, the fully matured phagosome maintains a highly acidic environment of approximately pH 5.5, along with increased destructive capacity [3].

2.4 The terminal phase: Pathogen clearance in phagolysosomes

Phagosome maturation culminates in the formation of phagolysosomes, characterized by a high concentration of degradative components [3]. These phagolysosomal compartments emerge from the fusion of fully matured phagosomes with lysosomes. This fusion process necessitates the migration of both fully matured phagosomal and lysosomal vesicles toward the microtubule-organizing center (MTOC) [47]. When Rab7 is bound to GTP, it facilitates the recruitment of additional proteins to late-stage phagosomes, including Rab-interacting lysosomal protein (RILP) and oxysterol-binding protein-related protein 1L (ORP1L) [62]. RILP, as a downstream effector of Rab7, acts as an adaptor for dynein, with its interaction with dynein being modulated by ORP1L. By promoting the mobilization of the dynein-dynactin motor complex, RILP enables the association between Rab7-enriched endocytic vesicles and tubulin-based intracellular scaffolds [63]. Subsequently, Rab7-containing late phagosomes are transported directionally along microtubules toward the negative terminus. This translocation is crucial for phagolysosome biogenesis, characterized by the fusion of Rab7-positive late phagosomes with lysosomal compartments [64]. These structures contain and secrete a diverse array of antimicrobial peptides that function optimally in acidic environments, along with various hydrolytic enzymes. The enzymatic repertoire includes glycosidases, DNases, cathepsins, proteases, lysozymes, and lipases [3,5]. The final phase of pathogen elimination occurs within phagolysosomes, which create an environment specifically designed for the destruction and breakdown of invading microorganisms.

3. Effector-mediated disruption of phagosomal maturation by *Mycobacterium tuberculosis*

As previously outlined, mycobacterial phagocytosis is an actin-mediated process. Notably, this actin polymerization occurs in a manner dependent on the Wiskott-Aldrich syndrome protein and SCAR Homolog (WASH), a multi-subunit complex that facilitates actin nucleation on endosomal membranes via the Arp 2/3 complex and protects mycobacteria from exposure to a bactericidal environment [65]. *Mtb*, a classical pathogen that consistently maintains its evolutionary pace, employs a diverse array of effector molecules to impede phagosome maturation at various stages. These effector molecules effectively disrupt the fusion between phagosomes and lysosomes while simultaneously interfering with the acidification of the phagosomal environment [66,67]. This section will examine the phagosomal damage resulting from phagosomal maturation arrest (PMA), caused by various mycobacterial effectors and mycolipids, which leads to PMA and thereby facilitates the intracellular survival of mycobacteria.

The repertoire of mycobacterial proteins responsible for disrupting phagosome-lysosome fusion includes ESX-1, SapM, Zmp1, PtpA, EsxG-EsxH, and EsxP, as depicted in Fig. 5.2. ESX-1, a type VII secretion system (T7SS), is central to mycobacterial virulence [68]. ESX-1 enhances pathogenicity through multiple mechanisms, including the activation of cytosolic signaling pathways, induction of macrophage death, and acceleration of tuberculous granuloma development. This acceleration occurs via MMP9 induction, which subsequently promotes mycobacterial growth [68–70]. The virulent ESX-1 secretory system of *Mtb* also functions to disrupt phagosome integrity and hinder its maturation process [68]. The membrane-disrupting function of ESX-1 has been attributed to its primary secreted component, ESAT-6 [71]. An independent line of investigation utilized purified recombinant ESAT-6 to elucidate its potential role in membrane lysis. In its purified state, ESAT-6 demonstrated lytic activity against artificial lipid bilayers, liposomes, and erythrocytes. Conversely, its co-secreted counterpart CFP-10 did not exhibit such effects, suggesting that ESAT-6 functions as a conventional pore-forming toxin [68,72,73]. Recent evidence indicates that ESAT-6 at its C-terminal tail comprises highly conserved residues, specifically methionine 83 (M83) and methionine 93 (M93). Consequently, modifications at these sites (such as M83I and M93T) result in a significant reduction in phagosomal membrane damage and attenuated virulence, despite these variants still being secreted in substantial quantities.

This observation suggests that the structural integrity of the C-terminus is essential for ESAT-6's functionality, extending beyond its role in secretion alone. The C-terminal region of ESAT-6 potentially contributes to its oligomerization or interactions with other proteins. Oligomerization entails the aggregation of multiple ESAT-6 molecules to form a functional unit. This assembled structure may be essential for generating pores or inducing disruptions in the phagosomal membrane. The C-terminus could facilitate the arrangement or stabilization of these oligomers, thereby enhancing their capacity to compromise membrane integrity. In vitro experiments utilizing acidified liposomes, a model system for investigating membrane disruption, have demonstrated that ESAT-6 exhibits the capacity to induce liposomal lysis. Remarkably, C-terminal variants such as M93T retained their ability to rupture acidified liposomes, indicating core membranolytic activity remains unaffected. Nevertheless, these mutations did not exhibit a correlation with in-vivo phagosomal damage, highlighting the interaction between the C-terminal end of ESAT-6 and other mycobacterial factors or host cell constituents within the phagosomal environment [74]. Although the precise mechanism remains elusive, ESAT-6 may contribute to stabilization within the membrane, potentially augmenting its membrane-damaging properties.

Recent research has elucidated that SapM, a secreted acid phosphatase, facilitates phagosomal permeabilization through the ESX-1 secretion system. Notably, SapM has also been shown to arrest phagosome maturation independently of ESX-1. The primary substrate of SapM is PI3P, a lipid located on the cytosolic side of the phagosomal membrane. SapM dephosphorylates PI3P, a critical process for

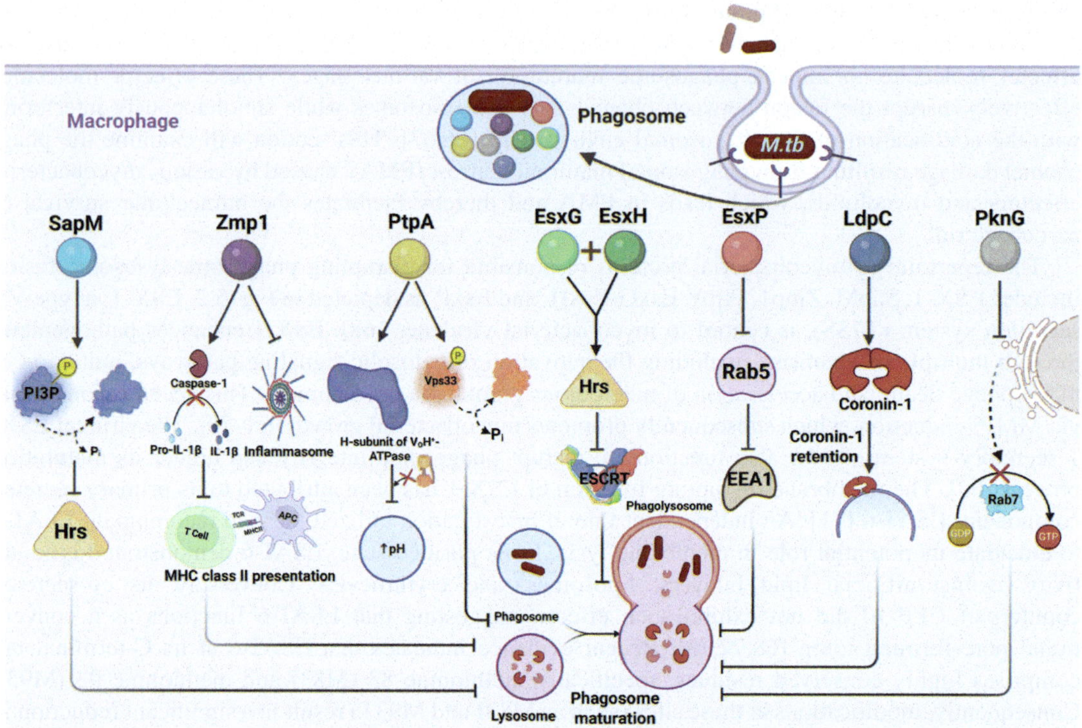

FIGURE 5.2 Mycobacterial effector-mediated inhibition of phagosome maturation in infected macrophages.

Mycobacterium tuberculosis (*M. tb*) produces a series of effector proteins that disrupt host cell trafficking and signaling pathways to suppress phagosome maturation and support intracellular persistence. SapM dephosphorylates phosphatidylinositol 3-phosphate (PI3P), inhibiting the recruitment of host proteins such as hepatocyte growth factor-regulated tyrosine kinase substrate (Hgs/Hrs) and preventing phagosomal trafficking. Zmp1 suppresses inflammasome activation and interleukin 1-beta (IL-1β) secretion, disrupting major histocompatibility (MHC) class II antigen presentation and T cell activation. PtpA dephosphorylates host VPS33 late endosome and lysosome associated (Vps33) and suppresses assembly of V-ATPase on the phagosome membrane, increasing intra-phagosomal pH and disrupting lysosomal fusion. EsxG and EsxH, which are members of the ESAT-6 secretion system 3 (ESX-3) secretion system, prevent endosomal sorting complex required for transport (ESCRT) complex recruitment by Hrs, inhibiting endosomal sorting and maturation. EsxP interferes with Ras-related protein Rab-5A (Rab5) activity and its downstream effector early endosome antigen 1 (EEA1), thus inhibiting early phagosome progression. LdpC facilitates Coronin-1 (CORO1A) retention, which prevents phagolysosome formation. PknG, mycobacterial serine/threonine kinase, prevents activation of Ras-related protein Rab-7A (Rab7) through GDP-GTP exchange blocking, inhibiting late endosomal maturation.

phagosome maturation. PI3P is instrumental in recruiting proteins such as EEA1 and hepatocyte growth factor-regulated tyrosine kinase substrate (Hrs), which are essential for the fusion of early phagosomes with late endosomes and lysosomes. By dephosphorylating PI3P, SapM disrupts these fusion events, thereby inhibiting the maturation of the phagosome into a phagolysosome. ΔsapM mutants in both *Mtb* and BCG demonstrate a defect in PMA, which correlates with a significant increase in localization to acidified phagosomes. This observation suggests that SapM can access the macrophage cytosol independently of ESX-1. Furthermore, the PMA defect in ΔsapM mutants results in attenuated growth within macrophages. Consequently, SapM is identified as a crucial effector molecule capable of functioning independently of the ESX-1 secretion system, revealing a novel strategy employed by *Mtb* to manipulate host cell processes and ensure its intracellular persistence and proliferation.

Zmp1, a putative Zn^{2+} metalloprotease, demonstrates significant immunogenicity and is secreted under conditions that simulate granuloma stress. It is involved in modulating macrophage innate defense by suppressing inflammasome activation and inhibiting IL-1β processing in a caspase1-dependent manner, thereby arresting phagosome maturation [75,76]. Notably, the *Mtb* strain deficient in Zmp1 expression exhibited an inability to inhibit IL-1β processing and phagosome maturation, leading to reduced survival in both macrophages and mice compared to the wild-type strain. This highlights the potential role of Zmp1 in suppressing inflammasome activation and halting phagosome maturation, contributing to the virulence of *Mtb*. Given that phagosomal vesicle maturation restricts antigen presentation via the class-II MHC pathway, this metalloprotease is implicated in impaired mycobacterial antigen presentation. The deletion of Zmp1 enhances MHC class II-mediated mycobacterial antigen presentation and increases the immunogenic potential of BCG in both in vitro and in vivo conditions, suggesting an inhibitory role of Zmp1 in the downregulation of MHC class II-mediated presentation of mycobacterial antigens and modulation of adaptive immunity [76]. Thus, Zmp1 functions as a versatile mycobacterial effector, contributing to its persistence and propagation within macrophages via PMA.

PtpA, a low molecular weight tyrosine phosphatase, employs two primary mechanisms to inhibit the fusion of phagosomes with lysosomes [65]. Firstly, PtpA dephosphorylates VPS33B late endosome and lysosome associated (VPS33B), a host protein crucial to the phagosome-lysosome fusion process, thereby disrupting its normal function. This dephosphorylation impairs the maturation of phagosomes, which is essential for the degradation of internalized mycobacterial bacilli [75]. The second mechanism involves PtpA binding to the H subunit of the VoH+-ATPase complex, thereby obstructing its transport to the phagosome. This interaction inhibits the acidification of the phagosome, a necessary step for its maturation. By preventing both the maturation and acidification of phagosomes, PtpA enables the bacteria to evade the host's degradative pathways, thereby contributing to the persistence and pathogenicity of *Mtb* [66].

Mtb employs T7SS effectors, specifically EsxG in conjunction with EsxH, to target and disrupt the host's ESCRT machinery. This disruption impairs the trafficking of mycobacteria to lysosomes, thereby promoting intracellular survival. EsxH, in association with EsxG, binds to Hrs and disrupts ESCRT function, which impairs phagosome maturation and prevents the delivery of *Mtb* to lysosomes, thus enhancing bacterial survival within macrophages [67]. Recent findings indicate that EsxG-EsxH respond to membrane damage independently of Ca^{2+} and ESCRT-III recruitment, suggesting their capability to detect damaged membranes. EsxG and EsxH antagonize the mobilization of ESCRT-III to damaged sites within endolysosomal compartments, as evidenced by the enhanced recruitment of ESCRT-III to phagosomes in the ΔesxH mutant strain [68]. Consequently, *Mtb* utilizes components of its secretory system to damage host phagosomes and evade the endomembrane repair response mediated by the ESCRT machinery.

EsxP, a T7SS effector molecule, plays a crucial role in promoting mycobacterial survival within host cells through a series of complex mechanisms. Initially, EsxP inhibits the expression of early phagosome markers Rab5 and EEA1, thereby preventing the maturation of phagosomes into lysosomes. This inhibition facilitates the escape of *Mtb* from phagosomes into the cytosol, thus avoiding the harmful acidic environment of lysosomes. Once *Mtb* enters the cytoplasm, it is detected by host cytoplasmic sensors, leading to the activation of the stimulator of interferon genes (STING) pathway, which subsequently recruits and phosphorylates TANK-binding kinase 1 (TBK1). The phosphorylation of TBK1 promotes the transcription of interferon-β (IFNβ), a critical factor that enhances bacterial survival by modulating the host immune response. In addition to blocking phagosome maturation, EsxP inhibits autophagy, a vital cellular process that degrades cytoplasmic components to limit the intracellular invasion of the TB pathogen. EsxP obstructs the LC3I to LC3II transition and promotes the accumulation of the autophagy substrate sequestosome1 (SQSTM1/p62). This autophagy inhibition is mediated through the EsxP-dependent activation of the STING/TBK1 pathway. The expression levels of autophagy-related proteins calcium binding and coiled-coil domain 2 (CALCOCO2/NDP52), SQSTM1 and LC3 are significantly decreased in the presence of EsxP, further indicating the suppression of autophagy Fig. 5.2.

4. SecA2-secreted effectors in phagosome maturation arrest

The mycobacterial effectors secreted via the SecA2 pathway, such as LdpC and PknG, also inhibit phagosome maturation. Lipoamide dehydrogenase (LdpC), released through the SecA2 pathway, plays a crucial role in the metabolism of branched-chain amino acids [69]. The cytosolic localization of LdpC enables its interaction with the host's CORO1A protein, resulting in its sustained presence on the membrane of the MCV, thereby impeding the maturation process of the phagosome [70,71]. Similarly, PknG (serine/threonine kinase) is another SecA2 pathway secretory molecule that localizes to the host's cytosol, where it continues to obstruct phagosome maturation [71–73]. Primarily, PknG targets the host's Rab GTPase Rab7L1 (known as Rab29 in mice) and, at the Golgi interface, interferes with the GDP-to-GTP exchange necessary for the activation of Rab7L1. Therefore, it prevents Rab7L1-GTP trafficking to the MCV, contributing to the inhibition of phagosome maturation [77]. Other mycobacterial effectors, such as mag 24-1 (orthologue of PE_PGRS30 in *M. marinum*), exhibit significant upregulation in the mycobacterial bacilli residing within the granuloma, and deletion of mag24-1 leads to a reduction in the virulence of mycobacteria in both ex vivo and in vivo models [78]. Furthermore, Mag 24-1 contributes to the mycobacteria's ability to impede phagosome maturation, likely through the exclusion of the VoH+-ATPase from the MCV [79].

5. Structural and functional complexity of the *Mycobacterium tuberculosis* cell wall

The current model delineates a tripartite structure for the cell envelope of *Mtb*, comprising an external layer (referred to as the capsule in pathogenic species), a core cell wall, and a conventional plasma membrane composed of phospholipids [80]. The essential components of the cell wall core include the mycomembrane, peptidoglycan (PG), and arabinogalactan (AG), with PG and AG being covalently bonded [81–83]. The mycomembrane is characterized by an asymmetrical dual-layer structure: the inner layer consists of α-alkyl, β-hydroxy long-chain mycolic acids covalently linked to AG, while the outer layer is composed of free and non-covalently associated lipids and glycolipids. These include trehalose monomycolates (TMM), trehalose dimycolate (TDM), glycerol monomycolates, glucose monomycolates, phthiocerol dimycocerosates (PDIM), poly-acylated threaloses (PAT), sulfolipids (SL), phosphatidylinositol mannosides (PIM), phenolic glycolipids (PGL), and mannose-capped lipoarabinomannans (Man-LAM). These outer leaflet mycolipids play a critical role in enhancing the TB pathogen's ability to modulate the immune response of its host [81,83,84], as illustrated in Fig. 5.3.

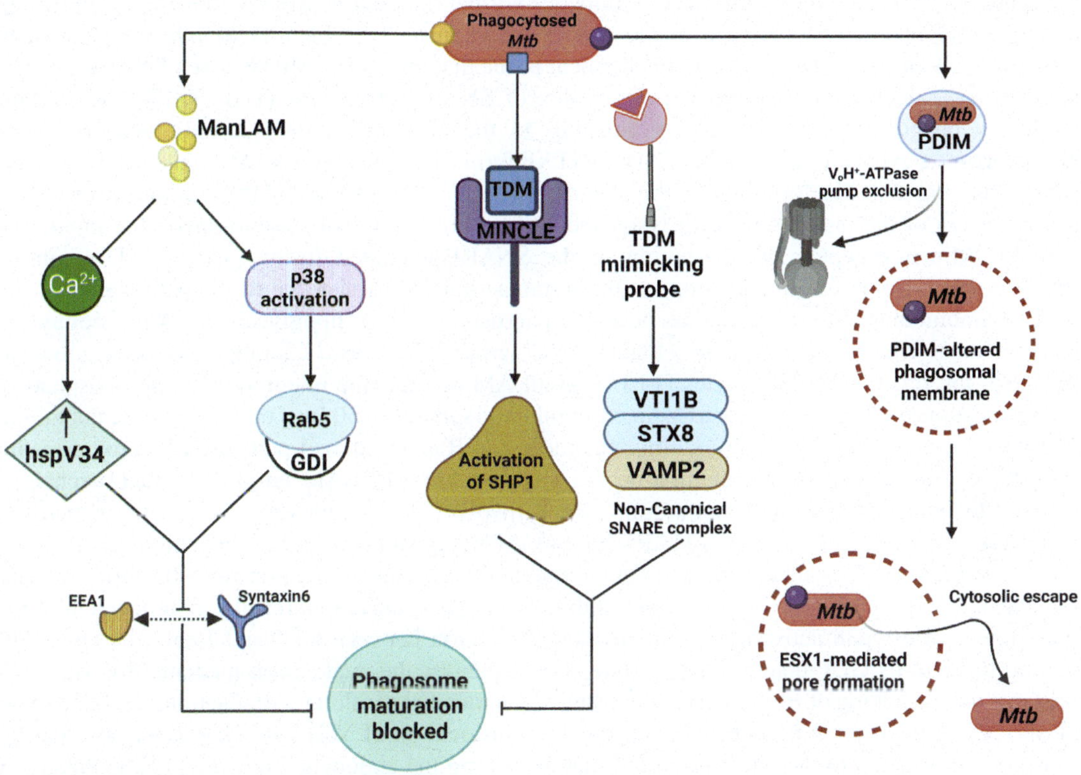

FIGURE 5.3 Mycolipids in phagosome maturation arrest and escape of *Mycobacterium tuberculosis* into the cytosol.

Mycobacterium tuberculosis (*Mtb*) employs mycolipids to circumvent host immune defenses. Mannosylated lipoarabinomannan (ManLAM) disrupts calcium signaling and p38 mitogen-activated protein kinase (MAPK) activation, thereby inhibiting phagosome maturation. Trehalose dimycolate (TDM) activates the macrophage-inducible C-type lectin (MINCLE)-SHP1 signaling pathway and forms a non-canonical soluble N-ethylmaleimide-sensitive factor-attachment protein receptors (SNARE) complex, which impedes vesicle fusion. Additionally, phthiocerol dimycocerosate (PDIM) alters the phagosomal membrane, preventing the inclusion of vacuolar-ATPase (V-ATPase) and inhibiting acidification, while also facilitating ESAT secretion system 1 (ESX-1)-mediated escape into the cytosol.

TDM, commonly referred to as the cord factor, is a critical and predominant mycolipid known for its high toxicity and granuloma-inducing potential [85]. The unique colony morphology of virulent mycobacterial strains is attributed to TDM [86]. As the principal lipid present on the surface of *Mtb*, TDM demonstrates the remarkable capacity to transition between distinct biological activities, depending on its physical structure. This adaptability suggests that TDM may influence TB pathogenesis in various ways, potentially affecting the disease's progression across different infection stages [85]. Notably, TDM enhances the intramacrophage survival of the TB pathogen by interfering with phagosome maturation processes [84]. The interaction of TDM with the Monocyte-Inducible C-type Lectin (Mincle) receptor results in the activation of SH2-domain-containing inositol polyphosphate 5′ phosphatase (SHP-1), which subsequently impedes normal maturation progression [87], as illustrated in Fig. 5.3. Recent studies have revealed a novel mechanism by which TDM inhibits phagosome formation, involving a TDM-mimicking chemical probe that interacts with the host's SNARE proteins in a non-canonical manner. Typically, a complex of SNARE proteins VTI1B, STX8, and VAMP8 facilitates phagosome-lysosome fusion, leading to the elimination of ingested microorganisms. However, upon exposure to TDM, VTI1B and STX8 form a complex with VAMP2 instead of VAMP8, as shown in Fig. 5.3. The formation of this non-canonical VTI1B-STX8-VAMP2 complex prevents the proper fusion of the phagosome with lysosomes [88]. Through a dual strategy involving interaction with the Mincle receptor and disruption of the SNARE-mediated fusion process, TDM inhibits phagosome-lysosome fusion and supports the intramacrophage persistence of the pathogen.

Lipoarabinomannan, a surface-associated macroamphiphilic lipoglycan of *Mtb*, commonly referred to as ManLAM due to its characteristic terminal mannose capping, has been shown to modulate the function of phagocytes [13,14]. ManLAM disrupts the integration of the phosphatidylinositol 3-kinase hVPS34 and its associated regulatory subunit p150 into the phagosome containing *Mtb* by inhibiting both Ca^{2+}/calmodulin-dependent and Rab5-mediated pathways, thereby arresting the progression of phagosomal maturation [13,18]. Notably, ManLAM employs a dual-mechanism strategy to impair phagosomal maturation, which involves the attenuation of Ca^{2+} level increases alongside the stimulation of the p38 MAP kinase (MAPK) signaling cascade. The inhibition of Ca^{2+} is unique to ManLAM, as uncapped lipoarabinomannan from non-pathogenic mycobacteria does not impede the increase in Ca^{2+} ions. The inhibition of Ca^{2+} elevation results in reduced hVPS34 recruitment, while p38 activation facilitates the association between GDI and Rab5, thereby maintaining Rab5 in its GDP-bound inactive state. Consequently, the dual action mediated by ManLAM prevents the trafficking of EEA1 within the phagosome membrane, along with the trans-Golgi network (TGN) SNARE Syntaxin 6, a critical step for the translocation of VoH^+-ATPase and lysosomal hydrolases (cathepsins) to the phagosome, ultimately halting the endocytic pathway [13]. Collectively, these mechanisms highlight the multifaceted strategies employed by ManLAM in modulating phagosomal transport and maturation, thereby affirming its central role in the pathogenesis of *Mtb*.

The category of mycolipids includes PDIM and their glycosylated variants, the phenol glycolipids (PGLs). PDIMs are exclusively synthesized by pathogenic strains within the *Mtb* complex. These lipids serve as critical mediators in the interaction between the pathogen and the host. Mycolipids facilitate mycobacterial proliferation by inhibiting the acidification of the phagosome. Recent studies have identified Rv3167c, a transcriptional repressor, as a regulator of PDIM expression. The interaction between Rc3167c and a palindromic DNA sequence located in the intergenic region between its promoter and the adjacent PDIM operon results in the repression of genes responsible for PDIM synthesis and transport. The deletion of Rv3167c is associated with increased expression of the PDIM operon and enhanced PDIM production, which facilitates the phagosomal escape of *Mtb*. Notably, PDIM and EsxA function in a coordinated manner as 'mycobacterial saboteurs' to breach the phagosomal barrier. PDIM-mediated alterations in the biophysical properties of the phagosomal membrane compromise its integrity, rendering it more susceptible to EsxA-mediated pore formation, thereby facilitating the escape of *Mtb* from the phagosome. With its multifaceted influence on phagosomal escape and other aspects, PDIM emerges as a key lipid virulence determinant in mycobacterial pathogenesis, underpinning its survival and dissemination.

5.1 Xenophagy and LC3-associated phagocytosis in host response to phagosomal damage

Autophagy is a critical innate immune defense mechanism that targets *Mtb*-containing phagosomes for degradation in macrophages [89]. It functions as a lysosomal degradative process that enhances phagosomal maturation, restricting intracellular *Mtb* growth. Thus, autophagy activation due to phagosome damage is not only a containment measure but also a vital immune surveillance mechanism. In the context of phagosomal damage by *Mtb*, autophagy can be triggered through two main pathways: xenophagy, which is activated following the detection of ESX-1-mediated phagosomal damage, and LC3-associated phagocytosis (LAP), which occurs independently of phagosomal damage [89]. Notably, the activation of both LAP and xenophagy pathways leads to acidification, lysosomal fusion, and ultimately the elimination of the mycobacteria [89]. The LAP process is activated upon phagocytosis of bacteria following the stimulation of cell surface PRRs, such that LC3 binds to the phagosomal membrane, a phagosome-like structure referred to as a LAPosome [90]. Interestingly, *Mtb* RNA within extracellular vesicles from infected macrophages, in combination with IFN-β, can also induce LAP through RNA sensor RIG-I/mitochondrial antiviral signaling protein (RIG-I/MAVS) signaling cascade. *Mtb's* ESX-1 secretion system induces phagosomal membrane rupture, allowing for the sensing of PAMPs by the cytosolic sensor cGAS, triggering STING-dependent xenophagy mediated by ubiquitin signaling Fig. 5.4 [89,91]. The E3 ubiquitin ligases Parkin (PRKN) and SMAD specific E3 ubiquitin protein ligase 1 (Smurf1) are required for this ubiquitin-tagged autophagic response against *Mtb* [92,93]. After the bacterial phagosome is ubiquitinated, autophagy receptors like SQSTM1 and CALCOCO2 bind to the ubiquitin marks and, via their LC3-interacting domains, enable the recruitment of these phagosomes to autophagosomes, resulting in their destruction in lysosomes Table 5.1 [91].

5.2 PknG, SapM, and CpsA: Coordinated disruption of host antimicrobial defenses

PknG functions as a 'moonlighting mycobacterial effector' by inducing autophagy while simultaneously inhibiting autophagosome formation. PknG competitively binds to the pleckstrin homology (PH) domain of AKT through its C-terminal region (amino acids 569–750). This competitive binding inhibits the phosphorylation and activation of AKT, which subsequently suppresses the activation of mTOR and the induction of autophagy. PknG directly interacts with the host small GTPase RAB14 via its tetratricopeptide repeat (TPR) domain. This interaction prevents RAB14-GTP hydrolysis, maintaining RAB14 in its active GTP-bound state. PknG inhibits RAB14 activation by sequestering its

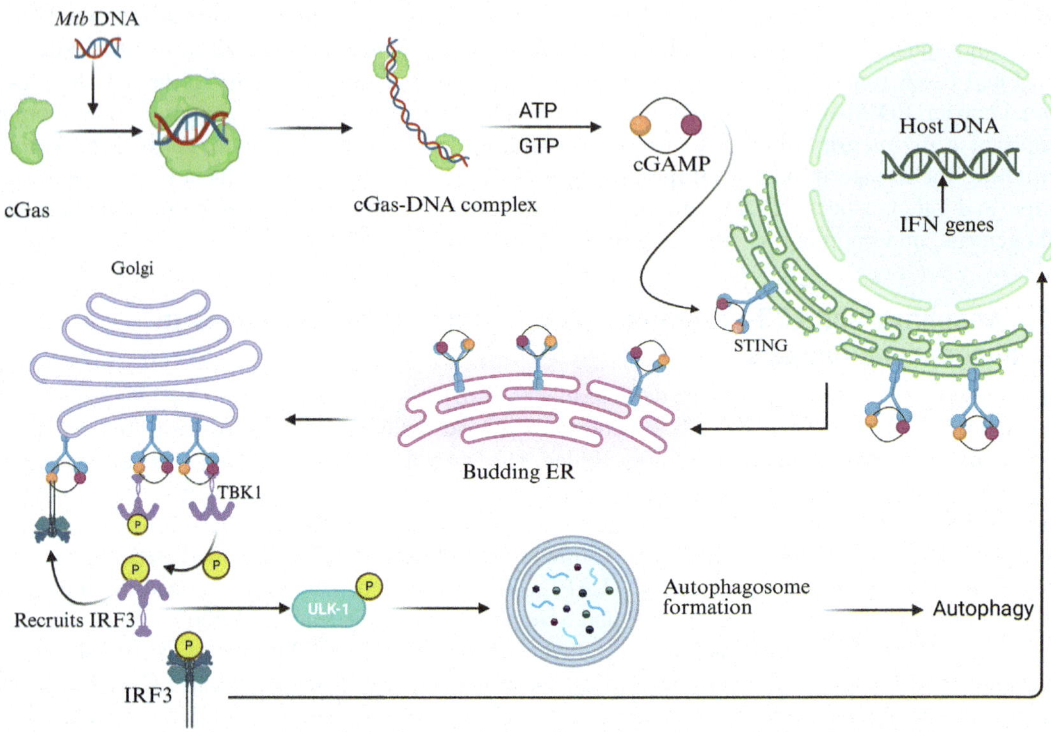

FIGURE 5.4 The cGAS-STING pathway: activation of the cGAS–STING pathway and its role in autophagy induction.

Cytosolic DNA of endogenous or pathogenic origin is sensed by cyclic GMP-AMP synthase (cGAS), which initiates the formation of cGAS-DNA liquid-phase droplets. These biomolecular condensates catalyze the synthesis of 2′3′-cyclic GMP-AMP (cGAMP) from ATP and GTP. The cGAMP molecule subsequently binds to the adaptor protein stimulator of interferon genes (STING), triggering its translocation from the endoplasmic reticulum (ER) to the Golgi and post-Golgi compartments. During this translocation, STING facilitates the recruitment and activation (autophosphorylation) of TANK binding kinase 1 (TBK1), which in turn phosphorylates interferon regulated factor 3 (IRF3). Phosphorylated IRF3 translocates to the nucleus to promote the transcription of type I interferons (IFN-I) and proinflammatory cytokines. Concurrently, TBK1 phosphorylates autophagy regulators such as unc-51 like autophagy activating kinase 1 (ULK1) and optineurin (OPTN), thereby initiating autophagosome formation and linking DNA sensing to autophagy activation.

GTP-bound state and blocking hydrolysis. It also inhibits the action of the GTPase-activating protein TBC1D1 inhibiting TBC1D4's GAP activity and obstructs the maturation of autophagosomes. Notably, SapM and PknG, working synergistically to impede the maturation process of both phagosomes and autophagosomes. Together, they maintain phagosomes and autophagosomes in an immature state, preventing their fusion with lysosomes and promoting *Mtb*'s intracellular survival and replication. *Mtb* prevents the conversion of autophagosomes to autophagolysosomes, resulting in increased levels of autophagy flux. Through their synergistic inhibition of endosomal trafficking and autophagic flux, PknG and SapM sustain a non-degradative intracellular niche, allowing *Mtb* to evade immune surveillance and maintain replication within host macrophages. CpsA, a protein containing the LytR-CpsA-Psr (LCP) domain, plays a critical role in safeguarding *Mtb* from the host's innate immune defenses. CpsA impedes the recruitment of NADPH oxidase components (gp91phox, p40phox, and p47phox) to the phagosomes harboring *Mtb*, as evidenced by the ΔcpsA mutant, which exhibited increased recruitment of NADPH oxidase to the phagosomes. This recruitment results in elevated ROS production and subsequent bacterial eradication [94]. LAP involves the recruitment of LC3 to phagosomes, facilitating their fusion with lysosomes and subsequent bacterial degradation. CpsA obstructs the association of LC3 with *Mtb*-containing phagosomes, thereby inhibiting LAP. This is demonstrated by the increased colocalization with LC3 and lysosomal markers (LAMP1) in the ΔcpsA mutant, indicating enhanced lysosomal trafficking and bacterial clearance [94].

6. Phagosomal integrity and host surveillance mechanisms

The damage to phagosomes, lysosomes, or the endo-lysosomal compartment can be highly dangerous to the cell, resulting in the release of ions and protons, imbalancing the cellular microenvironment. This can directly damage the cellular components and activate cell death pathways and inflammatory signals. Phagosomal membrane integrity is crucial for the proper containment and degradation of internalized pathogens. The detection of phagosomal damage represents a critical component of innate immunity, as many pathogens attempt to compromise phagosomal membranes to escape degradation. For instance, in the case of *Mtb,* the disruption of phagosome damages the phagosomal integrity and leads the pathogenic H37Rv to enter the cytosol and access the lumen. The consequences of ESX-1-induced phagosome damage and cytosolic access by pathogenic *Mtb* can lead to activation of several cell death pathways [95]. This section discusses several sophisticated surveillance mechanisms that detect breaches in phagosomal membrane integrity and coordinate appropriate cellular responses.

6.1 Galectins as first responders to phagosomal damage

Galectins have emerged as an essential initial sensors of phagosomal damage. Phagosomal damage activates galectins (Gals), which are carbohydrate-binding proteins located in the cytosol. Galectins are specialized in recognizing β-galactoside glycans of proteins and lipids. Under normal conditions, these glycans are confined to the lumen of phagosomes and are inaccessible to cytosolic proteins. Upon membrane damage, galectins can access and bind to these previously concealed glycans, serving as molecular indicators of membrane integrity loss. Galectins and ubiquitin (Ub) generate a "danger signal" that recruits autophagy receptors and mediates downstream degradation Table 5.1. The autophagy receptor (p62/NBR1 autophagy cargo receptor (NBR1), etc.) interacts with Gals3, 8, 9, and ubiquitinated cargo, as well as the lipidated LC3 present on the phagophore membrane, and fuses with lysosomes to degrade the tagged cargos. Gal-8 exerts an inhibitory effect on the mTOR, a key regulator of autophagy, while Gal-3 modulates the ESCRT machinery, facilitating the repair of lysosomal membrane damage. These findings suggest that cytosolic Gals are integral players in the response to lysosomal damage, a process essential for the preservation of lysosomal homeostasis. This reveals the broader functional significance of Gal proteins in maintaining cellular integrity under stress conditions. Research identifies Gals, particularly Gal-8, as the principal direct sensors of phagosomal damage. Studies demonstrate that Gal-3, -8, and -9 are all employed to the same *Mtb* population that colocalizes with selective autophagy markers (ubiquitin, SQSTM1, and LC3). The association between Gal-8 and Tax1 binding protein 1 (TAX1BP1) is novel and specific. The effective recruitment of TAX1BP1 to *Mtb* depends on its binding to Gal-8 as well as its ubiquitin-binding domain. This suggests a complex mechanism whereby Gal-8 recognizes damaged phagosomes and subsequently recruits TAX1BP1, which in turn interacts with ubiquitinated proteins to target the bacteria to selective autophagy.

6.2 ALG-2–ALIX: Calcium-sensing duo in damage detection

Calcium signaling constitutes a vital mechanism for detecting phagosomal damage. Initially, the phagophore is characterized by a high concentration of calcium sourced from the extracellular environment, establishing a gradient between the phagosomal lumen and the cytosol. Recent investigations utilizing advanced calcium-sensing probes have elucidated the dynamic nature of calcium movements across phagosomal membranes. Studies employing fluo4-coupled phagocytic targets have demonstrated that the initially elevated calcium concentration (millimolar range) within newly formed phagosomes diminishes to cytosolic levels (approximately 100 nM) within minutes following phagosomal closure. This reduction occurs via calcium-permeable channels integrated into the phagosomal membrane during phagocytosis, which effectively channel the phagosomal calcium into the surrounding cytosol [96]. Under normal physiological conditions, calcium efflux transpires in a regulated manner across the phagosomal lumen; however, when phagosomal membrane damage is detected, calcium ions leak uncontrollably into the cytosol. This unregulated calcium efflux serves as a crucial indicator of membrane damage. Cells can detect damage through various mechanisms, with the identification of calcium leakage serving as the primary detection method alongside Gal-mediated detection. Several calcium-binding proteins function as sensors to detect calcium leakage. Among these, the apoptosis-linked gene-2 (ALG-2) protein and its binding partner ALIX (ALG-2-interacting protein X) have emerged as particularly significant in damage sensing. ALIX has been reported to detect lysosomal and phagosomal damage specifically through sensing calcium leakage, a process facilitated by its calcium-binding partner ALG-2. The ALG-2-ALIX detection system relies on the successful interaction of calcium that leaks from the damaged membrane. Upon binding with calcium, ALG-2 undergoes conformational changes that enhance its interaction with ALIX, promoting the recruitment of this protein complex to damaged membranes [97]. The subsequent recruitment of the ESCRT-III repair complex to damaged membranes is contingent upon this calcium-dependent interaction [98].

6.3 ATG8 E3-like ligases in membrane damage surveillance

A recent report has suggested the crucial role of autophagy ATG8 E3-like ligase in the identification of damaged phagophores and the coordination of repair mechanisms. These ligases are instrumental in recruiting the ESCRT complex to the affected region. Membrane disruption results in the leakage of protons and calcium into the cytosol, a process facilitated by ATG8 E3-like ligase. The escape of protons from the compromised membrane initiates the V-ATPase-mediated recruitment of the ATG16L1-ATG5-ATG12 E3-like complex, whereas calcium leakage prompts the sphingomyelin-mediated recruitment of the TECPR1-ATG5-ATG12 E3-like complex. This mechanism ensures the accurate detection of membrane damage, irrespective of the specific ions that escape through the breach, and constitutes a critical component of the damage response cascade [98].

6.4 cGAS-STING axis: A double-edged sword in TB immunity

cGAS is characterized by two DNA-binding domains and a catalytic nucleotidyltransferase domain within its structural framework, which are essential for its role in immune signaling [99]. In the absence of bacterial or viral DNA, cGAS remains inactive; however, upon binding to intracellular bacterial DNA, it forms a complex that induces a conformational change at the enzyme's active site, facilitating the synthesis of cyclic GMP-AMP (cGAMP) through the strategic coordination of GTP and ATP substrates [100]. This cGAMP is distinguished by unique phosphodiester linkages: one between the 5′C-phosphate of AMP and the 2′C-OH of GMP, and the other between the 3′C-OH of AMP and the 5′C-phosphate of GMP. Cyclization catalyzes the formation of 2′3′-cGAMP, characterized by non-canonical 2′-5′ and 3′-5′ phosphodiester bonds. This distinctive second messenger binds to STING, an ER membrane adaptor protein, inducing structural reorganization that activates STING. STING exits the ER and translocates to the Golgi apparatus via clathrin-ordered protein II (COP-II)-mediated transport. At the Golgi, STING undergoes palmitoylation and clusters at the trans-Golgi network (TGN), creating a platform for TBK1 recruitment. STING directly binds TBK1, inducing its auto-phosphorylation and activation Fig. 5.4. Active TBK1 phosphorylates interferon regulated factor 3 (IRF3) at critical C-terminal residues (e.g., Ser386, Ser396), relieving autoinhibition and enabling dimerization. Phosphorylated IRF3, in homodimeric form, localizes to the nucleus, binding to interferon-stimulated response elements (ISREs) to initiate transcription of type I interferons (e.g., IFN-β) and antiviral genes [101]. Activated TBK1 also phosphorylates key autophagy regulators, such as Optineurin (OPTN) and Unc-41-like kinase 1 (ULK1), which are essential for autophagosome formation [102].

During *Mtb* infection, the production of host interferons also depends on the ESX-1 secretion system, which leads to phagosomal membrane damage. *Mtb*'s DNA is detected by host cytosolic sensors like cGAS, which immediately leads to the activation of STING and subsequently activates the STING-TBK1-IRF3 signaling axis, driving type I interferon (IFN-I) production, a paradoxical host defense mechanism that *Mtb* exploits for immune evasion [103] as shown in Fig. 5.4. However, mice with deleted IRF3 are resistant to infection caused by *Mtb*, underscoring IFN-1's detrimental role in TB pathogenesis. The phagosome disruption, mediated by ESX-1, is a mechanism of pathogenesis of *Mtb*, whereby cytosolic cGAS detects escaped *Mtb* DNA, recruiting ubiquitin ligases (PRKN, Smurf1) and adaptors (SQSTM1, CALCO2) to tag bacteria for xenophagy [104].

6.5 NLRs as guardians of phagosomal integrity in TB

Nucleotide oligomerization domain (NOD)-like receptors (NLRs) constitute a crucial class of intracellular receptors primarily responsible for recognizing microbial components and detecting disruptions in cellular homeostasis, including phagosomal damage. Within the NLR family, nucleotide-binding oligomerization domain (NOD1), NOD2, and the NLR family pyrin domain containing 3 (NLRP3) inflammasome are crucial in identifying breaches in phagosomal membranes. NOD1 and NOD2 recognize bacterial PG fragments, whereas NLRP3 detects physicochemical disruptions, such as ion flux and ROS, collectively initiating antimicrobial and inflammatory responses. Among NLRs, NLRP3 is particularly significant in the context of phagosomal damage. Both NOD1 and NOD2 possess cytosolic receptors with caspase-associated recruitment domain (CARD) domains that facilitate downstream nuclear factor kappa B (NF-κB) signaling. NOD1 contains a single CARD for downstream signaling, while NOD2 features two tandem CARD domains that enhance interaction with receptor-interacting serine/threonine kinase 2 (RIP2), a critical adaptor for NF-κB activation. Additionally, both NOD1 and NOD2 contain a Nucleotide-Binding Domain (NBD) that functions as a molecular switch through ATP binding and hydrolysis, enabling ligand-induced oligomerization and signal amplification. ATP binding stabilizes the active conformation, promoting the assembly of signaling complexes. The structural configuration of NLRs allows them to function as a sophisticated intracellular surveillance system, detecting various signs of cellular stress, including phagosomal damage. The relationship between phagosomal damage and NLRP3 activation is particularly evident in *Mtb* infection. Research indicates that *Mtb* induces caspase-1/NLRP3/gasdermin D (GSDMD)-mediated pyroptosis in macrophages through dual mechanisms: (1) contact-dependent plasma membrane damage during bacterial engulfment and (2) cytosolic membrane disruption following phagosomal rupture. These injuries induce K^+ efflux and activate the NLRP3 inflammasome, leading to IL-1β secretion and GSDMD pore formation, culminating in inflammatory cell death. This pyroptotic process facilitates bacterial dissemination to neighboring cells, highlighting *Mtb*'s exploitation of host cell death pathways for propagation. NLRs function as versatile sentinels, detecting both pathogen-derived signals and DAMPs associated with phagosomal compromise. These receptors respond to broader cellular stress, exemplified by ER stress-induced activation of NOD1 and NOD2, which triggers pro-inflammatory signaling apart from microbial sensing. This dual capacity to recognize both pathogen-associated and sterile cellular disturbances demonstrates NLRs' role as integrators of diverse stress signals, bridging organelle-specific damage to systemic immune responses. TLRs, the most significantly categorized member of PRRs, are not directly involved in detecting phagosomal membrane damage. Instead, they identify conserved microbial motifs such as PAMPs within phagosomes, initiating downstream signaling cascades that orchestrate antimicrobial responses. This distinction is critical: TLRs function as molecular sentinels for pathogen-derived ligands rather than sensors of compartment integrity. Phagosomal TLRs recognize intracellular pathogens and induce PRR signaling, and several intracellular pathogens, including *Mtb*, escape the phagosomes and gain access to the host lumen.

6.6 Ubiquitin codes: Decoding damage on phagosomal membranes

Ubiquitination is integral to the detection of membrane damage, particularly within phagosomes. Phagosomal injury triggers ubiquitination, which tags Ub chains, marking them for degradation via lysosomal or proteasome-mediated pathways. Ub, a compact protein comprising 76 amino acids, is involved in mediating autophagy, maintaining cellular homeostasis, and Ub-mediated proteasomal degradation. The cellular degradation pathway is determined by the topology of poly-Ub chains: K63-linked ubiquitination marks substrates for autophagic clearance, whereas K48-linked chains target proteins for proteasomal degradation. Selective autophagy receptors, including SQSTM1, NBR1, and CALCOCO2, bind these Ub tags and link them to LC3, a structural component of autophagosomal membranes. This receptor-mediated tethering ensures precise cargo recruitment, facilitating autophagosome encapsulation and subsequent lysosomal destruction. Maintaining phagosomal integrity through repair, replacement, or de novo synthesis is essential for cellular homeostasis [105]. The level of polyubiquitinated protein is elevated in the phagosomal membrane compared to the total cell lysate (TCL). Activation of macrophages by IFN-γ significantly increases the ubiquitination of phagosomal proteins. Notably, the level of ubiquitination in maturing phagosomes remains constant in both untreated and IFN-γ-treated cells, suggesting that polyubiquitination serves a function beyond protein degradation. A member of the E3 Ub ligase family, Ring Finger Protein 115 (RNF115), has been identified as a potential regulator of phagosomal functions. IFN-γ activation promotes RNF115 accumulation at phagosomal membranes, where its E3 Ub ligase activity inhibits maturation. Depletion of RNF115 or disruption of its catalytic function accelerates phagolysosome formation and amplifies proinflammatory cytokine responses to bacterial infection Table 5.1 [106].

6.7 TRIM proteins: Orchestrators of phagosomal defense

TRIM proteins, particularly TRIM16, are crucial in detecting phagosomal damage and activating subsequent signaling pathways. TRIM proteins belong to the extensive E3 Ub ligase family, which is integral to cellular processes such as autophagy, immunity, and protein quality control. In response to phagosomal damage, TRIM16 directly interacts with Gal, specifically Gal-3, to initiate a protective immune response. The TRIM16-Gal-8 complex constitutes a sophisticated surveillance mechanism that monitors phagosomal membrane integrity and responds to breaches in the barrier between pathogen and host cytosol. The sequential recruitment of components to damaged phagosomes provides insights into the temporal dynamics of damage detection. Initially, Gal-3 is recruited to damaged phagosomes harboring *Mtb*, followed by TRIM16, with Ub accumulation occurring subsequently. This sequence suggests that Gal-3 serves as the initial sensor of membrane damage, with TRIM16 acting as an effector that links damage recognition to downstream ubiquitination and autophagic responses. The temporal organization of this response ensures that damaged compartments are rapidly identified and addressed before pathogens can fully exploit the compromised barrier. Research has demonstrated that TRIM16 utilizes Gal-3 to detect such damaged phagosomes and lysosomes. The Gal-3-TRIM16 complex is specifically recruited to damaged phagosomes containing wild-type *Mtb* but not to phagosomes containing ESX-1 mutant bacteria (ΔESAT-6 *Mtb*) that cannot permeabilize the phagosomal membrane, confirming that recruitment is damage-dependent. This specificity demonstrates the precision of the TRIM16-Gal-3 system in identifying compromised phagosomes [107]. Recent research has identified another TRIM family protein, TRIM32, as involved in the response to phagosomal damage. TRIM32 plays a significant role in enhancing the autophagic destruction of *Mtb*, positioning it as a viable candidate for host-directed therapeutic (HDT) strategies against TB. Experimental knockdown of TRIM32 in THP-1 macrophages exacerbates intracellular *Mtb* proliferation, linked to three key deficits: impaired pathogen ubiquitination, reducing tagging for autophagy; diminished recruitment of autophagy adaptors CALCOCO2 and regulators BECN1 to bacterial foci; and suppressed autophagosome biogenesis, hindering bacterial containment. This triad of defects highlights TRIM32's centrality in coordinating Ub-dependent xenophagy to restrict *Mtb* growth. The identification of TRIM32 as a player in anti-mycobacterial defense suggests that multiple TRIM proteins may be involved in phagosomal damage detection and response, potentially targeting different subsets of the bacterial population or responding to different aspects of phagosomal damage Table 5.1 [108].

7. Mending the breach: Host repair mechanisms against phagosomal damage

The integrity of the phagosome is essential for the host immune system to effectively eliminate invading pathogens such as *Mtb*, while also regulating critical physiological processes including tissue regeneration, wound healing, and the maintenance of cellular homeostasis. However, the phagosomal membrane is frequently subjected to damage due to factors such as the mechanical stress associated with phagosome maturation [3,109], oxidative attacks by host defenses, or direct interference by invading pathogens [5]. Intracellular pathogens employ specialized secretion systems to release effector proteins that facilitate the breaching of the host's phagosomal membrane, allowing the import of secreted factors into the host cytoplasm to alter cellular trafficking and create a replicative niche for their proliferation. *Mtb* gains access to the host cytosol through the ESX-I T7SS, which is crucial for its intracellular survival and pathogenesis [20,110,111]. Once *Mtb* impairs host membrane repair mechanisms it enhances its survival and replication within infected macrophages. In response, the host employs various mechanisms to counteract these pathogen-induced disruptions. A primary repair mechanism involves lipid and membrane remodeling, where ESCRT proteins are rapidly recruited to seal perforations, particularly those caused by bacterial toxins [10]. Vesicular trafficking also plays a role in phagosomal repair, with small vesicles from various organelles, including endosomes, the endoplasmic reticulum, or the Golgi apparatus, being swiftly delivered to damaged sites to restore membrane integrity through the coordinated action of Rab GTPases [21,112,113], SNARE proteins, and tethering factors [114]. Another significant repair pathway is lysophagy, a specialized form of autophagy that selectively removes damaged lysosomal membrane fragments while simultaneously providing fresh membrane components for repair Table 5.2. Autophagy adaptors such as SQSTM1 and CALCOCO2 mediate the recognition and tagging of damaged sites for clearance and repair. Beyond autophagy, the Ub-mediated damage-sensing mechanism serves as an additional defense layer, where Ub ligases tag the damaged phagosomal membrane or targeted degradation cargos via autophagy [22,23].

The cell also initiates a rapid, calcium influx-driven repair response, which activates annexins and synaptotagmins to promote membrane resealing, particularly in response to pore-forming toxins [115,116]. The efficacy of these repair pathways in detecting and repairing phagosomal damage is fundamental to maintaining immune homeostasis and controlling pathogens. An effective repair mechanism prevents intracellular pathogens from escaping into the cytosol, where they could evade lysosomal degradation and manipulate host signaling pathways. Unresolved membrane damage can lead to excessive inflammation, increasing the risk of chronic immune activation and tissue damage. Any failure in the repair mechanisms can result in extensive immune dysregulation, pathogen dissemination, and, in extreme cases, cell death through necrotic or pyroptotic pathways, thereby exacerbating infection severity.

7.1 ESCRT-III at the frontline: Host repair response to *Mycobacterium tuberculosis*-induced phagosomal damage

The ESCRT-III machinery is involved in membrane repair processes following lysosomal damage and is recruited to *Mtb* phagosomes in an ESX-1-dependent manner. Key components of ESCRT-III

Table 5.2 The cGAS-STING pathway: activation of the cGAS–STING pathway and Its role in autophagy induction.

Counter actions applied via host against mycobacteria mediated modulation	Mechanisms	Key players
1. ESAT-6 secretion system 1 (ESX-1)-induced membrane disruption	*Mycobacterium tuberculosis (Mtb)* uses the ESX-1 (type VII) secretion system (T7SS) to perforate the phagosomal membrane, accessing host cytosol and altering trafficking.	ESX-1, early secreted antigenic target 6 (ESAT-6), culture filtrate protein 10 (CFP-10)
2. Endosomal sorting complex required for transport III (ESCRT-III)-driven membrane resealing	ESCRT-III components rapidly recruited to seal membrane lesions, especially at sites of ESX-1-mediated phagosomal injury.	Charged multivesicular body protein 1A (CHMP1A), CHMP4B, CHMP1B, ESCRT-III
3. Organelle-derived vesicular repair pathway	Vesicles from endoplasmic reticulum (ER), golgi, and endosomes fuse with damaged phagosome to restore integrity via membrane delivery.	Ras-related GTP-binding protein (Rab) GTPases, soluble N-ethylmale-imide-sensitive factor-attachment protein receptors (SNAREs), tethering factors
4. Autophagy-lysophagy clearance circuit	Damaged membranes are targeted via selective autophagy pathways; lysophagy removes damaged regions, supplying fresh membranes.	LC3, sequestome 1 (SQSTM1), CalCOCO2 (calcium binding and coiled-coil domain 2), ATG7, optineurin (OPTN)
5. Ubiquitin (Ub) triggered damage sensing axis	Ub ligases tag damaged membranes/ cargos to signal for degradation or repair via autophagy and ESCRT pathways.	TrafE (TRAF-like), ALG-2 interacting protein X (ALIX), vacuolar protein sorting 32 (Vps32), Vps4
6. Ca^{2+}-activated membrane patching	Rapid influx of Ca^{2+} upon membrane damage activates membrane-sealing proteins to rapidly restore phagosomal integrity.	Annexins, synaptotagmins
7. Glycan exposure surveillance system	Cytosolic exposure of luminal glycans recruits galectins (Gals) and tripartite motif containing (TRIM) E3 Ub ligases that tag membranes for xenophagy and help ESCRT recruitment.	Gal-3, Gal-8, Gal-9, TRIM16/17/20, SQSTM1, CALCOCO2, OPTN
8. Cytosolic DNA sensing via cyclic GMP-AMP synthase- stimulator of interferon response cGAMP interactor (cGAS–STING) axis	Sensing of cytosolic *Mtb* DNA activates CGAS-STING signaling, triggering interferon (IFN) response and autophagy recruitment.	cGAS, STING, TANK binding kinase 1 (TBK1), interferon regulatory factor 3 (IRF3), interleukin 6 (IL-6), tumor necrosis factor alfa (TNFα)

Continued

Table 5.2 The cGAS-STING pathway: activation of the cGAS–STING pathway and Its role in autophagy induction.—cont'd

Counter actions applied via host against mycobacteria mediated modulation	Mechanisms	Key players
9. ER-linked lipid exchange repair pathway	ER-derived oxysterol binding protein 8 (OSBP8) is recruited to mycobacteria containing vacuoles (MCVs) to rebalance phosphatidylinositol-4-phosphate (PI4P) and help restore damaged membranes, especially under ESX-1 influence.	OSBP8, PI4P
10. Ras-related GTP-binding protein 20 (Rab20)-guided membrane reinforcement	Rab20 modulates phagosome integrity and restricts *Mtb* replication by promoting phagosome maturation and membrane repair.	Rab20

namely charged multivesicular body protein 1A (CHMP1A), CHMP1B, and CHMP4B, play a significant role in facilitating the translocation of ESCRT-III to the damaged lysosomal membrane. In a study, the localization of CHMP1A was examined in bone-marrow-derived macrophages (BMDMs) infected with wild-type *Mtb* H37Rv. At 3 hours post-infection, prior to extensive phagosomal disruption, 85.2% of wild-type bacilli exhibited punctate CHMP1A. Further analysis of CHMP4B recruitment revealed extensive colocalization with the infected phagosomes, displaying a punctate pattern similar to that of CHMP1A accumulation at sites adjacent to bacterial localization. This pattern resembles the response to lysosomal damage induced by the lysosomotropic agent LLOME (l-leucyl–L-leucine O-methyl ester), suggesting that ESX-1 activity contributes not only to phagosomal recruitment but also to broader endomembrane remodeling in infected cells. Collectively, these findings establish ESX-1-mediated phagosomal membrane damage as a critical trigger for ESCRT-III-dependent repair pathways, highlighting a striking similarity between the host response to *Mtb*-induced phagosomal injury and canonical ESCRT-III-mediated lysosomal damage repair. This reinforces the notion that ESCRT-III serves as a key cellular defense mechanism against bacterial virulence strategies Table 5.2.

7.2 Autophagy interrupted: LC3 dynamics in *Mycobacterium tuberculosis*-containing vacuoles

Autophagy represents a vital innate defense mechanism employed by host cells to target invading pathogens, directing their degradation within autophagolysosomes [13]. The host cell utilizes the recruitment of the key autophagy regulator LC3 [117] to pathogen-containing vacuoles, facilitating phagosome maturation [90], fusion with lysosomes, and ultimately, bacterial clearance through lysosomal acidification. However, mycobacteria counteract this defense through complex mechanisms [24,118,119], thereby evading autophagic destruction. Real-time single-cell tracking of LC3 trafficking to *Mtb*-containing vacuoles revealed that although 70% of these vacuoles exhibited LC3 recruitment, only 30% retained it, indicating a transient and reversible autophagic response. Furthermore, those vacuoles that recruited LC3 failed to undergo acidification, suggesting that LC3 localization does not guarantee phagosomal maturation. The recruitment of LC3 was correlated with Gal-3 and OSBP1, markers of membrane damage, yet ATG7 knockdown did not impair phagosome membrane repair, indicating that autophagy is not directly responsible for membrane restoration. The mutual exclusion and alteration of LC3 and lysosomal markers suggest novel crosstalk between autophagy and membrane integrity, highlighting the complexity of host-pathogen interactions [25].

7.3 Triple threat: Coordinated host signaling against *Mycobacterium tuberculosis*-induced damage

The interaction of immune signaling pathways, including Gals, TRIM family proteins, and CGAS-STING signaling, plays a crucial role in recognizing and responding to *Mtb*-induced lysosomal damage. This interaction facilitates bacterial clearance through a selective autophagy process known as xenophagy and an inflammatory immune response. These pathways work synergistically to detect damage, initiate membrane repair, mark *Mtb* for degradation, and trigger an antimicrobial immune response [120]. The ESX-1 secretion system-mediated phagosomal membrane damage by *Mtb* recruits Gal-3, Gal-8, and Gal-9, followed by the recruitment of TRIM16, TRIM17, and TRIM20. These function as E3 Ub ligases, tagging the damaged phagosomal membrane for selective autophagy and degradation. TRIM16 and TRIM20 act as scaffolds for autophagy-related proteins, recruiting key autophagy adapters, which subsequently facilitate the recruitment of LC3 to the damaged sites. The activation of TRIM proteins further aids in recruiting the ESCRT-III machinery to seal phagosomal ruptures, preventing excessive cytosolic exposure to *Mtb* bacterial agents. However, persistent or excessive *Mtb*-mediated membrane damage that escapes repair is ultimately targeted for xenophagy [120]. Additionally, CGAS is activated, generating cGAMP, which binds and activates STING, a key mediator of innate immune signaling. This activation leads to the production of type I interferons (IFN-α/β) and inflammatory cytokines such as IL-6 and TNF-α, enhancing the host's antimicrobial defense and promoting the recruitment of autophagic machinery, thereby reinforcing bacterial degradation. The combined action of Gals, TRIM proteins, and CGAS-STING effectively restricts *Mtb* growth by either degrading the bacteria via autophagy or activating inflammatory immune responses to contain its replication. Notably, cells deficient in any of these components exhibited a higher intracellular *Mtb* burden, highlighting their crucial role in host defense Table 5.2 [120].

7.4 TrafE at the crossroads: Linking ESCRT and autophagy in membrane repair

The study explored the role of the evolutionarily conserved TRAF-like E3 Ub ligase TrafE in coordinating the ESCRT and autophagy pathways in response to endolysosomal damage caused by *M. marinum*, a close relative of *Mtb*. The strong recruitment of TrafE to the disrupted intracellular compartments indicated its involvement in recognizing and responding to membrane damage, thereby facilitating the intersection of the ESCRT and autophagy pathways. TrafE orchestrated the recruitment of ESCRT subunits ALIX, Vps32, and Vps4, which was significantly compromised in the absence of

TrafE. This impairment led to defective ESCRT-mediated and autophagy-mediated endolysosomal membrane repair, highlighting the critical role of TrafE in maintaining cellular integrity during pathogenic attacks [26].

7.5 OSBP8-mediated repair: A host strategy against *Mycobacterium tuberculosis*-induced membrane damage

The endomembrane is frequently damaged by pathogens such as *Mtb*, which exploit this damage to access the host cell cytosol and circumvent its innate immune defenses. In response to this mycobacterial-induced damage, the host activates repair mechanisms to retain the pathogen within the cytosol. A notable defense strategy employed by the host involves the ER-Golgi protein oxysterol-binding protein (OSBP) and its *Dictyostelium discoideum* homolog OSBP8, which are recruited to the MCV. This recruitment occurs under the influence of the ESX-1 secretion system, indicating that their mobilization is linked to membrane injury. Mycobacterial infection results in the hyperaccumulation of PI4P on MCV, a reduction in the host cell's lysosomal and degenerative potential, and decreased cell viability. OSBP8 trafficking in response to mycobacterium-induced membrane damage balances PI4P levels, restoring the MCVs through ER-dependent repair mechanisms [27].

7.6 Rab20: Guardian of phagosomal integrity in *Mycobacterium tuberculosis* infection

Another host protein, Rab20, plays a crucial role in controlling mycobacterial replication within infected macrophages. *Mtb* replicates inside host macrophages by residing in specialized compartments and employs strategies to inhibit phagosomal maturation [18,28], thereby favoring its prolonged persistence within the replicative niche [29,30,121]. To counteract this *Mtb*-mediated phagosomal modulation, the host transiently associates Rab20 with *Mtb*-containing phagosomes, which is further upregulated by IFN-γ stimulation. Rab20 translocation assists phagosomes in maintaining their spaciousness and membrane integrity, essential for phagosomal maturation and bacterial clearance Table 5.2. When Rab20 is depleted, phagosomes become smaller and compromised, facilitating mycobacterial proliferation within infected macrophages, further demonstrating its importance in combating disease progression [122].

7.7 Host's lipid-mediated phagosome membrane repair

Inhibition of phagosome maturation, acidification, membrane actin assembly, and fusion of phagosome with lysosome is among the critical strategies exploited by pathogenic mycobacterium species for their survival against host defences. To counteract mycobacterial infections, the host employ specific lipids such as sphingosine (Sph), ceramide (Cer), and sphingosine-1-phosphate (SIP) to promote actin nucleation on the mycobacteria containing phagosomal membrane, resulting in notable activation of the lipid-mediated actin assembly activation, facilitating the phagosomal maturation process, characterized by enhanced lysosome fusion and acidification, resulting to a significant reduction in the viability of *M.tb* and *M. avium* [123].

8. VPS18: A gatekeeper of phagosome integrity against *Mycobacterium tuberculosis*

Clustered regularly interspaced palindromic repeats (CRISPR)-based genome-wide screening has identified VPS18 as a crucial host factor in repairing phagosomal membranes damaged by *Mtb*. Shortly after infection, VPS18 co-localizes with *Mtb* in macrophages. Macrophages lacking VPS18 exhibit increased phagosomal membrane damage, as evidenced by a higher co-localization of *Mtb* with endosomal damage markers Gal-3 and Gal-9 [12]. Furthermore, vps18-knockout cells show impaired maturation of *Mtb*-containing phagosomes, indicated by decreased colocalization with the lysosomal marker lysosomal associated membrane protein 1 (LAMP1), leading to reduced exposure to lysosomal degradative enzymes and antimicrobial peptides. Additionally, mycobacterial phagosomes maintain a less acidic environment in VPS18-deficient macrophages. Both the impaired maturation and altered pH of the mycobacterial phagosome likely create a favorable environment for *Mtb*'s sustained survival and replication. Moreover, this reduced acidification compromises the effectiveness of the antibiotic pyrazinamide, which requires an acidic environment for optimal activity. Nevertheless, VPS18 is essential for repairing *Mtb*-induced phagosomal membrane damage, maintaining phagosome integrity, and ensuring effective antibiotic treatment [12,124], as depicted in Fig. 5.5. Thus, modulating phagosomal membrane integrity could be a potential strategy to enhance the efficacy of tuberculosis treatment.

8.1 CISH disarms phagosomes to aid *Mycobacterium tuberculosis* survival

Genome-wide RNA interference (RNAi) screening identified cytokine-inducible SH2-containing protein (CISH) as a critical host factor that is significantly upregulated in response to *Mtb* infection [17]. CISH is a member of the SOCS (Suppressor of Cytokine Signaling) family, which generally functions as a negative regulator of cytokine signaling, particularly within the Janus kinase/signal transducer and activator of transcription (JAK/STAT) pathway [125]. Under normal conditions, cytokines such as IL-2, IL-3, IL-5, and factors like granulocyte-macrophage colony-stimulating factor (GM-CSF) induce CISH expression in a JAK/STAT-dependent manner. The binding of cytokines to their receptors activates the JAK/STAT pathway, leading to the phosphorylation and activation of STAT proteins. These proteins then dimerize and move to the nucleus, where they bind to the promoter regions of SOCS genes, including CISH, initiating their transcription and expression. CISH protein subsequently acts as a negative feedback regulator by binding to phosphorylated cytokine receptors and JAKs, marking them for ubiquitination and proteasomal degradation, thus attenuating cytokine signaling to prevent immune overactivation. Concurrently, *Mtb* infection induces GM-CSF secretion, which triggers autocrine activation of STAT5 signaling, resulting in CISH overexpression. In the absence of intracellular bacteria, CISH activity is primarily cytoplasmic and not associated with specific organelles. However, *Mtb*-induced GM-CSF facilitates CISH recruitment and retention at the MCVs, where it represses phagosomal acidification by targeting ATPase H+ transporting V1 subunit A (ATP6V1A) for ubiquitination, as depicted in Fig. 5.6. The CISH protein contains a C-terminal SOCS domain (a dual-binding domain) that enables direct interaction with ATP6V1A. Once bound to ATP6V1A, the SOCS domain facilitates the recruitment of cullin-really interesting new gene (CUL-RING) E3 Ub ligase, which tags ATP6V1A with Ub molecules. The ubiquitination of ATP6V1A

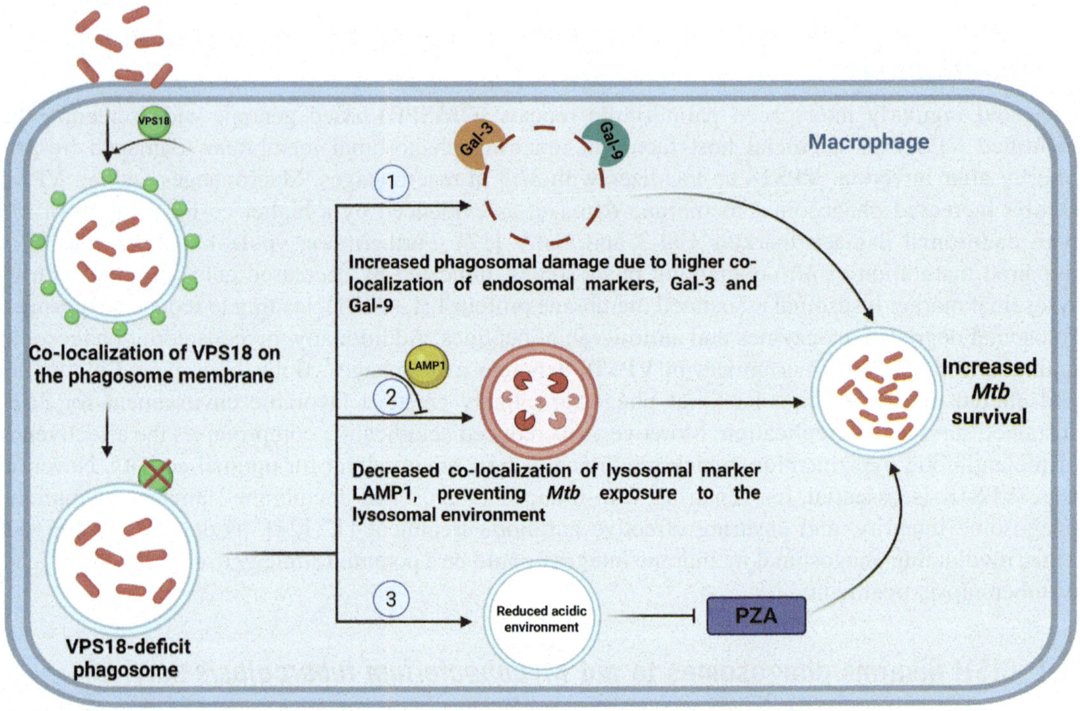

FIGURE 5.5 Role of VPS18 during phagosomal maturation and survival of *Mycobacterium tuberculosis*.

In macrophages, VPS18 core subunit of CORVET and HOPS complexes (VPS18) is localized on the phagosome membrane and plays a crucial role in phagosomal maturation. The loss of VPS18, a core subunit of the CORVET and HOPS complexes, results in: (1) increased phagosomal damage, as indicated by the heightened recruitment of endosomal markers Galectin-3 (Gal-3) and Gal-9, (2) diminished interaction with the lysosomal marker lysosomal associated membrane protein 1 (LAMP1), thereby reducing the exposure of *Mycobacterium tuberculosis* (*Mtb*) to degradative environments, and (3) a less acidic pH, which decreases the efficacy of pyrazinamide (PZA), anti-tuberculosis (anti-TB) drug. Collectively, these alterations enhance the intracellular survival of *Mtb*.

earmarks it for proteasomal degradation, leading to reduced activity of the VoH+-ATPase proton pump and inhibition of phagosomal acidification [17]. This highlights a strategic immune evasion mechanism and positions the CISH–ATP6V1A axis as a potential target for HDTs.

8.2 Coronin-1A: A mycobacterial ally in phagosome retention

CORO1A, also referred to as tryptophan/aspartate-containing coat protein (TACO or P57), is a member of the highly conserved coronin protein family and is ubiquitously present in eukaryotic cells. In phagocytic cells, viable bacilli actively engage with and sustain trimeric CORO1A on the phagosomal membrane, thereby enhancing their intracellular survival by inducing the Ca^{2+}/calcineurin

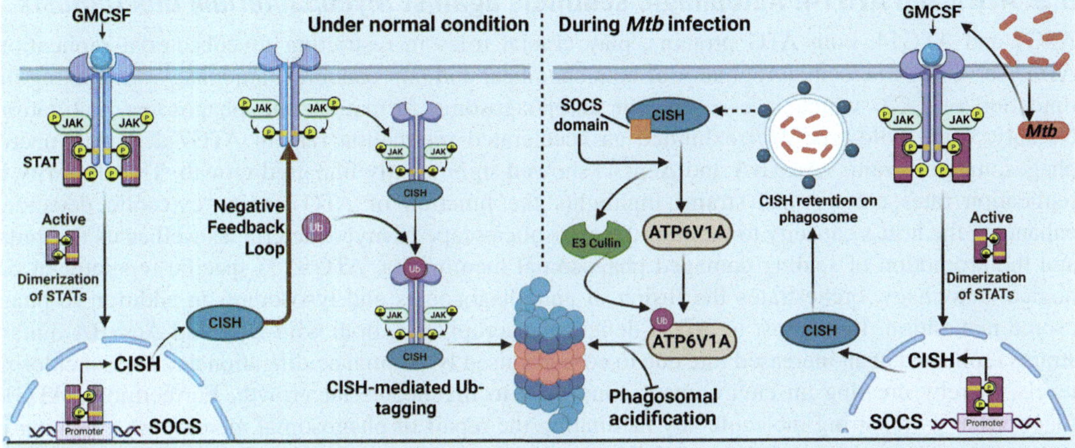

FIGURE 5.6 CISH disrupts GM-CSF-activated phagosome maturation and acidification.

The activation of the granulocyte-macrophage colony stimulating factor (GM-CSF) receptor typically initiates the Janus kinase-signal transducer and activator of transcription (JAK-STAT) signaling cascade, culminating in the dimerization of STAT and the induction of the suppressor of cytokine signaling (SOCS) family gene cytokine-inducible SH2-containing protein (CISH). CISH operates within a negative feedback loop by targeting JAKs for ubiquitination and subsequent degradation, thereby attenuating the pathway. During infection with *Mycobacterium tuberculosis* (*Mtb*), CISH localizes to the phagosome and interacts with the vacuolar ATPase subunit ATPase H+ transporting V1 subunit A (ATP6V1A). Through its SOCS domain and the recruitment of the E3 Ub ligase cullin (CUL) complex, CISH facilitates the ubiquitination of ATP6V1A, thereby hindering phagosomal acidification. This disruption aids the survival of *Mtb* within macrophages by impairing normal phagosomal maturation.

pathway, which circumvents delivery to degradative lysosomal compartments. Bovine CORO1A (bCORO1A) comprises a tryptophan-aspartate (WD) repeat domain and a coiled-coil domain. Mycobacterial infection leads to the upregulation of bCORO1A expression, and it interacts with mycobacterial lipoamide dehydrogenase C (LpdC) via its WD repeat domain. This interaction sequesters bCORO1A on the phagosomes, inhibiting their fusion with lysosomes and thereby facilitating mycobacterial survival within macrophages. Furthermore, the WD repeat domain of bCORO1A mediates a reduction in cytosolic Ca^{2+} levels. The decreased calcium levels result in reduced activation of calmodulin-dependent kinase II (CaMKII), a critical factor for phagosome-lysosome fusion [126]. Consequently, targeting bCORO1A and its mammalian homologs, particularly its WD repeat domain, could represent a strategy to enhance host resistance against TB.

8.3 ATG7 and ATG14: Autophagic sentinels against *Mycobacterium tuberculosis*

ATG7 and ATG14, both ATG proteins, play crucial roles in restricting mycobacterial replication. ATG7 serves as an essential regulator of both canonical and non-canonical autophagy, facilitating the lipidation of LC3, which is essential for autophagosome formation and phagosome maturation. Notably, while wild-type *Mtb* exhibited an accelerated replication rate in ATG7-deficient macrophages, mutant strains (ΔesxBA and ΔcpsA) showed significantly impaired growth. This disparity in replication rates among *Mtb* strains highlights the function of ATG7 as a "cytosolic detector," enhancing the host's capacity to monitor the cytosolic escape of mycobacteria, as well as its recapture and the promotion of sealing damaged phagosomal membranes. ATG14, a specific regulator of canonical autophagy, orchestrates the fusion of autophagosomes and lysosomes, in addition to phagosome maturation. In contrast to ATG7-deficient macrophages, both wild-type and ΔesxBA mutant strains replicated at an increased rate due to compromised lysosomal acidification and altered chlorine levels, thereby creating an environment conducive to mycobacterial growth. Furthermore, ATG14 functions as a "repairing molecule" by facilitating the repair of phagosomal membranes, resulting in increased accumulation of Gal-3 and LC3-associated tubulovesicular structures (LC-TVS), thereby preventing the cytosolic escape of *Mtb* [11]. These findings reveal the distinct roles of autophagy proteins and their potential in developing HDTs that could enhance the host's ability to curb *Mtb* by promoting phagosomal maturation.

9. Conclusions and perspectives

The phagosome plays important roles in the innate immune defense against intracellular pathogens such as *Mtb*, functioning not only as a degradative compartment but also as a central platform for immune signaling and molecular interactions. *Mtb* initiates its manipulation of the host at the early stages of infection by disrupting phagosomal maturation through interference with PI3P signaling, Rab GTPases, SNAREs, and tethering complexes. Effector molecules such as SapM, ManLAM, and PknG alter vesicular dynamics and immune signaling to delay acidification and lysosome fusion. The pathogen further compromises host defenses by inducing phagosomal membrane disruption via the ESX-1 secretion system and PDIM, facilitating controlled cytosolic release and modulation of the host immune response. In response to *Mtb*-induced phagosomal damage, the host employs molecular surveillance mechanisms involving Gal-3, Gal-8, and Gal-9, which bind to glycans exposed on damaged phagosomal membranes and recruit TRIM proteins (TRIM16, TRIM20) that scaffold selective autophagy adapters such as SQSTM1/p62, CALCOCO2/NDP52, and OPTN. These processes support xenophagy while also activating ESCRT-III to reseal membranes. Upon detection of damaged phagosomes, the host initiates rapid membrane repair by mobilizing ESCRT-III subunits (CHMP1A, CHMP4B) to sites of ESX-1-induced damage. Membrane restoration is facilitated by vesicular trafficking from the ER, Golgi, and endosomes, regulated by Rab GTPases, SNAREs, and tethering proteins. Ca^{2+} influx-activated responses, including annexins and synaptotagmins, also expedite membrane repair, whereas lysophagy selectively degrades irreparably damaged phagosomal fragments. Additionally, cGAS–STING signaling detects cytosolic *Mtb* DNA and activates TBK1-IRF3-dependent type I interferon and proinflammatory cytokine production, enhancing immune signaling and autophagic targeting. Concurrently, the ubiquitin-dependent repair network, exemplified by TrafE in *M. marinum*, bridges autophagy and ESCRT pathways, stabilizing the membrane. The OSBP8-dependent ER-Golgi repair system regulates PI4P accumulation in MCVs under ESX-1 regulation, rescuing endomembrane composition and function.

Furthermore, repair and reversal mechanisms are fundamental to the host's defensive strategies. Autophagy-related responses exhibit both functional potential and complexity. Autophagy proteins ATG7 and ATG14 play crucial roles, with the former acting as a cytosolic sensor of *Mtb* and facilitating LC3 lipidation, while the latter maintains membrane integrity and aids in lysosome fusion through specialized tubulovesicular structures. Host lipid mediators such as sphingosine, ceramide, and sphingosine-1-phosphate actively promote actin polymerization along the phagosome membrane, enabling maturation, lysosome fusion, and acidification, ultimately reducing the viability of *Mtb* and *Mycobacterium avium*. Other host molecules, such as Rab20, when transiently recruited to the damaged phagosome, are responsible for maintaining membrane integrity and maturation capability. A deficiency in Rab20 results in dysfunctional phagosomes and increased bacterial burden. CRISPR-identified VPS18 is critical for lysosomal maturation and acidification of *Mtb* phagosomes. The absence of Vps18 not only exacerbates phagosomal damage and bacterial survival but also diminishes the efficacy of acid-dependent drugs like pyrazinamide. As TB continues to pose a global health challenge, exacerbated by rising antimicrobial resistance and complex host-pathogen interactions, the insights presented in this chapter offer promising avenues for intervention. By modulating host pathways responsible for detecting, repairing, and reversing phagosomal damage, innovative HDTs can be developed. These therapies, when used in conjunction with traditional treatments, have the potential to restore phagosomal function and thereby facilitate the elimination of the TB pathogen.

Acknowledgments

Onaiza Ansari and Syeda Rushna sincerely acknowledge the Hamdard National Foundation for awarding a fellowship that supported the conduct of doctoral degree.

References

[1] Sun-Wada GH, Tabata H, Kawamura N, Aoyama M, Wada Y. Direct recruitment of H+-ATPase from lysosomes for phagosomal acidification. J Cell Sci 2009;122(14):2504–13. https://doi.org/10.1242/jcs.050443Japan.

[2] López de Armentia M, Amaya C, Colombo M. Rab GTPases and the autophagy pathway: bacterial targets for a suitable biogenesis and trafficking of their own vacuoles. Cells 2016;5(1):11. https://doi.org/10.3390/cells5010011.

[3] Lee HJ, Woo Y, Hahn TW, Jung YM, Jung YJ. Formation and maturation of the phagosome: a key mechanism in innate immunity against intracellular bacterial infection. Microorganisms 2020;8(9):1–22. https://doi.org/10.3390/microorganisms8091298.

[4] Omotade TO, Roy CR. Manipulation of host cell organelles by intracellular pathogens. Microbiol Spectr 2019;7(2). https://doi.org/10.1128/microbiolspec.BAI-0022-2019.

[5] Pauwels AM, Trost M, Beyaert R, Hoffmann E. Patterns, receptors, and signals: regulation of phagosome maturation. Trends Immunol 2017;38(6):407–22. https://doi.org/10.1016/j.it.2017.03.006.
[6] Xander C, Rajagopalan S, Jacobs WR, Braunstein M. The SapM phosphatase can arrest phagosome maturation in an ESX-1 independent manner in Mycobacterium tuberculosis and BCG. Infect Immun 2024; 92(7). https://doi.org/10.1128/iai.00217-24.
[7] Isberg RR, O'Connor TJ, Heidtman M. The Legionella pneumophila replication vacuole: making a cosy niche inside host cells. Nat Rev Microbiol 2009;7(1):13–24. https://doi.org/10.1038/nrmicro1967.
[8] Petit TJP, Lebreton A. Adaptations of intracellular bacteria to vacuolar or cytosolic niches. Trends Microbiol 2022;30(8):736–48. https://doi.org/10.1016/j.tim.2022.01.015.
[9] Bajunaid W, Haidar-Ahmad N, Kottarampatel AH, Manigat FO, Silué N, Tchagang CF, Tomaro K, Campbell-Valois F-X. The T3SS of Shigella: expression, structure, function, and role in vacuole escape. Microorganisms 2020;8(12):1933. https://doi.org/10.3390/microorganisms8121933.
[10] López-Jiménez AT, Cardenal-Muñoz E, Leuba F, Gerstenmaier L, Barisch C, Hagedorn M, King JS, Soldati T, Hawn TR. The ESCRT and autophagy machineries cooperate to repair ESX-1-dependent damage at the Mycobacterium-containing vacuole but have opposite impact on containing the infection. PLoS Pathog 2018;14(12):e1007501. https://doi.org/10.1371/journal.ppat.1007501.
[11] Aylan B, Bernard EM, Pellegrino E, Botella L, Fearns A, Athanasiadi N, Bussi C, Santucci P, Gutierrez MG. ATG7 and ATG14 restrict cytosolic and phagosomal Mycobacterium tuberculosis replication in human macrophages. Nat Microbiol 2023;8(5):803–18. https://doi.org/10.1038/s41564-023-01335-9.
[12] Jani C, Jain N, Marsh AK, Uchil P, Doan T, Hudspith M, Glover OT, Baskir ZR, Boucau J, Root DE, van der Wel NN, Doench JG, Barczak AK. VPS18 contributes to phagosome membrane integrity in Mycobacterium tuberculosis –infected macrophages. Sci Adv 2025;11(5). https://doi.org/10.1126/sciadv.adr6166.
[13] Chua J, Vergne I, Master S, Deretic V. A tale of two lipids: Mycobacterium tuberculosis phagosome maturation arrest. Curr Opin Microbiol 2004;7(1):71–7. https://doi.org/10.1016/j.mib.2003.12.011.
[14] Bussi C, Gutierrez MG. Mycobacterium tuberculosis infection of host cells in space and time. FEMS (Fed Eur Microbiol Soc) Microbiol Rev 2019;43(4):341–61. https://doi.org/10.1093/femsre/fuz006.
[15] Mohammadnabi N, Shamseddin J, Emadi M, Bodaghi AB, Varseh M, Shariati A, Rezaei M, Dastranj M, Farahani A. Mycobacterium tuberculosis: the mechanism of pathogenicity, immune responses, and diagnostic challenges. J Clin Lab Anal 2024;38(23). https://doi.org/10.1002/jcla.25122.
[16] Nguyen TA, Croon S, Rijkers G. Mycobacterium tuberculosis: escape room world champion. Pneumonia 2019;11(1). https://doi.org/10.1186/s41479-019-0060-2.
[17] Queval CJ, Song OR, Carralot JP, Saliou JM, Bongiovanni A, Deloison G, Deboosère N, Jouny S, Iantomasi R, Delorme V, Debrie AS, Park SJ, Gouveia JC, Tomavo S, Brosch R, Yoshimura A, Yeramian E, Brodin P. Mycobacterium tuberculosis controls phagosomal acidification by targeting CISH-mediated signaling. Cell Rep 2017;20(13):3188–98. https://doi.org/10.1016/j.celrep.2017.08.101.
[18] Vergne I, Chua J, Singh SB, Deretic V. Cell biology of Mycobacterium tuberculosis phagosome. Annu Rev Cell Dev Biol 2004;20:367–94. https://doi.org/10.1146/annurev.cellbio.20.010403.114015.
[19] Welin A, Raffetseder J, Eklund D, Stendahl O, Lerm M. Importance of phagosomal functionality for growth restriction of *Mycobacterium tuberculosis* in primary human macrophages. J Innate Immun 2011;3(5): 508–18. https://doi.org/10.1159/000325297.
[20] van der Wel N, Hava D, Houben D, Fluitsma D, van Zon M, Pierson J, Brenner M, Peters PJ, tuberculosis M, leprae M. Translocate from the phagolysosome to the cytosol in myeloid cells. Cell 2007; 129(7):1287–98. https://doi.org/10.1016/j.cell.2007.05.059.
[21] Menaceur C, Dusailly O, Gosselet F, Fenart L, Saint-Pol J, Trafficking V. A mechanism controlled by cascade activation of Rab proteins: focus on Rab27. Biology 2023;12(12):1530. https://doi.org/10.3390/biology12121530.

[22] Corkery DP, Castro-Gonzalez S, Knyazeva A, Herzog LK, Wu YW. An ATG12-ATG5-TECPR1 E3-like complex regulates unconventional LC3 lipidation at damaged lysosomes. EMBO Rep 2023;24(9). https://doi.org/10.15252/embr.202356841.

[23] Papadopoulos C, Kravic B, Meyer H. Repair or lysophagy: dealing with damaged lysosomes. J Mol Biol 2020;432(1):231–9. https://doi.org/10.1016/j.jmb.2019.08.010.

[24] Queval CJ, Brosch R, Simeone R. The macrophage: a disputed fortress in the battle against Mycobacterium tuberculosis. Front Microbiol 2017;8. https://doi.org/10.3389/fmicb.2017.02284.

[25] Augenstreich J, Phan AT, Allen CNS, Poddar A, Chen H, Srinivasan L, Briken V, Salgame P. Dynamic interplay of autophagy and membrane repair during Mycobacterium tuberculosis Infection. PLoS Pathog 2025;21(1):e1012830. https://doi.org/10.1371/journal.ppat.1012830.

[26] Raykov L, Mottet M, Nitschke J, Soldati T. A TRAF-like E3 ubiquitin ligase TrafE coordinates ESCRT and autophagy in endolysosomal damage response and cell-autonomous immunity to Mycobacterium marinum. eLife 2023;12. https://doi.org/10.7554/elife.85727.

[27] Anand A, Mazur A-C, Rosell-Arevalo P, Franzkoch R, Breitsprecher L, Listian SA, Hüttel SV, Müller D, Schäfer DG, Vormittag S, Hilbi H, Maniak M, Gutierrez MG, Barisch C, Roy CR. ER-dependent membrane repair of mycobacteria-induced vacuole damage. mBio 2023;14(5). https://doi.org/10.1128/mbio.00943-23.

[28] Deretic V, Fratti RA. Mycobacterium tuberculosis phagosome. Mol Microbiol 1999;31(6):1603–9. https://doi.org/10.1046/j.1365-2958.1999.01279.x.

[29] Barisch C, Paschke P, Hagedorn M, Maniak M, Soldati T. Lipid droplet dynamics at early stages of Mycobacterium marinum infection in D ictyostelium. Cell Microbiol 2015;17(9):1332–49. https://doi.org/10.1111/cmi.12437.

[30] Bermudez LE, Goodman J. Mycobacterium tuberculosis invades and replicates within type II alveolar cells. Infect Immun 1996;64(4):1400–6. https://doi.org/10.1128/iai.64.4.1400-1406.1996.

[31] Mwandumba HC, Russell DG, Nyirenda MH, Anderson J, White SA, Molyneux ME, Squire SB. Mycobacterium tuberculosis resides in nonacidified vacuoles in endocytically competent alveolar macrophages from patients with tuberculosis and HIV infection. J Immunol 2004;172(7):4592–8. https://doi.org/10.4049/jimmunol.172.7.4592.

[32] Upadhyay S, Mittal E, Philips JA. Tuberculosis and the art of macrophage manipulation. Pathog Dis 2018;76(4). https://doi.org/10.1093/femspd/fty037.

[33] Vergne I, Chua J, Deretic V. Mycobacterium tuberculosis phagosome maturation arrest: selective targeting of PI3P-dependent membrane trafficking. Traffic 2003;4(9):600–6. https://doi.org/10.1034/j.1600-0854.2003.00120.x.

[34] Clemens DL, Horwitz MA. Characterization of the Mycobacterium tuberculosis phagosome and evidence that phagosomal maturation is inhibited. J Exp Med 1995;181(1):257–70. https://doi.org/10.1084/jem.181.1.257.

[35] Sturgill-Koszycki S, Schaible UE, Russell DG. Mycobacterium-containing phagosomes are accessible to early endosomes and reflect a transitional state in normal phagosome biogenesis. EMBO J 1996;15(24):6960–8. https://doi.org/10.1002/j.1460-2075.1996.tb01088.x.

[36] Clemens DL, Horwitz MA. The Mycobacterium tuberculosis phagosome interacts with early endosomes and is accessible to exogenously administered transferrin. J Exp Med 1996;184(4):1349–55. https://doi.org/10.1084/jem.184.4.1349.

[37] Vergne I, Chua J, Lee HH, Lucas M, Belisle J, Deretic V. Mechanism of phagolysosome biogenesis block by viable Mycobacterium tuberculosis. Proc Natl Acad Sci U S A 2005;102(11):4033–8. https://doi.org/10.1073/pnas.0409716102.

[38] Hackam DJ, Rotstein OD, Zhang WJ, Demaurex N, Woodside M, Tsai O, Grinstein S. Regulation of phagosomal acidification. Differential targeting of Na +/H + exchangers, Na +/K +-ATPases, and vacuolar-type H +-ATPases. J Biol Chem 1997;272(47):29810–20. https://doi.org/10.1074/jbc.272.47.29810.

[39] Mukherjee S, Ghosh RN, Maxfield FR. Endocytosis. Physiol Rev 1997;77(3):759–803. https://doi.org/10.1152/physrev.1997.77.3.759.

[40] Liu D, Wang J, Xu Z, Chen X, Jiao Xin'an. Phagocytosis: strategies for macrophages to hunt Mycobacterium tuberculosis. One Health Adv 2024;2(1). https://doi.org/10.1186/s44280-024-00065-9.

[41] Westman J, Grinstein S. Determinants of phagosomal pH during host-pathogen interactions. Front Cell Dev Biol 2021;8. https://doi.org/10.3389/fcell.2020.624958.

[42] Graves AR, Curran PK, Smith CL, Mindell JA. The Cl-/H+ antiporter ClC-7 is the primary chloride permeation pathway in lysosomes. Nature 2008;453(7196):788–92. https://doi.org/10.1038/nature06907.

[43] Jin X, Zhang Y, Alharbi A, Hanbashi A, Alhoshani A, Parrington J. Targeting two-pore channels: current progress and future challenges. Trends Pharmacol Sci 2020;41(8):582–94. https://doi.org/10.1016/j.tips.2020.06.002.

[44] Nauseef WM. The phagocyte NOX2 NADPH oxidase in microbial killing and cell signaling. Curr Opin Immunol 2019;60:130–40. https://doi.org/10.1016/j.coi.2019.05.006.

[45] Wang G. Chloride flux in phagocytes. Immunol Rev 2016;273(1):219–31. https://doi.org/10.1111/imr.12438.

[46] Pungerčar JR, Caglič D, Sajid M, Dolinar M, Vasiljeva O, Požgan U, Turk D, Bogyo M, Turk V, Turk B. Autocatalytic processing of procathepsin B is triggered by proenzyme activity. FEBS J 2009;276(3):660–8. https://doi.org/10.1111/j.1742-4658.2008.06815.x.

[47] Gutierrez MG. Functional role(s) of phagosomal Rab GTPases. Small GTPases 2013;4(3). https://doi.org/10.4161/sgtp.25604.

[48] Weigert R, Yeung AC, Li J, Donaldson JG. Rab22a regulates the recycling of membrane proteins internalized independently of clathrin. Mol Biol Cell 2004;15(8):3758–70. https://doi.org/10.1091/mbc.E04-04-0342.

[49] Zhu H, Liang Z, Li G, Linstedt A. Rabex-5 is a Rab22 effector and mediates a Rab22-Rab5 signaling cascade in endocytosis. Mol Biol Cell 2009;20(22):4720–9. https://doi.org/10.1091/mbc.e09-06-0453.

[50] Kälin S, Hirschmann DT, Buser DP, Spiess M. Rabaptin5 is recruited to endosomes by Rab4 and Rabex5 to regulate endosome maturation. J Cell Sci 2015;128(22):4126–37. https://doi.org/10.1242/jcs.174664.

[51] Jeschke A, Haas A. Deciphering the roles of phosphoinositide lipids in phagolysosome biogenesis. Commun Integr Biol 2016;9(3):e1174798. https://doi.org/10.1080/19420889.2016.1174798.

[52] Mills IG, Urbé S, Clague MJ. Relationships between EEA1 binding partners and their role in endosome fusion. J Cell Sci 2001;114(10):1959–65. https://doi.org/10.1242/jcs.114.10.1959.

[53] Peplowska K, Markgraf DF, Ostrowicz CW, Bange G, Ungermann C. The CORVET tethering complex interacts with the yeast Rab5 homolog Vps21 and is involved in endo-lysosomal biogenesis. Dev Cell 2007;12(5):739–50. https://doi.org/10.1016/j.devcel.2007.03.006.

[54] Solinger JA, Spang A. Tethering complexes in the endocytic pathway: CORVET and HOPS. FEBS J 2013;280(12):2743–57. https://doi.org/10.1111/febs.12151.

[55] Lobingier BT, Merz AJ. Sec1/Munc18 protein Vps33 binds to SNARE domains and the quaternary SNARE complex. Mol Biol Cell 2012;23(23):4611–22. https://doi.org/10.1091/mbc.E12-05-0343.

[56] Kinchen JM, Ravichandran KS. Identification of two evolutionarily conserved genes regulating processing of engulfed apoptotic cells. Nature 2010;464(7289):778–82. https://doi.org/10.1038/nature08853.

[57] Nordmann M, Cabrera M, Perz A, Bröcker C, Ostrowicz C, Engelbrecht-Vandré S, Ungermann C. The Mon1-Ccz1 complex is the GEF of the late endosomal Rab7 homolog Ypt7. Curr Biol 2010;20(18):1654–9. https://doi.org/10.1016/j.cub.2010.08.002.

[58] Poteryaev D, Datta S, Ackema K, Zerial M, Spang A. Identification of the switch in early-to-late endosome transition. Cell 2010;141(3):497–508. https://doi.org/10.1016/j.cell.2010.03.011.

[59] Kinchen JM, Ravichandran KS. Phagosome maturation: going through the acid test. Nat Rev Mol Cell Biol 2008;9(10):781–95. https://doi.org/10.1038/nrm2515.

[60] Becken U, Jeschke A, Veltman K, Haas A. Cell-free fusion of bacteria-containing phagosomes with endocytic compartments. Proc Natl Acad Sci 2010;107(48):20726–31. https://doi.org/10.1073/pnas.1007295107.

[61] Huynh KK, Eskelinen EL, Scott CC, Malevanets A, Saftig P, Grinstein S. LAMP proteins are required for fusion of lysosomes with phagosomes. EMBO J 2007;26(2):313–24. https://doi.org/10.1038/sj.emboj.7601511.

[62] Ma X, Liu K, Li J, Li H, Li J, Liu Y, Yang C, Liang H. A non-canonical GTPase interaction enables ORP1L-Rab7-RILP complex formation and late endosome positioning. J Biol Chem 2018;293(36):14155–64. https://doi.org/10.1074/jbc.ra118.001854.

[63] Harrison RE, Bucci C, Vieira OV, Schroer TA, Grinstein S. Phagosomes fuse with late endosomes and/or lysosomes by extension of membrane protrusions along microtubules: role of Rab7 and RILP. Mol Cell Biol 2003;23(18):6494–506. https://doi.org/10.1128/MCB.23.18.6494-6506.2003.

[64] Jordens I, Fernandez-Borja M, Marsman M, Dusseljee S, Janssen L, Calafat J, Janssen H, Wubbolts R, Neefjes J. The Rab7 effector protein RILP controls lysosomal transport by inducing the recruitment of dynein-dynactin motors. Curr Biol 2001;11(21):1680–5. https://doi.org/10.1016/s0960-9822(01)00531-0.

[65] Wang J, Ge P, Qiang L, Tian F, Zhao D, Chai Q, Zhu M, Zhou R, Meng G, Iwakura Y, Gao GF, Liu CH. The mycobacterial phosphatase PtpA regulates the expression of host genes and promotes cell proliferation. Nat Commun 2017;8(1). https://doi.org/10.1038/s41467-017-00279 z.

[66] Zhang Q-ao, Ma S, Li P, Xie J. The dynamics of Mycobacterium tuberculosis phagosome and the fate of infection. Cell Signal 2023;108:110715. https://doi.org/10.1016/j.cellsig.2023.110715.

[67] Mehra A, Zahra A, Thompson V, Sirisaengtaksin N, Wells A, Porto M, Köster S, Penberthy K, Kubota Y, Dricot A, Rogan D, Vidal M, Hill DE, Bean AJ, Philips JA, Sassetti CM. Mycobacterium tuberculosis type VII secreted effector EsxH targets host ESCRT to impair trafficking. PLoS Pathog 2013;9(10):e1003734. https://doi.org/10.1371/journal.ppat.1003734.

[68] Mittal E, Skowyra ML, Uwase G, Tinaztepe E, Mehra A, Köster S, Hanson PI, Philips JA. Mycobacterium tuberculosis type VII secretion system effectors differentially impact the ESCRT endomembrane damage response. mBio 2018;9(6). https://doi.org/10.1128/mBio.01765-18.

[69] Venugopal A, Bryk R, Shi S, Rhee K, Rath P, Schnappinger D, Ehrt S, Nathan C. Virulence of Mycobacterium tuberculosis depends on lipoamide dehydrogenase, a member of three multienzyme complexes. Cell Host Microbe 2011;9(1):21–31. https://doi.org/10.1016/j.chom.2010.12.004.

[70] Deghmane A-E, Soualhine H, Bach H, Sendide K, Itoh S, Tam A, Noubir S, Talal A, Lo R, Toyoshima S, Av-Gay Y, Hmama Z. Lipoamide dehydrogenase mediates retention of coronin-1 on BCG vacuoles, leading to arrest in phagosome maturation. J Cell Sci 2007;120(16):2796–806. https://doi.org/10.1242/jcs.006221.

[71] Zulauf KE, Sullivan JT, Braunstein M, Behr MA. The SecA2 pathway of Mycobacterium tuberculosis exports effectors that work in concert to arrest phagosome and autophagosome maturation. PLoS Pathog 2018;14(4):e1007011. https://doi.org/10.1371/journal.ppat.1007011

[72] Walburger A, Koul A, Ferrari G, Nguyen L, Prescianotto-Baschong C, Huygen K, Klebl B, Thompson C, Bacher G, Pieters J. Protein kinase G from pathogenic mycobacteria promotes survival within macrophages. Science 2004;304(5678):1800–4. https://doi.org/10.1126/science.1099384.

[73] van der Woude AD, Stoop EJM, Stiess M, Wang S, Ummels R, van Stempvoort G, Piersma SR, Cascioferro A, Jiménez CR, Houben ENG, Luirink J, Pieters J, van der Sar AM, Bitter W. Analysis of secA2-dependent substrates in mycobacterium marinum identifies protein kinase G (PknG) as a virulence effector. Cell Microbiol 2014;16(2):280–95. https://doi.org/10.1111/cmi.12221.

[74] Osman MM, Shanahan JK, Chu F, Takaki KK, Pinckert ML, Pagan AJ, Brosch R, Conrad WH, Ramakrishnan L. The C terminus of the mycobacterium ESX-1 secretion system substrate ESAT-6 is required for phagosomal membrane damage and virulence. Proc Natl Acad Sci U S A 2022;119(11). https://doi.org/10.1073/pnas.2122161119.

[75] Naeem MA, Ahmad W, Tyagi R, Akram Q, Younus M, Liu X. Stealth strategies of mycobacterium tuberculosis for immune evasion. Curr Issues Mol Biol 2021;41:597–616. https://doi.org/10.21775/CIMB.041.597.

[76] Pal R, Bisht MK, Mukhopadhyay S. Secretory proteins of Mycobacterium tuberculosis and their roles in modulation of host immune responses: focus on therapeutic targets. FEBS J 2022;289(14):4146–71. https://doi.org/10.1111/febs.16369.
[77] Pradhan G, Shrivastva R, Mukhopadhyay S. Mycobacterial PknG targets the Rab7l1 signaling pathway to inhibit phagosome–lysosome fusion. J Immunol 2018;201(5):1421–33. https://doi.org/10.4049/jimmunol.1800530.
[78] Ramakrishnan L, Federspiel NA, Falkow S. Granuloma-specific expression of Mycobacterium virulence proteins from the glycine-rich PE-PGRS family. Science 2000;288(5470):1436–9. https://doi.org/10.1126/science.288.5470.1436.
[79] Hagedorn M, Soldati T. Flotillin and RacH modulate the intracellular immunity of Dictyostelium to Mycobacterium marinum infection. Cell Microbiol 2007;9(11):2716–33. https://doi.org/10.1111/j.1462-5822.2007.00993.x.
[80] Chiaradia L, Lefebvre C, Parra J, Marcoux J, Burlet-Schiltz O, Etienne G, Tropis M, Daffé M. Dissecting the mycobacterial cell envelope and defining the composition of the native mycomembrane. Sci Rep 2017;7(1). https://doi.org/10.1038/s41598-017-12718-4.
[81] Gago G, Diacovich L, Gramajo H. Lipid metabolism and its implication in mycobacteria–host interaction. Curr Opin Microbiol 2018;41:36–42. https://doi.org/10.1016/j.mib.2017.11.020.
[82] Queiroz A, Riley LW. Bacterial immunostat: Mycobacterium tuberculosis lipids and their role in the host immune response. Rev Soc Bras Med Trop 2017;50(1):9–18. https://doi.org/10.1590/0037-8682-0230-2016.
[83] Simeone R, Sayes F, Lawarée E, Brosch R. Breaching the phagosome, the case of the tuberculosis agent. Cell Microbiol 2021;23(7). https://doi.org/10.1111/cmi.13344.
[84] Augenstreich J, Briken V. Host cell targets of released lipid and secreted protein effectors of Mycobacterium tuberculosis. Front Cell Infect Microbiol 2020;10. https://doi.org/10.3389/fcimb.2020.595029.
[85] Hunter RL, Olsen M, Jagannath C, Actor JK. Trehalose 6,6′-dimycolate and lipid in the pathogenesis of caseating granulomas of tuberculosis in mice. Am J Pathol 2006;168(4):1249–61. https://doi.org/10.2353/ajpath.2006.050848.
[86] Hunter RL, Venkataprasad N, Olsen MR. The role of trehalose dimycolate (cord factor) on morphology of virulent M. tuberculosis in vitro. Tuberculosis 2006;86(5):349–56. https://doi.org/10.1016/j.tube.2005.08.017.
[87] Uribe-Querol E, Rosales C. Control of phagocytosis by microbial pathogens. Front Immunol 2017;8. https://doi.org/10.3389/fimmu.2017.01368.
[88] Santamaria C, Biegas KJ, Kim CY, Cabral J, Lee JR, Swarts BM, Siegrist MS. Trehalose dimycolate inhibits phagosome maturation and promotes intracellular M. tuberculosis growth via noncanonical SNARE interaction. bioRxiv 2024. https://doi.org/10.1101/2024.12.16.627577.
[89] Bah A, Sanicas M, Nigou J, Guilhot C, Astarie-Dequeker C, Vergne I. The lipid virulence factors of Mycobacterium tuberculosis exert multilayered control over autophagy-related pathways in infected human macrophages. Cells 2020;9(3):666. https://doi.org/10.3390/cells9030666.
[90] Upadhyay S, Philips JA. LC3-associated phagocytosis: host defense and microbial response. Curr Opin Immunol 2019;60:81–90. https://doi.org/10.1016/j.coi.2019.04.012.
[91] Kim JK, Silwal P, Jo EK. Host-pathogen dialogues in autophagy, apoptosis, and necrosis during mycobacterial infection. Immune Netw 2020;20(5):1–15. https://doi.org/10.4110/in.2020.20.e37.
[92] Franco LH, Nair VR, Scharn CR, Xavier RJ, Torrealba JR, Shiloh MU, Levine B. The ubiquitin ligase Smurf1 functions in selective autophagy of Mycobacterium tuberculosis and anti-tuberculous host defense. Cell Host Microbe 2017;21(1):59–72. https://doi.org/10.1016/j.chom.2016.11.002.
[93] Manzanillo PS, Ayres JS, Watson RO, Collins AC, Souza G, Rae CS, Schneider DS, Nakamura K, Shiloh MU, Cox JS. The ubiquitin ligase parkin mediates resistance to intracellular pathogens. Nature 2013;501(7468):512–6. https://doi.org/10.1038/nature12566.

[94] Köster S, Upadhyay S, Chandra P, Papavinasasundaram K, Yang G, Hassan A, Grigsby SJ, Mittal E, Park HS, Jones V, Hsu FF, Jackson M, Sassetti CM, Philips JA. Mycobacterium tuberculosis is protected from NADPH oxidase and LC3-associated phagocytosis by the LCP protein CpsA. Proc Natl Acad Sci U S A 2017;114(41):E8711. https://doi.org/10.1073/pnas.1707792114.
[95] Wong KW. The role of ESX-1 in Mycobacterium tuberculosis pathogenesis. Microbiol Spectr 2017;5(3). https://doi.org/10.1128/microbiolspec.TBTB2-0001-2015.
[96] Dewitt S, Green J, Laffafian I, Lewis KJ, Hallett MB. Intraphagosomal free Ca2+ changes during phagocytosis. Int J Mol Sci 2024;25(8). https://doi.org/10.3390/ijms25084254.
[97] Duran J, Salinas JE, Wheaton Rping, Poolsup S, Allers L, Rosas-Lemus M, Chen L, Cheng Q, Pu J, Salemi M, Phinney B, Ivanov P, Lystad AH, Bhaskar K, Rajaiya J, Perkins DJ, Jia J. Calcium signaling from damaged lysosomes induces cytoprotective stress granules. EMBO J 2024;43(24):6410–43. https://doi.org/10.1038/s44318-024-00292-1.
[98] Corkery DP, Li S, Wijayatunga D, Herzog LK, Knyazeva A, Wu YW. ESCRT recruitment to damaged lysosomes is dependent on the ATG8 E3-like ligases. bioRxiv 2024. https://doi.org/10.1101/2024.04.30.591897.
[99] Sun L, Wu J, Du F, Chen X, Chen ZJ. Cyclic GMP-AMP synthase is a cytosolic DNA sensor that activates the type I interferon pathway. Science 2013;339(6121):786–91. https://doi.org/10.1126/science.1232458.
[100] Civril F, Deimling T, De Oliveira Mann CC, Ablasser A, Moldt M, Witte G, Hornung V, Hopfner KP. Structural mechanism of cytosolic DNA sensing by cGAS. Nature 2013;498(7454):332–7. https://doi.org/10.1038/nature12305.
[101] Tanaka Y, Chen ZJ. STING specifies IRF3 phosphorylation by TBK1 in the cytosolic DNA signaling pathway. Sci Signal 2012;5(214). https://doi.org/10.1126/scisignal.2002521.
[102] Oakes JA, Davies MC, Collins MO. TBK1: a new player in ALS linking autophagy and neuroinflammation. Mol Brain 2017;10(1):1–10. https://doi.org/10.1186/s13041-017-0287-x.
[103] Manzanillo PS, Shiloh MU, Portnoy DA, Cox JS. Mycobacterium tuberculosis activates the DNA-dependent cytosolic surveillance pathway within macrophages. Cell Host Microbe 2012;11(5):469–80. https://doi.org/10.1016/j.chom.2012.03.007.
[104] Hu W, Chan H, Lu L, Wong KT, Wong SH, Li MX, Xiao ZG, Cho CH, Gin T, Chan MTV, Wu WKK, Zhang L. Autophagy in intracellular bacterial infection. Semin Cell Dev Biol 2020;101:41–50. https://doi.org/10.1016/j.semcdb.2019.07.014.
[105] Shariq M, Khan MF, Raj R, Ahsan N, Kumar P. PRKAA2, MTOR, and TFEB in the regulation of lysosomal damage response and autophagy. J Mol Med 2024;102(3):287–311. https://doi.org/10.1007/s00109-023-02411-7.
[106] Bilkei Gorzo O, Heunis T, Marín-Rubio JL, Cianfanelli FR, Raymond BBA, Inns J, Fabrikova D, Peltier J, Oakley F, Schmid R, Härtlova A, Trost M. The E3 ubiquitin ligase RNF115 regulates phagosome maturation and host response to bacterial infection. EMBO J 2022;41(23). https://doi.org/10.15252/embj.2021108970.
[107] Chauhan S, Kumar S, Jain A, Ponpuak M, Mudd MH, Kimura T, Choi SW, Peters R, Mandell M, Bruun J-A, Johansen T, Deretic V. TRIMs and galectins globally cooperate and TRIM16 and galectin-3 Co-direct autophagy in endomembrane damage homeostasis. Dev Cell 2016;39(1):13–27. https://doi.org/10.1016/j.devcel.2016.08.003.
[108] Romagnoli A, Di Rienzo M, Petruccioli E, Fusco C, Palucci I, Micale L, Mazza T, Delogu G, Merla G, Goletti D, Piacentini M, Fimia GM. The ubiquitin ligase TRIM32 promotes the autophagic response to Mycobacterium tuberculosis infection in macrophages. Cell Death Dis 2023;14(8). https://doi.org/10.1038/s41419-023-06026-1.
[109] Levin-Konigsberg R, Mantegazza AR. A guide to measuring phagosomal dynamics. FEBS J 2021;288(5):1412–33. https://doi.org/10.1111/febs.15506.

[110] McDonough KA, Kress Y, Bloom BR. Pathogenesis of tuberculosis: interaction of Mycobacterium tuberculosis with macrophages. Infect Immun 1993;61(7):2763–73. https://doi.org/10.1128/iai.61.7.2763-2773.1993.

[111] Watson RO, Manzanillo PS, Cox JS. Extracellular M. tuberculosis DNA targets bacteria for autophagy by activating the host DNA-sensing pathway. Cell 2012;150(4):803–15. https://doi.org/10.1016/j.cell.2012.06.040.

[112] Fratti RA, Backer JM, Gruenberg J, Corvera S, Deretic V. Role of phosphatidylinositol 3-kinase and Rab5 effectors in phagosomal biogenesis and mycobacterial phagosome maturation arrest. JCB (J Cell Biol) 2001;154(3):631–44. https://doi.org/10.1083/jcb.200106049.

[113] Guichard A, Nizet V, Bier E. RAB11-mediated trafficking in host–pathogen interactions. Nat Rev Microbiol 2014;12(9):624–34. https://doi.org/10.1038/nrmicro3325.

[114] Berón W, Alvarez-Dominguez C, Mayorga L, Stahl PD. Membrane trafficking along the phagocytic pathway. Trends Cell Biol 1995;5(3):100–4. https://doi.org/10.1016/S0962-8924(00)88958-8.

[115] Cheng X, Zhang X, Yu L, Xu H. Calcium signaling in membrane repair. Semin Cell Dev Biol 2015;45: 24–31. https://doi.org/10.1016/j.semcdb.2015.10.031.

[116] Nakamura M, Parkhurst SM. Calcium influx rapidly establishes distinct spatial recruitments of Annexins to cell wounds. Genetics 2024;227(4). https://doi.org/10.1093/genetics/iyae101.

[117] Deretic V. Autophagy in inflammation, infection, and immunometabolism. Immunity 2021;54(3):437–53. https://doi.org/10.1016/j.immuni.2021.01.018.

[118] Armstrong JA, Hart PD. Phagosome-lysosome interactions in cultured macrophages infected with virulent tubercle bacilli. Reversal of the usual nonfusion pattern and observations on bacterial survival. J Exp Med 1975;142(1):1–16. https://doi.org/10.1084/jem.142.1.1.

[119] Chandra P, Grigsby SJ, Philips JA. Immune evasion and provocation by Mycobacterium tuberculosis. Nat Rev Microbiol 2022;20(12):750–66. https://doi.org/10.1038/s41579-022-00763-4.

[120] Malik AA, Shariq M, Sheikh JA, Zarin S, Ahuja Y, Fayaz H, Alam A, Ehtesham NZ, Hasnain SE. Activation of the lysosomal damage response and selective autophagy: the coordinated actions of galectins, TRIM proteins, and CGAS-STING1 in providing immunity against Mycobacterium tuberculosis. Crit Rev Microbiol 2025;51(1):108–27. https://doi.org/10.1080/1040841X.2024.2321494.

[121] Moreira AL, Wang J, Tsenova-Berkova L, Hellmann W, Freedman VH, Kaplan G. Sequestration of Mycobacterium tuberculosis in tight vacuoles in vivo in lung macrophages of mice infected by the respiratory route. Infect Immun 1997;65(1):305–8. https://doi.org/10.1128/iai.65.1.305-308.1997.

[122] Schnettger L, Rodgers A, Repnik U, Lai RP, Pei G, Verdoes M, Wilkinson RJ, Young DB, Gutierrez MG. A Rab20-dependent membrane trafficking pathway controls M. tuberculosis replication by regulating phagosome spaciousness and integrity. Cell Host Microbe 2017;21(5):619. https://doi.org/10.1016/j.chom.2017.04.004.

[123] Anes E, Kühnel MP, Bos E, Moniz-Pereira J, Habermann A, Griffiths G. Selected lipids activate phagosome actin assembly and maturation resulting in killing of pathogenic mycobacteria. Nat Cell Biol 2003;5(9): 793–802. https://doi.org/10.1038/ncb1036.

[124] Jani C, Marsh A, Uchil P, Jain N, Baskir ZR, Glover OT, Root DE, Doench JG, Barczak AK. Vps18 contributes to phagosome membrane integrity in Mycobacterium tuberculosis-infected macrophages. bioRxiv 2023. https://doi.org/10.1101/2023.10.01.560397.

[125] Yoshimura A, Suzuki M, Sakaguchi R, Hanada T, Yasukawa H. SOCS, inflammation, and autoimmunity. Front Immunol 2012;3. https://doi.org/10.3389/fimmu.2012.00020.

[126] Yang J, Li Z, Li A, Liu Y, Zhang X, Zhang Y, Gao Y. The tryptophan-aspartate (WD) repeat domain of bovine Coronin-1A promotes mycobacterial survival by inhibiting calcium signaling-mediated phagosome-lysosome fusion. Vet Res 2025;56(1):33. https://doi.org/10.1186/s13567-025-01471-6.

CHAPTER

6

Autophagy and *Mycobacterium tuberculosis*: Host defense mechanisms versus pathogen evasion

Yashika Ahuja[1], Waseem Ali[1], Haleema Fayaz[1], Anwar Alam[1, 2, 3], Nasreen Z. Ehtesham[1] and Seyed E. Hasnain[1, 4]

[1]*Department of Life Science, School of Bioscience and Technology, Sharda University, Greater Noida, Uttar Pradesh, India;* [2]*Department of Biotechnology, School of Bioscience and Technology, Sharda University, Greater Noida, Uttar Pradesh, India;* [3]*DST-FIST, Sharda University, Greater Noida, Uttar Pradesh, India;* [4]*Department of Biochemical Engineering and Biotechnology, IIT Delhi, New Delhi, India*

Highlights

1. Autophagy is a key innate defense against *Mycobacterium tuberculosis* (*M. tb*).
2. *M. tb* evades autophagy to persist intracellularly.
3. Host directed therapies that boost autophagy enhance *M. tb* clearance.
4. Vitamin D and repurposed drugs show potential as autophagy adjuncts.
5. Challenges include complex regulation, pathogen adaptation, and host variability.
6. Future focus: selective autophagy modulators and personalized strategies.
7. Combining autophagy enhancers with TB drugs may improve efficacy and shorten therapy.

Abbreviations

AAA+ ATPases associated with diverse cellular activities
ABC ATP binding Cassette
AICAR 5-Aminoimidazole-4-carboxamide ribonucleotide
AKT Protein Kinase B
ALPL Alkaline phosphatase, biomineralization associated
AMPK AMP activated protein kinase
ATG Autophagy related gene
ATP Adenosine triphosphate
BCG Bacille Calmette-Guérin
BCL2L11 BCL2 Like 11 (Bim)
BECN1 Beclin-1
CALCOCO2 Calcium binding and coiled-coil domain 2 (NDP52)

Navigating Tuberculosis. https://doi.org/10.1016/B978-0-443-27735-1.00015-0

CMA Chaperone mediated autophagy
CtpF Calcium transporting P2A ATPase
DNA Deoxyribonucleic acid
Eis Enhanced intracellular survival protein
EPG5 Ectopic P granules 5 autophagy tethering factor
ERRα Estrogen related receptor alpha
ESAT-6 Early secreted antigenic target of 6 kDa
ESX-1 ESAT-6 secretion system-1
FUNDC1 FUN14 domain containing 1
GABA Gamma aminobutyric acid
HIV Human immunodeficiency virus
HOPS Homotypic fusion and protein sorting complex
HSV-1 Herpes simplex virus 1
IKKβ Inhibitor of nuclear factor kappa B kinase subunit beta
IL-10 Interleukin-10
IL-1β Interleukin-1 Beta
IMQ Imiquimod
iNOS Inducible nitric oxide synthase
IRGM Immunity related GTPase M
LAM Lipoarabinomannan
LAP LC3-associated phagocytosis
LC3 Microtubule associated protein 1A/1B-light chain 3
LL-37 Cathelicidin antimicrobial peptide
MAC Mycobacterium avium complex
ManLAM Mannose capped lipoarabinomannan
MAP1LC3 Microtubule associated protein 1 light chain 3
MAPK Mitogen activated protein kinase
MDR Multidrug resistant
MHC Major histocompatibility complex
miRNA MicroRNA
mTOR Mechanistic target of rapamycin
mTORC1 mTOR complex 1
NBR1 NBR1 autophagy cargo receptor
NF-κB Nuclear factor kappa B
NLRP3 NLR family pyrin domain containing 3
NOD2 Nucleotide binding oligomerization domain-containing protein 2
NR1D1 Nuclear receptor subfamily 1, group D, member 1
NTM Nontuberculous mycobacteria
OPTN Optineurin
PBA Phenylbutyrate
PE/PPE Proline glutamate/proline proline glutamate (Mycobacterial protein family)
PI3K Phosphatidylinositol 3-kinase
PI3KC3 Phosphatidylinositol 3-kinase catalytic subunit type 3
PI3P Phosphatidylinositol 3-phosphate
PPARα Peroxisome proliferator activated receptor alpha
PRKAA2 Protein kinase AMP activated catalytic subunit alpha 2
PRKN Parkin RBR E3 ubiquitin protein ligase
PRR Pattern recognition receptor
PTPA/PTPB Protein tyrosine phosphatase A/B

Rab1A/RAB14 Ras related protein 1A/14
ROS Reactive oxygen species
rpS30 Ribosomal protein S30
RUBCN Rubicon autophagy regulator
SBL Soybean lectin
SIRT1/SIRT3 Sirtuin 1/3
SMURF1 SMAD specific E3 Ubiquitin Protein Ligase 1
SNARE Soluble N-ethylmaleimide sensitive factor attachment protein receptor
SNP Single nucleotide polymorphism
SOCS Suppressor of cytokine signaling
SQSTM1 Sequestosome 1 (p62)
STING Stimulator of interferon genes
STX/SNAP25 Syntaxin/Synaptosome associated protein 25
T1R Type 1 reversal reaction (Leprosy)
TAX1BP1 Tax1-binding protein 1
TB Tuberculosis
TBC1D4 TBC1 domain family member 4
TCA Tricarboxylic acid cycle
TFEB Transcription factor EB
TLR Toll like receptor
TNF-α Tumor necrosis factor alpha
TOLLIP Toll interacting protein
TRIM Tripartite motif containing protein
ULK1 Unc-51 Like autophagy activating kinase 1
V-ATPase Vacuolar type H+-ATPase
VAMP Vesicle associated membrane protein
XDR Extensively drug resistant
Zmp1 Zinc metalloprotease 1

1. Introduction

Tuberculosis (TB) constitutes a significant global health challenge, with epidemiological data from 2023 reporting approximately 10.8 million incidences and 1.25 million associated deaths [1]. The causative agent, *Mycobacterium tuberculosis* (*M. tb*), is notable for its ability to persist intracellularly within macrophages, primarily by establishing a latent state [2]. Additionally, the capacity to develop resistance to antitubercular drugs presents a considerable clinical obstacle [3]. The survival strategies of *M. tb* rely on modulating host immune responses, evading macrophage mediated destruction, and circumventing autophagic degradation (Fig. 6.1) [4,5].

Autophagy, a conserved homeostatic process, is responsible for the selective degradation of pathogen containing cargos and the recycling of cellular components, thereby playing a critical role in host defense and adaptation to exogenous stress. It represents a crucial innate immune mechanism that directly targets intracellular *M. tb* for degradation, thereby forming a central axis in the host pathogen interaction. A key strategy employed by *M. tb* to subvert this process involves the early arrest of phagosome maturation following internalization by macrophages [6]. This disruption prevents the fusion of the phagosome with lysosomes, thereby protecting the pathogen from the harsh degradative environment of the phagolysosome and preventing bacterial elimination [7,8]. Recent research has demonstrated that the exogenous induction of autophagy in *M. tb* infected macrophages can effectively counteract this blockade. Autophagy stimulation restores phagosome maturation, facilitating the

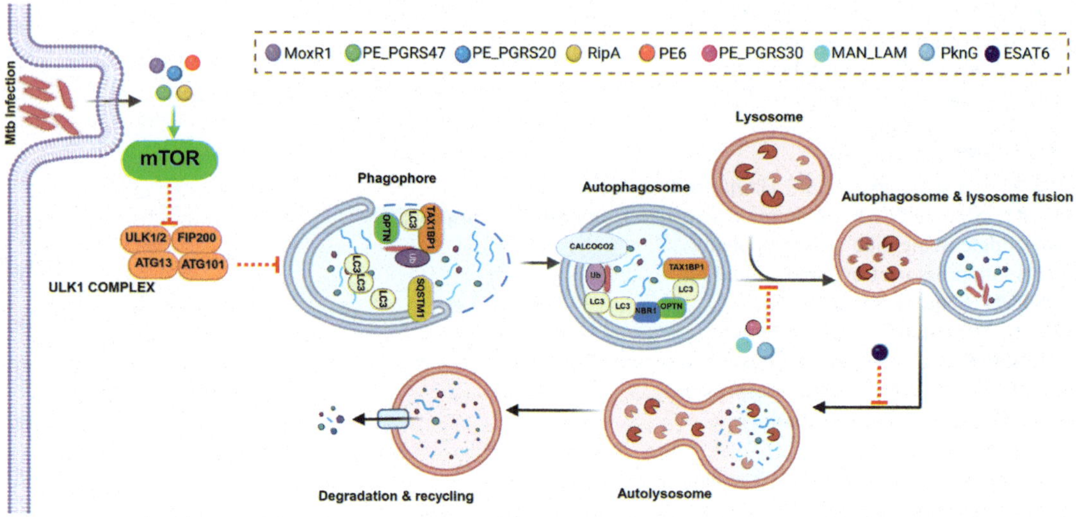

FIGURE 6.1 Inhibition of host autophagy by *Mycobacterium tuberculosis* (*M. tb*) proteins.

This figure illustrates how *Mycobacterium tuberculosis* (*M. tb*) disrupts distinct stages of the host autophagy pathway using various virulence factors. Mtb infection activates the mechanistic target of rapamycin (mTOR), a central negative regulator of autophagy. Activated mTOR inhibits the ULK1 complex thereby suppressing autophagy initiation. While multiple *M.tb* proteins, including MoxR1, PE_PGRS47, PE_PGRS20, RipA, and PE6, act upstream by enhancing mTOR activity to inhibit phagophore formation, others (MAN_LAM, PknG) inhibit autophagosome lysosome fusion or impede autolysosome maturation and function (ESAT-6). These targeted interruptions at multiple steps collectively enable *M. tb* to subvert host defense mechanisms and evade immune clearance ensuring intracellular survival. The color coded legend indicates the specific *M. tb* proteins and their respective points of interference in the autophagy pathway.

delivery of mycobacteria to degradative compartments and promoting bacterial clearance [9,10]. The process of autophagy involves the formation of double membraned vesicles or autophagosomes which encapsulate cytoplasmic material for degradation in lysosomes, significantly strengthening the innate immune response, particularly *via* the clearance of intracellular pathogens [11]. Autophagy is vital for controlling infections by various intracellular pathogens, including *M. tb*. It serves as an antimicrobial defense mechanism by targeting pathogens for lysosomal degradation, thereby limiting their replication. Autophagy can also activate adaptive immunity by promoting antigen presentation, thus bridging the innate and adaptive arms of the immune system [12]. This chapter explores the complex relationship between autophagy and *M. tb* infection. It examines the host's use of autophagy as a defense mechanism against *M. tb* and how *M. tb* has evolved various strategies to evade this process.

2. Autophagy: Mechanisms and function in host defense

Autophagy is a fundamental and evolutionarily conserved process crucial for maintaining cellular homeostasis and defending against infections [13]. It is categorized into macroautophagy, microautophagy, and chaperone mediated autophagy (CMA), each contributing uniquely to immune defense. **Macroautophagy**, the most well studied form, involves the formation of autophagosomes that engulf cytoplasmic material and subsequently fuse with lysosomes for degradation [14]. This process is meticulously regulated by a series of key molecules, including unc-51 like autophagy activating kinase 1 (ULK1), beclin-1 (BECN1), phosphatidylinositol 3-kinase complex (PI3KC3), microtubule associated protein 1 light chain 3 (MAP1LC3/LC3), and autophagy related (ATG) proteins, which coordinate the stages of autophagy [15]. Cellular stress, such as nutrient starvation or low ATP levels, initiates autophagy, a critical degradation pathway. Under nutrient rich conditions, the mechanistic target of rapamycin kinase complex 1 (mTORC1) inhibits the ULK1 complex, thereby suppressing autophagy. However, when nutrients are scarce, mTORC1 inactivation permits ULK1 to initiate autophagy. This regulatory interplay between mTORC1 and ULK1 is essential for controlling autophagy initiation, as evidenced by research demonstrating ULK1's capacity to inhibit mTORC1 signaling and promote Raptor phosphorylation [16]. Concurrently, low ATP levels activate protein kinase AMP activated catalytic subunit alpha 2 (PRKAA2/AMPK), which subsequently phosphorylates and activates ULK1. The active ULK1 complex then phosphorylates BECN1 within the PI3KC3-C1 complex, leading to the secretion of phosphatidyl inositol 3 phosphate (PI3P), an essential molecule for nucleating the phagophore. This membrane, originating from endoplasmic reticulum (ER) or Golgi, elongates through the action of ATG proteins (ATG7, ATG3, ATG4, ATG5, ATG12) and LC3 lipidation. The presence of LC3 facilitates cargo recruitment through various adaptor proteins, including optineurin (OPTN), Tax1 binding protein 1 (TAX1BP1), NBR1 autophagy cargo receptor (NBR1), toll interacting protein (TOLLIP), FUN14 domain containing 1 (FUNDC1), SMAD specific E3 ubiquitin protein ligase 1 (SMURF1), tripartite motif containing (TRIM) and baculoviral IAP repeat containing 6 (BIRC6/BRUCE). Subsequently, ATG9 containing vesicles fuse to close the phagophore, forming the double membrane autophagosome. The final step involves autophagolysosomal fusion, a process mediated by soluble N-ethylmaleimide-sensitive factor attachment protein receptor (SNARE) proteins (ST8 alpha-N-acetylneuraminide alpha-2,8-sialyltransferase 2 (ST8SIA2/STX), vesicle associated membrane protein 1 (VAMP), synaptosome associated protein 25 (SNAP25/SNAP), ectopic P-granules 5 autophagy tethering factor (EPG5), alkaline phosphatase, biomineralization associated (ALPL/HOPS), pleckstrin homology and RUN domain containing M1 (PLEKHM1), and tectonin beta-propeller repeat-containing 1 (TECPR1), culminating in the formation of the autolysosome where cargo degradation occurs *via* hydrolytic enzymes [17].

Microautophagy, by contrast, facilitates basal cellular turnover through direct invaginations of the lysosomal membrane, whereas CMA selectively degrades proteins by recognizing chaperone tagged substrates and transporting them to lysosomes *via* lysosomal associated membrane protein 2 (LAMP2) [18]. Although these pathways have traditionally been considered independent, emerging research indicates functional crosstalk, particularly in response to cellular stress, revealing the complexity of autophagy regulation [19]. Beyond its role in maintaining homeostasis, autophagy is essential for host defense, particularly in controlling intracellular pathogens such as Mycobacteria. Upon infection, macrophages phagocytose *M. tb*, enclosing it within a phagosome, which under normal conditions fuses with lysosomes for bacterial degradation. However, *M. tb* actively inhibits phagosome lysosome fusion by deploying multiple virulence factors such as ESAT-6 and Zmp1, allowing intracellular survival [20,21]. Autophagy counteracts this immune evasion by tagging intracellular *M. tb* through ubiquitin binding autophagy receptors such as sequestosome1 (SQSTM1/p62) and calcium binding and coiled coil domain 2 (CALCOCO2/NDP52), which recruit LC3 and facilitate xenophagy mediated clearance [22,23]. The significance of xenophagy in TB immunity is highlighted by genetic studies demonstrating that polymorphisms in autophagy related genes (ATG5, ATG16L1) correlate with susceptibility to *M. tb* infection, suggesting that impaired autophagy acts as a determinant of disease progression [24]. These findings raise critical questions about the feasibility of therapeutic autophagy modulation, particularly given that excessive or dysregulated autophagy could lead to unintended immune suppression. These findings not only point out that autophagy is crucial for host defense against intracellular pathogens, such as *M. tb*, but also suggest the critical consideration of therapeutically modulating autophagy, as enhanced and ectopic activation might inadvertently compromise immune homeostasis.

Autophagy serves as a critical link between innate and adaptive immunity by modulating antigen presentation. Microbial components that undergo degradation are processed and presented *via* MHC-II in turn activating $CD4^+$ T cells, while cross presentation through MHC class I enhances $CD8^+$ cytotoxic responses [25]. Nonetheless, the extent to which autophagy contributes to antigen processing relative to conventional proteasomal degradation remains debatable. Some studies propose that autophagy augments antigen cross presentation in dendritic cells, whereas others suggest pathogen specific variations in antigen routing mechanisms [26]. This complexity reveals the necessity for further research to ascertain whether modulating autophagy could enhance vaccine efficacy, particularly against intracellular pathogens. The autophagic process interacts with pattern recognition receptors (PRRs) such as toll like receptors (TLRs) and nucleotide binding oligomerization domain containing 2 (NOD2), which detect microbial components and induce nuclear factor kappa B (NF-κB) activation and mitogen activated protein kinase (MAPK) pathways, resulting in production of proinflammatory cytokines like tumor necrosis factor alfa (TNF-α) and interleukin 1 beta (IL-1β) [27–29]. Furthermore, autophagy regulates inflammasome activation, particularly the NLR family pyrin domain containing 3 (NLRP3) inflammasome, which activates caspase-1, resulting in IL-1β maturation and secretion, a key inflammatory response in bacterial infections [30]. However, the role of autophagy in inflammasome regulation is context dependent, while it can enhance cytokine responses during infection, excessive autophagy may also degrade inflammasome components, thereby dampening immune activation [31]. This dual role suggests that modulating autophagy for therapeutic purposes necessitates precise control to avoid immune suppression.

In case of viral infections, autophagy exhibits both antiviral and pro viral roles. While it restricts viral replication in pathogens like human immune deficiency virus (HIV), Influenza A, and severe acute respiratory syndrome coronavirus 2 (SARS-CoV-2), certain viruses, including herpes simplex virus 1 (HSV-1) and hepatitis C virus (HCV), actively inhibit autophagy to evade host immunity [32,33]. This viral subversion highlights the challenge of employing autophagy in antiviral therapy, as indiscriminate enhancement of autophagy could paradoxically benefit some viruses. Given its multifaceted role, autophagy is being investigated as a therapeutic target in other infectious diseases as well. Pharmacological agents such as mTOR inhibitors (rapamycin), vitamin D3, and statins have demonstrated potential in enhancing xenophagy and bacterial clearance, yet their effects on global immune responses remain incompletely understood [34]. A significant challenge lies in ensuring pathogen specific autophagy activation without disrupting homeostatic autophagic functions essential for cell survival. Developing precision targeted autophagy modulators and exploring potential synergies with antimicrobial therapies, particularly in antibiotic resistant infections, may thus prove extremely beneficial.

3. Harnessing host autophagy: Repurposed drugs and natural compounds for tuberculosis control

Novel cutting edge strategies leveraging host directed therapy (HDT) to induce autophagy represent a promising avenue for the effective and rapid treatment of drug-resistant tuberculosis, including multi drug resistant (MDR) and extensively drug resistant (XDR) tuberculosis. Several repurposed drugs have exhibited antimicrobial activity against mycobacteria by inducing autophagy. Vitamin D enhances the expression of cathelicidin, an antimicrobial peptide that facilitates autophagosome maturation and mycobacterial eradication. In human monocytes and macrophages, vitamin D enhances the activation of antimicrobial autophagy and elimination of intracellular *M. tb* through induction of cathelicidin and autophagy gene activation. Phenylbutyrate (PBA) synergizes with vitamin D to stimulate LL-37 expression, thereby reinforcing autophagy activation in macrophages. The combination of PBA and vitamin D3 elicits a favorable immunomodulatory response in peripheral blood mononuclear cells (PBMC), effectively suppressing intracellular inflammation. Notably, the therapeutic outcomes of vitamin D supplementation may vary depending on individual differences as well as experimental parameters [34–36]. Other repurposed drugs that induce autophagy include rapamycin, metformin, valproic acid, gefitinib, and fluoxetine (Fig. 6.2).

A range of small molecules, including SMER18, SMER28, GABA, NSC 18725, AICAR, and SRT 1720, have been evaluated for their potential to induce autophagy and inhibit mycobacterial replication, yielding promising results in various cell lines and primary cells. Remarkably, plant derived compounds also exhibit efficacy in activating autophagy against mycobacterial infections. Resveratrol, a stilbene derivative, induces autophagy and facilitates the fusion of lysosome phagosome complexes by activating sirtuin 1 (SIRT1) deacetylase. Baicalin, a flavone glycoside, promotes autophagy by inhibiting the AKT serine/threonine kinase/phosphatidyl inositol 3 kinase/mTOR (Akt/PI3K/mTOR) pathway. Pasakbumin A, an eurycomanone compound from *Eurycoma longifolia*, suppresses *M. tb* replication. Soybean lectin (SBL) enhances autophagy and restricts the proliferation of *M. bovis* BCG and *M. smegmatis* mc2 155. Epigallocatechin-3-gallate (EGCG) from green tea extract enhances autophagic flux and inhibits intracellular mycobacterial proliferation. Honokiol, a polyphenol, activates SIRT3, inducing the PPARα transcription factor and upregulating genes essential for autophagic flux [37–40].

Nuclear receptors (NRs) also contribute to autophagy induction and antimicrobial defense during mycobacterial infection. Nuclear receptor subfamily 1, group D, member 1 (NR1D1) enhances autophagic flux and antimycobacterial effects in human macrophages by upregulating transcription factor EB (TFEB) and increasing lysosome and autophagosome numbers. NR1D1, along with peroxisome proliferator activated receptor alpha (PPARα), activates autophagy and lysosomal biogenesis [41]. The orphan nuclear receptor estrogen related receptor α (ERRα), a vital regulator of metabolic homeostasis, activates autophagy and boosts antimycobacterial responses by altering key autophagy genes and proteins transcriptionally and posttranslationally. Estrogen related receptor alpha (ESRRA/ERRα) regulates ATGs such as ATG5, ATG6, and ATG16L1. PPARα also promotes host innate immunity and lipid degradation *via* autophagy through transcriptional activation of ATGs. Additionally, imiquimod (IMQ), a nucleoside analogue of imidazoquinolinone, triggers TLR7, leading to increased autophagy and reduced *M. tb* infection *via* enhanced reactive oxygen species (ROS) production [42].

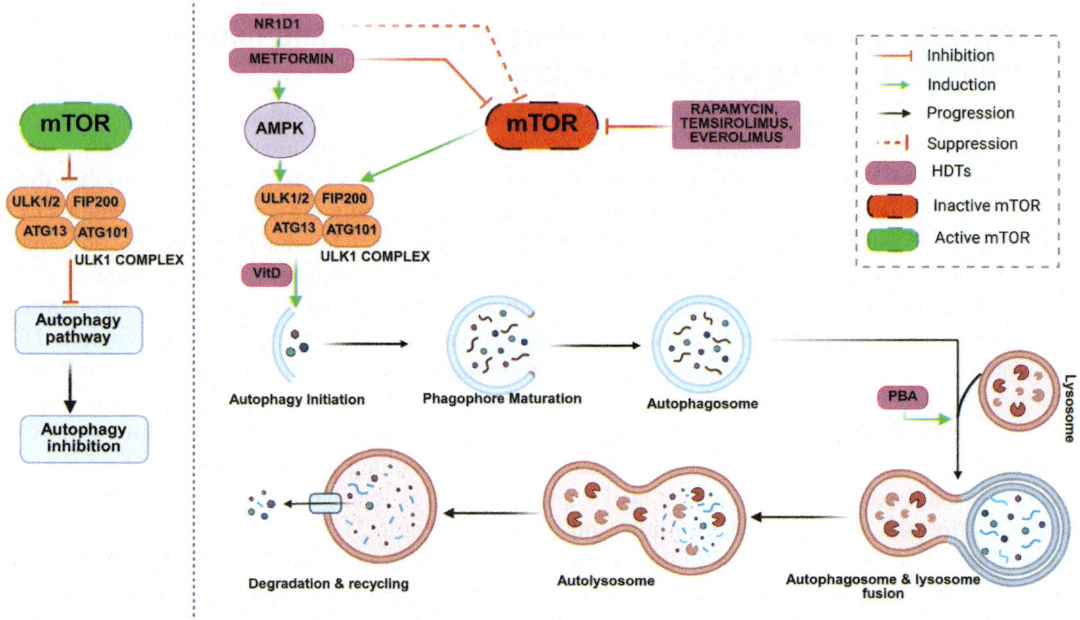

FIGURE 6.2 Modulation of autophagy by host directed therapeutics (HDTs).

The figure illustrates the canonical autophagy pathway and the specific stages influenced by various host directed therapeutics (HDTs). Autophagy is initiated *via* the ULK1 complex, comprising ULK1/2, ATG13, FIP200, and ATG101, which is activated by AMPK and inhibited by mTOR. In the left panel, mTOR activation blocks the ULK1 complex, thereby suppressing autophagy initiation. In the central panel, metformin and NR1D1 act in a dual manner both by stimulating AMPK and supressing mtor activation, promoting ULK1 activation and enhancing autophagy initiation. Vitamin D also contributes to autophagy induction by supporting early phagophore formation. The lower panel illustrates the progression of autophagy enabling the degradation and recycling of intracellular components. PBA promotes autophagosome–lysosome fusion, a critical step in autophagic maturation. This action helps restore autophagic flux and alleviate blockages in infected or stressed cells, particularly in the context of *M. tb* infection.

Notably, the activation of autophagy to combat *M. tb* has been achieved through the use of specific autophagy activating agents, such as clionamines [43]. Vaccine based strategies represent another novel approach, wherein autophagy inducing mycobacteria, such as *M. smegmatis*, are utilized to enhance the efficacy of BCG vaccination [44]. Furthermore, the combination of autophagy inducing agents with conventional anti TB drugs shows promise for improving treatment outcomes, as evidenced by vitamin D receptor signaling or AMPK activation [34,45]. Finally, research suggests that enhancing lysosomal activity through autophagy induction can augment the host cells' ability to eliminate bacteria, with ubiquitin derived peptides playing a role in mycobacterial eradication within the lysosomal compartment [46].

4. Evasion strategies of *Mycobacterium tuberculosis* and host countermeasures

M. tb has developed advanced mechanisms to circumvent autophagic processes. Utilizing a mechanism dependent on ROS, *M. tb* employs its Eis protein to inhibit autophagy and diminish macrophage cell death [47]. Similarly, the ESAT-6 protein, secreted *via* the ESX-1 type VII secretion system (T7SS), disrupts the late stage autophagy in human dendritic cells [48]. A comprehensive analysis of autophagy's role in mycobacterial infections is imperative, given its critical function in restricting *M. tb*'s proliferation within host cells. Autophagy also manifests in more specialized forms, such as selective autophagy and LC3-associated phagocytosis (LAP). Xenophagy specifically targets intracellular infections. Both the conventional (canonical) and alternative (noncanonical) autophagy pathways are vital for host defense and the innate immune system's response to intracellular infections. In the context of mycobacterial infections, xenophagy is activated when bacteria enter the cytoplasm through the ESX-1 T7SS. The successful eradication of *M. tb* relies on autophagy receptors like SQSTM1 and CALCOCO2, as well as the stimulator of interferon response cGAMP interactor (STING) dependent cytosolic process [49,50]. Ubiquitination, supported by proteins such as tripartite motif containing 16 (TRIM16) and ubiquitin ligases like Parkin (PRKN) and SMAD specific E3 ubiquitin protein ligase 1 (SMURF1), plays a crucial role in xenophagy [51–53]. Moreover, by modulating AMPK and interaction with autophagic components, immunity related GTPase M (IRGM) in mice and its human counterpart, are pivotal for the elimination of mycobacteria through autophagy. Although the exact role of IRGM in selective autophagy during mycobacterial infections remains unclear, evidence indicates that autophagy contributes to the regulation of *M. tb* growth, despite studies questioning the importance of certain ATG genes in host resistance. LAP is characterized by the conjugation of LC3 to phagosomal membranes during the phagocytosis of microorganisms or apoptotic cells. ATG proteins, such as LC3 and BECN1, are recruited to the phagosome to facilitate its maturation and pathogen elimination. Unlike conventional autophagy, LAP is activated by TLRs and involves proteins such as BECN1, ATG5, ATG7, and ATG3, rather than traditional autophagic machinery [54]. Rubicon autophagy regulator (RUBCN) serves as a major regulator of LAP, promoting the decoration of phagocytosed cargo with PI3P. This mechanism is essential for generating ROS and lipidating LC3-II on the LAPosomal membrane. While *M. tb* activates PRRs, the role of LAP in modulating antimicrobial and inflammatory responses during mycobacterial infection remains ambiguous, possibly due to *M. tb*'s efforts to evade LAP induced effects.

5. Species-specific modulation of host autophagy by *Mycobacterium tuberculosis* and nontuberculous mycobacteria

Autophagy is integral to host defense against mycobacterial infections; however, its induction varies markedly among different Mycobacterium species. Both pathogenic and nonpathogenic mycobacteria activate mTOR signaling, yet non-pathogenic species robustly induce autophagy, whereas pathogenic non-tuberculous Mycobacteria (NTMs) frequently manipulate or evade autophagic clearance [55]. For instance, *M. smegmatis*, a model for pathogenic mycobacteria, necessitates significantly higher doses of mTOR inhibitors such as rapamycin and Torin for clearance, indicating the involvement of a noncanonical autophagy pathway. Notably, prior research has demonstrated that treating *M. tb* infected macrophages with *M. smegmatis* enhances bacterial clearance by stimulating autophagy, suggesting potential therapeutic applications of autophagy inducing NTMs. Despite its recognized role in *M. tb* immunity, the role of autophagy in NTM infections remains insufficiently explored. Nonetheless, emerging genetic studies suggest that polymorphisms in ATG genes such as NOD2, PARK2, IRGM, and ATG16L1 influence susceptibility to NTM diseases [56]. Specifically, a single nucleotide polymorphism (SNP) in PARK2 is associated with higher susceptibility to *M. ulcerans*, while an SNP in NOD2 is linked to accelerated disease progression. Conversely, ATG16L1 variants appear to confer protection against severe *M. ulcerans* infection. Additionally, mutations in inducible nitric oxide synthase (iNOS) and IFN-γ have been linked to heightened susceptibility to *M. ulcerans*, leprosy, and TB [57–59]. The virulence of *M. ulcerans* is largely attributed to mycolactone, a cytotoxic macrolide that induces autophagy while impairing autophagic flux, thereby allowing bacterial persistence. Mycolactone further inhibits mTOR, leading to the activation of BCL2 like 11 (BCL2L11/Bim), a proapoptotic protein, *via* an alternative mTOR dependent pathway. This suggests that modulating mTOR activity could potentially suppress apoptosis and restore autophagic control over infection. Similarly, *M. abscessus* and *M. fortuitum* exist in rough (R) and smooth (S) variants, with the R variant being more virulent due to its ability to block autophagosome lysosome fusion, thereby enhancing intracellular survival [56]. Recent studies highlight how NTMs manipulate autophagy related pathways to evade immune clearance. While total lipids from pathogenic as well as nonpathogenic mycobacteria can enhance autophagy, differences in lipoarabinomannan (LAM) capping significantly influence immune responses. PI-LAM (phosphatidyl-myo-inositol LAM) from rapid growing mycobacteria (RGM) triggers autophagy and apoptosis *via* TLR2, whereas Man-LAM (mannose-capped LAM) from pathogenic mycobacteria does not induce autophagy [60–62]. *M. marinum* also exhibits dual behavior, where it induces autophagy but simultaneously inhibits autophagic flux, thereby increasing its intracellular survival [63–65]. These findings suggest that NTMs employ diverse strategies to modulate autophagy, either suppressing or exploiting it to establish infection.

Another challenge posed by NTMs is their capacity to form biofilms and extracellular bacterial environments, which complicate treatment by diminishing antibiotic efficacy and fostering drug resistance [66]. Many mycobacteria responsible for ulcerative diseases exhibit inherent antibiotic resistance, complicating clinical management. Recent studies indicate that autophagy based HDTs may augment antibiotic treatment by preventing biofilm formation and limiting bacterial persistence. These findings suggest the potential therapeutic value of autophagy modulation in managing chronic NTM infections.

Autophagy also plays an indispensable role in the dissemination of *M. leprae*. Macrophages from patients with paucibacillary leprosy demonstrate enhanced autophagic control, whereas those from multibacillary leprosy patients exhibit impaired autophagy regulation, attributed to elevated levels of interleukin-10 (IL-10), a cytokine known to inhibit autophagy. Furthermore, in type 1 reversal reactions (T1R), patients display dysregulated expression of ATG genes, overactivation of the mTOR, and increased activity of the NLRP3 inflammasome, resulting in elevated secretion of IL-1β. These observations suggest that proautophagic therapies may enhance leprosy treatment by modulating immune responses and mitigating inflammatory damage [56].

Given that different mycobacterial species employ distinct virulence mechanisms, their ability to modulate host autophagy varies considerably. While some NTMs elicit robust autophagic responses, other species, including *M. tb*, appear to have evolved strategies to suppress or evade autophagic degradation. However, significant gaps remain in understanding species specific autophagy regulation, highlighting the need for further comparative studies. Recent studies have investigated the distinct immune and autophagic reactions induced by *M. tb* and four NTMs, *M. abscessus*, *M. smegmatis*, *M.*

intracellulare, and *M. avium*, in THP-1 macrophages. The findings of this study revealed that while *M. smegmatis* and *M. abscessus* strongly induced autophagy, infections with Mycobacterium avium complex (MAC) species and *M. tb* resulted in significantly lower levels of autophagy induction, suggesting a pathogen specific modulation of autophagic pathways [67]. A comparative study evaluating autophagy induction across multiple mycobacterial species, including *M. tb*, BCG, and NTMs found significant variation in autophagic responses (Table 6.1). Notably, *M. kansasii* induced minimal autophagy, whereas *M. smegmatis* and *M. fortuitum* exhibited more pronounced autophagic activation [55]. These findings suggest that mycobacterial species regulate host autophagy in a species -dependent manner, potentially influencing their ability to persist or be cleared within macrophages.

6. Mechanistic insights into *Mycobacterium tuberculosis*-mediated subversion of host autophagy

M. tb has evolved highly sophisticated mechanisms to subvert host autophagy, facilitating its intracellular survival and persistence within macrophages. These strategies primarily involve the inhibition of phagosome maturation, disruption of autophagy signaling pathways, suppression of ATG gene expression, and modulation of immune responses [19]. By targeting key host cell processes, *M. tb* not only evades immune clearance but also establishes a niche for long term persistence and infection recurrence. A critical strategy used by *M. tb* to evade immune clearance is the arrest of phagosome maturation, which prevents the fusion of phagosomes with lysosomes. This inhibition prevents the acidic pH, hydrolytic enzymes, and antimicrobial peptides necessary for bacterial degradation from taking effect. The secreted acid phosphatase SapM is an important virulence factor responsible for preventing phagosome maturation. It dephosphorylates PI3P, an essential molecule required for phagosomal trafficking and fusion with lysosomes. Loss of SapM in mutant *M. tb* strains results in a failure to arrest phagosome maturation, leading to reduced bacterial survival in macrophages and attenuation of virulence in animal models. Additionally, SapM modulates dendritic cell recruitment and function, further subverting host immune responses (Table 6.2) [68].

ESAT-6, secreted *via* the ESX-1 T7SS, plays a dual role in disrupting phagosomal membranes and influencing macrophage polarization. By facilitating *M. tb*'s escape into the cytosol, ESAT-6 enables intracellular spread. Moreover, it initially induces a proinflammatory M1 macrophage phenotype, characterized by cytokines, which contribute to granuloma formation. Over time, ESAT-6 drives macrophages to adopt an antiinflammatory M2 phenotype, suppressing immune responses and promoting bacterial persistence. Additionally, ESAT-6 induces miR-155 expression, which inhibits suppressors of cytokine signaling (SOCS) and enhances apoptotic pathways in macrophages [48]. EspL, another component of the ESX-1 secretion system, directly inhibits autophagy by interacting with ULK1. EspL promotes the phosphorylation of ULK1 at Ser757, leading to its inactivation and suppression of autophagy. This results in impaired autophagic clearance, enhancing bacterial survival [69].

6.1 *Mycobacterium tuberculosis*-mediated suppression of autophagosome maturation

In addition to inhibiting the initiation of autophagy, *M. tb* actively disrupts the maturation of autophagosomes to prevent their degradation within lysosomes. Several virulence factors interfere with the

Table 6.1 Autophagy regulation in different mycobacterial species.

S. No.	Mycobacterial species	Autophagy regulation	Key mechanisms
1.	***Mycobacterium tuberculosis* (*Mtb*)**	Suppresses autophagy	Inhibits phagosome maturation and autophagosome lysosome fusion; modifies host immune responses to evade clearance
2.	***Mycobacterium. avium* complex (MAC)**	Weak autophagy induction	miR-125a-5p activates autophagy *via* signal transducer and activator of transcription 3 (STAT3) suppression; alpha-1-antitrypsin (AAT) treatment enhances autophagosome formation
3.	***Mycobacterium abscessus* (MABC)**	Suppresses autophagy (R variant)/moderate induction (S variant)	R variant blocks autophagic flux; S variant promotes phagosome cytosol communication but reduces autophagy induction
4.	**M. marinum**	Induces autophagy but inhibits autophagic flux	ESAT secretion system 1 (ESX-1) secretion system triggers autophagosome formation while blocking xenophagic degradation; selective autophagy *via* DNA damage regulated autophagy modulator 1 (DRAM1) and sequestome 1 (SQSTM1/p62)
5.	***Mycobacterium smegmatis***	Strong autophagy induction	Induces autophagy independent of membrane damage and ubiquitination; vitamin D3 enhances bacterial clearance
6.	***Mycobacterium ulcerans***	Induces autophagy but disrupts flux	Mycolactone inhibits mechanistic target of rapamycin kinase (mTOR) and increases autophagosome formation; autophagy related 16 like 1 (ATG16L1) polymorphisms linked to disease severity
7.	***Mycobacterium kansasii***	Minimal autophagy induction	Weakly induces autophagy compared to *Mycobacterium smegmatis* and *Mycobacterium fortuitum*
8.	***Mycobacterium fortuitum***	Moderate autophagy induction	Induces autophagy but with weaker response compared to *Mycobacterium smegmatis*
9.	***Mycobacterium terrae***	Induces autophagy	Interleukin 17A (IL-17A) and interleukin 17F (IL-17F) activate autophagosome formation and enhance bacterial clearance

Many mycobacterial species possess specialized mechanisms to regulate autophagy within host cells, impacting their survival and immune evasion. This table lists well studied examples including Mycobacterium tuberculosis, M. bovis BCG, and various non-tuberculous mycobacteria (NTMs) such as M. smegmatis, M. marinum, and M. avium. Species differ significantly in their ability to induce or inhibit host autophagy through factors like PE/PPE proteins and secretion systems. Understanding these diverse regulatory strategies not only reveals fundamental host pathogen interactions but also identifies new therapeutic and vaccine targets for mycobacterial diseases.

host's lysosomal trafficking machinery, thereby creating an intracellular environment conducive to bacterial persistence. One of the virulence factors, PknG, a serine threonine kinase, plays a dual role in modulating autophagy flux. It promotes the initiation of autophagy by inhibiting AKT activation, while simultaneously obstructing autophagosome maturation through interaction with RAB14, a member of the RAS oncogene family (RAB14) and phosphorylation of TBC1 domain family member 4 (TBC1D4). This dual function ensures that the autophagic pathway is arrested before bacterial degradation, thereby favoring *M. tb* persistence [70]. Beyond its role in autophagy, PknG enhances bacterial antibiotic tolerance and supports survival under hypoxic conditions, rendering it an attractive target for drug development [70].

The PE/PPE protein family, including members such as PE_PGRS20 and PE_PGRS47, inhibits autophagy by interacting with RAB1A, a small GTPase involved in autophagy initiation. This interaction prevents the formation of autophagosomes, thereby impeding bacterial clearance. Additionally, other PE/PPE proteins increase mTOR activity while reducing the production of proinflammatory cytokines, further reinforcing bacterial persistence [71]. Furthermore, Mannosecapped lipoarabinomannan (ManLAM) inhibits phagosome maturation and suppresses autophagy-inducing signals from IFN-γ, a key cytokine involved in host defense. By disrupting IFN-γ signaling, ManLAM enables *M. tb* to evade host immune responses [72,73]. Cumulatively, these strategies highlight the multimodal mechanisms adopted by *M.tb* to disrupt the process of autophagy at multiple stages, from its initiation to the degradation process (Fig. 6.1). By targeting essential host signaling pathways and manipulating autophagy related molecules, *M. tb* maximally establishes a permissive intracellular environment favoring its persistence and survival, thereby emphasizing the utmost need for therapeutics that can reinstate autophagic function and counteract bacterial immune evasion (Table 6.2).

6.2 Signal subversion: How *Mycobacterium tuberculosis* hijacks host kinases and autophagy regulators for intracellular survival

M.tb has developed advanced strategies to modulate host's autophagy by subverting host kinase and autophagy regulators, which are central in orchestrating several critical processes such as autophagy initiation, phagosome maturation, and inflammatory signaling. In this context, RipA, a peptidoglycan hydrolase, inhibits autophagy by disrupting the activity of ULK1, a key initiating kinase in the autophagy pathway. It activates the PI3K-AKT-mTORC1 signaling cascade, which subsequently suppresses ULK1. Given that ULK1 is indispensable for initiating the formation of the autophagosome, its inactivation by RipA effectively halts the initiation of autophagy, allowing the pathogen to evade host cellular defences and promote its intracellular survival [74]. Furthermore, MoxR1, an AAA+ family protein acting as chaperone, inhibits autophagy by activating the PI3K-AKT-mTOR cascade. It phosphorylates ULK1 at Ser757, leading to autophagy suppression. Additionally, MoxR1 induces metabolic reprogramming to dampen host defense mechanism [75]. In addition, the mycobacterial tyrosine phosphatases PtpA and PtpB prevent phagosomal acidification by blocking the recruitment of V-ATPase, a proton pump essential for phagosome acidification and pathogen degradation. These phosphatases specifically target host signaling pathways despite the absence of tyrosine kinases in *M. tb*. Thus, neutralization of PtpA with targeted antibodies has been shown to reduce bacterial proliferation in macrophages, making it a potential therapeutic target [76,77]. Also, CtpF, a calcium transporting P2A ATPase, disrupts host calcium homeostasis by effluxing calcium from macrophages. This leads to reduced mTOR activation and inhibition of autophagy, further promoting bacterial survival [78]. By disrupting multiple nodes of autophagy and phagosome maturation

Table 6.2 Genes involved in mycobacterial manipulation of autophagy.

S. No.	Gene/Rv number	Origin (host/ *Mtb*)	Effect on autophagy	Mechanism of action
1.	**SapM**	*Mtb*	Inhibits phagosome maturation	Dephosphorylates phosphatidylinositol-3-phosphate (PI3P), blocking phagosome lysosome fusion
2.	**PknG**	*Mtb*	Blocks autophagosome maturation	Interacts with Rab associated binding 14 (RAB14), phosphorylates TBC1 domain family member 4 (TBC1D4
3.	**EspL**	*Mtb*	Inhibits autophagy initiation	Promotes phosphorylation of unc-51 like autophagy activating kinase 1 (ULK1) at Ser757, inactivating it
4.	**PDIM**	*Mtb*	Inhibits LC3-associated phagocytosis (LAP)	Prevents nicotinamide adenine dinucleotide phosphate hydrogen (NADPH) oxidase recruitment to phagosomes
5.	**PE_PGRS20**	*Mtb*	Inhibits autophagy initiation	Interacts with Rab associated binding 1 A (Rab1A), preventing autophagosome formation
6.	**PE_PGRS47**	*Mtb*	Inhibits autophagy initiation	Similar to PE_PGRS20, disrupts Rab1A mediated autophagy
7.	**Rv0790c**	*Mtb*	Suppresses early stage autophagy	Enhances mechanistic target of rapamycin kinase (mTOR) activity, phosphorylating ULK1 and ribosomal protein S6 Kinase (P70S6K)
8.	**Rv2347c**	*Mtb*	Blocks phagosome/ autophagosome maturation	Activates stimulator of interferon response cGAMP interactor 1/ TANK binding kinase 1 (STING/ TBK1) pathway
9.	**MoxR1**	*Mtb*	Inhibits autophagy	Activates phosphoinositide-3-kinase - AKT serine/threonine kinase 1 mTOR (PI3K-AKT-mTOR) pathway, phosphorylates ULK1
10.	**AcpM**	*Mtb*	Suppresses autophagy gene expression	Inhibits transcription factor EB (TFEB) activation, blocking lysosomal biogenesis
11.	**MtbPRT**	*Mtb*	Inhibits autophagy related gene expression	Induces histone hypermethylation at autophagy related 5 (Atg5) and Atg7 promoters
12.	**Rv3034c**	*Mtb*	Regulates pexophagy	Modulates mechanistic target of rapamycin kinase complex 1 (mTORC1) and peroxisome proliferator activated receptor gamma (PPAR-γ) signaling

Table 6.2 Genes involved in mycobacterial manipulation of autophagy.—cont'd

S. No.	Gene/Rv number	Origin (host/ *Mtb*)	Effect on autophagy	Mechanism of action
13.	**LprE**	*Mtb*	Inhibits autophagy and antimicrobial peptides	Downregulates microtubule associated protein 1 light chain 3 (MAP1LC3/LC3), Atg5, beclin-1 (BECN1), suppresses cathelicidin
14.	**Kxd1**	*Mtb*	Suppresses lysosome repositioning	Blocks delivery of lysosomes to phagosomes
15.	**Plekhm2**	*Mtb*	Like KxDL motif containing 1 (Kxd1)	Prevents lysosomal delivery to *Mtb* containing phagosomes
16.	**CtpF**	*Mtb*	Inhibits autophagy *via* calcium modulation	Effluxes calcium, disrupting mTOR activation
17.	**BCG_2432c**	*Mtb*	Inhibits autophagosome lysosome fusion	Prevents degradation of bacterial antigens
18.	**Eis**	*Mtb*	Suppresses reactive oxygen species (ROS) dependent autophagy	Inhibits reactive oxygen species (ROS) production
19.	**PtpA**	*Mtb*	Disrupts phagosomal acidification	Blocks vacuolar ATPase (V-ATPase) recruitment to phagosomes
20.	**PtpB**	*Mtb*	Inhibits lysosomal function	Disrupts phagosome maturation *via* phosphatase activity
21.	**miR-20a-3p**	Host (manipulated by *Mtb*)	Suppresses autophagy	Targets I-Kappa-B Kinase b (IKKβ), downregulates ATG7 and autophagy related 16 like 1 (ATG16L1)
22.	**PPE51**	*Mtb*	Inhibits autophagy	Blocks toll like receptor 2 (TLR-2) signaling, suppressing extracellular regulated kinase 1/2 (ERK1/2) activation
23.	**RipA**	*Mtb*	Suppresses autophagy initiation	Activates PI3K-AKT-mTOR pathway, blocking ULK1
24.	**Coronin-1a**	Host (manipulated by *Mtb*)	Prevents phagosome lysosome fusion	Activates calcineurin, inhibiting lysosomal trafficking
25.	**ManLAM**	*Mtb*	Inhibits phagosome maturation	Blocks interferon amma (IFN-γ) signaling, preventing autophagy induction

This table lists key mycobacterial genes known to manipulate host cell autophagy, affecting both bacterial survival and immune response. Genes such as PE_PGRS20, PE_PGRS47, Eis, SapM, and autophagy related genes like ATG6, ATG7, and ATG13 play distinctive roles in either inducing or inhibiting autophagy pathways. Their mechanisms include modulation of signaling cascades, evasion of lysosomal degradation, and alteration of host gene expression. Understanding these genes provides insight into host pathogen interactions and may help in the development of new therapeutic interventions targeting autophagy in mycobacterial diseases.

pathways, *M. tb* not only strengthens its intramacrophage survival but also establish a persistent chronic stage infection, highlighting the therapeutic potential of restoring host signaling integrity as a countermeasure against TB.

6.3 Epigenetic silencing and miRNA modulation: How *Mycobacterium tuberculosis* suppresses host autophagy at the transcriptional level

Beyond its direct interference with the host signaling pathways, *M. tb* also employs epigenetic strategies that downregulates the expression of ATG genes, thus modulating autophagic degradation. The enzyme *M. tb* PRT, a phosphoribosyl transferase, facilitates histone hypermethylation at the promoters of critical ATG genes, including ATG5 and ATG7. This epigenetic alteration results in the suppression of gene expression, thereby inhibiting autophagy [79]. Furthermore, *M. tb* modulates host microRNAs (miRNAs) to inhibit autophagy. The upregulation of miR-20a-3p leads to the inactivation of inhibitor of nuclear factor kappa B kinase subunit beta (IKKβ), which impairs NF-κB activation and autophagy by downregulating ATG7 and ATG16L1. Several other microRNAs, including miR-30a, miR-33, and miR-1958, are involved in autophagy repression through posttranscriptional regulation of essential ATG genes that play roles in the autophagic process initiation and development [80,81].

7. *Mycobacterium tuberculosis* manipulation of granulomas and host immunity

Granuloma formation serves as a critical human defense mechanism to contain *M. tb* infection, characterized by a core of dormant or semidormant *M. tb* surrounded by lymphocytes, macrophages, and other immune cells [82]. Despite this defensive arrangement, *M. tb* alters the granuloma's architecture to establish a latent state, thereby evading immune detection while maintaining the reactivation potential upon compromised state of the host's immune system. In order to maintain the granulomas as progressive niche, the TB bacilli secretes sulfatides, which inhibit the fusion of phagosomes with lysosomes and impede phagosome maturation *via* the ESX-1 secretion pathway, thereby enhancing bacterial persistence [83]. *M. tb* prolongs its intracellular survival by downregulating IFN-γ responses, thereby inhibiting effective antibacterial activity. Furthermore, the bacterium disrupts autophagic initiation by interfering with the recruitment of BECN1, a key regulator of autophagy [23]. Recent studies show that the ESX-1 T7SS also contributes to the inhibition of autophagosome development, thereby hindering efficient bacterial degradation [84,85]. *M. tb* can manipulate host lipid metabolism by inducing miR-33 and miR-33*, which inhibit autophagy and promote lipid body formation. These lipid bodies provide a nutrient source for the bacteria, facilitating their persistence within the host [86]. These additional evasion strategies underscore the complex tactics employed by *M. tb* to survive within the host, contributing to its persistence and the challenges associated with TB treatment.

8. Navigating the promise and pitfalls of autophagy adjunctive therapies in TB

Autophagy adjunctive therapeutics have emerged as a promising HDT strategy for TB, enhancing the host's ability to clear *M. tb* while potentially reducing treatment duration and addressing drug resistant strains. Despite encouraging preclinical findings, several challenges hinder their clinical translation, including variable efficacy in trials, complex regulation of autophagy pathways, pathogen adaptation, host variability, and drug delivery limitations. Additionally, uncertainties regarding long term safety, optimal dosing, and drug interactions with standard TB treatments must be addressed before autophagy modulating therapies can be integrated into clinical practice. Gefitinib, fluoxetine, loperamide, and verapamil are among the repositioned drugs showing promise in preclinical models, yet they require rigorous validation through well structured clinical trials (Fig. 6.2). Notably, an ongoing phase II trial (NCT 2968927) is evaluating the autophagy enhancing effects of everolimus and auranofin in TB treatment, while a completed study (NCT 2774993) investigated doxycycline's impact on pulmonary TB [87,88]. Another challenge in evaluating autophagy adjunctive therapeutics is drug delivery and patient selection. Targeted drug delivery remains poorly understood, and most autophagy inducing agents lack specificity for TB infected cells, enhancing the risk of nonspecificity and toxicity. Furthermore, treatment failures due to drug resistance, poor adherence, and adverse reactions necessitate careful patient selection for adjunctive therapy trials. Biomarker driven stratification within personalized medicine frameworks offers significant potential for identifying patient subsets most likely to respond positively to therapeutics targeting.

9. Balancing immunity and adaptation: The double-edged role of autophagy in tuberculosis therapy

Autophagy is regulated by several pathways, including mTOR, AMPK, and PI3K-AKT, which play roles in metabolism, immune signaling, and cellular homeostasis. The challenge lies in selectively enhancing autophagy without disrupting these vital processes. Current autophagy inducing agents, such as rapamycin, metformin, and resveratrol, exhibit broad systemic effects that can lead to neurodegeneration, immunosuppression, or tumor progression. The development of highly selective autophagy modulators that specifically target *M. tb* infected macrophages is critical. While autophagy enhances bacterial clearance, excessive activation may promote intracellular *M. tb* survival or cause tissue damage. Mouse models have provided critical insights into autophagy's role in TB, but species specific differences limit direct clinical translation. For example, the vitamin D-cathelicidin axis, which plays a key role in autophagy mediated bacterial killing in humans, is not functional in murine models. This reveals the need for human relevant models in preclinical studies. Phagolysosomal acidification may contribute to direct *M. tb* killing through lysosomal enzymes that degrade bacterial components or through antimicrobial peptides, such as ribosomal protein S30 (rpS30), which have been implicated in autophagy mediated *M. tb* clearance. Long term autophagy induction may exert selective pressure on *M. tb*, potentially leading to pathogen adaptation and resistance. Continuous exposure to autophagy enhancing agents might favor *M. tb* strains that evade autophagic clearance, necessitating combination therapies to counteract potential resistance mechanisms.

10. Future directions and concluding remarks

Despite considerable progress in understanding the role of autophagy in TB, numerous questions remain unresolved. Future research should aim to further elucidate the mechanisms through which *M. tb* manipulates host cell autophagic pathways and identify specific molecular targets for intervention. Moreover, investigating how autophagy intersects with other host immune pathways, such as the inflammasome and cell death mechanisms, could provide deeper insights into the host pathogen interaction. Furthermore, developing strategies to enhance autophagic responses in the host, either through pharmacological agents or genetic manipulation, may offer new therapeutic approaches to treat both latent and active TB. Given the increasing threat of MDR and XDR *M. tb*, novel therapeutic strategies are essential for combating this global health crisis. Additionally, the role of autophagy in TB latency and reactivation warrants further investigation. Understanding how autophagy regulates the transition between latent and active TB could provide valuable insights for preventing reactivation and controlling the spread of the disease. While autophagy adjunctive therapies hold significant potential for improving TB treatment outcomes, several challenges must be addressed before their widespread clinical adoption. Further investigations should focus on developing highly selective autophagy modulators which curtail offtarget effects while effectively targeting *M. tb* infected macrophages. Given the variability in patient responses due to genetic, immunological, and metabolic differences, personalized treatment strategies should be explored to optimize therapy for specific patient subgroups. Additionally, combining autophagy enhancing agents with standard TB drugs could improve treatment efficacy, but further studies are required to understand potential drug-drug interactions and optimize dosing regimens. Investigating novel antimicrobial peptides and lysosomal pathways involved in direct *M. tb* clearance could lead to the identification of new therapeutic targets which enhance the antibacterial effects of autophagy. Moreover, significant advancements in drug delivery technologies are needed to ensure that autophagy inducing agents reach infected tissues with minimal systemic toxicity. Large scale clinical trials will be crucial in determining the long term safety, efficacy, and feasibility of integrating autophagy targeted therapies into existing TB treatment regimens. By addressing these challenges, autophagy based HDTs may become a viable addition to TB treatment strategies, particularly in combating drug resistant TB and reducing the duration of antibiotic therapy. Advancing this field requires a multidisciplinary approach, integrating molecular research, clinical trials, and personalized medicine to fully harness the therapeutic potential of autophagy in TB treatment.

Future research on autophagy based HDTs is anticipated to enhance our comprehension of host defense mechanisms against TB and facilitate the development of innovative therapeutic strategies. Key areas of focus should include the identification of novel antimicrobial peptides, the elucidation of interactions between autophagy and other host immune pathways, and the optimization of drug delivery methods, all of which are crucial for advancing autophagy adjunctive therapies. Additionally, forthcoming clinical trials should emphasize personalized treatment approaches, considering patient specific factors such as vitamin D status, genetic polymorphisms, and geographic influences on TB susceptibility. Ultimately, a more thorough understanding of autophagy mediated host defense and its therapeutic potential will be pivotal in enhancing TB treatment outcomes and addressing the global challenge of drug resistant TB.

References

[1] 2024 Global tuberculosis report.

[2] Shariq M, Sheikh JA, Quadir N, Sharma N, Hasnain SE, Ehtesham NZ. COVID-19 and tuberculosis: the double whammy of respiratory pathogens. Eur Respir Rev 2022;31:210264. https://doi.org/10.1183/16000617.0264-2021.

[3] Naqvi N, Ahuja Y, Zarin S, Alam A, Ali W, Shariq M, Hasnain SE, Ehtesham NZ. BCG's role in strengthening immune responses: implications for tuberculosis and comorbid diseases. Infect Genet Evol 2025;127:105703. https://doi.org/10.1016/j.meegid.2024.105703.

[4] Upadhyay S, Mittal E, Philips JA. Tuberculosis and the art of macrophage manipulation. Pathog Dis 2018; 76. https://doi.org/10.1093/femspd/fty037.

[5] Vu A, Glassman I, Campbell G, Yeganyan S, Nguyen J, Shin A, Venketaraman V. Host cell death and modulation of immune response against Mycobacterium tuberculosis infection. Int J Mol Sci 2024.

[6] Manjunath P, Ahmad J, Samal J, Rani A, Sheikh JA, Zarin S, Ahuja Y, Alam A, Hasnain SE, Ehtesham NZ. Expression of a unique M. tuberculosis DNA MTase Rv1509 in M. smegmatis alters the gene expression pattern and enhances virulence. Front Microbiol 2024;15:1344857. https://doi.org/10.3389/fmicb.2024.1344857.

[7] Vergne I, Gilleron M, Nigou J. Manipulation of the endocytic pathway and phagocyte functions by Mycobacterium tuberculosis lipoarabinomannan. Front Cell Infect Microbiol 2014;4:187. https://doi.org/10.3389/fcimb.2014.00187.

[8] Voskuil MI, Bartek IL, Visconti K, Schoolnik GK. The response of mycobacterium tuberculosis to reactive oxygen and nitrogen species. Front Microbiol 2011;2:105. https://doi.org/10.3389/fmicb.2011.00105.

[9] Deretic V, Singh S, Master S, Harris J, Roberts E, Kyei G, Davis A, De Haro S, Naylor J, Lee H-H, et al. Mycobacterium tuberculosis inhibition of phagolysosome biogenesis and autophagy as a host defence mechanism. Cell Microbiol [Internet] 2006;8:719–27. https://doi.org/10.1111/j.1462-5822.2006.00705.x.

[10] Gutierrez MG, Master SS, Singh SB, Taylor GA, Colombo MI, Deretic V. Autophagy is a defense mechanism inhibiting BCG and Mycobacterium tuberculosis survival in infected macrophages. Cell 2004;119: 753–66. https://doi.org/10.1016/j.cell.2004.11.038.

[11] Bento CF, Puri C, Moreau K, Rubinsztein DC. The role of membrane-trafficking small GTPases in the regulation of autophagy. J Cell Sci 2013:1059–69.

[12] Sumpter R, Levine B. Selective autophagy and viruses. Autophagy. Taylor and Francis Inc; 2011. p. 260–5.

[13] Levine B, Kroemer G. Biological functions of autophagy genes: a disease perspective. Cell 2019;176:11–42. https://doi.org/10.1016/j.cell.2018.09.048.

[14] Erratum: in brief: nuclear envelope. Nat Rev Mol Cell Biol 2008;9. https://doi.org/10.1038/nrm2445.

[15] Cao Z, Tian K, Ran Y, Zhou H, Zhou L, Ding Y, Tang X. Beclin-1: a therapeutic target at the intersection of autophagy, immunotherapy, and cancer treatment. Front Immunol 2024;15. https://doi.org/10.3389/fimmu.2024.1506426.

[16] Dunlop EA, Hunt DK, Acosta-Jaquez HA, Fingar DC, Tee AR. ULK1 inhibits mTORC1 signaling, promotes multisite Raptor phosphorylation and hinders substrate binding. Autophagy 2011;7:737–47. https://doi.org/10.4161/auto.7.7.15491.

[17] Park J-M, Jung CH, Seo M, Otto NM, Grunwald D, Kim KH, Moriarity B, Kim Y-M, Starker C, Nho RS, et al. The ULK1 complex mediates MTORC1 signaling to the autophagy initiation machinery via binding and phosphorylating ATG14. Autophagy 2016;12:547–64. https://doi.org/10.1080/15548627.2016.1140293.

[18] Marston AL, Amon A. Meiosis: cell-cycle controls shuffle and deal. Nat Rev Mol Cell Biol 2004;5:983–97. https://doi.org/10.1038/nrm1526.

[19] Sharma N, Shariq M, Quadir N, Singh J, Sheikh JA, Hasnain SE, Ehtesham NZ. Mycobacterium tuberculosis protein PE6 (Rv0335c), a novel TLR4 agonist, evokes an inflammatory response and modulates the cell death pathways in macrophages to enhance intracellular survival. Front Immunol 2021;12:696491. https://doi.org/10.3389/fimmu.2021.696491.

[20] Behura A, Das M, Kumar A, Naik L, Mishra A, Manna D, Patel S, Mishra A, Singh R, Dhiman R. ESAT-6 impedes IL-18 mediated phagosome lysosome fusion via microRNA-30a upon Calcimycin treatment in mycobacteria infected macrophages. Int Immunopharmacol 2021;101:108319. https://doi.org/10.1016/j.intimp.2021.108319.

[21] Ferraris DM, Sbardella D, Petrera A, Marini S, Amstutz B, Coletta M, Sander P, Rizzi M. Crystal structure of Mycobacterium tuberculosis zinc-dependent metalloprotease-1 (Zmp1), a metalloprotease involved in pathogenicity. J Biol Chem 2011;286:32475–82. https://doi.org/10.1074/jbc.M111.271809.
[22] Cooney R, Baker J, Brain O, Danis B, Pichulik T, Allan P, Ferguson DJP, Campbell BJ, Jewell D, Simmons A. NOD2 stimulation induces autophagy in dendritic cells influencing bacterial handling and antigen presentation. Nat Med 2010;16:90–7. https://doi.org/10.1038/nm.2069.
[23] Shariq M, Quadir N, Alam A, Zarin S, Sheikh JA, Sharma N, Samal J, Ahmad U, Kumari I, Hasnain SE, et al. The exploitation of host autophagy and ubiquitin machinery by Mycobacterium tuberculosis in shaping immune responses and host defense during infection. Autophagy 2023;19:3–23. https://doi.org/10.1080/15548627.2021.2021495.
[24] Deretic V. Autophagy in inflammation, infection, and immunometabolism. Immunity 2021;54:437–53. https://doi.org/10.1016/j.immuni.2021.01.018.
[25] Münz C. Antigen processing for MHC class II presentation via autophagy. Front Immunol 2012;3:9. https://doi.org/10.3389/fimmu.2012.00009.
[26] Münz C. Autophagy beyond intracellular MHC class II antigen presentation. Trends Immunol 2016;37:755–63. https://doi.org/10.1016/j.it.2016.08.017.
[27] Chen L, Deng H, Cui H, Fang J, Zuo Z, Deng J, Li Y, Wang X, Zhao L. Inflammatory responses and inflammation-associated diseases in organs. Oncotarget. Impact Journals LLC; 2018. p. 7204–18.
[28] Delgado M, Singh S, De Haro S, Master S, Ponpuak M, Dinkins C, Ornatowski W, Vergne I, Deretic V. Autophagy and pattern recognition receptors in innate immunity. Immunol Rev 2009:189–202.
[29] Li D, Wu M. Pattern recognition receptors in health and diseases. Signal Transduct Target Ther 2021.
[30] Kelley N, Jeltema D, Duan Y, He Y. The NLRP3 inflammasome: an overview of mechanisms of activation and regulation. Int J Mol Sci. MDPI AG 2019.
[31] Gan H, Ma Q, Hao W, Yang N, Chen ZS, Deng L, Chen J. Targeting autophagy to counteract neuroinflammation: a novel antidepressant strategy. Pharmacol Res 2024.
[32] Chen T, Tu S, Ding L, Jin M, Chen H, Zhou H. The role of autophagy in viral infections. J Biomed Sci 2023.
[33] Shariq M, Malik AA, Sheikh JA, Hasnain SE, Ehtesham NZ. Regulation of autophagy by SARS-CoV-2: the multifunctional contributions of ORF3a. J Med Virol 2023;95. https://doi.org/10.1002/jmv.28959.
[34] Bento CF, Empadinhas N, Mendes V. Autophagy in the fight against tuberculosis. DNA Cell Biol 2015;34:228–42. https://doi.org/10.1089/dna.2014.2745.
[35] Mily A, Rekha RS, Kamal SMM, Arifuzzaman ASM, Rahim Z, Khan L, Haq MA, Zaman K, Bergman P, Brighenti S, et al. Significant effects of oral phenylbutyrate and Vitamin D3 adjunctive therapy in pulmonary tuberculosis: a randomized controlled trial. PLoS One 2015;10. https://doi.org/10.1371/journal.pone.0138340.
[36] Yuk JM, Shin DM, Lee HM, Yang CS, Jin HS, Kim KK, Lee ZW, Lee SH, Kim JM, Jo EK. Vitamin D3 induces autophagy in human monocytes/macrophages via cathelicidin. Cell Host Microbe 2009;6:231–43. https://doi.org/10.1016/j.chom.2009.08.004.
[37] Arora G, Gagandeep BA, Gosain TP, Shaliwal RP, Kidwai S, Singh P, Kandi SK, Dhiman R, Rawat DS, et al. NSC 18725, a Pyrazole derivative inhibits growth of intracellular Mycobacterium tuberculosis by induction of autophagy. Front Microbiol 2019;10:3051. https://doi.org/10.3389/fmicb.2019.03051.
[38] Bhandage AK, Barragan A. GABAergic signaling by cells of the immune system: more the rule than the exception. Cell Mol Life Sci 2021:5667–79.
[39] Floto RA, Sarkar S, Perlstein EO, Kampmann B, Schreiber SL, Rubinsztein DC. Small molecule enhancers of rapamycin-induced TOR inhibition promote autophagy, reduce toxicity in Huntington's disease models and enhance killing of mycobacteria by macrophages (Autophagy). Autophagy. Taylor and Francis Inc; 2007. p. 620–2.

[40] Yang CS, Kim JJ, Lee HM, Jin HS, Lee SH, Park JH, Kim SJ, Kim JM, Han YM, Lee MS, et al. The AMPK-PPARGC1A pathway is required for antimicrobial host defense through activation of autophagy. Autophagy 2014;10:785–802. https://doi.org/10.4161/auto.28072.

[41] Chandra V, Bhagyaraj E, Nanduri R, Ahuja N, Gupta P. NR1D1 ameliorates Mycobacterium tuberculosis clearance through regulation of autophagy. Autophagy 2015;11:1987–97. https://doi.org/10.1080/15548627.2015.1091140.

[42] Lee H-J, Kang S-J, Woo Y, Hahn T-W, Ko H-J, Jung Y-J. TLR7 stimulation with imiquimod induces selective autophagy and controls Mycobacterium tuberculosis growth in mouse macrophages. Front Microbiol 2020; 11:1684. https://doi.org/10.3389/fmicb.2020.01684.

[43] Persaud R, Li SC, Chao JD, Forestieri R, Donohue E, Balgi AD, Zheng X, Chao JT, Yashiroda Y, Yoshimura M, et al. Clionamines stimulate autophagy, inhibit Mycobacterium tuberculosis survival in macrophages, and target Pik1. Cell Chem Biol 2022;29:870–882.e11. https://doi.org/10.1016/j.chembiol.2021.07.017.

[44] Gonzalez-Orozco M, Strong EJ, Paroha R, Lee S. Reversing BCG-mediated autophagy inhibition and mycobacterial survival to improve vaccine efficacy. BMC Immunol 2022;23. https://doi.org/10.1186/s12865-022-00518-z.

[45] Paik S, Kim JK, Chung C, Jo EK. Autophagy: a new strategy for host-directed therapy of tuberculosis. Virulence 2019:448–59.

[46] Alonso S, Pethe K, Russell DG, Purdy GE. Lysosomal killing of Mycobacterium mediated by ubiquitin-derived peptides is enhanced by autophagy. Proc Natl Acad Sci U S A 2007;104:6031–6. https://doi.org/10.1073/pnas.0700036104.

[47] Shin DM, Jeon BY, Lee HM, Jin HS, Yuk JM, Song CH, Lee SH, Lee ZW, Cho SN, Kim JM, et al. Mycobacterium tuberculosis Eis regulates autophagy, inflammation, and cell death through redox-dependent signaling. PLoS Pathog 2010;6. https://doi.org/10.1371/journal.ppat.1001230.

[48] Passos BBS, Araújo-Pereira M, Vinhaes CL, Amaral EP, Andrade BB. The role of ESAT-6 in tuberculosis immunopathology. Front Immunol 2024.

[49] Shariq M, Quadir N, Sheikh JA, Singh AK, Bishai WR, Ehtesham NZ, Hasnain SE. Post translational modifications in tuberculosis: ubiquitination paradox. Autophagy 2021;17:814–7. https://doi.org/10.1080/15548627.2020.1850009.

[50] Malik AA, Shariq M, Sheikh JA, Fayaz H, Srivastava G, Thakuri D, Ahuja Y, Ali S, Alam A, Ehtesham NZ, et al. Regulation of type I interferon and autophagy in immunity against Mycobacterium tuberculosis: role of CGAS and STING1. Adv Biol 2024;8:e2400174. https://doi.org/10.1002/adbi.202400174.

[51] Kim JK, Silwal P, Jo EK. Host-pathogen dialogues in autophagy, apoptosis, and necrosis during mycobacterial infection. Immune Netw 2020:1–15.

[52] Tripathi-Giesgen I, Behrends C, Alpi AF. The ubiquitin ligation machinery in the defense against bacterial pathogens. EMBO Rep. 2021;22. https://doi.org/10.15252/embr.202152864.

[53] Malik AA, Shariq M, Sheikh JA, Zarin S, Ahuja Y, Fayaz H, Alam A, Ehtesham NZ, Hasnain SE. Activation of the lysosomal damage response and selective autophagy: the coordinated actions of galectins, TRIM proteins, and CGAS-STING1 in providing immunity against Mycobacterium tuberculosis. Crit Rev Microbiol 2025;51:108–27. https://doi.org/10.1080/1040841X.2024.2321494.

[54] Peña-Martinez C, Rickman AD, Heckmann BL. Beyond autophagy: LC3-associated phagocytosis and endocytosis. Sci Adv 2022.

[55] Zullo AJ, Lee S. Mycobacterial induction of autophagy varies by species and occurs independently of mammalian target of rapamycin inhibition. J Biol Chem 2012;287:12668–78. https://doi.org/10.1074/jbc.M111.320135.

[56] Strong EJ, Lee S. Targeting autophagy as a strategy for developing new vaccines and host-directed therapeutics against mycobacteria. Front Microbiol 2021.

[57] Bibert S, Bratschi MW, Aboagye SY, Collinet E, Scherr N, Yeboah-Manu D, Beuret C, Pluschke G, Bochud P-Y. Susceptibility to Mycobacterium ulcerans disease (Buruli ulcer) is associated with IFNG and iNOS gene polymorphisms. Front Microbiol 2017;8. https://doi.org/10.3389/fmicb.2017.01903.

[58] Capela C, Dossou AD, Silva-Gomes R, Sopoh GE, Makoutode M, Menino JF, Fraga AG, Cunha C, Carvalho A, Rodrigues F, et al. Genetic variation in autophagy-related genes influences the risk and phenotype of Buruli ulcer. PLoS Negl Trop Dis 2016;10:e0004671. https://doi.org/10.1371/journal.pntd.0004671.

[59] Manry J, Vincent QB, Johnson C, Chrabieh M, Lorenzo L, Theodorou I, Ardant M-F, Marion E, Chauty A, Marsollier L, et al. Genome-wide association study of Buruli ulcer in rural Benin highlights role of two LncRNAs and the autophagy pathway. Commun Biol 2020;3:177. https://doi.org/10.1038/s42003-020-0920-6.

[60] Bohsali A, Abdalla H, Velmurugan K, Briken V. The non-pathogenic mycobacteria M. smegmatis and M. fortuitum induce rapid host cell apoptosis via a caspase-3 and TNF dependent pathway. BMC Microbiol 2010;10:237. https://doi.org/10.1186/1471-2180-10-237.

[61] Shui W, Petzold CJ, Redding A, Liu J, Pitcher A, Sheu L, Hsieh T, Keasling JD, Bertozzi CR. Organelle membrane proteomics reveals differential influence of mycobacterial lipoglycans on macrophage phagosome maturation and autophagosome accumulation. J Proteome Res 2011;10:339–48. https://doi.org/10.1021/pr100688h.

[62] Singh B, Saqib M, Chakraborty A, Bhaskar S. Lipoarabinomannan from Mycobacterium indicus pranii shows immunostimulatory activity and induces autophagy in macrophages. PLoS One 2019;14. https://doi.org/10.1371/journal.pone.0224239.

[63] Kim SW, Subhadra B, Whang J, Back YW, Bae HS, Kim HJ, Choi CH. Clinical Mycobacterium abscessus strain inhibits autophagy flux and promotes its growth in murine macrophages. Pathog Dis 2017;75. https://doi.org/10.1093/femspd/ftx107.

[64] Lerena MC, Colombo MI. Mycobacterium marinum induces a marked LC3 recruitment to its containing phagosome that depends on a functional ESX-1 secretion system. Cell Microbiol 2011;13:814–35. https://doi.org/10.1111/j.1462-5822.2011.01581.x.

[65] Oliveira FM, Marinho FV, Oliveira SC, Resende DP, Junqueira-Kipnis AP, Kipnis A. *Mycobacterium abscessus* subsp. *massiliense* expressing bacterioferritin have improved resistance to stressful conditions. J Appl Microbiol 2020;128:1802–13. https://doi.org/10.1111/jam.14585.

[66] Greendyke R, Byrd TF. Differential antibiotic susceptibility of *Mycobacterium abscessus* variants in biofilms and macrophages compared to that of planktonic bacteria. Antimicrob Agents Chemother 2008;52:2019–26. https://doi.org/10.1128/AAC.00986-07.

[67] Feng Z, Bai X, Wang T, Garcia C, Bai A, Li L, Honda JR, Nie X, Chan ED. Differential responses by human macrophages to infection with Mycobacterium tuberculosis and non-tuberculous mycobacteria. Front Microbiol 2020;11. https://doi.org/10.3389/fmicb.2020.00116.

[68] Zhang W, Dong C, Xiong S. Mycobacterial SapM hampers host autophagy initiation for intracellular bacillary survival via dephosphorylating Raptor. iScience 2024;27. https://doi.org/10.1016/j.isci.2024.109671.

[69] Cai L, Lei Y, Xie T, Liu Y, Fan Y, Yang B, Dong S, Cao G, Chen X. Overexpression of EspL inhibits autophagy and antigen presentation to promote the intracellular survival of Mycobacterium tuberculosis avirulent strains. Anim Dis 2024;4. https://doi.org/10.1186/s44149-024-00128-9.

[70] Ge P, Lei Z, Yu Y, Lu Z, Qiang L, Chai Q, Zhang Y, Zhao D, Li B, Pang Y, et al. Tuberculosis PknG manipulates host autophagy flux to promote pathogen intracellular survival. Autophagy 2022;18:576–94. https://doi.org/10.1080/15548627.2021.1938912.

[71] Strong EJ, Ng TW, Porcelli SA, Lee S. Mycobacterium tuberculosis PE_PGRS20 and PE_PGRS47 proteins inhibit autophagy by interaction with Rab1A. mSphere 2021;6. https://doi.org/10.1128/msphere.00549-21.

[72] Fratti RA, Chua J, Vergne I, Deretic V. Mycobacterium tuberculosis glycosylated phosphatidylinositol causes phagosome maturation arrest. Proc Natl Acad Sci U S A 2003;100:5437–42. https://doi.org/10.1073/pnas.0737613100.

[73] Selvapandiyan A, Puri N, Kumar P, Alam A, Ehtesham NZ, Griffin G, Hasnain SE. Zooming in on common immune evasion mechanisms of pathogens in phagolysosomes: potential broad-spectrum therapeutic targets against infectious diseases. FEMS Microbiol Rev 2023;47. https://doi.org/10.1093/femsre/fuac041.

[74] Shariq M, Quadir N, Sharma N, Singh J, Sheikh JA, Khubaib M, Hasnain SE, Ehtesham NZ. Mycobacterium tuberculosis RipA dampens TLR4-mediated host protective response using a multi-pronged approach involving autophagy, apoptosis, metabolic repurposing, and immune modulation. Front Immunol 2021;12. https://doi.org/10.3389/fimmu.2021.636644.

[75] Quadir N, Shariq M, Sheikh JA, Singh J, Sharma N, Hasnain SE, Ehtesham NZ. Mycobacterium tuberculosis protein MoxR1 enhances virulence by inhibiting host cell death pathways and disrupting cellular bio-energetics. Virulence 2023;14. https://doi.org/10.1080/21505594.2023.2180230.

[76] Nasiri MJ, Venketaraman V. Advances in host-pathogen interactions in tuberculosis: emerging strategies for therapeutic intervention. Int J Mol Sci [Internet] 2025:1621. https://doi.org/10.3390/ijms26041621.

[77] Wong D, Bach H, Sun J, Hmama Z, Av-Gay Y. Mycobacterium tuberculosis protein tyrosine phosphatase (PtpA) excludes host vacuolar-H +-ATPase to inhibit phagosome acidification. Proc Natl Acad Sci U S A. 2011;108:19371–6. https://doi.org/10.1073/pnas.1109201108.

[78] Garg R, Borbora SM, Bansia H, Rao S, Singh P, Verma R, Balaji KN, Nagaraja V. Mycobacterium tuberculosis calcium pump CtpF modulates the autophagosome in an mTOR-dependent manner. Front Cell Infect Microbiol 2020;10. https://doi.org/10.3389/fcimb.2020.00461.

[79] Sengupta S, Nayak B, Meuli M, Sander P, Mishra S, Sonawane A. Mycobacterium tuberculosis Phosphoribosyltransferase promotes bacterial survival in macrophages by inducing histone hypermethylation in autophagy-related genes. Front Cell Infect Microbiol 2021;11. https://doi.org/10.3389/fcimb.2021.676456.

[80] Behrouzi A, Alimohammadi M, Nafari AH, Yousefi MH, Rad FR, Vaziri F, Siadat SD. The role of host mirnas on mycobacterium tuberculosis. ExRNA 2019.

[81] Guo L, Zhao J, Qu Y, Yin R, Gao Q, Ding S, Zhang Y, Wei J, Xu G. microRNA-20a inhibits autophagic process by targeting ATG7 and ATG16l1 and favors mycobacterial survival in macrophage cells. Front Cell Infect Microbiol 2016;6. https://doi.org/10.3389/fcimb.2016.00134.

[82] Barry CE, Boshoff HI, Dartois V, Dick T, Ehrt S, Flynn JA, Schnappinger D, Wilkinson RJ, Young D. The spectrum of latent tuberculosis: rethinking the biology and intervention strategies. Nat Rev Microbiol 2009: 845–55.

[83] Pieters J. Mycobacterium tuberculosis and the macrophage: maintaining a balance. Cell Host Microbe 2008: 399–407.

[84] Romagnoli A, Etna MP, Giacomini E, Pardini M, Remoli ME, Corazzari M, Falasca L, Goletti D, Gafa V, Simeone R, et al. ESX-1 dependent impairment of autophagic flux by Mycobacterium tuberculosisin human dendritic cells. Autophagy 2012;8:1357–70. https://doi.org/10.4161/auto.20881.

[85] Malik AA, Shariq M, Sheikh JA, Jaiswal U, Fayaz H, Shrivastava G, Ehtesham NZ, Hasnain SE. Mechanisms of immune evasion by Mycobacterium tuberculosis: the impact of T7SS and cell wall lipids on host defenses. Crit Rev Biochem Mol Biol 2024;59:310–36. https://doi.org/10.1080/10409238.2024.2411264.

[86] Ouimet M, Koster S, Sakowski E, Ramkhelawon B, van Solingen C, Oldebeken S, Karunakaran D, Portal-Celhay C, Sheedy FJ, Ray TD, et al. Mycobacterium tuberculosis induces the miR-33 locus to reprogram autophagy and host lipid metabolism. Nat Immunol 2016;17:677–86. https://doi.org/10.1038/ni.3434.

[87] Miow QH, Vallejo AF, Wang Y, Hong JM, Bai C, Teo FS, Wang AD, Loh HR, Tan TZ, Ding Y, et al. Doxycycline host-directed therapy in human pulmonary tuberculosis. J Clin Investig 2021;131. https://doi.org/10.1172/JCI141895.

[88] Adikesavalu H, Gopalaswamy R, Kumar A, Ranganathan UD, Shanmugam S. Autophagy induction as a host-directed therapeutic strategy against Mycobacterium tuberculosis infection. Medicina (Kaunas) 2021; 57. https://doi.org/10.3390/medicina57060522.

CHAPTER

7

Unraveling NCoR1's role in Mycobacterium tuberculosis pathogenicity

Subhasish Prusty, Sourya Prakash Nayak, R. Divya and Sunil K. Raghav

Immuno-genomics & Systems Biology Laboratory, BRIC-Institute of Life Sciences, Bhubaneswar, Odisha, India

Highlights

1. NCoR1 is a key regulatory protein in host defense against *Mycobacterium tuberculosis* infection.
2. NCoR1 modulates autophagy and lysosomal biogenesis genes, and plays a complex role in regulating autophagy.
3. NCoR1 manipulates myeloid cell metabolism through oxidative phosphorylation (OXPHOS) during *Mtb* infection.
4. NCoR1 depletion in macrophages and mice leads to enhanced *Mtb* burden.
5. NCoR1 regulates *Mtb* infection by preserving the balance between energy levels and the AMPK-mTOR-TFEB axis.
6. NCoR1 is a unique candidate for host-pathogen interactions that can be exploited for TB therapy.
7. Understanding NCoR1's role in non-myeloid cells is required to unveil new host-pathogen interactions.

Abbreviations

AMP Adenosine monophosphate
AMPK AMP activated protein kinase
AMs Alveolar macrophages
AP1 Activator protein 1
ATG Autophagy related genes
ATP Adenosine triphosphate
BCG Bacillus Calmette-Guérin
Bcl B cell lymphoma
BMDMs Bone marrow-derived macrophages
BPaLM Bedaquiline, Pretomanid, Linezolid, Moxifloxacin
CD Cluster of differentiation
cDC Conventional dendritic cells
CEBPα CCAAT enhancer-binding protein alpha
CF10 Culture filtrate protein 10
CFU Colony-forming units

Navigating Tuberculosis. https://doi.org/10.1016/B978-0-443-27735-1.00001-0

cGAS-STING Cyclic GMP-AMP synthase stimulator of interferon genes
cGMP Cyclic Guanosine monophosphate- Adenosine monophosphate
c-JUN Cellular JUN or Jun proto-oncogene AP-1 transcription factor subunit
CLR C-type lectin receptors
CNS Central nervous system
COVID-19 Coronavirus disease 2019
CPLA2γ Cytosolic phospholipase A2γ
CTLA-4 Cytotoxic T-lymphocyte associated protein 4
DC Dendritic cells
DC-SIGN DC-specific intracellular adhesion molecule-3-grabbing non-integrin
ESAT6 Early secreted antigenic target 6 kDa
ESX-1 ESAT-6 secretion system-1
FAO Fatty acid oxidation
FOXP3 Forkhead box P3
GLUT Glucose transporters
GMP Guanosine monophosphate
GrpE GroP-like protein E
GSK3-α Glycogen synthase kinase 3 alpha
H37Rv H37 Reference virulent strain
HDAC Histone deacetylases
HDT Host-directed therapy
HIF-1α Hypoxia inducible factor 1 subunit a
HIV Human Immunodeficiency Virus
HSP Heat shock proteins
IEC Intestinal epithelial cell
IFN Interferon
IL Interleukin
ILC Innate lymphoid cells
IM Interstitial macrophages
iNKT Invariant natural killer cells
IRF Interferon regulatory factor
ISG Interferon stimulatory genes
KD Knockdown
LAM Lipoarabinomannan
LAMP1 Lysosomal-associated membrane protein 1
LC3 Microtubule-associated protein 1 light chain 3
LM Lipomannan
LpqH Lipoprotein H
LpqT Lipoprotein T
LprE Lipoprotein E
LPS Lipopolysaccharide
LXR Liver X receptor
LXRβ Liver X-receptor β
LysM Cre FvB
MDM Monocyte-derived macrophages
MDR Multidrug resistance
MHC-II Major histocompatibility II
miR30 Micro RNA 30
MPT83 Mycobacterium Protein Tuberculosis 83

Mtb *Mycobacterium tuberculosis*
mTOR Mammalian target of Rapamycin
mTORC mTOR complex
MyeKO Myeloid Knockout
NADPH Nicotinamide adenine dinucleotide phosphate hydrogen
NCoR1 Nuclear Receptor Co-repressor 1
NCoR1fl/fl Nuclear receptor corepressor 1 floxed/floxed
NCR Natural cytotoxicity receptor
NF-κB Nuclear factor kappa
NK Natural killer cells
NLR Nod-like receptors
NO Nitric oxide
NOD2 Nucleotide-binding oligomerization domain containing 2
NR Nuclear receptors
OXPHOS Oxidative phosphorylation
PAMP Pathogen-associated molecular patterns
PCR Polymerase Chain Reaction
PD Periodontitis
pDC Plasmacytoid dendritic cells
Pdl1 Programmed death-1
PE_PGRS33 Proline-Glutamate/Polymorphic GC-Rich Sequence Protein 33
PI3K-AKT Phosphoinositide 3 kinase-AKT serine/threonine kinase
PIM Phosphatidylinositol mannoside
PPARγ Peroxisome proliferator-activated receptor gamma
PPE32 Proline-Proline-Glutamate Protein 32
PRKAA2 Phosphorylated protein kinase AMP-activated catalytic subunit alpha 2
PRR Pattern recognition receptors
PtpA Protein phosphatase 2a
RAR Retinoic acid receptors
ROS Reactive oxygen species
RpfB Resuscitation-promoting factor B
RpfE Resuscitation-promoting factor E
Rv0652 Reference Variant 0652
Rv288c Reference Variant 288c
Rv3529c Reference Variant 3529c
SeV Sendai Virus
SGLT-1 Solute carrier family 5 member 1
shRNA Short hairpin RNA
SIRT1 Sirtuin 1
SJL- Swiss Jim Lambert
SMRT Silencing mediator of retinoic acid and thyroid hormone
SNH Severe neonatal hyperbilirubinemia
TB Tuberculosis
TCR T cell receptors
TFEB Transcription factor EB
Th1 T-helper type 1 cells
THP-1 Tohoku Hospital Pediatrics-1
THR Thyroid hormone receptor
TLR Toll-like receptors

TNF Tumor-necrosis factor
Tregs Regulatory T cells
TST Tuberculin skin tests
v-ATPase Vacuolar ATPase
WHO World Health Organization

1. Introduction

Mycobacterium tuberculosis (*Mtb*) causes tuberculosis (TB) in humans. According to the World Health Organization (WHO), more than 10.8 million people are infected with TB, of which 1.25 million died in the year 2023. TB has surpassed coronavirus disease 2019 (COVID-19) to become a leading cause of death worldwide. Malnutrition, lack of awareness, unavailability of effective vaccines with long-term memory, complex host-pathogen interactions, and multidrug resistance (MDR) are the main reasons for the increasing global TB burden [1–3]. When an infected person coughs, aerosol-containing bacilli are released into the environment, which, when inhaled by a healthy individual, causes TB. Within the respiratory tract, bacilli encounter the patrolling police of the lungs, alveolar macrophages (AMs). Unfortunately, the cell designated to kill these bacteria becomes the harboring house for the pathogen. By secreting virulence factors, intracellular pathogens such as *Mtb* prevent the fusion of phagosomes and lysosomes and survive in the host's hostile environment [4]. It replicates inside AMs and is released into the alveolar environment following cell death. The fate of *Mtb* depends on the host's immune system. If the immune system is strong, it attempts to clear the infection by restraining bacterial growth and dissemination, recruiting multiple layers of immune cells around the infected macrophages, and forming granulomas. However, *Mtb* takes advantage of granulomas by manipulating the host immune response to create a favorable environment for survival and chronic infections [5,6]. This is also known as latent TB infection, which can last up to several years until the granuloma ruptures to release the bacteria for active infection. Around 5%–10% of these latent TB patients with human immunodeficiency virus (HIV), cancer, or immune-suppressed diseases are at a high risk for the development of active or primary TB.

For the last 100 years, Bacillus Calmette-Guérin (BCG) has been the only licensed TB vaccine available, which only protects infants up to 10–12 years [7]. Currently, all probable vaccine candidates are in different phases of clinical trials, and many of them, those that showed effective protection during preclinical trials, failed to do so in clinical trials. In addition, the anti-TB treatment regimen is very long, and MDR TB strains (isoniazid- and rifampicin-resistant) are a major threat to global health [8]. Although increasing the use of antibiotics (BPaLM TB therapy) is not a long-term solution to fight MDR-TB, host factors and proteins may be crucial in creating new approaches to lower TB burden, which are sometimes referred to as host-directed therapies (HDTs). HDTs can be performed in combination with traditional anti-TB drugs to improve the efficacy of TB treatment. HDTs target immune pathways and metabolic processes to boost the host's ability to control infections. For example, oxysophocarpine, a quinolizidine alkaloid, helps reduce lung granuloma formation in mice by augmenting the host's antioxidant defenses and regulating excessive inflammation [9]. Resveratrol is another natural compound that can control *Mtb* growth by initiating autophagy and phagosome-lysosome fusion through sirtuin 1 (SIRT1) [10]. Similarly, several host cell proteins that are regulated during *Mtb* pathogenesis must be identified. NCoR1 is a global transcriptional repressor and highly conserved protein whose expression is tightly modulated during *Mtb* pathogenesis. By comprehending the function of NCoR1, we can correlate host-*Mtb* interactions and formulate a distinctive TB therapy regimen.

2. Mycobacterium tuberculosis

Mtb causes TB in humans, which belongs to the family *Mycobacteriaceae*, order *Actinomycetales,* and class *Actinobacteria* [11]. *Mtb* is a non-motile, obligate, and aerobic rod-shaped bacterium with a distinct waxy cell wall composed of glycolipids (peptidoglycans, mycolic acids, lipomannan (LM), and lipoarabinomannan (LAM)), which play a crucial role in host-pathogen interactions, including phagosome maturation arrest and immune evasion [12]. This modification renders bacteria resistant to many common antibiotics. Although the primary infection site is the lungs, extrapulmonary TB is also seen in vital organs, such as the kidneys, bones, spine, intestines, and eyes. Despite the slow growth of *Mtb,* it is highly infectious and can lead to fatality if left untreated. Its doubling time is almost 24 hours, and in solid culture, distinct rough colonies are formed. It can be visualized using Ziehl-Neelsen acid-fast staining. Although new polymerase chain reaction (PCR)-based tests are available for the detection of TB, the Mantoux and tuberculin skin tests (TSTs) are commonly used. Patients with active TB have common symptoms, such as prolonged cough, chest pain, appetite loss, weakness, fatigue, weight loss, fever, chills, and night sweats.

2.1 *Mycobacterium tuberculosis* pathogenesis

Understanding consecutive events during *Mtb* infection is essential for dissecting host-pathogen interactions. In the lungs, *Mtb* interacts with alveolar lining fluid, a mixture of surfactant proteins and hydrolases that facilitates its uptake by AMs. In addition to AMs, immune cells, such as dendritic cells (DCs) and neutrophils, are susceptible to infection [13]. *Mtb* accesses host resources such as iron and fatty acids and manipulates the host machinery to encounter minimal oxidative stress for survival. Following AM infection, *Mtb* spreads to other macrophage populations, such as interstitial macrophages (IMs) and monocyte-derived macrophages (MDMs) [14].

2.2 Mechanism of *Mycobacterium tuberculosis* recognition

Various cells interact with *Mtb* during the disease progression. Although numerous experiments have been conducted, significant gaps remain in our understanding of host-pathogen interactions during *Mtb* infection. Innate immune cells detect *Mtb* via pattern recognition receptors (PRRs), such as toll-like receptors (TLRs), nod-like receptors (NLRs), C-type lectin receptors (CLRs), and mannose receptors. Lung-resident macrophages and DCs recognize *Mtb* components using receptors such as TLR2, TLR4, Dectin-1, DC-SIGN, and nucleotide-binding oligomerization domain-containing 2 (NOD2). *Mtb* infection regulates various processes in host cells, including apoptosis, necrosis, necroptosis, autophagy, and cytokine release [15–17].

TLR2 recognizes mycobacterial glycolipids, such as LAM and LM, and glycoproteins, such as phosphatidylinositol mannoside (PIM) [18,19]. *Mtb* antigens MPT83, PE_PGRS33, LpqH, and PPE32 trigger apoptosis, whereas antigens Rv3529c and LpqT inhibit apoptosis in phagocytes such as macrophages via TLR2. Similarly, a few lipoproteins, such as LpqT and LpqH, hinder major histocompatibility II (MHC-II) and pro-inflammatory responses, leading to immune evasion. In contrast,

TLR2 also recognizes MPT83, another lipoprotein that facilitates macrophage apoptosis, thereby facilitating bacterial clearance [20]. It also modulates phagosomal maturation and oxidative stress in host cells [21]. TLR4 detects acylated LM, Rv0652, Rv288c, GrpE, RpfB, RpfE, and heat shock proteins (HSPs) to mediate Th1 responses by DCs [22]. TLR4-mediated signals maintain equilibrium between apoptosis and necrosis upon *Mtb* infection [23]. Blocking the TLR4 receptor using HTA125 and an endotoxin agonist (E5531) reduces phagocytosis and pro-inflammatory responses, leading to bacterial survival [24].

Dectin-1 receptors are found on various immune cells, such as macrophages, DCs, neutrophils, and T cells. Although it primarily detects β-glucans in fungal pathogens, it also assists the immune cells in *Mtb* detection. It helps immune cells secrete tumor necrosis factor (TNF), interleukin 6 (IL-6), and IL-12 [25,26]. In addition, DC-SIGN (CD206), which is present on DCs, is involved in *Mtb* infection. It can act as both a PRR and an adhesion receptor, which can help in DC migration and T cell interactions. DC-SIGN binds to Man-LAM, LM, and α-glucan present on *Mtb* to initiate an anti-inflammatory response via IL-10 production [27]. Studies have shown that *Mtb* induces interleukin 1 beta (IL-1β) production through NOD2 and suppresses inflammasome activation [28].

2.3 Role of immune cells during *Mycobacterium tuberculosis* infection

AMs form the primary innate defense mechanism against *Mtb* and play a crucial role in pathogen elimination. In addition to macrophages, other immune cells play different roles during *Mtb* infection. Neutrophils are the first cell type to be assigned from the vasculature to the site of infection. They quickly phagocytose the pathogen and secrete antibacterial peptides and pro-inflammatory cytokines to fight the infection [29]. Eosinophils are involved in various immune responses related to allergens, parasites, fibrosis, tissue repair, and remodeling. Few studies have investigated the interaction between eosinophils and AMs during *Mtb* infection in mice [5]. Type I interferon (IFN) signaling recruits inflammatory monocytes during *Mtb* infection. Inflammatory monocytes play a crucial role in pathogenesis by differentiating into phagocytic macrophages or antigen-presenting DCs. They also produce inflammatory cytokines, such as TNF, IL-1β, and reactive oxygen species (ROS), which can lead to inflammation [30]. DCs detect *Mtb* via their TLR9 receptors, followed by phagocytosis. It then processes the antigens and presents them to $CD8^+$ T cells using MHC-I molecules. This leads to enhanced pathogen recognition and recruitment of innate immune cells such as monocytes and neutrophils for bacterial clearance. Both classical and monocyte-derived DCs effectively prime $CD4^+$ T lymphocytes during pathogenesis [31]. Recent studies have revealed that natural killer (NK) cells recognize *Mtb* via TLR2 and Nkp44 (a member of the natural cytotoxicity receptor (NCR)) binding [32]. NK cells reduce pathogen survival by inducing cytotoxicity in infected cells and modulating cytokine signaling in granulomas [27,33].

Invariant natural killer T (iNKT) cells are a special type of lymphocyte that can show features of both innate and adaptive immune cells. During *Mtb* infection, their number increases significantly to provide protection against *Mtb* infection. T cell receptors (TCRs) facilitate the activation of iNKT cells in response to bacterial antigens, followed by the release of IL-10 and TNF and the recruitment of various innate immune cells to the site of infection [34]. Innate lymphoid cells (ILCs) accumulate in the lungs of TB patients [35,36]. A recent report suggested the differentiation of ILC precursor to ILC1, which helps the production of sustained interferon-gamma (IFN-γ), resulting in Mycobacterial clearance [37]. $CD4^+$ and $CD8^+$ T cells play a critical role in *Mtb* pathogenesis in the adaptive response. $CD4^+$ Th1 cells and $CD8^+$ T cells secrete IFN-γ, which controls the pathogen burden in mice and humans [38].

2.3.1 Mycobacterium tuberculosis-macrophage interaction

Macrophages, the lung's first line of defense, reside in airways and parenchyma where they clear pathogens and, upon *Mtb* infection, may undergo necrosis, apoptosis, or survive [39]. *Mtb* triggers apoptosis and antimycobacterial activity in human macrophages through cytosolic phospholipase A2γ (CPLA2γ), which releases arachidonic acid from membrane phospholipids. Arachidonic acid and its metabolites act as secondary messengers in TNF-induced apoptosis, generating ROS and activating sphingomyelinase to produce ceramide, which drives cell death [40,41]. Macrophage population polarization into M1 (pro-inflammatory, Th1-dependent) or M2 (anti-inflammatory, Th2-dependent) subtypes is crucial for *Mtb* immunity. M1 macrophages, driven by factors such as IFN-γ and TNF, exhibit high antigen presentation, nitric oxide (NO) production, and cytokine secretion, enabling a strong defense against intracellular pathogens. In contrast, M2-type macrophages induced by IL-4 and IL-10 are less effective in controlling infection and promoting tissue repair [12]. Macrophages detect pathogen-associated molecular patterns (PAMPs) through various PRRs, including TLRs, CLRs, NLRs, and cyclic GMP-AMP synthase-stimulator of interferon response cGAMP interactor (cGAS-STING), which activates antimicrobial pathways and modulates intracellular trafficking.

Upon internalization, macrophages employ multiple mechanisms to eliminate *Mtb*, including nutrient restriction, heavy metal intoxication, antimicrobial peptide production, the generation of ROS/nitrogen species, and phagosome acidification. The infection triggers a metabolic shift to glycolysis (Warburg effect), providing adenosine triphosphate (ATP) and nicotinamide adenine dinucleotide phosphate hydrogen (NADPH) to support antimicrobial responses, thereby restricting *Mtb* in macrophages, mice, and non-human primates [14]. Although AMs are not typically considered to leave their alveolar niche, recent findings show that bacteria-containing AMs can traverse the airway epithelium and reach the interstitium. This substantiates previous observations that AM depletion reduces the bacterial load, suggesting that AM facilitates *Mtb* transmission to monocytes, where bacterial replication occurs [42]. The uptake of large *Mtb* aggregates leads to a higher and faster frequency of cell death than the uptake of smaller aggregates. Once the host cell dies, *Mtb* rapidly replicates inside dead macrophages, regardless of macrophage type or prior IFN-γ exposure. Bacterial growth within dead cells is faster than in the extracellular space [43]. Additionally, NK cells support macrophages during the immune response [44].

2.4 Granuloma formation during *Mycobacterium tuberculosis* pathogenesis

Despite the recruitment of various immune cells and an increase in inflammatory responses, *Mtb* persists inside the host. Granuloma formation plays a crucial role in controlling TB infection. Granulomas are composed of multiple layers of immune cells that restrict pathogen replication and dissemination. Within granulomas, *Mtb* survives as a latent infection and persists for several years. Inside the granuloma, *Mtb* embraces a dormant, metabolically altered, constrained state, leading to latent TB that may be reactivated under various conditions (immune suppression, HIV, etc.). *Mtb* also exploits granuloma as it creates a barrier to antibiotics and determines disease outcomes, with TNF and IL-6 maintaining their structures [45]. Active TB granulomas are caseous and necrotic, whereas mature granulomas are calcified, sterile, and organized. $CD4^+$ T cell dysfunction during HIV infection or TNF inhibition impairs granuloma formation, increasing the risk of bacterial dissemination and active TB infection [14]. Similarly, *Mtb* induces macrophage differentiation into foam cells via mycolic acid production, which contributes to caseous necrosis and cavitation [46]. Granuloma varies by infection stage: classical granuloma in both active and latent TB features a necrotic center; non-necrotizing granuloma, mainly macrophages with few lymphocytes, occurs in active disease; and fibrotic granuloma, mostly fibroblasts, is found in latent TB. *Mtb* remains in a metabolically altered state in the central hypoxic zone of the granuloma during latent infection but replicates in peripheral oxygenated areas during active TB [39].

2.5 Evasion mechanisms of *Mycobacterium tuberculosis*

The mycobacterial protein PtpA binds to vacuolar ATPase (v-ATPase) and inhibits the efficient recruitment of phagosomes. It attenuates phagosome acidification and promotes bacterial survival [47]. Some *Mtb* strains develop favorable cytoplasmic conditions to survive within endocytic vesicles by activating phospholipase A2 in host cells. *Mtb* inhibits phagosome maturation by preventing acidification and maintaining a near-neutral pH (~6.2) within the phagosomes. This is achieved by altering phagosome composition and using specific cell wall molecules as barriers, thus protecting the pathogen from an acidic environment. *Mtb* employs multiple strategies to evade autophagy-mediated immune clearance. Inhibiting phagosome-lysosome fusion is a key mechanism for preventing pathogen degradation in macrophages. While autophagy-related (ATG) genes contribute to host immunity, *Mtb* evades degradation by inhibiting autophagolysosome formation through ESAT-6 secretion system 1 (ESX-1), thereby creating a protected niche for intracellular survival [48]. The ESX-1 system secretes proteins, such as the early secreted antigenic target 6 kDa (ESAT-6) and culture filtrate protein 10 (CFP-10), which disrupt the host cell phagosomal membrane, leading to bacterial escape. Additionally, this system induces host cell death (macrophage apoptosis) and inhibits antigen presentation [49]. Multiple studies have also suggested that virulence factors such as miR-30A regulate macrophage apoptosis and inhibit autophagy, facilitating the intracellular survival of *Mtb* [39]. *Mtb* inhibition of apoptosis and autophagy promotes immune evasion and latent infection [39]. Although apoptosis suppresses the intracellular replication of *Mtb*, virulent strains evade apoptosis through various effectors, thereby enhancing their virulence. However, apoptosis during later stages of infection contributes to *Mtb* dissemination [15]. *Mtb* evades lysosomal breakdown by escaping phagocytosis in the cytoplasm [50]. *Mtb* can also block host apoptosis via PtpA-mediated dephosphorylation of glycogen synthase kinase 3 alpha (GSK3-α), thereby enhancing its survival [51]. *Mtb* alters phosphatidylinositol and Rab proteins in phagosomes to prevent autophagic clearance. *Mtb* further suppresses autophagy by modulating IL-6 and LprE protein expression, which impairs vitamin D3-mediated immune response.

Cellular immunity driven by IFN-γ-producing T cells activates macrophages to produce NO, which subsequently targets *Mtb*. However, the adaptive immune response to *Mtb* in the lungs is notably delayed, allowing bacteria to grow unchecked for many days, remaining confined but persistent within the host. Studies in transgenic mice expressing TCRs specific for *Mtb* antigens (Ag85B and ESAT-6) highlight the slow development of T cell responses, which are initiated in the mediastinal lymph nodes after DCs transport *Mtb*. Delayed adaptive immunity is a key mechanism by which *Mtb* evades immune elimination, in contrast to the faster responses observed in viral and other bacterial infections [52]. These mechanisms enable *Mtb* to evade both innate and adaptive immune responses [44].

3. NCoR1: A key player in cellular function

Nuclear receptors (NR) are a superfamily of ligand-dependent transcription factors that control gene expression to regulate development, metabolism, and immunity [53,54]. NCoR1, an essential transcriptional corepressor, is involved in cellular homeostasis and regulates numerous physiological processes by interacting with transcription factors and nuclear receptors. NCoR1 creates transcriptionally repressive chromatin states that silence gene expression by attracting histone deacetylases (HDACs) to target genes.

3.1 Chromatin remodeling and transcriptional regulation

The fundamental role of NCoR1 is to serve as a scaffold for the formation of transcriptional repressor complexes. The deacetylation of histone tails is catalyzed by the recruitment of HDACs, which results in chromatin condensation and inhibition of gene transcription. This regulation is critical for maintaining cellular identity and for responding to environmental and developmental cues. It acts as a transcriptional checkpoint, repressing inflammatory gene expression by interacting with the dephosphorylated Jun proto-oncogene AP-1 transcription factor subunit (c-Jun) bound to target promoters in the absence of an NR ligand. Upon stimulation with innate inflammatory signals, such as lipopolysaccharide (LPS), structural alterations in the ligand-binding domain of NR diminish its affinity for NCoR1 while enhancing coactivator recruitment, thereby driving the transcriptional activation of pro-inflammatory genes [53]. NCoR1 also regulates transcriptional responses to hormone signals by coordinating with NRs, including retinoic acid receptors (RAR) and thyroid hormone receptors (THR) [55].

3.2 Metabolism

NCoR1 significantly influenced metabolic processes by regulating genes associated with mitochondrial biogenesis, lipid metabolism, and energy expenditure. Lipid and glucose homeostasis regulate the activity of transcription factors such as liver X receptor (LXR) and peroxisome proliferator-activated receptor gamma (PPARγ). Through the efficient regulation of energy production and storage, this metabolic control allows cells to adapt to fluctuating energy requirements. Its depletion in intestinal cells alters glucose metabolism by inhibiting carbohydrate transporters such as solute carrier family 5 member 1 (SGLT-1), Glucose transporter 2 (GLUT-2), and GLUT-5 [56]. NCoR1 controls metabolic processes such as fatty acid oxidation (FAO) and glycolysis to maintain a fine balance between inflammatory and tolerogenic responses in both murine and human DCs [57]. To investigate NCoR1's role in thyroid hormone receptor (THR) regulation, Hollenberg et al. created a mutated version of NCoR1 (NCoR1ΔID) lacking key nuclear receptor interaction domains. Mice with this disrupted binding domain exhibit lower cholesterol levels in the liver and increased production of alternative bile acids. These fewer bile acids are less effective at binding fats and sterols in the intestine, leading to the reduced absorption of these compounds [58].

3.3 Immune regulation

NCoR1 strongly controls immune responses by maintaining a fine balance between tolerogenic and inflammatory gene signatures, which are essential for cellular homeostasis. Targeted NCoR1 deletions in immune cells, the liver, adipose tissue, and muscle reveal its regulatory role in immune metabolism [53,56].

3.3.1 NCoR1 is identified as a key master regulator of the tolerogenic program

NCoR1 is a key player in regulating the immune tolerance observed in antigen-presenting cells, such as DCs. To prevent excessive immune activation by modulating gene expression, NCoR1 promotes immune homeostasis in DCs [59]. Additionally, NCoR1 and the silencing mediator of retinoic acid and thyroid hormone (SMRT) play central roles in balancing inflammatory and tolerogenic responses [60]. NCoR1-depleted DCs exhibit a heightened ability to polarize Th lymphocytes into regulatory T cells (Tregs) both ex vivo and in vivo, which is associated with an exacerbated disease phenotype. Integrative genomic analyses revealed that NCoR1 strongly repressed genes associated with immune tolerance. Knockdown (KD) alters nuclear factor kappa B (NF-κB) activity in activated DCs, shifting their immune response toward tolerance. This underscores NCoR1's critical role in maintaining immune homeostasis by controlling tolerogenic gene expression and modulating immune response. Upon NCoR1 depletion, genes such as *Il-10*, Cd83 molecule (*Cd83*), cytotoxic T-lymphocyte-associated protein 4 (*Ctla4*), and programmed death-1 (*Pdl1*) are derepressed, leading to increased transcription [59].

The deletion of NCoR1 in adipocytes and macrophages exerts anti-inflammatory effects via peroxisome proliferator-activated receptor gamma (PPAR-γ) and LXR, while disruption of Bcl6 transcription repressor (Bcl6)-SMRT/NCoR1 interactions in the bone marrow induces severe atherosclerosis, highlighting the tissue- and cell-specific roles of NCoR1 in regulation [61]. NF-κB and other transcriptional activators that regulate the expression of proinflammatory cytokines and other immune effectors are suppressed. Maintaining this regulatory balance is essential to avoid excessive inflammation, which can result in autoimmune diseases and tissue damage [62]. Furthermore, NCoR1 plays a crucial role in immunological homeostasis by regulating T cell differentiation and macrophage polarization [63,64].

3.3.2 Antiviral immunity

NCoR1 also plays a role in various immune functions related to antiviral response. Plasmacytoid DCs (pDCs) induced a strong type I IFN response, whereas classical or conventional type 1 (cDC1) DCs triggered a mild response. NCoR1 directly controls TLR-specific enhancer activity and gene expression in cDC1 cells. NCoR1 depletion boosts TLR9-specific gene expression, leading to robust TLR9 response. In contrast, TLR3-mediated antiviral genes are suppressed by reduced IRF3 activation and binding to ISGs, leading to lower expression. NCoR1-depleted cDC1 cells had weaker antiviral responses, reduced Sendai Virus (Sev) clearance, and lower $CD8^+$ T cell cytotoxicity upon TLR3 activation [65,66].

3.4 Cellular differentiation and development

The significance of NCoR1 in developmental processes is highlighted by its role in regulating pathways required for cellular differentiation. By suppressing genes that regulate neuronal differentiation, osteogenesis, and adipogenesis, NCoR1 guarantees the accurate temporal and spatial regulation of these processes [67]. NCoR1 plays an essential role in forkhead box P3 $(FOXP3)^+$ Treg differentiation and effector function. Hence, NCoR1-ablated Tregs failed to protect the mice from colitis-induced inflammation. It also controls cholesterol homeostasis independent of LXRβ [63]. NCoR1 and

SMRT interact with transcription factors and control genes that regulate B cell development. Deleting any of the transcriptional repressors leads to impaired B cell differentiation, resulting in an increased risk of leukemia [68]. In addition, NCoR1 is involved in central nervous system (CNS) development in mice, and its depletion impairs erythropoiesis [69].

3.5 Cellular health and pathology

The loss and dysregulation of NCoR1 have major consequences on cellular and systemic health. NCoR1 demonstrates complex regulatory mechanisms across diseases, such as autism spectrum disorder, type 2 diabetes, and atherosclerosis, by modulating transcription factors in a cell-type-specific manner [70]. Conditional gene targeting and transgenic mouse models expressing mutant forms of NCoR1 have been used to investigate their functions in adult tissues and whole organisms. Pathway analysis of NCoR1-depleted naive $CD4^+$ T cells revealed upregulation of cholesterol biosynthesis pathways, indicating that NCoR1 plays a crucial role in regulating cholesterol metabolism in these cells [71]. In a mouse model of diet-induced obesity, myeloid cell-specific ablation of NCoR1 paradoxically suppressed inflammation and enhanced insulin sensitivity [72].

Deletion of NCoR1 in neonatal mice has been reported to induce intestinal epithelial cell (IEC) proliferation, a process that is absent in adults, suggesting a cause of severe neonatal hyperbilirubinemia (SNH) during the developmental period [73]. The absence of NCoR1 enhances osteoclast and neutrophil formation in periodontitis (PD) by transcriptionally regulating the expression of CCAAT enhancer-binding protein alpha (CEBPα) via PPARγ, thereby promoting osteoclast differentiation. These findings reveal a critical role for NCoR1 in the pathophysiology of PD [70]. NCoR1 regulates colonic stem cell proliferation and secretory differentiation [61]. Jepsen et al. demonstrated that complete NCoR1 knockout is lethal and impairs thymocyte development [53]. Zhang et al. further revealed that NCoR1 ablation in male SJL mice hindered fenofibrate-induced IFN-γ repression, potentially explaining the sex-specific differences in Th1 responses [77]. NCoR1 is a key regulator of muscle cells [74]. NCoR1 deficiency in macrophages augments myocardial infarction and neointimal hyperplasia in mice [74].

4. Role of NCoR1 in *Mycobacterium tuberculosis* pathogenesis

As discussed earlier, NCoR1 has profound functions in metabolism and immune regulation, which encouraged us to investigate the function of NCoR1 during *Mtb* infection in myeloid cells. Additionally, *Mtb* rewires host metabolism and immune functions for survival inside the host, which makes NCoR1 an important candidate for TB therapy.

4.1 NCoR1 expression during the early stages of *Mycobacterium tuberculosis* infection

NCoR1 plays a crucial role in shaping the host immune response during *Mtb* infections. NCoR1 levels have been investigated in TB patients in clinical settings. Whole blood gene expression profiling of patients with TB revealed reduced NCoR1 transcript levels during active infection compared with healthy controls. This reduction was evident during the initial 6 months and was restored after 12 months of anti-TB treatment. Experimental evidence indicates that macrophages and DCs exhibit a dynamic shift in the kinetics of

NCoR1 expression. In THP-1-derived macrophages (mo-MΦ) infected with H37Rv (a pathogenic *Mtb* strain), NCoR1 expression increased during the first 24 hours and then gradually decreased after 48 hours. This change in expression implies the importance of early response mechanisms against *Mtb* infection. Additionally, immunofluorescence and western blot analyses revealed that NCoR1 expression peaked at 12 hours post-infection, followed by a reduction at subsequent time points. Similar NCoR1 dynamics were also observed in murine-derived type 1 conventional DCs at both the transcript and protein levels during infection with *Mtb* H37Rv and *Mycobacterium smegmatis* [75,76]. Thus, the importance of NCoR1 in the management of *Mtb* infection cannot be ignored.

4.2 NCoR1 depletion enhances *Mycobacterium tuberculosis* burden in myeloid cells

NCoR1 plays a pivotal role in regulating *Mtb* infection in myeloid cells such as macrophages and DCs. To validate this hypothesis, NCoR1 expression was depleted by lentivirus-mediated shRNA transduction in THP-1 mo-MΦ. H37Rv and *M. smegmatis* infection in these NCoR1 knockdown (KD) cells led to elevated *Mtb* infection, which was supported by both flow cytometry and colony-forming unit (CFU) assays. Since there was no difference in the phagocytosis rate during the early time points in these NCoR1-depleted cells, it was determined that NCoR1 plays a role that extends beyond the initial uptake of pathogens. Using myeloid-specific conditional knockout murine models (NCoR1MyeKO mice) generated by mating homozygous NCoR1fl/fl-CD11c-Cre C57BL6 mice with LysM-Cre FvB mice, the influence of NCoR1 is studied in innate immune responses. Bone marrow-derived macrophages (BMDMs) and peritoneal macrophages obtained from NCoR1MyeKO mice show increased susceptibility to *Mtb* infection. This indicated the regulatory role of NCoR1 in reducing bacterial proliferation in DCs and macrophages.

4.3 Enhanced *Mycobacterium tuberculosis* infection in NCoR1 MyeKO mice in vivo

During TB progression, NCoR1 plays a central role in regulating the host defense against *Mtb*, thereby maintaining immune homeostasis. During *Mtb* infection, the absence of NCoR1 alters the body weight dynamics, organ pathology, and bacterial load. In NCoR1 KD mice, there was a marked reduction in body weight during the early stages of infection, which gradually decreased over time. A similar trend was observed for the spleen size, where NCoR1 KD mice experienced splenomegaly, which is a primary indicator of systemic immune activation. As the disease progressed, the *Mtb* load in both the lungs and splenic tissue increased [75].

Significant changes in the immune cell population in the lungs and the spleen in the absence of NCoR1 demonstrate its role in regulating immune cell composition.. There was a substantial increase in the number of neutrophils, DCs, and eosinophils in the lungs and a marked decrease in alveolar and infiltrating macrophages. A higher infiltration of neutrophils, macrophages, and monocytes has been reported in splenic tissues with constant levels of DCs. This study focuses on the critical role of NCoR1 in the regulation of bacterial containment within myeloid cells. In contrast, the lymphocyte populations, mainly the B and T cell subtypes, remained unaffected, further strengthening the impact of NCoR1 deficiency on innate immune cell dynamics. Elevated levels of pro-inflammatory cytokines, such as IL-6 and IL-17A, in lung tissues maintain an inflammatory environment, which is substantiated by the cellular infiltration observed in the histology of the murine lungs [75].

4.4 NCoR1 as a potent modulator of autophagy during *Mycobacterium tuberculosis* infection

NCoR1 modulates autophagy and lysosome-mediated bacterial clearance by regulating the expression of various genes involved in this process. The modulation of autophagy by NCoR1 was confirmed by examining the expression of key ATG genes, such as *ATG3*, *ATG5*, and *ATG13*, as well as hypoxia-inducible factor 1 subunit alpha (HIF-1α), which possesses antimicrobial properties. Apart from ATG genes, NCoR1 also influences inflammatory mediators such as TNF and NF-κB, thus having a broad regulatory impact on immune mechanisms. An important mechanism in autophagy is the conversion of LC3-I to LC3-II to form autophagosomes, which degrade pathogens in the phagolysosomes. However, in cells with NCoR1 deficiency, autophagic flux is disrupted because of changes in the ratio of LC3-II to LC3-I. This was further confirmed by bafilomycin treatment, where NCoR1-depleted cells inhibited autophagosome-lysosome fusion in contrast to normal cells. This further strengthens NCoR1's role in maintaining autophagic balance. Regulation of autophagy by NCoR1 ensures the effective colocalization of LC3 with intracellular *Mtb*, facilitating bacterial entrapment and degradation. This confirmed the importance of NCoR1 in myeloid cells to provide a proper immune response against *Mtb* infection.

4.5 NCoR1 regulating the mTOR-TFEB axis for autophagy induction

Transcription factor EB (TFEB) plays a pivotal role in regulating genes involved in autophagy induction and maturation, and its effect on genes involved in autophagic processes. Recent findings have highlighted the interplay between NCoR1 and TFEB in myeloid cells, which regulate autophagy during *Mtb* infection. Elevated levels of NCoR1 and TFEB in CD11c+ and F4/80+ cells in *Mtb*-infected lungs confirmed their interactions in vivo. Overexpression of Flag-tagged TFEB in NCoR1 KD mo-MΦ cells restored LC3-II and LAMP1 levels, highlighting NCoR1's role in autophagy and lysosomal function through TFEB. Furthermore, the reduced bacterial infection upon TFEB overexpression in NCoR1 KD mo-MΦ cells focused on the role of NCoR1 in controlling *Mtb* infection.

The study is not only limited to infection scenarios but is also conducted under stress conditions such as starvation. In control mo-MΦ cells, NCoR1 expression increased along with TFEB and LC3-II to LC3-I ratio during starvation, but these markers were significantly reduced in NCoR1 KD cells. This indicated the involvement of NCoR1 in cellular homeostasis under stressful conditions. Mechanistic target of rapamycin kinase (mTOR), a well-known regulator of autophagy, regulates TFEB activity through its phosphorylation. mTOR signaling is upregulated in NCoR1-depleted cells, which further increases the phosphorylation of TFEB, leading to impairment of autophagy. Similar observations have been reported in BMDMs from NCoR1MyeKO mice, suggesting a potential link between NCoR1 and the mTOR-TFEB pathway. Furthermore, in NCoR1-deficient cells, TFEB activity and autophagy were restored after the inhibition of mTOR signaling. When these cells were treated with rapamycin, a potent mTOR inhibitor, mTOR phosphorylation was reduced, reversing the phosphorylation of TFEB and elevating the levels of LC3-II. This ultimately results in bacterial clearance through the induction of autophagy and lysosomal biogenesis (Fig. 7.1).

4.6 NCoR1 controls mTOR by curbing ATP-AMPK levels

NCoR1-depleted cells show alterations in the phosphoinositide 3 kinase-AKT serine/threonine kinase (PI3K-AKT)-mTOR pathway, which affects autophagic flux and energy homeostasis. The mTOR complex (mTORC1) is negatively regulated by the phosphorylated protein kinase AMP-activated catalytic subunit alpha 2 (PRKAA2/AMPK). The balance between mTOR and AMPK is perturbed during *Mtb* infection, leading to elevated mTOR activity and reduced AMPK activity in the macrophages. AMPK, being a key sensor of cellular ATP levels and is inversely related to mTOR, thereby altering the cellular energy profile. ATP levels were elevated in NCoR1-depleted macrophages during *Mtb* infection, indicating the involvement of NCoR1 in the metabolic pathways. This perturbation leads to increased OXPHOS and ATP production. In the context of autophagy and infection, the balance between AMPK and mTOR is necessary for the regulation of autophagy during infection. When signaling pathways are disrupted during the depletion of

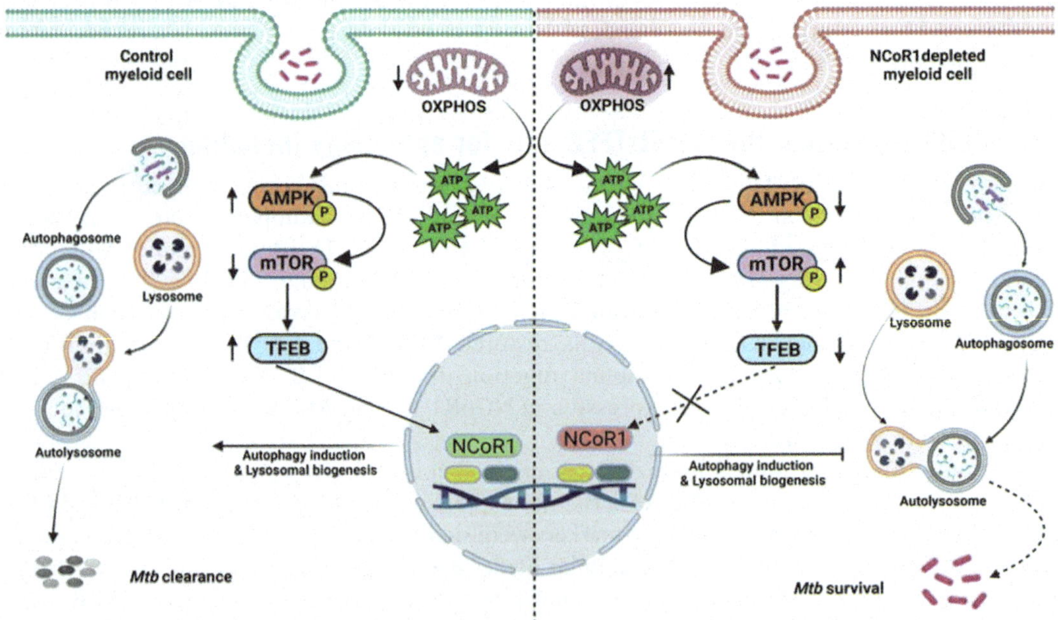

FIGURE 7.1 Schematic Illustration Explaining *Mtb* Survival in NCoR1-Depleted Myeloid Cells.

NCoR1 depletion leads to enhanced OXPHOS and ATP production. mTOR inhibits the translocation of TFEB into the nucleus, followed by the inhibition of genes related to autophagy induction and lysosomal biogenesis, thereby promoting the survival of Mycobacteria in myeloid cells. *ATP*, adenosine triphosphate; *mTOR*, mechanistic target of rapamycin kinase; *NCoR1*, Nuclear receptor corepressor 1; *OXPHOS*, oxidative phosphorylation; *TFEB*, transcription factor EB.

NCoR1, the autophagic process is impaired, further increasing the bacterial burden. Metformin reduces mTOR activity, increases autophagic flux, and reduces bacterial infection. Cellular ATP levels are altered by pharmacological interventions such as antimycin A, which indirectly affects autophagy through energy production. The use of antimycin A reduces ATP levels, further enhancing AMPK function, leading to a reduction in mTOR activation, and finally reducing infection levels. This regulatory loop ensures the optimal induction of TFEB, driving the formation of autophagolysosomes required for pathogen clearance. When NCoR1 was overexpressed in NCoR1-depleted macrophages, the AMPK/mTOR signaling pathway was rescued, enhancing the ability of the cells to clear infection through autophagy.

5. Conclusions and perspectives

Experimental observations have shown that NCoR1 is involved in fine-tuning autophagosome formation to control *Mtb* infection via the AMPK-mTOR-TFEB signaling axis. Enhanced expression of NCoR1 in macrophages and DCs during early pathogenesis plays a major role in host defense. NCoR1 depletion hinders pathogen clearance by establishing an equilibrium in the AMPK-mTOR signaling, leading to activation of TFEB function. In the absence of NCoR1, TFEB enhanced *Mtb* survival. By contrast, NCoR1-depleted macrophages can eliminate bacteria via autophagy by overexpressing TFEB. Overall, NCoR1 is crucial for controlling *Mtb* infection in myeloid cells such as macrophages. NCoR1 is also involved in various cellular functions, such as regulating inflammatory and tolerogenic responses, antiviral responses, and metabolism in immune cells. Given its central role in immune function, NCoR1 has emerged as a potential target for therapeutic strategies to modulate host immune and metabolic processes to minimize the ability of the pathogen to evade host defense mechanisms. Further investigations on the role of NCoR1 and other non-myeloid cells are required to unveil new unexplored areas of host-pathogen interactions.

Acknowledgments

The authors express their sincere gratitude to Dr. Viplov Biswas and Dr. Kaushik Sen for their pioneering research and foundational contributions to understanding the role of NCoR1 in *Mycobacterium tuberculosis* pathogenesis, which provided the basis for this book chapter. We are deeply thankful to the BRIC Institute of Life Sciences (ILS) for providing access to its advanced research infrastructure, including the state-of-the-art BSL-3 and ABSL3 facilities for conducting *Mtb*-related experiments and animal house facilities for animal studies. We extend our appreciation to all lab members, colleagues, and collaborators for their invaluable support, technical assistance, and constructive discussions throughout this research. Finally, we would like to express our heartfelt thanks to Prof. S.E Hasnain, Prof. N.Z. Ehtesham and Dr. M. Shariq for the opportunity and encouragement to contribute to this book chapter.

References

[1] Moliva JI, Turner J, Torrelles JB. Immune responses to bacillus Calmette-Guérin vaccination: why do they fail to protect against mycobacterium tuberculosis? Front Immunol 2017;8. https://doi.org/10.3389/fimmu.2017.00407.

[2] Heidary M, Shirani M, Moradi M, Goudarzi M, Pouriran R, Rezaeian T, Khoshnood S. Tuberculosis challenges: resistance, co-infection, diagnosis, and treatment. Eur J Microbiol Immunol 2022;12(1):1–17. https://doi.org/10.1556/1886.2021.00021.

[3] Ockenga J, Fuhse K, Chatterjee S, Malykh R, Rippin H, Pirlich M, Yedilbayev A, Wickramasinghe K, Barazzoni R. Tuberculosis and malnutrition: the European perspective. Clin Nutr 2023;42(4):486–92. https://doi.org/10.1016/j.clnu.2023.01.016.

[4] Mohareer K, Asalla S, Banerjee S. Cell death at the cross roads of host-pathogen interaction in Mycobacterium tuberculosis infection. Tuberculosis 2018;113:99–121. https://doi.org/10.1016/j.tube.2018.09.007.

[5] Bohrer AC, Castro E, Tocheny CE, Assmann M, Schwarz B, Bohrnsen E, Makiya MA, Legrand F, Hilligan KL, Baker PJ, Torres-Juarez F, Hu Z, Ma H, Wang L, Niu L, Wen Z, Lee SH, Kamenyeva O, Kauffman KD, Donato M, Sher A, Barber DL, Via LE, Scriba TJ, Khatri P, Song Y, Wong K-W, Bosio CM, Klion AD, Mayer-Barber KD. Rapid GPR183-mediated recruitment of eosinophils to the lung after Mycobacterium tuberculosis infection. Cell Rep 2022;40(4):111144. https://doi.org/10.1016/j.celrep.2022.111144.

[6] Cronan MR. In the thick of it: formation of the tuberculous granuloma and its effects on host and therapeutic responses. Front Immunol 2022;13. https://doi.org/10.3389/fimmu.2022.820134.

[7] Nguipdop-Djomo P, Heldal E, Rodrigues LC, Abubakar I, Mangtani P. Duration of BCG protection against tuberculosis and change in effectiveness with time since vaccination in Norway: a retrospective population-based cohort study. Lancet Infect Dis 2016;16(2):219–26. https://doi.org/10.1016/s1473-3099(15)00400-4.

[8] Dheda K, Mirzayev F, Cirillo DM, Udwadia Z, Dooley KE, Chang KC, Omar SV, Reuter A, Perumal T, Horsburgh CR, Murray M, Lange C. Multidrug-resistant tuberculosis. Nat Rev Dis Primers 2024;10(1). https://doi.org/10.1038/s41572-024-00504-2.

[9] Huang X, Lowrie DB, Fan X-Y, Hu Z. Natural products in anti-tuberculosis host-directed therapy. Biomed Pharmacother 2024;171:116087. https://doi.org/10.1016/j.biopha.2023.116087.

[10] Jeong, Lee, Jung. Host-directed therapies for tuberculosis. Pathogens 2022;11(11).

[11] Gordon SV, Parish T. Microbe profile: Mycobacterium tuberculosis: humanity's deadly microbial foe. Microbiology 2018;164(4):437–9. https://doi.org/10.1099/mic.0.000601.

[12] Bo H, Moure UAE, Yang Y, Pan J, Li L, Wang M, Ke X, Cui H. Mycobacterium tuberculosis-macrophage interaction: molecular updates. Front Cell Infect Microbiol 2023;13. https://doi.org/10.3389/fcimb.2023.1062963.

[13] Lerner TR, Borel S, Gutierrez MG. The innate immune response in human tuberculosis. Cell Microbiol 2015;17(9):1277–85. https://doi.org/10.1111/cmi.12480.

[14] Chandra P, Grigsby SJ, Philips JA. Immune evasion and provocation by Mycobacterium tuberculosis. Nat Rev Microbiol 2022;20(12):750–66. https://doi.org/10.1038/s41579-022-00763-4.

[15] Kim JK, Silwal P, Jo EK. Host-pathogen dialogues in autophagy, apoptosis, and necrosis during mycobacterial infection. Immune Netw 2020;20(5):1–15. https://doi.org/10.4110/in.2020.20.e37.

[16] Sankar, Mishra BB. Early innate cell interactions with Mycobacterium tuberculosis in protection and pathology of tuberculosis.

[17] Nisa, Kipper FC, Panigrahy, Tiwari S, Kupz, Subbian. Different modalities of host cell death and their impact on Mycobacterium tuberculosis infection. Am J Physiol Cell Physiol 2022;323(5):1444.

[18] Kim JS, Kim YR, Yang CS. Latest comprehensive knowledge of the crosstalk between TLR signaling and mycobacteria and the antigens driving the process. J Microbiol Biotechnol 2019;29(10):1506–21. https://doi.org/10.4014/jmb.1908.08057.

[19] Turner J, Torrelles JB. Mannose-capped lipoarabinomannan in Mycobacterium tuberculosis pathogenesis. Pathog Dis 2018;76(4). https://doi.org/10.1093/femspd/fty026.

[20] Wang L, Zuo M, Chen H, Liu S, Wu X, Cui Z, Yang H, Liu H, Ge B. Mycobacterium tuberculosis lipoprotein MPT83 induces apoptosis of infected macrophages by activating the TLR2/p38/COX-2 signaling pathway. J Immunol 2017;198(12):4772–80. https://doi.org/10.4049/jimmunol.1700030.

[21] Su H, Zhu S, Zhu L, Huang W, Wang H, Zhang Z, Xu Y. Recombinant lipoprotein Rv1016c derived from Mycobacterium tuberculosis is a TLR-2 ligand that induces macrophages apoptosis and inhibits MHC II antigen processing. Front Cell Infect Microbiol 2016;6. https://doi.org/10.3389/fcimb.2016.00147.

[22] Dowling JK, Mansell A. Toll-like receptors: the swiss army knife of immunity and vaccine development. Clin Transl Immunol 2016;5(5). https://doi.org/10.1038/cti.2016.22.

[23] Pattanaik KP, Sengupta S, Jit BP, Kotak R, Sonawane A. Host-mycobacteria conflict: immune responses of the host vs. the mycobacteria TLR2 and TLR4 ligands and concomitant host-directed therapy. Microbiol Res 2022;264. https://doi.org/10.1016/j.micres.2022.127153.

[24] Lv J, He X, Wang H, Wang Z, Kelly GT, Wang X, Chen Y, Wang T, Qian Z. TLR4-NOX2 axis regulates the phagocytosis and killing of Mycobacterium tuberculosis by macrophages. BMC Pulm Med 2017;17(1). https://doi.org/10.1186/s12890-017-0517-0.

[25] Thompson A, Orr SJ. Emerging IL-12 family cytokines in the fight against fungal infections. Cytokine 2018; 111:398–407. https://doi.org/10.1016/j.cyto.2018.05.019.

[26] Pedro ARV, Lima T, Fróis-Martins R, Leal B, Ramos IC, Martins EG, Cabrita ARJ, Fonseca AJM, Maia MRG, Vilanova M, Correia A. Dectin-1-Mediated production of pro-inflammatory cytokines induced by yeast β-glucans in bovine monocytes. Front Immunol 2021;12. https://doi.org/10.3389/fimmu.2021.689879.

[27] Garand M, Goodier M, Owolabi O, Donkor S, Kampmann B, Sutherland JS. Functional and phenotypic changes of natural killer cells in whole blood during Mycobacterium tuberculosis infection and disease. Front Immunol 2018;9. https://doi.org/10.3389/fimmu.2018.00257.

[28] Silvério D, Gonçalves R, Appelberg R, Saraiva M. Advances on the role and applications of interleukin-1 in tuberculosis. mBio 2021;12(6). https://doi.org/10.1128/mBio.03134-21.

[29] Alcantara G, Nguyen P, Venketaraman, et al. Neutrophils in Mycobacterium tuberculosis. Vaccines 2023;11(3).

[30] Mijiritsky E, Gardin C, Ferroni L, Lacza Z, Zavan B. Albumin-impregnated bone granules modulate the interactions between mesenchymal stem cells and monocytes under in vitro inflammatory conditions. Mater Sci Eng C 2020;110:110678. https://doi.org/10.1016/j.msec.2020.110678.

[31] Khan N, Vidyarthi A, Pahari S, Agrewala JN. Distinct strategies employed by dendritic cells and macrophages in restricting Mycobacterium tuberculosis infection: different philosophies but same desire. Int Rev Immunol 2016;35(5):386–98. https://doi.org/10.3109/08830185.2015.1015718.

[32] Abebe F. Immunological basis of early clearance of Mycobacterium tuberculosis infection: the role of natural killer cells. Clin Exp Immunol 2021;204(1):32–40. https://doi.org/10.1111/cei.13565.

[33] Allen M, Bailey C, Cahatol I, Dodge L, Yim J, Kassissa C, Luong J, Kasko S, Pandya S, Venketaraman V. Mechanisms of control of Mycobacterium tuberculosis by NK cells: role of glutathione. Front Immunol 2015;6. https://doi.org/10.3389/fimmu.2015.00508.

[34] Ruibal P, Voogd L, Joosten SA, Ottenhoff THM. The role of donor-unrestricted T-cells, innate lymphoid cells, and NK cells in anti-mycobacterial immunity. Immunol Rev 2021;301(1):30–47. https://doi.org/10.1111/imr.12948.

[35] Ardain A, Domingo-Gonzalez R, Das S, Kazer SW, Howard NC, Singh A, Ahmed M, Nhamoyebonde S, Rangel-Moreno J, Ogongo P, Lu L, Ramsuran D, Garcia-Hernandez Mde la L, Ulland TK, Darby M, Park E, Karim F, Melocchi L, Madansein R, Dullabh KJ, Dunlap M, Marin-Agudelo N, Ebihara T, Ndung'u T, Kaushal D, Pym AS, Kolls JK, Steyn A, Zúñiga J, Horsnell W, Yokoyama WM, Shalek AK, Kløverpris HN, Colonna M, Leslie A, Khader SA. Group 3 innate lymphoid cells mediate early protective immunity against tuberculosis. Nature 2019;570(7762):528–32. https://doi.org/10.1038/s41586-019-1276-2.

[36] Pan L, Chen X, Liu X, Qiu W, Liu Y, Jiang W, Yang Z, Mou Y, Xu W, Li X, Ge H, Zhu H. Innate lymphoid cells exhibited IL-17-expressing phenotype in active tuberculosis disease. BMC Pulm Med 2021;21(1). https://doi.org/10.1186/s12890-021-01678-1.

[37] Corral D, Charton A, Krauss MZ, Blanquart E, Levillain F, Lefrançais E, Sneperger T, Vahlas Z, Girard J-P, Eberl G, Poquet Y, Guéry J-C, Argüello RJ, Belkaid Y, Mayer-Barber KD, Hepworth MR, Neyrolles O, Hudrisier D. ILC precursors differentiate into metabolically distinct ILC1-like cells during Mycobacterium tuberculosis infection. Cell Rep 2022;39(3):110715. https://doi.org/10.1016/j.celrep.2022.110715.

[38] Tambunan BA, Priyanto H, Nugraha J. Soedarsono, CD4+ and CD8+ T-cells expressing interferon gamma in active pulmonary tuberculosis patients. Afr J Infect Dis 2018;12(1):49–53. https://doi.org/10.2101/Ajid.12v1S.6.
[39] Zhai W, Wu F, Zhang Y, Fu Y, Liu Z. The immune escape mechanisms of Mycobacterium Tuberculosis. Int J Mol Sci 2019;20(2). https://doi.org/10.3390/ijms20020340.
[40] Ueda N. A rheostat of ceramide and sphingosine-1-phosphate as a determinant of oxidative stress-mediated kidney injury. Int J Mol Sci 2022;23(7):4010. https://doi.org/10.3390/ijms23074010.
[41] Yura Y, Masui A, Hamada M. Inhibitors of ceramide- and sphingosine-metabolizing enzymes as sensitizers in radiotherapy and chemotherapy for head and neck squamous cell carcinoma. Cancers 2020;12(8):2062. https://doi.org/10.3390/cancers12082062.
[42] Cohen SB, Gern BH, Delahaye JL, Adams KN, Plumlee CR, Winkler JK, Sherman DR, Gerner MY, Urdahl KB. Alveolar macrophages provide an early Mycobacterium tuberculosis niche and initiate dissemination. Cell Host Microbe 2018;24(3):439. https://doi.org/10.1016/j.chom.2018.08.001.
[43] Mahamed D, Boulle M, Ganga Y, Mc Arthur C, Skroch S, Oom L, Catinas O, Pillay K, Naicker M, Rampersad S, Mathonsi C, Hunter J, Sreejit G, Pym AS, Lustig G, Sigal A. Intracellular growth of Mycobacterium tuberculosis after macrophage cell death leads to serial killing of host cells. eLife 2017;6. https://doi.org/10.7554/eLife.22028.
[44] Zhuang L, Yang L, Li L, Ye Z, Gong W. Mycobacterium tuberculosis: immune response, biomarkers, and therapeutic intervention. MedComm 2024;5(1). https://doi.org/10.1002/mco2.419.
[45] Shukla S, Bhardwaj N, Singh A. Drug resistance in Mycobacterium tuberculosis: an evolutionary perspective and its adaptation to the lung microenvironment. Microbe 2024;5:100189. https://doi.org/10.1016/j.microb.2024.100189.
[46] Yang J, Zhang L, Qiao W, Luo Y. Mycobacterium tuberculosis: pathogenesis and therapeutic targets. MedComm 2023;4(5). https://doi.org/10.1002/mco2.353.
[47] Pal R, Kumar Bisht M, Mukhopadhyay S. Secretory proteins of Mycobacterium tuberculosis and their roles in modulation of host immune responses: focus on therapeutic targets. FEBS J 2022;289(14):4146–71. https://doi.org/10.1111/febs.16369.
[48] Chai Q, Wang L, Liu CH, Ge B. New insights into the evasion of host innate immunity by Mycobacterium tuberculosis. Cell Mol Immunol 2020;17(9):901–13. https://doi.org/10.1038/s41423-020-0502-z.
[49] Vu A, Glassman I, Campbell G, Yeganyan S, Nguyen J, Shin A, Venketaraman V. Host cell death and modulation of immune response against Mycobacterium tuberculosis infection. Int J Mol Sci 2024;25(11):6255. https://doi.org/10.3390/ijms25116255.
[50] Chen Z, Wang T, Liu Z, Zhang G, Wang J, Feng S, Liang J. Inhibition of autophagy by MiR-30A induced by Mycobacteria tuberculosis as a possible mechanism of immune escape in human macrophages. Jpn J Infect Dis 2015;68(5):420–4. https://doi.org/10.7883/yoken.JJID.2014.466.
[51] Rankine-Wilson LI, Shapira T, Emani CS, Av-Gay Y. From infection niche to therapeutic target: the intracellular lifestyle of Mycobacterium tuberculosis. Microbiology (United Kingdom) 2021;167(4). https://doi.org/10.1099/MIC.0.001041.
[52] Korb VC, Chuturgoon AA, Moodley D. Mycobacterium tuberculosis: manipulator of protective immunity. Int J Mol Sci 2016;17(3). https://doi.org/10.3390/ijms17030131.
[53] Geiger MA, Guillaumon AT, Paneni F, Matter CM, Stein S. Role of the nuclear receptor corepressor 1 (NCOR1) in atherosclerosis and associated immunometabolic diseases. Front Immunol 2020;11. https://doi.org/10.3389/fimmu.2020.569358.
[54] Müller, Hainberger, Stolz, Ellmeier. NCOR1—a new player on the field of T cell development. J Leukoc Biol 2018;104(6).
[55] Shimizu H, Astapova I, Ye F, Bilban M, Cohen RN, Hollenberg AN. NCoR1 and SMRT play unique roles in thyroid hormone action in vivo. Mol Cell Biol 2015;35(3):555–65. https://doi.org/10.1128/MCB.01208-14.

[56] Ritter MJ, Amano I, van der Spek AH, Gower AC, Undeutsch HJ, Rodrigues VAP, Daniel HE, Hollenberg AN. Nuclear receptor corepressors NCOR1 and SMRT regulate metabolism via intestinal regulation of carbohydrate transport. Endocrinology 2024;165(9). https://doi.org/10.1210/endocr/bqae100.

[57] Sen K, Pati R, Jha A, Mishra GP, Prusty S, Chaudhary S, Swetalika S, Podder S, Sen A, Swain M, Kumar Nanda R, Raghav SK. NCoR1 controls immune tolerance in conventional dendritic cells by fine-tuning glycolysis and fatty acid oxidation. Redox Biol 2023;59:102575. https://doi.org/10.1016/j.redox.2022.102575.

[58] Vella KR, Hollenberg AN. The actions of thyroid hormone signaling in the nucleus. Mol Cell Endocrinol 2017;458:127–35. https://doi.org/10.1016/j.mce.2017.03.001.

[59] Ahad A, Stevanin M, Smita S, Mishra GP, Gupta D, Waszak S, Sarkar UA, Basak S, Gupta B, Acha-Orbea H, Raghav SK. NCoR1: putting the brakes on the dendritic cell immune tolerance. iScience 2019;19:996–1011. https://doi.org/10.1016/j.isci.2019.08.024.

[60] Jha A, Ahad A, Mishra GP, Sen K, Smita S, Minz AP, Biswas VK, Tripathy A, Senapati S, Gupta B, Acha-Orbea H, Raghav SK. SMRT and NCoR1 fine-tune inflammatory versus tolerogenic balance in dendritic cells by differentially regulating STAT3 signaling. Front Immunol 2022;13. https://doi.org/10.3389/fimmu.2022.910705.

[61] Mennillo E, Yang X, Paszek M, Auwerx J, Benner C, Chen S. NCoR1 protects mice from dextran sodium sulfate–induced colitis by guarding colonic crypt cells from luminal insult. Cell Mol Gastroenterol Hepatol 2020;10(1):133–47. https://doi.org/10.1016/j.jcmgh.2020.01.014.

[62] Liu CH, Liu H, Ge B. Innate immunity in tuberculosis: host defense vs pathogen evasion. Cell Mol Immunol 2017;14(12):963–75. https://doi.org/10.1038/cmi.2017.88.

[63] Stolz V, Silva Rde Fe, Rica R, Zhu C, Preglej T, Hamminger P, Hainberger D, Alteneder M, Müller L, Waldherr M, Waltenberger D, Hladik A, Agerer B, Schuster M, Frey T, Krausgruber T, Knapp S, Campbell C, Schmetterer K, Trauner M, Bergthaler A, Bock C, Boucheron N, Ellmeier W. Nuclear receptor corepressor 1 controls regulatory T cell subset differentiation and effector function. eLife 2024;13. https://doi.org/10.7554/elife.78738.

[64] Hou C, Yan L, Sun K, Zhou T, Zou Y, Xiong W, Duan S-Z. Nuclear receptor corepressor 1 deficiency exacerbates asthma by modulating macrophage polarization. Cell Death Discov 2023;9(1). https://doi.org/10.1038/s41420-023-01724-3.

[65] Ahad A, Smita S, Mishra GP, Biswas VK, Sen K, Gupta B, Garcin D, Acha-Orbea H, Raghav SK. NCoR1 fine-tunes type-I IFN response in cDC1 dendritic cells by directly regulating Myd88-IRF7 axis under TLR9. Eur J Immunol 2020;50(12):1959–75. https://doi.org/10.1002/eji.202048566.

[66] Mishra GP, Jha A, Ahad A, Sen K, Sen A, Podder S, Prusty S, Biswas VK, Gupta B, Kumar Raghav S. Epigenomics of conventional type-I dendritic cells depicted preferential control of TLR9 versus TLR3 response by NCoR1 through differential IRF3 activation. Cell Mol Life Sci 2022;79(8). https://doi.org/10.1007/s00018-022-04424-w.

[67] Paluvai H, Shanmukha KD, Tyedmers J, Backs J. Insights into the function of HDAC3 and NCoR1/NCoR2 co-repressor complex in metabolic diseases. Front Mol Biosci 2023;10. https://doi.org/10.3389/fmolb.2023.1190094.

[68] Lee RD, Knutson TP, Munro SA, Miller JT, Heltemes-Harris LM, Mullighan CG, Jepsen K, Farrar MA. Nuclear corepressors NCOR1/NCOR2 regulate B cell development, maintain genomic integrity and prevent transformation. Nat Immunol 2022;23(12):1763–76. https://doi.org/10.1038/s41590-022-01343-7.

[69] Wan X, Liu L, Zhou P, Hui X, He Q, Yu F, Zhang W, Ding X, Yuan X, Zhang N, Zhao Y, Zhu R, Liu Y, Hao P, Auwerx J, Song X, Leng Q, Zhang Y. The nuclear receptor corepressor NCoR1 regulates hematopoiesis and leukemogenesis in vivo. Blood Adv 2019;3(4):644–57. https://doi.org/10.1182/bloodadvances.2018022756.

[70] Ma XX, Meng XQ, Wang YL, Liu Y, Shi XR, Shao S, Duan SZ, Lu HX. Ncor1 deficiency promotes osteoclastogenesis and exacerbates periodontitis. J Dent Res 2023;102(1):72–81. https://doi.org/10.1177/00220345221116927.

[71] Hainberger D, Stolz V, Zhu C, Schuster M, Müller L, Hamminger P, Rica R, Waltenberger D, Alteneder M, Krausgruber T, Hladik A, Knapp S, Bock C, Trauner M, Farrar MA, Ellmeier W. NCOR1 orchestrates transcriptional landscapes and effector functions of CD4+ T cells. Front Immunol 2020;11. https://doi.org/10.3389/fimmu.2020.00579.

[72] Shimizu H, Horibata Y, Amano I, Ritter MJ, Ohyama M, Domae M, Sugimoto H, Hollenberg AN. 7260 Nuclear Corepressor 1 (NCoR1) plays important role in the regulation of insulin gene transcription in mouse Min6 pancreatic beta cells. J Endocr Soc 2024;8(Suppl. 1). https://doi.org/10.1210/jendso/bvae163.870.

[73] Chen S, Lu W, Yueh M-F, Rettenmeier E, Liu M, Miles P, Auwerx J, Yu RT, Evans RM, Wang K, Karin M, Tukey RH. Intestinal NCoR1, a regulator of epithelial cell maturation, controls neonatal hyperbilirubinemia. Proc Natl Acad Sci 2017;114(8). https://doi.org/10.1073/pnas.1700232114.

[74] Du LJ, Sun JY, Zhang WC, Liu Y, Liu Y, Lin WZ, Liu T, Zhu H, Wang YL, Shao S, Zhou LJ, Chen BY, Lu H, Li RG, Jia F, Duan SZ. NCOR1 maintains the homeostasis of vascular smooth muscle cells and protects against aortic aneurysm. Cell Death Differ 2023;30(3):618–31. https://doi.org/10.1038/s41418-022-01065-1.

[75] Biswas VK, Sen K, Ahad A, Ghosh A, Verma S, Pati R, Prusty S, Nayak SP, Podder S, Kumar D, Gupta B, Raghav SK, Gutierrez MG. NCoR1 controls Mycobacterium tuberculosis growth in myeloid cells by regulating the AMPK-mTOR-TFEB axis. PLoS Biol 2023;21(8):e3002231. https://doi.org/10.1371/journal.pbio.3002231.

[76] Sen K, Biswas VK, Ghosh A, Prusty S, Nayak SP, Podder S, Gupta B, Raghav SK. NCoR1: a key player regulating mycobacterium tuberculosis pathogenesis. Autophagy 2024;20(3):697–8. https://doi.org/10.1080/15548627.2023.2277583.

[77] Zhang M, et al. Antagonizing peroxisome proliferator–activated receptor α activity selectively enhances Th1 immunity in male mice. J. Immunol. 2015;195(11):5189–202. https://doi.org/10.4049/jimmunol.1500449.

CHAPTER

Division: An eternal rhythm of cells

8

Yogita Kapoor[1,2,4], Amit Chakraborty[1,2] and Vinay Kumar Nandicoori[1,2,3]

[1]*CSIR-Centre for Cellular and Molecular Biology, Hyderabad, Telangana, India;* [2]*Academy of Scientific and Innovative Research (AcSIR), Ghaziabad, Uttar Pradesh, India;* [3]*BRIC-National Institute of Immunology, New Delhi, India;* [4]*Ragon Institute of MGH, MIT and Harvard University, Cambridge, MA, United States*

Highlights

1. Cell division is a fundamental process that unifies all life on our planet.
2. Bacteria possess a unique set of proteins that drive various division processes.
3. Mycobacteria grow asymmetrically in a polar fashion and V-snap into two daughter cells.
4. Mycobacteria contain typical SEDS proteins and a unique DivIVA-domain-containing protein called Wag31.
5. Wag31 aids polar elongation in mycobacteria by recruiting cell wall synthesis machinery.
6. SepIVA is another essential DivIVA-domain-containing protein involved in mycobacterial cell division.
7. Mycobacterial cell division is regulated at both transcriptional and post-translational levels.

Abbreviations

ACCase acetyl-CoA carboxylase
AG arabinogalactan
Crescentin intermediate filament homolog
CwsA cell wall synthesis protein A
Divisome multi-protein cell division machinery
EmbCAB EmbCAB operon (*embC, embA and embB*)
EmbR EmbR transcription factor
FASI fatty acid synthase complex I
FASII fatty acid synthase complex II
FasR fatty acid synthase regulator
FtsA Filamenting temperature-sensitive mutant A
FtsK DNA translocase and cell division protein
FtsZ Filamenting temperature-sensitive mutant Z (tubulin-like GTPase)
IMD intracellular membrane domain
MA mycolic acid
MabR mycolic acid biosynthesis regulator
MAPc mycolylarabinogalactan–peptidoglycan complex
MinC, MinD, MinE, MinJ min system centering proteins

Navigating Tuberculosis. https://doi.org/10.1016/B978-0-443-27735-1.00012-5

MM mycomembrane
MreB bacterial actin homolog
MtrA MtrA sensor kinase
MtrB MtrB response regulator
MurG lipid II synthase
NBS Noc-binding site
NO nucleoid occlusion
OM outer membrane
PbpB penicillin-binding protein B
PG peptidoglycan
PlrA polar growth regulator
RpfB resuscitation-promoting factor B
SEDS shape, elongation, division, and sporulation proteins
STPK serine/threonine protein kinase
TCS two-component signaling system
Wag31 DivIVA-domain-containing protein Wag31
ZipA ZipA (membrane anchor)

1. Cell division: An overview

One of the core principles of physiology is "The cell theory," which unifies all life on the planet through a simple cellular process called division. The cell theory suggests that all biological entities are composed of cells. Rudolf Virchow, a German pathologist, stated "omnis cellula e cellula," meaning that all cells arise from preexisting cells [1]. It was in the nineteenth century that scientists discovered all cells reproduce by dividing into two, making cell division a fundamental aspect of cell biology. In simple terms, cell division implies the division of a parental cell into two daughter cells. But why do cells divide? According to the "Origins of Life," which proposes that life as an independent entity depends on its constituents, life in any form (such as a cell) results from the inherent connectivity of its parts, which it tries to maintain to respond effectively to environmental stimuli. If a cell merely grew larger, its parts would eventually lose their connection, causing the cell to lose its identity. To restore this, cells undergo division, which help them re-establish their connectivity. This is also the basis of the "Sizer theory" which suggests that cells undergo division after attaining a certain size. Cell size is of utmost importance to both unicellular (for fitness) and multicellular organisms (ability to function in an integrated manner) [2]. Another theory, i.e., the "Timer theory" proposes that cells divide after a fixed amount of time has elapsed since their birth. And finally, according to the "Adder theory," cells divide after incrementing their size by a fixed amount [3]. Cell biologists around the globe believe that these models of growth are not mutually exclusive. One system could undergo a combination of either of them for a continuum.

Cell division is the final outcome of the cell cycle process. The cell cycle includes various stages, starting with the synthesis of multiple proteins and other metabolites that support cell growth, followed by replication of the genetic material, and ending with cytokinesis, which divides the cell's cytoplasm into two. These processes occur in a sequential order and are tightly regulated in terms of timing and location. A change in the sequence of any event or misregulation of any step can lead to defective cell division or prevent it entirely, which is detrimental to the organism. In prokaryotes, defective cell division affects overall fitness and survival, while in higher organisms, abnormal cell division can cause diseases such as cancer and chromosomal disorders. Therefore, the process of cell division is closely connected with several regulatory systems that ensure accurate segregation of genetic material and proper division of mother cells into two daughter cells.

2. Cell division in bacteria

Bacteria are unicellular, membrane-bound organisms that need to divide to survive. All bacteria have conserved mechanisms to accomplish this feat, in some cases several times per hour. They duplicate their chromosomes, position the division machinery accurately at the mid-cell, and partition their contents precisely, all the while maintaining the integrity of their cell envelope to withstand turgor pressure. Bacterial cell division involves a plethora of essential division proteins and accessory proteins that function in a concerted fashion to ensure successful division.

Bacteria have homologs of eukaryotic cytoskeletal elements that assist in division and related processes, such as FtsZ: a tubulin homolog that polymerizes to form a ring marking the division site [4–7]; MreB: an actin homolog that supports cell shape and organelle maintenance [8–10], chromosome segregation [11,12], and cellular motility [13]; and Crescentin: a homolog of intermediate filaments that also helps maintain cell shape [14]. Additionally, they possess MinD and ParA proteins, which are unique to bacterial cytoskeletons and aid in chromosome partitioning and divisome placement [15,16].

Bacterial division begins with the placement of a Z-ring composed of FtsZ filaments at the mid-cell, which acts as a scaffold for the assembly of many proteins that together form "divisome" and "septasome" components. The nano-machineries recruited by the FtsZ ring viz. septasome and divisome assist in cell wall synthesis at the septum and its constriction to produce two cells. The stages of bacterial cell division are briefly described below (Fig. 8.1).

(i) FtsZ anchorage to the membrane
(ii) FtsZ ring polymerization
(iii) Assembly of the septasome
(iv) Septation

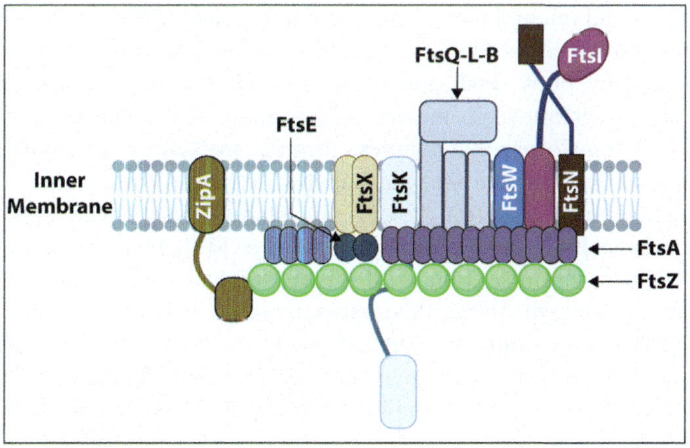

FIGURE 8.1 Assembly of the Divisome.

FtsA and ZipA tether FtsZ to the bacterial membrane to form a proto-ring which is later joined by FtsEX complex and the DNA translocon FtsK, which has both membranes bound N- and C-terminal cytosolic domains. Subsequently, FtsQ-FtsL-FtsB form a ternary complex and recruit septal synthases such as FtsW to the divisome. At the end, FtsN joins the complex activating FtsW.

2.1 FtsZ anchorage to the membrane

FtsZ, the eukaryotic tubulin homolog, cannot bind directly to the membrane. Instead, it interacts with FtsA (an actin homolog) and ZipA (a protein that interacts with FtsZ), which anchor it to the membrane [17,18]. Both FtsA and ZipA are membrane-associated proteins that bind to the C-terminal of FtsZ [19,20]. FtsA, the actin homolog, possesses ATPase activity and can form filaments [21]. These three proteins together constitute what is known as a "proto-ring," which prepares the stage for the recruitment of other proteins [22]. This proto-ring is later joined by FtsE and FtsX proteins, part of the ABC membrane transporter superfamily, with FtsE exhibiting ATPase activity and FtsX being an integral membrane protein [23]. The two proteins together coordinate peptidoglycan (PG) synthesis and hydrolysis at the septum.

2.2 FtsZ ring polymerization

Cells contain roughly 10,000–20,000 molecules of FtsZ. Being the tubulin homolog, it has GTPase activity, which it uses to power the formation of straight or gently curved FtsZ protofilaments [24,25]. These protofilaments come together to form an annular ring linked to the inner cytoplasmic membrane. FtsZ molecules in the ring are highly dynamic, with a half-time of 10–30 seconds for turnover as observed in Fluorescence Recovery After Photobleaching (FRAP) experiments [26]. As mentioned above, a treadmilling motion has been observed for the FtsZ ring, in which FtsZ molecules dissociate from one end and associate at the other, producing an apparent directional motion of the ring at a speed of ~30–40 nm/s observed in vivo [27]. Since GTP hydrolysis powers FtsZ polymerization, mutants defective in GTPase activity also affect FtsZ treadmilling but do not completely abolish it, highlighting the roles of other, yet uncharacterized mechanisms that drive FtsZ treadmilling. FtsZ-interacting proteins and its membrane anchors have mixed effects on FtsZ dynamics. While proteins such as ZapA, promote bundling and increase lateral interactions between FtsZ filaments, negatively impacting the FtsZ ring dynamics [28], FtsA co-treadmills with FtsZ filaments and enhances its dynamism [29].

2.3 Assembly of the septasome

Upon FtsZ ring assembly, several peripheral and integral membrane proteins localize at the proto-ring in a sequential manner, FtsK arrives first, followed by FtsQ, FtsL, and FtsB, which form a ternary complex called FtsQBL. This is succeeded by FtsW, FtsI, and FtsN [30–34]. The order has been deduced from numerous localization studies performed in mutant backgrounds of the Fts proteins. FtsK, the first to arrive at the site, is a DNA translocase that couples chromosome segregation to cell division [35,36]. It has an N-terminal membrane domain that participates in cell division and a C-terminal cytoplasmic tail involved in nucleoid segregation [34,37,38]. FtsK interacts with FtsQ, FtsL, and FtsI, in addition to FtsZ, as shown by bacterial two-hybrid assays [39]. FtsQ, a bitopic membrane protein, interacts with FtsB via its C-terminal domain [40]. FtsB and FtsL interact through both their periplasmic and transmembrane regions; the dimer then interacts with FtsQ in a 2:2:2 ratio [41,42]. This transmembrane complex links early cell division proteins to late cell division proteins. The complex is activated by FtsN, which promotes septal PG synthesis [30,33]. FtsA interacts with both FtsN and the FtsQBL complex to recruit them to the septum [43]. FtsB and FtsL also interact with highly conserved FtsW and FtsI, essential septal cell wall synthases. FtsW, belonging to the SEDS family, functions as a transglycosylase, while FtsI is its cognate transpeptidase; together, they form septal PG [32,44]. As previously mentioned, the activity of FtsW and FtsI is triggered by FtsN, which further recruits more FtsN, creating a positive feedback loop that accelerates inward PG synthesis [45].

2.4 Septation

Septation is the final process that leads to the scission of the membrane, separating the sister cells into two, each with a copy of the genetic material. Amidases, specifically AmiB and AmiC, localize at the septum to cleave the membrane by digesting the PG cross-linking peptide chains [46–51]. The last Fts protein to localize, FtsN, is responsible for recruiting AmiC, whereas AmiB depends on FtsX for its localization [46,51–53]. Successful cell constriction not only requires PG hydrolysis but also mechanical force for splitting the cell. The exact molecular players responsible for the latter process remain to be identified. Current models for septation focus on FtsZ-based constriction. Many reports support the hypothesis that FtsZ rings act like actomyosin rings that generate enough force upon contraction to overcome the internal turgor pressure of a bacterial cell and pull the septal cell wall inward, causing septation.

3. Regulators of bacterial cell division

The process of FtsZ ring formation is central to bacterial cell division. The assembly of the Z-ring at the mid-cell not only determines whether the cell will divide but also influences the size and shape of the daughter cells that would emerge from the mother cell. Therefore, this process is regulated both spatially and temporally. Bacteria, whether gram-negative or gram-positive, possess (a) the Min system and (b) the NOC system, which negatively regulate FtsZ ring polymerization to ensure its formation occurs only at the mid-cell. They also include (c) positive regulators of Z-ring placement.

3.1 The Min system

This divisome centering system was discovered over 30 years ago. The Min system includes three proteins—MinC, MinD, and MinE. MinC is the first protein that binds directly to FtsZ via its N- and C-terminal domains, restricting lateral interactions between FtsZ proto-filaments to prevent bundling [54,55]. MinD is shaped like a long funnel with its rim oriented toward the mid-cell. It exists in the cell in either ATP- or ADP-bound form. MinD-ATP associates with the membrane and recruits MinC there (at the base of the funnel), inhibiting FtsZ polymerization [56,57]. MinE is the third protein in the sequence that interacts with the MinC-MinD complex, binding to MinD-ATP and stimulating its ATP hydrolysis activity [58,59]. When bound to ADP, MinD cannot attach to the membrane and lodges off of it, diffusing into the cytosol. After ATP exchange, MinD-ATP relocates to the opposite pole, which is devoid of MinE. It then forms a funnel-like concentration gradient at the opposite pole, concentrating most of the MinC there to prevent FtsZ assembly at that location. This allows FtsZ to form a ring at the mid-cell. The MinCD complex at this pole is once again chased away by MinE, leading to its dissociation [60]. MinCDE proteins oscillate between the poles, preventing FtsZ ring formation [60,61]. In this system, MinC acts as the active negative regulator of FtsZ, while MinD and MinE are oscillators that regulate MinC's activity by restricting it to the cell poles and away from the mid-cell

[62]. Without this system, FtsZ would assemble at the poles, resulting in the formation of anucleoid mini-cells (Fig. 8.2A).

The Min system of gram-positive bacteria, such as *B. subtilis*, is slightly different. It doesn't contain the MinE protein; instead, it uses MinJ, which provides localization cues to the MinCD complex [58,63]. MinJ is recruited by another protein called DivIVA, which senses the negative membrane curvature at the poles and septum [64,65]. This makes DivIVA the spatial determinant of the MinCDJ complex in *B. subtilis* [66]. Unlike gram-negative bacteria, the MinCD of gram positives doesn't oscillate; instead, it forms a bipolar static gradient with the highest concentration at the poles and the lowest at the mid-cell, preventing ring formation at the poles (Fig. 8.2B) [67].

3.2 Nucleoid occlusion system

In several model organisms, it has been observed that cell division does not occur near the nucleoid due to active DNA-related processes, and this phenomenon is called nucleoid occlusion. The molecular mechanisms of this process differ between gram-negative and gram-positive bacteria. The former possess a DNA-binding protein called SlmA (a synthetic lethal allele of Min) that forms an increasing concentration gradient toward the pole and has more binding sites on the nucleoid near the Ori region

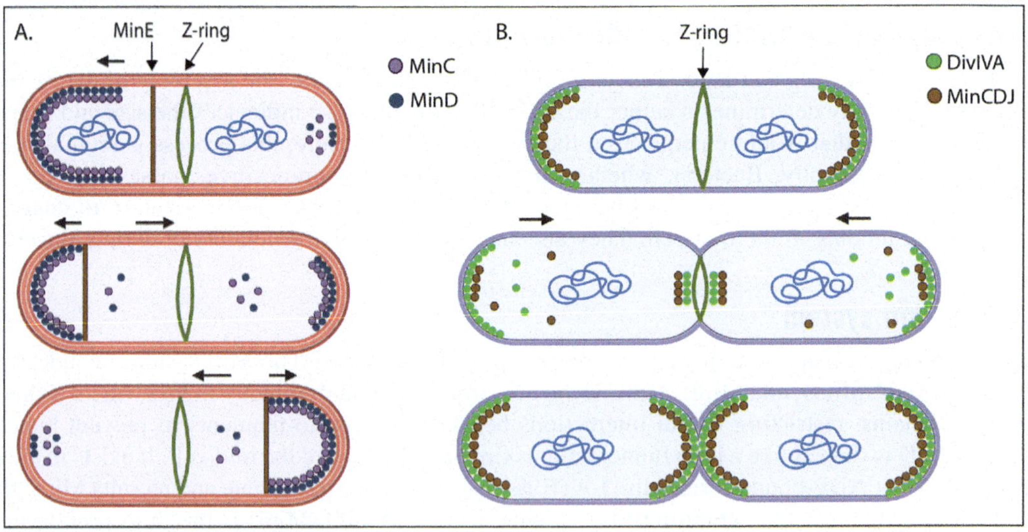

FIGURE 8.2 The Bacterial Min System.

(A) In *E. coli*, MinCD form a complex at the poles whose concentration is present in a decreasing gradient manner as it reaches the septum. This complex is chased by MinE which causes MinD hydrolysis and dissociates from the membrane, taking MinC along. The MinCD complex oscillates from one pole to another followed by MinE. Such a back-and-forth movement excludes the septum wherein FtsZ ring forms. (B) In *B. subtilis*, MinCD do not oscillate. The complex is held at the poles and septum with the help of MinJ which is recruited by DivIVA, making it the spatial determinant of the complex. In both the systems, MinCD complex prevents FtsZ ring formation at the poles.

(polar) [68]. SlmA also binds to FtsZ, preventing its assembly near the nucleoid (Fig. 8.3A) [69]. In contrast, gram-positive bacteria have Noc protein, which, although structurally different from SlmA, performs the same function of inhibiting FtsZ ring formation over the nucleoid [70]. However, they achieve this differently: SlmA protects the DNA by binding to specific sites, whereas Noc associates with the membrane via its N-terminal amphipathic helix, holding the DNA there and discouraging FtsZ ring formation over the nucleoid [71–74]. Deletion of these proteins results in misplacement of the FtsZ ring over unsegregated nucleoids. Noc has been found to have at least 70 binding sites in the *Bacillus subtilis* chromosome, called Noc binding sites, which are 14 bp palindromes. These sites are spread across the entire bacterial chromosome except for the Ter region (Fig. 8.3A).

3.3 Positive regulators of the FtsZ ring assembly

Many species also contain positive regulators of Z-ring assembly that assist in positioning the ring at the center of a bacterial cell. *Escherichia coli* has MatP protein that binds to regions flanking the Ter sites on DNA, which localize at the mid-cell, thereby aiding in Z-ring placement (Fig. 8.3B) [75]. It

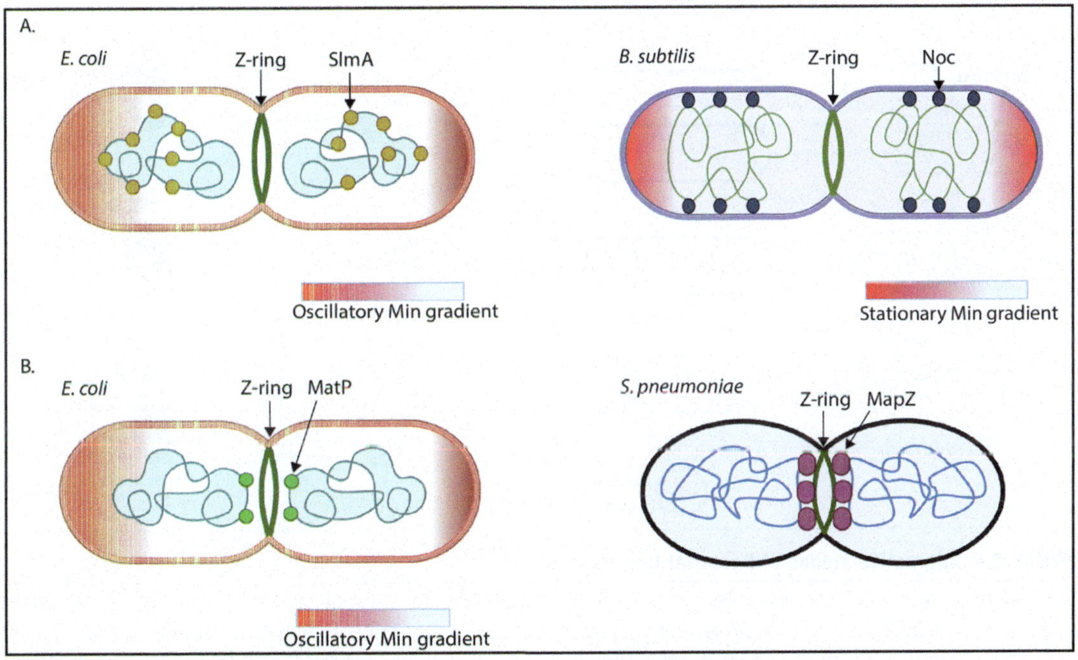

FIGURE 8.3 Positive and Negative Regulators of Bacterial FtsZ Ring Assembly.

(A) Apart from MinCDE system, SlmA protein in *E. coli* constitutes the NO system and prevents Z-ring formation over the nucleoid. (B) *B. subtilis* contains Noc protein that performs the same role of occluding bacterial chromosomes from the zone of FtsZ ring formation. (C) MatP acts as a positive regulator FtsZ ring formation in *E. coli* and helps in placement of Z-ring in the mid cell. (D) MapZ in *S. pneumoniae* is a positive regulator of Z-ring formation.

has also been found to interact with ZapB, an interactor of ZapA that promotes FtsZ-bundling [76–78]. Other positive regulators include MapZ, encoded by *Streptococci* (Fig. 8.3B) [79,80].

4. Alternative modes of bacterial cell division

Bacteria come in various shapes and sizes, as do their modes of reproduction! *E. coli*, *Caulobacter crescentus (C. crescentus)*, and *B. subtilis* are extensively studied model organisms because of the availability of genetic tools, ease of growth in laboratory conditions, presence of a plethora of microscopy-based markers, and their simple mode of cell division. Both *E. coli* and *B. subtilis* undergo binary fission, a mode of division in which the septum forms along the short axis of the cell, resulting in nearly symmetric division into two cells (Fig. 8.4A) [81]. Many non-model organisms reproduce through alternative methods to produce offspring. These methods include multiple fission, endospore formation, budding, and intracellular offspring production, which are discussed below.

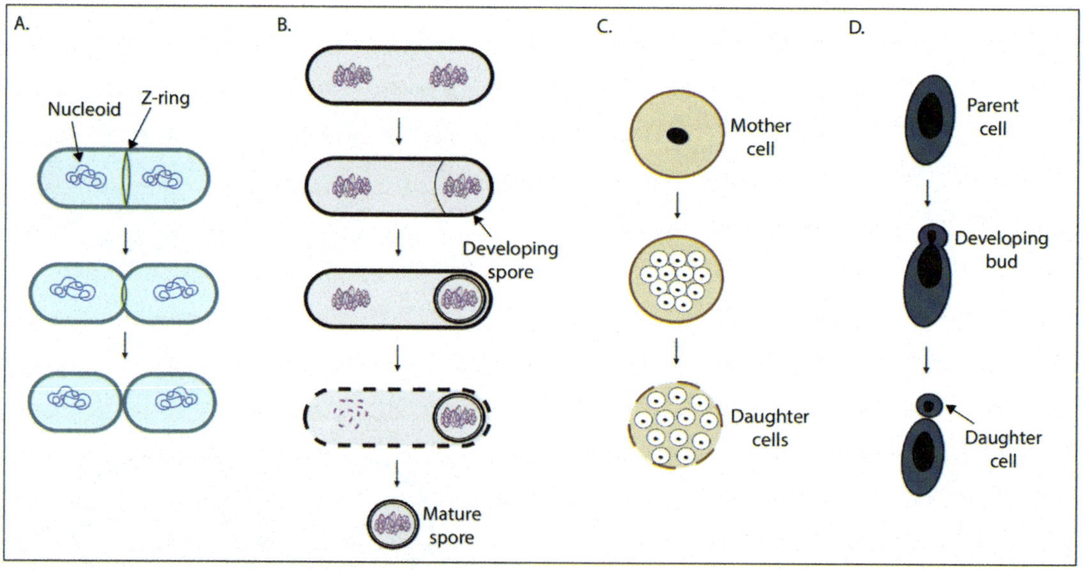

FIGURE 8.4 Alternative Modes of Bacterial Cell Division.

(A) In both *E. coli* and *B. subtilis*, binary fission is the most common form of cell division in which septum forms at the midcell leading to near symmetric division of the parent cell into two daughter cells. (B) Endospores are tough, nonreproducing structures formed by bacteria such as *B. subtilis* under nutritional stress conditions. Spore forms intracellularly within parent cell's body and survives even after the parent cell dies. It germinates once favorable condition for growth are restored. (C) Multiple fission is a modified form of binary fission where multiple progenies are produced from a single parent cell. This mode of reproduction is reported in Cyanobacteria, Actinobacteria, and Proteobacteria. (D) Cyanobacteria, Prosthecate proteobacteria, and Planctomycetes undergo budding where daughter cell develops as a small bud in parent cell. The bud later matures and becomes a separate individual.

4.1 Endospore formation

As the name suggests, this mode of reproduction leads to the generation of offspring intracellularly, i.e., within the body of the mother cell. It is a modified version of binary fission where the septum forms near both poles instead of at the mid-cell. Successful division occurs at the side of the old pole, leading to the production of a smaller forespore and a larger mother cell, each containing a copy of the duplicated nucleoid. After separation, the forespore is engulfed again by the mother cell, where it matures. Once sporulation is complete, the mother cell dies, and the spore continues to live on for very long periods, sometimes over tens of thousands of years, until environmental conditions trigger germination [82]. After germination, spores develop into cells that undergo regular vegetative cycle of growth cycle. This type of reproduction is observed in *B. subtilis* (Fig. 8.4B) [83–85].

4.2 Multiple fission

It is a form of reproduction triggered by nutritional stress that results in the production of multiple progenies from a single parent cell. The parent undergoes repeated rounds of genome duplication, producing multiple nucleoids that, along with some amount of cytoplasm, become enclosed by a membrane. This enlarged, multinucleated cell then divides either synchronously or undergoes multiple successive rounds of division, forming multiple daughter cells or spores (Fig. 8.4C). This reproductive mode has been reported in *Cyanobacteria*, *Actinobacteria*, and *Proteobacteria* so far. The progenies produced by this method are called baeocytes in the case of *Pleurocapsales*, a lineage of Cyanobacteria. Baeocytes are 1–2 μm in diameter and can grow up to 10 times their original size, i.e., ~30 μm in diameter, concomitant with multiple rounds of genetic material duplication during vegetative growth. The genetic material is segregated within the cytoplasm, ultimately leading to the fission of the mother cell into numerous (4–1000) baeocytes [86].

4.3 Budding

This mode of reproduction is adapted by various environmental bacteria, in which a small bud emerges on the mother cell and continues to grow until it attains the size of the mother cell, at which point it separates from the mother cell (Fig. 8.4D). Many low G + C bacteria, Cyanobacteria, Prosthecate Proteobacteria, and Planctomycetes undergo budding [87–89]. Bacteria such as *Caulobacter* and several species of Proteobacteria exist in two stages—a motile, swarmer cell and a sessile, stalked cell stage. The swarmer cell cannot replicate its genome. Once it differentiates into a prostheca, it replicates, and a bud emerges from the distal end of the stalk. The bud matures and acquires a flagellum, and disperses.

5. Cell division in Actinobacteria

Actinobacteria are high G+C, filamentous bacteria that are either aerobic, facultative aerobic, or anaerobic [90]. They are highly diverse and produce several metabolites of commercial importance, including antibiotics, agricultural, and other medicinal products [91]. They constitute one of the most diverse bacterial phyla and include several species such as *Streptomyces*, *Mycobacteria*, *Nocardia*, *Rhodococcus*, *Deinococcus*, *Corynebacterium*, *Bifidobacterium*, among others, with various shapes—coccoid, rods, and mycelium. Actinobacteria reproduce through different modes, such as binary fission and sporulation. They are often classified as gram-positive bacteria based on phylogeny, but their cell envelopes are multilayered, unlike gram-positives, which are usually monoderm. Their Peptidoglycan (PG) is attached to an Arabinogalactan (AG) layer, which is further covalently linked to Mycolic acids (MA), collectively known as the Mycolylarabinogalactan–peptidoglycan complex [92]. In addition to MA, they possess unique lipids forming an outer lipid layer. The MA layer, along with the outer membrane, is collectively called the Mycomembrane (MM) [93,94]. Species of the Mycobacterium genus have an additional capsule beyond the MM, comprising polysaccharides (mainly Lipomannan and its modified forms), proteins, and very small quantities of lipids.

Usually, rod-shaped bacteria such as *E. coli* and *B. subtilis* elongate by adding new cell wall material laterally along the wall [95,96]. MreB, the actin homolog, determines cell shape in these bacteria. However, rod-shaped bacteria of the actinobacteria phylum do not grow laterally; instead, they add new cell wall material at the cell poles [97,98]. They use an MreB-independent mechanism to maintain their shape. These bacteria grow in a polar fashion by targeting the cell wall synthesizing machinery precisely at the poles, with the help of adaptor proteins that self-assemble there and function to recruit the components of the elongasome. DivIVA is one such adaptor protein that orchestrates polar growth in actinobacteria [99–102]. It works with Scy and FilP to form a polarisome responsible for growth from the hyphal tips in *Streptomyces* spp. [103,104]. Actinobacteria produce unique cell division proteins, such as CrgA, found in *Streptomyces coelicolor* and *mycobacteria*, which interacts with FtsZ, FtsI, FtsQ, and PbpA, and is crucial for the localization of FtsI [105–107]. CrgA does not affect FtsZ ring assembly, but its deletion alongside CwsA, a *mycobacteria*-specific cell division protein, affects FtsZ ring dynamics and membrane integrity [108]. As previously mentioned, regulators of FtsZ ring assembly are absent from actinobacterial genomes, but *S. coelicolor* encodes SsgA and SsgB, which together assist in FtsZ ring assembly [109].

6. Cell division in mycobacteria

Mycobacteria belong to the actinobacteria phylum that elongates by polar division of cell wall material [110]. Cell division in mycobacteria can be broken down into various stages, such as elongation of the poles, synthesis of the septal cell wall (septation), and constriction of the septal membrane followed by its hydrolysis. Actinobacteria generally follow a "V-snapping" mode of division in which the daughter cells rapidly separate from one another [111,112]. They divide asymmetrically to produce daughter cells that are genotypically similar but phenotypically dissimilar (Fig. 8.5). This dissimilarity manifests in their growth rates, sizes, and cell wall composition, making the progeny population heterogeneous and resistant to antibiotics [113–116]. This heterogeneity may further influence bacterial survival inside the host by affecting the pathogen's ability to evade host-induced stresses. The heterogeneity is amplified in the host environment through cell wall remodeling and the switch to dormancy, which not only shapes the bacterial response to stressors produced by the host but also influences the host's immunological response [117–120].

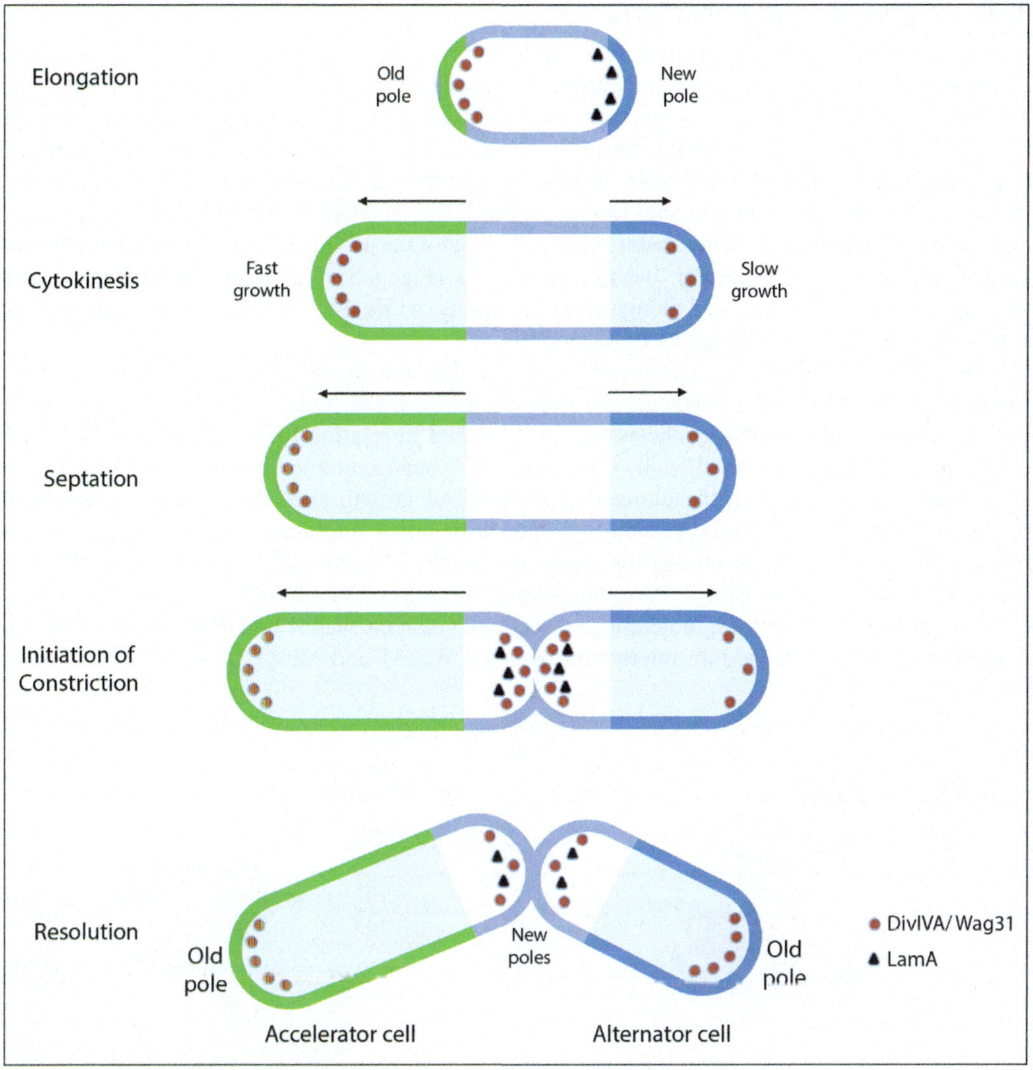

FIGURE 8.5 Unipolar Growth in Mycobacteria.

Mycobacteria elongate in a unipolar manner by adding new cell wall material at the poles. Each cell has two poles, an old pole, which was derived from the mother cell, and a new pole, acquired during the previous round of division. Mycobacteria grow by adding new cell wall material in polar fashion at just one pole, i.e., old pole, leading to the production of two asymmetric daughter cells. Once elongation is complete, cells transcend into cytokinesis stage, wherein the FtsZ ring assembly starts taking place and then septation occurs. Septation marks the formation of septal cell wall which is facilitated by PG synthases and hydrolases. This leads to constriction and physical separation of the daughter cells.

6.1 Polar growth of mycobacteria

Like other rod-shaped bacteria, mycobacterial cells also have two poles: an old pole, which is inherited from the mother cell, and a new pole, which was synthesized during the previous round of cell division. In mycobacteria, polar growth is asymmetric and results in the formation of unequally sized daughter cells [113–116]. Two models explain this process: one suggests that growth is unipolar, with the old pole elongating more than the new pole, or that the growth rates of both poles differ, producing two asymmetric daughter cells. The second model proposes that growth occurs equally at both poles, but the old pole's growth rate becomes faster shortly after cytokinesis [113–115]. Asymmetric division of mycobacteria generates accelerator and alternator cells (Fig. 8.5). The former inherits the old pole from the mother cell, is larger, and elongates more rapidly, while the latter inherits the new pole of the mother cell, is shorter, and elongates more slowly.

Polar growth of mycobacteria requires coordinated synthesis of various cell wall components, including PG, AG, MA, OM lipids, capsular polysaccharides, and proteins. Little is known about this complex process, and only PG synthesis has been studied in relation to cell division [121,122]. The synthesis mechanisms of other cell wall components have been examined in isolation [123–126]. Polar synthesis involves assembling the elongation complex at growth sites, which are the old poles in mycobacteria. Wag31, the DivIVA homolog, detects negative membrane curvature, self-assembles there, and forms scaffolds for elongation machinery at the old pole [102,110,127]. The PG precursor, Lipid II, is produced by MurG. Phosphomimetic versions of Wag31 have been shown to increase the activity of Lipid II synthases, indicating a direct link between Wag31 and PG synthesis (Fig. 8.6A). Our group recently established an interaction between Wag31 and MurG [128]. We also found that

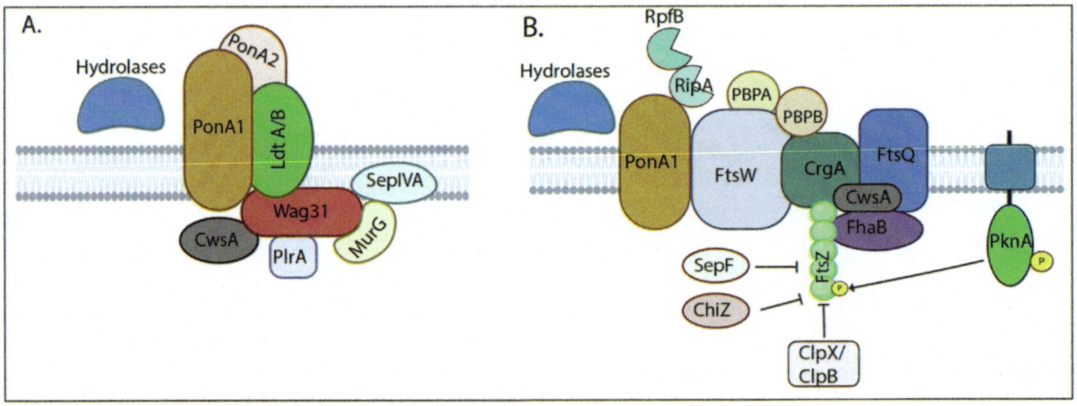

FIGURE 8.6 The Elongation and Division Complex of Mycobacteria.

(A) The elongation complex of mycobacteria comprises of PG synthases and hydrolases which are held at the poles with the help of structural proteins such as Wag31, which is stabilized by interaction with CwsA, SepIV A, MurG, and PlrA. (B) After the elongation is complete, proteins assemble at the septum to drive cell division, forming divisome complex. FtsZ forms a ring at the septum which is regulated by ChiZ, ClpX, Rv3660c, and SepF. The ring acts as a scaffold for the recruitment of various structural proteins including FtsW, FtsQ, CrgA, CwsA. The scaffold is also joined by PG synthases and hydrolases that together drive septum cell wall formation and cell separation.

Wag31 interacts with yet another DivIVA-domain-containing protein in mycobacteria called SepIVA (discussed below), which has been reported as a spatiotemporal regulator of MurG [128]. These interactions imply that these three proteins may form a complex to facilitate mycobacterial elongation. SepF, an essential component of the mycobacterial divisome that localizes in a ring-like structure, interacts with MurG and FtsZ to promote mycobacterial elongation and division (Fig. 8.6A) [129,130].

CwsA, which is cell wall synthesis protein and cell shape protein A, is involved in PG synthesis and it interacts with Wag31. The absence or overexpression of CwsA perturbs Wag31 polar and septal assembly, leading to severe morphological defects, further linking Wag31 to governing cell elongation in mycobacteria [108]. A recently described protein called PlrA or Polar Protein A, which helps in polar elongation, has also been linked to positioning Wag31 at the poles [102]. The absence of PlrA results in abnormal accumulation of Wag31 at the poles. Other components involved in PG remodeling, such as transglycosylases, transpeptidases, and hydrolysis enzymes like Rip hydrolases, have been shown to localize at the polar and septal sites. It is hypothesized that Wag31 may interact with them for recruitment to these locations (Fig. 8.6A).

6.2 Septation

After elongation is complete, the PG synthesis machinery moves to the septum, and other proteins involved in division become part of this complex to form what is known as the divisome [131]. The divisome consists of structural proteins, accessory proteins, cytoskeletal proteins, and PG remodelers, which are involved in septal wall build-up and hydrolysis. The determinants for division site selection are not known in mycobacteria. Studies suggest that a wide septum forms in the cell body, slightly away from the mid-cell position. One of the initial steps of septation is the formation of the FtsZ ring, guided by a MinD homolog in mycobacteria, Rv3660c. Rv3660c is inversely related to FtsZ—its overexpression causes a marked reduction in FtsZ levels and prevents its assembly. FtsZ is also regulated by an essential mycobacterial protein kinase, PknA, which mediates phosphorylation, decreasing FtsZ's GTPase activity and polymerization capacity [132]. Once the FtsZ ring assembles, structural components of the divisome are recruited to the scaffold. These include FhaA, FhaB, FtsQ, FtsW, CrgA, and CwsA [133]. FtsZ associates with FhaB and FtsQ in a ternary complex, as well as with FtsW and penicillin-binding protein B (PbpB) in another ternary complex (Fig. 8.6B). The FtsZ-FhaB-FtsQ complex is important for bacterial replication inside macrophages during oxidative stress, and the FtsZ-FtsW-PbpB complex is crucial for initiating septal PG synthesis in mycobacteria. Mycobacterial septal PG synthase FtsW has a long cytoplasmic tail, unlike *E. coli* or *B. subtilis*, which serves as an interaction hub for both FtsZ and PbpB [134]. PbpB, a septal PG synthase, interacts with Wag31, which recruits elongation machinery at the pole, linking elongation with division. PbpB is recruited by CrgA, an actinomycete-specific cell division protein [107]. CrgA also interacts with FtsQ, FtsZ, PbpA, and PbpB. Additionally, CrgA, a component of the divisome, interacts with CwsA, an elongasome component linking the two machineries (Fig. 8.6B). As expected, CrgA-null cells show defective cell length [107].

6.3 Division

Once the septal cell wall is ready, enzymes act on it to cleave the membrane, dividing the cells into two physically separate daughter cells. These enzymes are called PG hydrolases [122]. Unlike *E. coli,* mycobacteria do not encode amidases that cleave the cell wall. Instead, they possess Rip proteins—RipA and RipC [53,111]. RipA's function is well understood, while RipC's role is not adequately delineated. RipA works with another hydrolase, Resuscitation-promoting factor B (RpfB), and both collaborate to cleave PG [135]. PonA1 (PG synthase) also interacts with RipA. The dynamics of RipA-PonA1 and RipA-RpfB interactions regulate PG synthesis and hydrolysis, ensuring hydrolysis does not occur before synthesis is complete (Fig. 8.6B) [136].

7. Regulation of mycobacterial cell division

Environmental cues and host-induced stresses initiate signaling in bacteria that influence their physiology and morphology. Cell division and maintaining cell shape are essential to bacterial physiology. It involves multiple proteins that control cell division at both the transcriptional and post-translational levels.

7.1 Transcriptional regulation

Cell wall synthesis is coordinated with cell division, and several cell wall synthases are regulated by transcription factors. The EmbCAB operon, which encodes arabinosyltransferases that facilitate the incorporation of arabinose into AG and lipoarabinomannan, has been linked to resistance against the frontline TB drug ethambutol and is tightly controlled by the EmbR transcription factor [137]. Inside host macrophages, PknH levels increase, phosphorylating the FhaA domain of EmbR, which promotes its binding to the embCAB promoter and activates its expression. Overall, remodeling of the cell envelope confers resistance to ethambutol and enhances survival within the host (Fig. 8.7A) [138].

The production of α-MA, which forms the MM, is also regulated by two transcription factors: MabR, a mycolic acid biosynthesis regulator, and FasR, a fatty acid synthesis regulator [139,140]. Fatty acid synthase complexes I and II are involved in producing mycolic acids. While FASI causes de novo synthesis of fatty acids, FASII elongates them to form meromycolates. MabR is located upstream of the FASII operon and binds upstream of one of its genes, i.e., *fabD*. This binding results in downregulation of the transcription of fabD, acpM, kasA, kasB, and ultimately α-MA biosynthesis [140]. FasR, on the other hand, positively regulates α-MA production by binding to the fas promoter. As expected, loss of FasR decreases the overall MA content in the cell [139]. The opposing functions of MabR and FasR maintain a balance in α-MA production.

Two-component signaling systems (TCS), which are widely present in bacteria, also help regulate mycobacterial division. The MtrA-MtrB TCS is essential for cell envelope biogenesis and cell division. MtrA binds to the promoter regions of various cell division proteins, including MA synthases and PG hydrolases, thereby regulating their activity. Losing MtrB from the system results in cell bulging and chaining of cells, which is reflective of aberrant daughter-cell separation [141,142].

7.2 Post-translational regulation

Mycobacterium tuberculosis (Mtb) encodes 11 serine-threonine protein kinases (STPKs), of which two, PknA and PknB, are essential for cell growth and survival. PknA and PknB phosphorylate a plethora of proteins that are part of the elongasome and divisome [101,143]. PknA phosphorylates

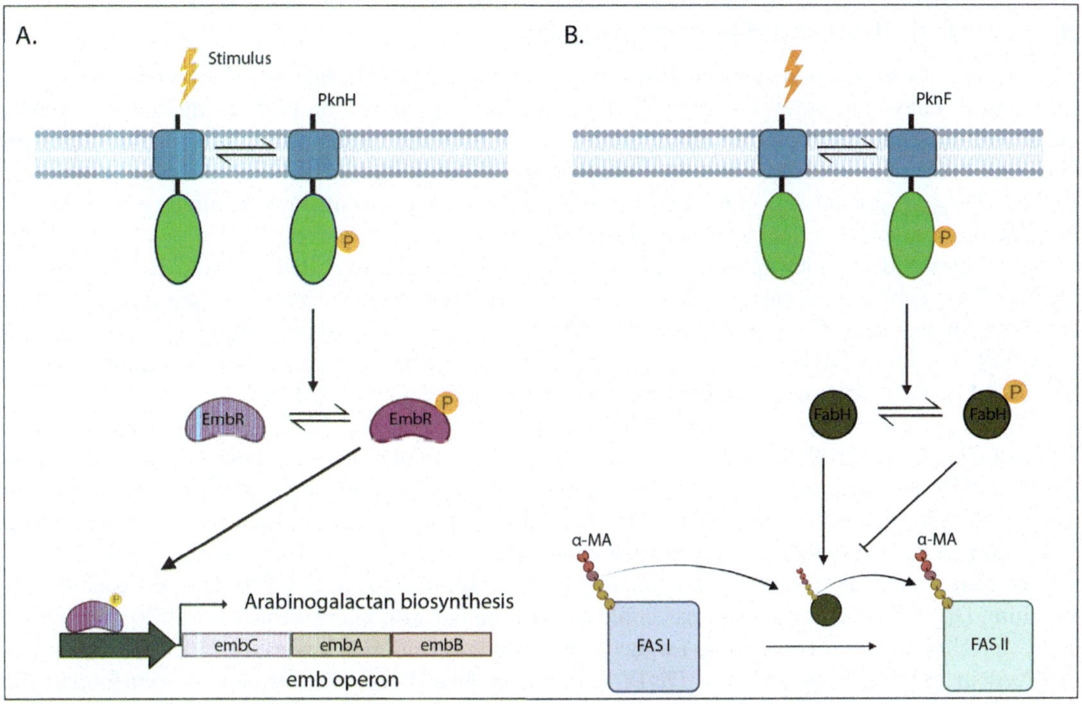

FIGURE 8.7 Regulation of Mycobacterial Cell Division.

(A) In the transcriptional mode of regulation, the transcription factor, EmbR, plays a crucial role in regulating AG synthesis. The STPK, PknH, phosphorylates EmbR, that increases its binding to the embCAB promoter which upregulates AG biosynthesis. (B) In the posttranslational mode of regulation, FabH, which transports α-MA from FASI to FASII, gets phosphorylated by PknF, which decreases its activity, eventually downregulating α-MA synthesis.

FtsZ, and as mentioned earlier, phosphorylated FtsZ has reduced polymerization activity [132]. FtsZ, together with FhaB and FtsQ, forms a ternary complex, and both FhaB and FtsQ become phosphorylated. FhaB phosphorylation is necessary to form this complex, but the biological significance of FtsQ phosphorylation remains unclear [144,144].

Elongation machinery proteins, such as adaptor protein Wag31, lipid II transporter MviN, and PG synthase PonA1, are also modified by STPKs [121]. The phosphorylation of Wag31 by PknA influences its localization and PG synthesis. MviN is phosphorylated by PknB, which is important for its interaction with FhaB [121]. Various synthases involved in α-MA and AG biosynthesis pathways also undergo phosphorylation. These include EmbR, FabD, FabH, MabA, InhA, KasAB, Pks13, and PcaA [144–146]. PknF-mediated phosphorylation of FabH, which channels mycolic acid precursors from FASI to FASII, attenuates its catalytic activity and downregulates mycolic acid biosynthesis (Fig. 8.7B).

8. DivIVA-domain-containing proteins

The successful cell division event requires the coordinated action of dozens of proteins that perform structural or enzymatic roles. Much of the focus has been primarily on proteins that carry out enzymatic functions like synthesizing DNA, proteins, and essential metabolites. Structural proteins that play a vital role in assembling such multi-enzyme complexes at the correct time and place remain less studied. They do not possess intrinsic enzymatic activity but contain protein-binding domains that facilitate protein–protein interactions to form macromolecular complexes.

Firmicutes and actinobacteria encode one such adaptor protein called DivIVA. It has also been identified in some Cyanobacteria, Deltaproteobacteria, Tenericutes, and Deinococcus [147–152]. DivIVA proteins were discovered more than 30 years ago while studying cell division in *B. subtilis* [153,154]. The name DivIVA was assigned to mutants defective in division. DivIVA mutants, such as Div IV-A1 and Div IVA-B1, exhibited improper septal placement [155,156].

DivIVA protein contains an N-terminal DivIVA domain and random coiled-coil domains in its C-terminal (Fig. 8.8A). Both domains are linked via a variable linker region whose length varies among species [157]. The basic unit of the DivIVA domain is a heptad, a sequence of seven amino acids that repeats and determines the length of the DivIVA domain. These seven amino acids are labeled from a to g and include hydrophobic residues such as Leucine, Isoleucine, or Valine at the "a" and "d" positions, while positions "e" and "g" contain hydrophilic amino acids (Fig. 8.8B). Due to the coiled-coil structure, DivIVA proteins can self-assemble to form higher oligomeric structures [158]. The coiled-coil C-terminal region is responsible for mediating oligomerization, whereas the N-terminal domain is involved in binding lipid vesicles. DivIVA tetramers have been observed in *B. subtilis* and *Mtb*

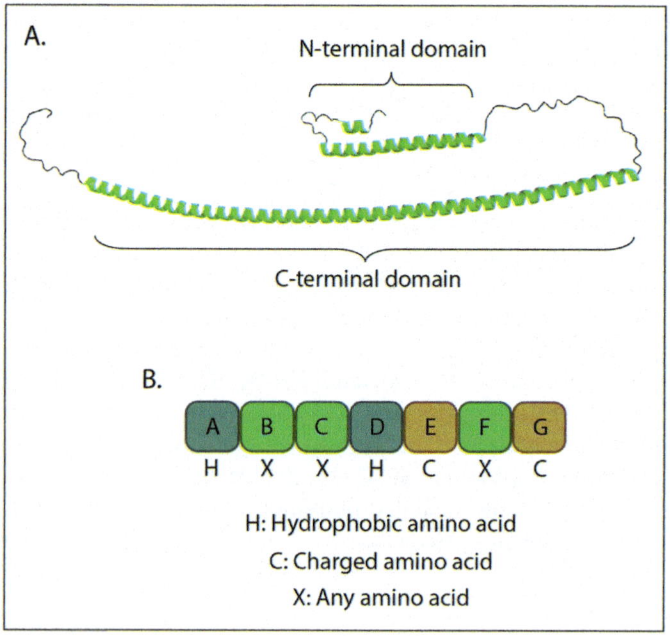

FIGURE 8.8 The DivIVA Protein of Mycobacteria.

(A) Structure of DivIVA domain containing protein Wag31 of mycobacteria. It contains an N- and C-terminal domain joined by a linker. DivIVA domain is present in the N-terminal region. (B) Organization of heptad repeat in the DivIVA domain. The amino acids are denoted from a to g.

(A) (Source: AlphaFold).

[158,159]. These proteins are peripheral membrane proteins that associate with the cytoplasmic side of the membrane, recognizing the negative curvature of the membrane present at poles and septa, and self-assembling at these sites. They serve as scaffolds to recruit various proteins to specific locations within the cell [157]. While the overall structure of DivIVA proteins is conserved across species, their functions vary. Multiple studies indicate that DivIVA proteins have species-specific roles, which are influenced by different interacting protein partners in various bacteria.

9. Divergent functions of DivIVA-domain-containing proteins

9.1 Firmicutes

B. subtilis is the preferred model for studying cell division in gram-positive bacteria, and DivIVA protein was first identified in it. The DivIVA protein in Bacillus has dual functions: positioning the septum and aiding chromosome segregation during sporulation [156]. Along with MinJ, it regulates the MinCD system to ensure correct septum placement at the mid-cell [58,63]. It forms rings adjacent to both sides of the FtsZ ring at mid-cell, preventing further FtsZ assembly [66]. The absence of DivIVA results in minicells that are devoid of nucleoid [155]. Additionally, DivIVA performs two other unrelated roles during sporulation and competence by interacting with different protein subsets within the cell. Its interactions with RacA and Spo0A, coupled with its localization at the cell pole, help anchor OriC to the pole during sporulation [127,160,161]. It facilitates asymmetric septum placement by localizing at the septum and sequestering SpoIIE, which enables compartment-specific gene expression in the mother cell and spore. On the other hand, its interaction with ComN influences competence, and the DivIVA-Maf interaction inhibits bacterial division [162–164].

Firmicutes phyla contain both rod-shaped and coccoid-shaped bacteria. The non-rod species of this phyla lack the conventional Min system and MreB protein. However, they contain only DivIVA protein, making it central for cell division, chromosome segregation, and cell shape maintenance [165]. In ovococcoid *Streptococcus pneumoniae*, DivIVA helps in cell elongation and separation by interacting with GpsB and EzrA, and in nucleoid segregation by interacting with ParB [166]. In *Streptococcus suis*, it is the determinant of division plane selection [167], while in *Enterococcus faecalis*, DivIVA plays all three roles [168]. In spherical firmicutes such as *Staphylococcus aureus*, DivIVA functions in DNA segregation. Interestingly, in *Listeria monocytogenes*, DivIVA imparts swarming motility to the cells and promotes secretion of autolysins like p60 and MurA via the SecA2 system. In the absence of DivIVA, the cells can neither swim nor divide, forming chains [64,169–171].

9.2 Actinobacteria

DivIVA is an essential adaptor protein in various species of Actinobacteria, including Streptomyces, Corynebacteria, and mycobacteria. It localizes at the tips of hyphae and branch sites to recruit elongation machinery and promote PG synthesis in *S. coelicolor* [99,100,172]. It interacts with other cytoskeletal elements such as Scy and FilP. Together with FilP, it facilitates hyphal growth in *Streptomyces venezuelae* [103,104,173,174]. Since *Corynebacterium* lack actin homologs, DivIVA plays a major role in maintaining cell shape. It interacts with RodA, a Lipid II flippase, to enhance PG synthesis at the poles and localizes at the cell pole and the site of septation [175,176]. In the case of mycobacteria, DivIVA protein is known as Wag31, which governs polar cell wall synthesis by recruiting elongation machinery at the pole. After septation, it localizes to the septum, likely promotes septal cell wall synthesis, although evidence for this remains circumstantial. It helps maintain the rod shape of mycobacterial cells, as the loss of Wag31 results in a morphological change from rod to round. It has also been shown to interact with components of the FAS complex to assist in synthesizing mycolate precursors [101,108,177,178].

10. Wag31 and SepIVA—The essential DivIVA-domain-containing proteins of mycobacteria

10.1 Wag31

Mycobacteria, which belong to the Actinomycetes phylum, encode a DivIVA protein called Wag31, which is essential for the organism's survival. Like other bacteria, the DivIVA in mycobacteria plays a vital role in polar elongation, maintaining morphology, and cell division. It localizes at the cell poles, more at the old pole than at the new pole, and the septum (Fig. 8.5). Consequently, in the absence of Wag31, the rod shape of the cells is lost, and they become spherical. The overexpression of Wag31 also disturbs cellular morphology, where cells bulge from one pole [179]. Structural studies of Wag31 based on crystallography reveal that the Wag31 tends to tetramerize, similar to DivIVA of *B. subtilis*. The N-terminal lipid-binding domain forms dimers that arrange in an antiparallel fashion to form tetramers, as seen in crystals and size-exclusion chromatography. Moreover, Wag31 polymerizes to form very long filaments, as long as 30 μm [180].

10.1.1 Role of Wag31 in cell envelope biogenesis

Wag31, being an adaptor protein, governs polar elongation by recruiting cell wall synthesis machinery to the old pole and septum, promoting polar and septal cell wall synthesis [110]. Co-immunoprecipitation analysis of Wag31 led to the identification of AccA3, AccD4, AccD5, Fas, InhA, and MmpL3 as its partners [178]. Acc proteins are components of the multi-enzymatic acetyl-CoA carboxylase (ACCase) complex and are involved in forming mycolate precursors that ultimately form long-chain α-MA [181]. α-MA, together with outer membrane lipids, constitute the MM, which is vital for the pathogen's survival inside the host [93,94]. Several proteins of the FASII complex, including InhA, MabA, KasA, and KasB, along with the MA transporter MmpL3, colocalize with Wag31 at the poles and septum [182]. Another study found Wag31 to be essential for the localization of MmpL3, which not only transports α-MA across the membrane but also regulates global cell wall metabolism during conditions of nutrient or energy [183]. These protein–protein interactions position Wag31 as a key effector in α-MA and lipid synthesis of the mycobacterial cell wall. Additionally, Wag31 depletion is linked to the loss of an intracellular membrane domain (IMD), a specialized membrane compartment in mycobacteria that is physically separate from the cell wall and serves as a major site of lipid synthesis [97]. Recently, our work identified a central role for Wag31 in tethering the membrane and maintaining lipid homeostasis in mycobacteria [128].

10.1.2 Role in chromosome segregation

ParA/B system proteins are known to facilitate proper nucleoid segregation in bacteria. Mycobacteria also encode ParA and ParB proteins. ParA was shown to colocalize with Wag31, and later, a direct interaction between them was observed through bacterial two-hybrid analysis [184]. Another study later implicated T (Threonine) 3 of ParA as the key residue mediating interaction with Wag31. Furthermore, disrupting ParA-DNA binding increased ParA-Wag31 binding, indicating that DNA and Wag31 compete for ParA to regulate chromosome segregation. Therefore, abolishing the ParA-Wag31 interaction impacts nucleoid segregation and cell elongation [185]. Recently, a novel ParA partner, PapM, was identified in a bacterial two-hybrid screen, which modulates the interaction between ParA and Wag31 [186]. ParA, PapM, and Wag31 form a tripartite complex regulated by the phosphorylation status of Wag31. Wild-type Wag31 binds to ParA, while the phosphorylated form reduces its affinity for ParA, recruiting more PapM to the tripartite complex.

10.2 SepIVA, the second essential DivIVA

SepIVA was identified in 2018 during the discovery of septal factors in mycobacteria [187,188]. FtsQ, a component of the divisome, interacts with FtsB and FtsL to form a ternary complex, FtsQBL, linking early cell division proteins to late ones. Two independent studies tagged FtsQ to explore its interacting partners and identified Msm2416 as a key interactor [167,187]. Bioinformatic analysis showed that Msm2416 contains a DivIVA domain [188]. Wag31 was previously the only known DivIVA-domain protein in mycobacteria. Comparing the primary structures of the two proteins revealed about 20% amino acid identity and 30% similarity with Wag31. Msm2416 was predicted to be essential for mycobacterial survival in vitro based on Himar Transposon mutagenesis [189]. The researchers created a proteasome degradation-controlled Msm2416 and found that its absence halted cell division without affecting polar elongation. Cells depleted of Msm2416 became filamentous, branched, and unviable, a phenotype typical of proteins involved in septation [190]. This indicated a role for the newly identified DivIVA-domain protein in septation, leading to its name as SepIVA.

Fluorescently tagged SepIVA exhibited a localization pattern characteristic of IMD proteins. It was found to be present at the septum, subpolar regions, and as patches along the membrane. Additionally, it showed colocalization with a known IMD marker, GlfT2, indicating its presence in the membrane and potential involvement in lipid biogenesis [167]. Another study from the group that identified SepIVA discovered that arginines in SepIVA undergo methylation [191]. This modification influences SepIVA's localization and distribution at the septum and pole. Moreover, disrupting arginine methylation affected PG synthesis and polar elongation. Additionally, SepIVA was also found to be important for MurG localization. Degradation of SepIVA delocalized MurG from the IMD to the cytosol as puncta. Methylation of the N-terminal domain of SepIVA also regulates MurG association to the pole. While unmethylated SepIVA promotes MurG localization at the poles, methylation increases its distribution at the septum. Interestingly, the C-terminal domain of SepIVA plays a role in cell division rather than polar elongation, unlike the N-terminal domain. Methylation mutants of arginines toward the C-terminal showed longer, branched cells with slower growth. The study established a role for SepIVA in elongation and division [191].

11. Concluding remark

Division is an intrinsic trait of all cells. Both intrinsic factors, such as bacterial cell shape, size, and internal environment, and extrinsic factors, like resource availability and environmental conditions (temperature, pH, light, water, oxygen, osmotic pressure, etc.), collectively influence whether a bacterial cell will undergo division and, if so, determine its mode. In this chapter, we discuss various modes of bacterial reproduction influenced by both internal and external factors. The process of cell division involves a sequence of events occurring over space and time, driven by numerous proteins forming multi-unit machineries that communicate to ensure successful completion and regulation of division. Several stages of this cycle and the protein machineries involved are described here, offering insight into the dynamics and complexity of the process. Although the roles of participating proteins and the phenomenon are broadly understood, gaps remain in our knowledge, which current omics techniques are striving to address. Further research to identify and characterize unknown molecular players will help resolve these uncertainties. Understanding this process becomes more relevant in pathogenic bacteria, as cell division relates directly to cell growth, a common target of antibiotics. Over the past decade, antimicrobial resistance has increased, a trend linked to mutations in cell division proteins, highlighting the need for discovering new, druggable targets. Emerging technologies such as super-resolution microscopy, electron microscopy, high-throughput omics, and advanced tools like CRISPRi systems for bacterial genome editing hold great promise for advancing the field. These innovations are pivotal in both fundamental and translational biology, enabling more precise dissection of molecular pathways and opening new therapeutic avenues for improving global public health.

References

[1] Ribatti D. Rudolf Virchow, the founder of cellular pathology. Rom J Morphol Embryol 2019;60(4):1381–2.

[2] Ginzberg MB, et al. Cell size sensing in animal cells coordinates anabolic growth rates and cell cycle progression to maintain cell size uniformity. eLife 2018;7.

[3] Cadart C, et al. Volume growth in animal cells is cell cycle dependent and shows additive fluctuations. eLife 2022;11.

[4] Beall B, Lutkenhaus J. FtsZ in Bacillus subtilis is required for vegetative septation and for asymmetric septation during sporulation. Genes Dev 1991;5(3):447–55.

[5] Bi E, et al. FtsZ and cell division. Res Microbiol 1991;142(2–3):249–52.

[6] Bi E, Lutkenhaus J. FtsZ ring structure associated with division in *Escherichia coli*. Nature 1991;354 (6349):161–4.

[7] Dai K, Lutkenhaus J. ftsZ is an essential cell division gene in *Escherichia coli*. J Bacteriol 1991;173(11): 3500–6.

[8] Doi M, et al. Determinations of the DNA sequence of the mreB gene and of the gene products of the mre region that function in formation of the rod shape of *Escherichia coli* cells. J Bacteriol 1988;170(10): 4619–24.

[9] Wachi M, et al. New mre genes mreC and mreD, responsible for formation of the rod shape of *Escherichia coli* cells. J Bacteriol 1989;171(12):6511–6.

[10] Wachi M, Matsuhashi M. Negative control of cell division by mreB, a gene that functions in determining the rod shape of *Escherichia coli* cells. J Bacteriol 1989;171(6):3123–7.

[11] Gitai Z, et al. MreB actin-mediated segregation of a specific region of a bacterial chromosome. Cell 2005; 120(3):329–41.

[12] Kruse T, et al. Dysfunctional MreB inhibits chromosome segregation in *Escherichia coli*. EMBO J 2003;22 (19):5283–92.

[13] Mauriello EM, et al. Bacterial motility complexes require the actin-like protein, MreB and the Ras homologue, MglA. EMBO J 2010;29(2):315–26.

[14] Ausmees N, Kuhn JR, Jacobs-Wagner C. The bacterial cytoskeleton: an intermediate filament-like function in cell shape. Cell 2003;115(6):705–13.
[15] Godfrin-Estevenon AM, Pasta F, Lane D. The parAB gene products of Pseudomonas putida exhibit partition activity in both P. putida and *Escherichia coli*. Mol Microbiol 2002;43(1):39–49.
[16] Kim HJ, et al. Partitioning of the linear chromosome during sporulation of Streptomyces coelicolor A3(2) involves an oriC-linked parAB locus. J Bacteriol 2000;182(5):1313–20.
[17] Hale CA, de Boer PA. Recruitment of ZipA to the septal ring of *Escherichia coli* is dependent on FtsZ and independent of FtsA. J Bacteriol 1999;181(1):167–76.
[18] Pichoff S, Lutkenhaus J. Unique and overlapping roles for ZipA and FtsA in septal ring assembly in *Escherichia coli*. EMBO J 2002;21(4):685–93.
[19] Buske PJ, Levin PA. Extreme C terminus of bacterial cytoskeletal protein FtsZ plays fundamental role in assembly independent of modulatory proteins. J Biol Chem 2012;287(14):10945–57.
[20] Ma X, et al. Interactions between heterologous FtsA and FtsZ proteins at the FtsZ ring. J Bacteriol 1997;179 (21):6788–97.
[21] Szwedziak P, et al. FtsA forms actin-like protofilaments. EMBO J 2012;31(10):2249–60.
[22] Rico AI, Krupka M, Vicente M. In the beginning, *Escherichia coli* assembled the proto-ring: an initial phase of division. J Biol Chem 2013;288(29):20830–6.
[23] Du S, et al. How FtsEX localizes to the Z ring and interacts with FtsA to regulate cell division. Mol Microbiol 2019;112(3):881–95.
[24] de Boer P, Crossley R, Rothfield L. The essential bacterial cell-division protein FtsZ is a GTPase. Nature 1992;359(6392):254–6.
[25] Wang X, Lutkenhaus J. The FtsZ protein of Bacillus subtilis is localized at the division site and has GTPase activity that is dependent upon FtsZ concentration. Mol Microbiol 1993;9(3):435–42.
[26] Stricker J, et al. Rapid assembly dynamics of the Escherichia coli FtsZ-ring demonstrated by fluorescence recovery after photobleaching. Proc Natl Acad Sci 2002;99(5):3171–5.
[27] McCausland JW, et al. Treadmilling FtsZ polymers drive the directional movement of sPG-synthesis enzymes via a Brownian ratchet mechanism. Nat Commun 2021;12(1):609.
[28] Caldas P, et al. Cooperative ordering of treadmilling filaments in cytoskeletal networks of FtsZ and its crosslinker ZapA. Nat Commun 2019;10(1):5744.
[29] Bisson-Filho AW, et al. Treadmilling by FtsZ filaments drives peptidoglycan synthesis and bacterial cell division. Science 2017;355(6326):739–43.
[30] Addinall SG, Cao C, Lutkenhaus J. FtsN, a late recruit to the septum in *Escherichia coli*. Mol Microbiol 1997;25(2):303–9.
[31] Buddelmeijer N, Beckwith J. A complex of the *Escherichia coli* cell division proteins FtsL, FtsB and FtsQ forms independently of its localization to the septal region. Mol Microbiol 2004;52(5):1315–27.
[32] Mercer KL, Weiss DS. The *Escherichia coli* cell division protein FtsW is required to recruit its cognate transpeptidase, FtsI (PBP3), to the division site. J Bacteriol 2002;184(4):904–12.
[33] Ursinus A, et al. Murein (peptidoglycan) binding property of the essential cell division protein FtsN from *Escherichia coli*. J Bacteriol 2004;186(20):6728–37.
[34] Wang L, Lutkenhaus J. FtsK is an essential cell division protein that is localized to the septum and induced as part of the SOS response. Mol Microbiol 1998;29(3):731–40.
[35] Aussel L, et al. FtsK Is a DNA motor protein that activates chromosome dimer resolution by switching the catalytic state of the XerC and XerD recombinases. Cell 2002;108(2):195–205.
[36] Liu G, Draper GC, Donachie WD. FtsK is a bifunctional protein involved in cell division and chromosome localization in *Escherichia coli*. Mol Microbiol 1998;29(3):893–903.
[37] Barre FX, et al. FtsK functions in the processing of a Holliday junction intermediate during bacterial chromosome segregation. Genes Dev 2000;14(23):2976–88.

[38] Draper GC, et al. Only the N-terminal domain of FtsK functions in cell division. J Bacteriol 1998;180(17):4621–7.
[39] Grenga L, et al. The *Escherichia coli* FtsK functional domains involved in its interaction with its divisome protein partners. FEMS Microbiol Lett 2008;287(2):163–7.
[40] Kureisaite-Ciziene D, et al. Structural analysis of the interaction between the bacterial cell division proteins FtsQ and FtsB. mBio 2018;9(5).
[41] Choi Y, et al. Structural insights into the FtsQ/FtsB/FtsL complex, a key component of the divisome. Sci Rep 2018;8(1):18061.
[42] Gonzalez MD, et al. Multiple interaction domains in FtsL, a protein component of the widely conserved bacterial FtsLBQ cell division complex. J Bacteriol 2010;192(11):2757–68.
[43] Busiek KK, et al. The early divisome protein FtsA interacts directly through its 1c subdomain with the cytoplasmic domain of the late divisome protein FtsN. J Bacteriol 2012;194(8):1989–2000.
[44] Wang L, et al. FtsI and FtsW are localized to the septum in *Escherichia coli*. J Bacteriol 1998;180(11):2810–6.
[45] Lyu Z, et al. FtsN maintains active septal cell wall synthesis by forming a processive complex with the septum-specific peptidoglycan synthases in *E. coli*. Nat Commun 2022;13(1):5751.
[46] Bernhardt TG, de Boer PA. The *Escherichia coli* amidase AmiC is a periplasmic septal ring component exported via the twin-arginine transport pathway. Mol Microbiol 2003;48(5):1171–82.
[47] Dubey A, Priyadarshini R. Amidase activity is essential for medial localization of AmiC in *Caulobacter crescentus*. Curr Genet 2018;64(3):661–75.
[48] Heidrich C, et al. Involvement of N-acetylmuramyl-L-alanine amidases in cell separation and antibiotic-induced autolysis of *Escherichia coli*. Mol Microbiol 2001;41(1):167–78.
[49] Uehara T, et al. Daughter cell separation is controlled by cytokinetic ring-activated cell wall hydrolysis. EMBO J 2010;29(8):1412–22.
[50] Yakhnina AA, McManus HR, Bernhardt TG. The cell wall amidase AmiB is essential for *Pseudomonas aeruginosa* cell division, drug resistance and viability. Mol Microbiol 2015;97(5):957–73.
[51] Yang DC, et al. An ATP-binding cassette transporter-like complex governs cell-wall hydrolysis at the bacterial cytokinetic ring. Proc Natl Acad Sci U S A 2011;108(45):E1052–60.
[52] Cook J, et al. Insights into bacterial cell division from a structure of EnvC bound to the FtsX periplasmic domain. Proc. Natl Acad Sci U S A 2020;117(45):28355–65.
[53] Mavrici D, et al. *Mycobacterium tuberculosis* FtsX extracellular domain activates the peptidoglycan hydrolase, RipC. Proc Natl Acad Sci U S A 2014;111(22):8037–42.
[54] Dajkovic A, et al. MinC spatially controls bacterial cytokinesis by antagonizing the scaffolding function of FtsZ. Curr Biol 2008;18(4):235–44.
[55] De Boer P, Crossley RE, Rothfield LI. Central role for the *Escherichia coli* minC gene product in two different cell division-inhibition systems. Proc Natl Acad Sci 1990;87(3):1129–33.
[56] De Boer P, et al. The MinD protein is a membrane ATPase required for the correct placement of the *Escherichia coli* division site. EMBO J 1991;10(13):4371–80.
[57] Huang J, Cao C, Lutkenhaus J. Interaction between FtsZ and inhibitors of cell division. J Bacteriol 1996;178(17):5080–5.
[58] Patrick JE, Kearns DB. MinJ (YvjD) is a topological determinant of cell division in Bacillus subtilis. Mol Microbiol 2008;70(5):1166–79.
[59] Shih YL, et al. The N-terminal amphipathic helix of the topological specificity factor MinE is associated with shaping membrane curvature. PLoS One 2011;6(6):e21425.
[60] Raskin DM, de Boer PA. Rapid pole-to-pole oscillation of a protein required for directing division to the middle of *Escherichia coli*. Proc Natl Acad Sci U S A 1999;96(9):4971–6.
[61] Juarez JR, Margolin W. Changes in the Min oscillation pattern before and after cell birth. J Bacteriol 2010;192(16):4134–42.

[62] Hu Z, Lutkenhaus J. Topological regulation of cell division in *Escherichia coli* involves rapid pole to pole oscillation of the division inhibitor MinC under the control of MinD and MinE. Mol Microbiol 1999;34(1): 82–90.

[63] Bramkamp M, et al. A novel component of the division-site selection system of Bacillus subtilis and a new mode of action for the division inhibitor MinCD. Mol Microbiol 2008;70(6):1556–69.

[64] Kaval KG, Rismondo J, Halbedel S. A function of DivIVA in Listeria monocytogenes division site selection. Mol Microbiol 2014;94(3):637–54.

[65] van Baarle S, et al. Protein-protein interaction domains of Bacillus subtilis DivIVA. J Bacteriol 2013;195 (5):1012–21.

[66] Eswaramoorthy P, et al. Cellular architecture mediates DivIVA ultrastructure and regulates min activity in Bacillus subtilis. mBio 2011;2(6).

[67] van Baarle S, Bramkamp M. The MinCDJ system in Bacillus subtilis prevents minicell formation by promoting divisome disassembly. PLoS One 2010;5(3):e9850.

[68] Bernhardt TG, De Boer PA. SlmA, a nucleoid-associated, FtsZ binding protein required for blocking septal ring assembly over chromosomes in *E. coli*. Mol Cell 2005;18(5):555–64.

[69] Tonthat NK, et al. Molecular mechanism by which the nucleoid occlusion factor, SlmA, keeps cytokinesis in check. EMBO J 2011;30(1):154–64.

[70] Wu LJ, Errington J. Coordination of cell division and chromosome segregation by a nucleoid occlusion protein in Bacillus subtilis. Cell 2004;117(7):915–25.

[71] Adams DW, Wu LJ, Errington J. Nucleoid occlusion protein Noc recruits DNA to the bacterial cell membrane. EMBO J 2015;34(4):491–501.

[72] Cho H, et al. Nucleoid occlusion factor SlmA is a DNA-activated FtsZ polymerization antagonist. Proc Natl Acad Sci U S A 2011;108(9):3773–8.

[73] Tonthat NK, et al. SlmA forms a higher-order structure on DNA that inhibits cytokinetic Z-ring formation over the nucleoid. Proc Natl Acad Sci U S A 2013;110(26):10586–91.

[74] Yu Y, et al. The division defect of a Bacillus subtilis minD noc double mutant can Be suppressed by spx-dependent and spx-independent mechanisms. J Bacteriol 2021;203(18):e0024921.

[75] Espéli O, et al. A MatP-divisome interaction coordinates chromosome segregation with cell division in *E. coli*. EMBO J 2012;31(14):3198–211.

[76] Buss J, et al. A multi-layered protein network stabilizes the *Escherichia coli* FtsZ-ring and modulates constriction dynamics. PLoS Genet 2015;11(4):e1005128.

[77] Galli E, Gerdes K. Spatial resolution of two bacterial cell division proteins: ZapA recruits ZapB to the inner face of the Z-ring. Mol Microbiol 2010;76(6):1514–26.

[78] Männik J, Bailey MW. Spatial coordination between chromosomes and cell division proteins in *Escherichia coli*. Front Microbiol 2015;6:306.

[79] Dresen M, et al. Identification and characterization of the cell division protein MapZ from Streptococcus suis. Microbiologyopen 2021;10(5):e1234.

[80] Feng Z, et al. Multi-functional regulator MapZ controls both positioning and timing of FtsZ polymerization. Biochem J 2019;476(10):1433–44.

[81] Donachie WD. Co-ordinate regulation of the *Escherichia coli* cell cycle or the cloud of unknowing. Mol Microbiol 2001;40(4).

[82] Nicholson WL, et al. Resistance of Bacillus endospores to extreme terrestrial and extraterrestrial environments. Microbiol Mol Biol Rev 2000;64(3):548–72.

[83] Eichenberger P, et al. The program of gene transcription for a single differentiating cell type during sporulation in Bacillus subtilis. PLoS Biol 2004;2(10):e328.

[84] Grossman AD. Integration of developmental signals and the initiation of sporulation in B. subtilis. Cell 1991;65(1):5–8.

[85] Stragier P. A gene Odyssey: exploring the genomes of endospore-forming bacteria. In: Sonenshein AL, Hoch JA, Losick R, editors. Bacillus subtilis and its closest relatives: from genes to cells. ASM Press; 2002.
[86] Waterbury JB, Stanier RY. Patterns of growth and development in pleurocapsalean cyanobacteria. Microbiol Rev 1978;42(1):2–44.
[87] Ajithkumar VP, et al. A novel filamentous Bacillus sp., strain NAF001, forming endospores and budding cells. Microbiology 2001;147(6):1415–23.
[88] Staley JT, et al. Bergey's manual of systematic bacteriology, vol. 3. Baltimore, MD: Williams & Wilkins; 1989.
[89] Waterbury J, Stanier R. Two unicellular cyanobacteria which reproduce by budding. Arch Microbiol 1977; 115:249–57.
[90] Barka EA, et al. Taxonomy, physiology, and natural products of actinobacteria. Microbiol Mol Biol Rev 2016;80(1):1–43.
[91] Rajivgandhi G, et al. Chapter 9—production of secondary metabolites from endophytic actinomycetes isolated from marine mangrove plants. In: Egamberdieva D, Parray JA, Davranov K, editors. Plant endophytes and secondary metabolites. Academic Press; 2024. p. 133–57.
[92] Brennan PJ. Structure, function, and biogenesis of the cell wall of *Mycobacterium tuberculosis*. Tuberculosis 2003;83(1–3):91–7.
[93] Chiaradia L, et al. Dissecting the mycobacterial cell envelope and defining the composition of the native mycomembrane. Sci Rep 2017;7(1):12807.
[94] Dahl JL. Electron microscopy analysis of *Mycobacterium tuberculosis* cell division. FEMS Microbiol Lett 2004;240(1):15–20.
[95] Egan AJ, Vollmer W. The physiology of bacterial cell division. Ann N Y Acad Sci 2013;1277:8–28.
[96] Gerding MA, et al. The trans-envelope Tol-Pal complex is part of the cell division machinery and required for proper outer-membrane invagination during cell constriction in *E. coli*. Mol Microbiol 2007;63(4): 1008–25.
[97] Garcia-Heredia A, et al. Peptidoglycan precursor synthesis along the sidewall of pole-growing mycobacteria. eLife 2018;7.
[98] Howell M, Brown PJB. Building the bacterial cell wall at the pole. Curr Opin Microbiol 2016;34:53–9.
[99] Flärdh K. Growth polarity and cell division in Streptomyces. Curr Opin Microbiol 2003;6(6):564–71.
[100] Hempel AM, et al. Assemblies of DivIVA mark sites for hyphal branching and can establish new zones of cell wall growth in Streptomyces coelicolor. J Bacteriol 2008;190(22):7579–83.
[101] Kang CM, et al. The *Mycobacterium tuberculosis* serine/threonine kinases PknA and PknB: substrate identification and regulation of cell shape. Genes Dev 2005;19(14):1692–704.
[102] Quintanilla SY, et al. PlrA (MSMEG_5223) is an essential polar growth regulator in Mycobacterium smegmatis. PLoS One 2023;18(1):e0280336.
[103] Fröjd MJ, Flärdh K. Apical assemblies of intermediate filament-like protein FilP are highly dynamic and affect polar growth determinant DivIVA in Streptomyces venezuelae. Mol Microbiol 2019;112(1):47–61.
[104] Holmes NA, et al. Coiled-coil protein Scy is a key component of a multiprotein assembly controlling polarized growth in Streptomyces. Proc Natl Acad Sci 2013;110(5):E397–406.
[105] Del Sol R, et al. Influence of CrgA on assembly of the cell division protein FtsZ during development of Streptomyces coelicolor. J Bacteriol 2006;188(4):1540–50.
[106] Del Sol R, et al. The product of a developmental gene, crgA, that coordinates reproductive growth in Streptomyces belongs to a novel family of small actinomycete-specific proteins. J Bacteriol 2003;185(22): 6678–85.
[107] Plocinski P, et al. Characterization of CrgA, a new partner of the *Mycobacterium tuberculosis* peptidoglycan polymerization complexes. J Bacteriol 2011;193(13):3246–56.
[108] Plocinski P, et al. *Mycobacterium tuberculosis* CwsA overproduction modulates cell division and cell wall synthesis. Tuberculosis 2013;93(Suppl. 1):S21–7.

[109] Willemse J, et al. Positive control of cell division: FtsZ is recruited by SsgB during sporulation of Streptomyces. Genes Dev 2011;25(1):89–99.
[110] Meniche X, et al. Subpolar addition of new cell wall is directed by DivIVA in mycobacteria. Proc Natl Acad Sci 2014;111(31):E3243–51.
[111] Hett EC, et al. A mycobacterial enzyme essential for cell division synergizes with resuscitation-promoting factor. PLoS Pathog 2008;4(2):e1000001.
[112] Zhou X, et al. Sequential assembly of the septal cell envelope prior to V snapping in Corynebacterium glutamicum. Nat Chem Biol 2019;15(3):221–31.
[113] Aldridge BB, et al. Asymmetry and aging of mycobacterial cells lead to variable growth and antibiotic susceptibility. Science 2012;335(6064):100–4.
[114] Joyce G, et al. Cell division site placement and asymmetric growth in mycobacteria. PLoS One 2012;7(9): e44582.
[115] Santi I, et al. Single-cell dynamics of the chromosome replication and cell division cycles in mycobacteria. Nat Commun 2013;4(1):2470.
[116] Singh B, et al. Asymmetric growth and division in Mycobacterium spp.: compensatory mechanisms for non-medial septa. Mol Microbiol 2013;88(1):64–76.
[117] Bhamidi S, et al. A bioanalytical method to determine the cell wall composition of *Mycobacterium tuberculosis* grown in vivo. Anal Biochem 2012;421(1):240–9.
[118] Jain M, et al. Lipidomics reveals control of *Mycobacterium tuberculosis* virulence lipids via metabolic coupling. Proc Natl Acad Sci U S A 2007;104(12):5133–8.
[119] Rachman H, et al. Unique transcriptome signature of *Mycobacterium tuberculosis* in pulmonary tuberculosis. Infect Immun 2006;74(2):1233–42.
[120] Seiler P, et al. Cell-wall alterations as an attribute of *Mycobacterium tuberculosis* in latent infection. J Infect Dis 2003;188(9):1326–31.
[121] Gee CL, et al. A phosphorylated pseudokinase complex controls cell wall synthesis in mycobacteria. Sci Signal 2012;5(208):ra7.
[122] Typas A, et al. From the regulation of peptidoglycan synthesis to bacterial growth and morphology. Nat Rev Microbiol 2011;10(2):123–36.
[123] Birch HL, et al. Biosynthesis of mycobacterial arabinogalactan: identification of a novel $\alpha(1\rightarrow3)$ arabinofuranosyltransferase. Mol Microbiol 2008;69(5):1191–206.
[124] Crick DC, Mahapatra S, Brennan PJ. Biosynthesis of the arabinogalactan-peptidoglycan complex of *Mycobacterium tuberculosis*. Glycobiology 2001;11(9):107R–18R.
[125] Portevin D, et al. A polyketide synthase catalyzes the last condensation step of mycolic acid biosynthesis in mycobacteria and related organisms. Proc Natl Acad Sci 2004;101(1):314–9.
[126] Wolucka BA, et al. Recognition of the lipid intermediate for arabinogalactan/arabinomannan biosynthesis and its relation to the mode of action of ethambutol on mycobacteria. J Biol Chem 1994;269(37):23328–35.
[127] Kloosterman TG, et al. Complex polar machinery required for proper chromosome segregation in vegetative and sporulating cells of Bacillus subtilis. Mol Microbiol 2016;101(2):333–50.
[128] Kapoor Y, et al. Wag31, a membrane tether, is crucial for lipid homeostasis in mycobacteria. bioRxiv 2024. https://doi.org/10.1101/2024.10.22.619650.
[129] Gola S, et al. The essential role of SepF in mycobacterial division. Mol Microbiol 2015;97(3):560–76.
[130] Gupta S, et al. Essential protein SepF of mycobacteria interacts with FtsZ and MurG to regulate cell growth and division. Microbiol (Read) 2015;161(8):1627–38.
[131] Van der Ploeg R, et al. Colocalization and interaction between elongasome and divisome during a preparative cell division phase in *E scherichia coli*. Mol Microbiol 2013;87(5):1074–87.
[132] Thakur M, Chakraborti PK. GTPase activity of mycobacterial FtsZ is impaired due to its transphosphorylation by the eukaryotic-type Ser/Thr kinase, PknA. J Biol Chem 2006;281(52):40107–13.

[133] Li Y, et al. FtsZ protofilaments use a hinge-opening mechanism for constrictive force generation. Science 2013;341(6144):392–5.
[134] Datta P, et al. Interaction between FtsW and penicillin-binding protein 3 (PBP3) directs PBP3 to mid-cell, controls cell septation and mediates the formation of a trimeric complex involving FtsZ, FtsW and PBP3 in mycobacteria. Mol Microbiol 2006;62(6):1655–73.
[135] Hett EC, et al. A partner for the resuscitation-promoting factors of *Mycobacterium tuberculosis*. Mol Microbiol 2007;66(3):658–68.
[136] Hett EC, Chao MC, Rubin EJ. Interaction and modulation of two antagonistic cell wall enzymes of mycobacteria. PLoS Pathog 2010;6(7):e1001020.
[137] Sharma K, et al. Transcriptional control of the mycobacterial embCAB operon by PknH through a regulatory protein, EmbR, in vivo. J Bacteriol 2006;188(8):2936–44.
[138] Telenti A, et al. The emb operon, a gene cluster of *Mycobacterium tuberculosis* involved in resistance to ethambutol. Nat Med 1997;3(5):567–70.
[139] Mondino S, Gago G, Gramajo H. Transcriptional regulation of fatty acid biosynthesis in mycobacteria. Mol Microbiol 2013;89(2):372–87.
[140] Salzman V, et al. Transcriptional regulation of lipid homeostasis in mycobacteria. Mol Microbiol 2010;78 (1):64–77.
[141] Plocinska R, et al. Septal localization of the *Mycobacterium tuberculosis* MtrB sensor kinase promotes MtrA regulon expression. J Biol Chem 2012;287(28):23887–99.
[142] Rajagopalan M, et al. *Mycobacterium tuberculosis* origin of replication and the promoter for immunodominant secreted antigen 85B are the targets of MtrA, the essential response regulator. J Biol Chem 2010;285 (21):15816–27.
[143] Prisic S, et al. Extensive phosphorylation with overlapping specificity by *Mycobacterium tuberculosis* serine/threonine protein kinases. Proc Natl Acad Sci U S A 2010;107(16):7521–6.
[144] Molle V, et al. Phosphorylation of InhA inhibits mycolic acid biosynthesis and growth of *Mycobacterium tuberculosis*. Mol Microbiol 2010;78(6):1591–605.
[145] Corrales RM, et al. Phosphorylation of mycobacterial PcaA inhibits mycolic acid cyclopropanation: consequences for intracellular survival and for phagosome maturation block. J Biol Chem 2012;287(31): 26187–99.
[146] Khan S, et al. Phosphorylation of enoyl-acyl carrier protein reductase InhA impacts mycobacterial growth and survival. J Biol Chem 2010;285(48):37860–71.
[147] Akiyama T, Inouye S, Komano T. Novel developmental genes, fruCD, of Myxococcus xanthus: involvement of a cell division protein in multicellular development. J Bacteriol 2003;185(11):3317–24.
[148] Danchin A, Fang G. Unknown unknowns: essential genes in quest for function. Microb Biotechnol 2016;9 (5):530–40.
[149] Lambert C, et al. The first bite–profiling the predatosome in the bacterial pathogen Bdellovibrio. PLoS One 2010;5(1):e8599.
[150] Maurya GK, et al. Phosphorylation of FtsZ and FtsA by a DNA damage-responsive ser/thr protein kinase affects their functional interactions in Deinococcus radiodurans. mSphere 2018;3(4).
[151] Maurya GK, Modi K, Misra HS. Divisome and segrosome components of Deinococcus radiodurans interact through cell division regulatory proteins. Microbiol (Read) 2016;162(8):1321–34.
[152] Miyagishima SY, Wolk CP, Osteryoung KW. Identification of cyanobacterial cell division genes by comparative and mutational analyses. Mol Microbiol 2005;56(1):126–43.
[153] Mendelson NH. Cell division suppression in the Bacillus subtilis div IC-A1 minicell-producing mutant. J Bacteriol 1975;121(3):1166–72.
[154] Van Alstyne D, Simon MI. Division mutants of Bacillus subtilis: isolation and PBS1 transduction of division-specific markers. J Bacteriol 1971;108(3):1366–79.
[155] Cha JH, Stewart GC. The divIVA minicell locus of Bacillus subtilis. J Bacteriol 1997;179(5):1671–83.

[156] Edwards DH, Errington J. The Bacillus subtilis DivIVA protein targets to the division septum and controls the site specificity of cell division. Mol Microbiol 1997;24(5):905–15.

[157] Oliva MA, et al. Features critical for membrane binding revealed by DivIVA crystal structure. EMBO J 2010;29(12):1988–2001.

[158] Stahlberg H, et al. Oligomeric structure of the Bacillus subtilis cell division protein DivIVA determined by transmission electron microscopy. Mol Microbiol 2004;52(5):1281–90.

[159] Muchová KN, et al. Oligomerization of the Bacillus subtilis division protein DivIVA. Microbiol (Read) 2002;148(Pt 3):807–13.

[160] Ben-Yehuda S, Rudner DZ, Losick R. RacA, a bacterial protein that anchors chromosomes to the cell poles. Science 2003;299(5606):532–6.

[161] Wu LJ, Errington J. RacA and the Soj-Spo0J system combine to effect polar chromosome segregation in sporulating Bacillus subtilis. Mol Microbiol 2003;49(6):1463–75.

[162] Briley Jr K, et al. Maf acts downstream of ComGA to arrest cell division in competent cells of B. subtilis. Mol Microbiol 2011;81(1):23–39.

[163] dos Santos VT, Bisson-Filho AW, Gueiros-Filho FJ. DivIVA-mediated polar localization of ComN, a posttranscriptional regulator of Bacillus subtilis. J Bacteriol 2012;194(14):3661–9.

[164] Ogura M, Tanaka T. The Bacillus subtilis late competence operon comE is transcriptionally regulated by yutB and under post-transcription initiation control by comN (yrzD). J Bacteriol 2009;191(3):949–58.

[165] Pinho MG, Kjos M, Veening JW. How to get (a)round: mechanisms controlling growth and division of coccoid bacteria. Nat Rev Microbiol 2013;11(9):601–14.

[166] Beilharz K, et al. Control of cell division in Streptococcus pneumoniae by the conserved Ser/Thr protein kinase StkP. Proc Natl Acad Sci U S A 2012;109(15):E905–13.

[167] Ni H, et al. Streptococcus suis DivIVA protein is a substrate of ser/thr kinase STK and involved in cell division regulation. Front Cell Infect Microbiol 2018;8:85.

[168] Ramirez-Arcos S, et al. *Enterococcus faecalis* divIVA: an essential gene involved in cell division, cell growth and chromosome segregation. Microbiol (Read) 2005;151(Pt 5):1381–93.

[169] Halbedel S, et al. DivIVA affects secretion of virulence-related autolysins in Listeria monocytogenes. Mol Microbiol 2012;83(4):821–39.

[170] Kaval KG, et al. Genetic dissection of DivIVA functions in Listeria monocytogenes. J Bacteriol 2017;199 (24).

[171] Valenčíková R, et al. Clostridial DivIVA and MinD interact in the absence of MinJ. Anaerobe 2018;50: 22–31.

[172] Richards DM, et al. Mechanistic basis of branch-site selection in filamentous bacteria. PLoS Comput Biol 2012;8(3):e1002423.

[173] Ditkowski B, et al. Dynamic interplay of ParA with the polarity protein, Scy, coordinates the growth with chromosome segregation in Streptomyces coelicolor. Open Biol 2013;3(3):130006.

[174] Donovan C, et al. A synthetic *Escherichia coli* system identifies a conserved origin tethering factor in Actinobacteria. Mol Microbiol 2012;84(1):105–16.

[175] Donovan C, et al. Chromosome segregation impacts on cell growth and division site selection in Corynebacterium glutamicum. PLoS One 2013;8(2):e55078.

[176] Donovan C, et al. Subcellular localization and characterization of the ParAB system from Corynebacterium glutamicum. J Bacteriol 2010;192(13):3441–51.

[177] Singh V, et al. Identification of aminopyrimidine-sulfonamides as potent modulators of Wag31-mediated cell elongation in mycobacteria. Mol Microbiol 2017;103(1):13–25.

[178] Xu WX, et al. The Wag31 protein interacts with AccA3 and coordinates cell wall lipid permeability and lipophilic drug resistance in Mycobacterium smegmatis. Biochem Biophys Res Commun 2014;448(3): 255–60.

[179] Kapoor Y, et al. Wag31, a membrane tether, is crucial for lipid homeostasis in mycobacteria. eLife 2025;14.

[180] Choukate K, Chaudhuri B. Structural basis of self-assembly in the lipid-binding domain of mycobacterial polar growth factor Wag31. IUCrJ 2020;7(Pt 4):767–76.
[181] Gago G, et al. Fatty acid biosynthesis in actinomycetes. FEMS Microbiol Rev 2011;35(3):475–97.
[182] Carel C, et al. *Mycobacterium tuberculosis* proteins involved in mycolic acid synthesis and transport localize dynamically to the old growing pole and septum. PLoS One 2014;9(5):e97148.
[183] Arejan NH, et al. MmpL3, Wag31, and PlrA are involved in coordinating polar growth with peptidoglycan metabolism and nutrient availability. J Bacteriol 2024;206(10):e0020424.
[184] Ginda K, et al. ParA of Mycobacterium smegmatis co-ordinates chromosome segregation with the cell cycle and interacts with the polar growth determinant DivIVA. Mol Microbiol 2013;87(5):998–1012.
[185] Pióro M, et al. Competition between DivIVA and the nucleoid for ParA binding promotes segrosome separation and modulates mycobacterial cell elongation. Mol Microbiol 2019;111(1):204–20.
[186] Matusiak I, et al. The interplay between the polar growth determinant DivIVA, the segregation protein ParA, and their novel interaction partner PapM controls the Mycobacterium smegmatis cell cycle by modulation of DivIVA subcellular distribution. Microbiol Spectr 2023;11(6):e0175223.
[187] Jain P, et al. Delineating FtsQ-mediated regulation of cell division in *Mycobacterium tuberculosis.* J Biol Chem 2018;293(32):12331–49.
[188] Wu KJ, et al. Characterization of conserved and novel septal factors in Mycobacterium smegmatis. J Bacteriol 2018;200(6). p. e00649-17.
[189] DeJesus MA, et al. Comprehensive essentiality analysis of the *Mycobacterium tuberculosis* genome via saturating transposon mutagenesis. mBio 2017;8(1). https://doi.org/10.1128/mbio.02133-16.
[190] Pickford H, et al. A mycobacterial DivIVA domain-containing protein involved in cell length and septation. Microbiol (Read) 2020;166(9):817–25.
[191] Freeman AH, et al. Arginine methylation sites on SepIVA help balance elongation and septation in Mycobacterium smegmatis. Mol Microbiol 2023;119(2):208–23.

CHAPTER

9

Programmed cell death during *Mycobacterium tuberculosis* infection: The roles of pyroptosis and necroptosis in host defense

Krishnaveni Mohareer and Sharmistha Banerjee

Department of Biochemistry, School of Life Sciences, University of Hyderabad, Hyderabad, Telangana, India

Highlights

1. *Mycobacterium tuberculosis* (*Mtb*) modulates programmed cell death pathways to evade immune responses.
2. Pyroptosis and necroptosis are pro-inflammatory forms of programmed cell death triggered by *Mtb* infection.
3. *Mtb* factors can induce and inhibit pyroptosis and necroptosis in the host macrophages.
4. The interplay between different cell death modalities shapes the balance between the host defense and pathology.
5. The bacterial load and timing of cell death influence the trajectory of disease progression.
6. Targeting cell death modulators holds promise for developing novel host-directed therapies for tuberculosis (TB).
7. Understanding the complex roles of pyroptosis and necroptosis in TB may reveal novel therapeutic targets.

Abbreviations

ASC apoptosis-associated speck-like protein containing a CARD (caspase recruitment domain)
ATP adenosine triphosphate
CARD caspase recruitment domain
CNS central nervous system
DAMP damage-associated molecular pattern
DFNA5 deafness, autosomal dominant 5 (gene)
ESAT-6 early secreted antigenic target-6 kDa
FAK focal adhesion kinase
GRIM-19 gene associated with retinoid-interferon-induced mortality-19
GSDMD gasdermin D
IL-18 interleukin-18
IL-1β interleukin-1 beta
LAM lipoarabinomannan
LPS lipopolysaccharide

Navigating Tuberculosis. https://doi.org/10.1016/B978-0-443-27735-1.00010-1

LUBAC linear ubiquitin chain assembly complex
MLKL mixed lineage kinase domain-like
MOI multiplicity of infection
Mtb *mycobacterium tuberculosis*
NAD+ nicotinamide adenine dinucleotide
NETosis neutrophil extracellular trap formation
NF-κB nuclear factor kappa-beta
NLRP3 NOD-like receptor pyrin domain-containing protein 3
NOD nucleotide-binding oligomerization domain
PAMP pathogen-associated molecular pattern
PCD programmed cell death
PDIM phthiocerol dimycocerosate
PE/PPE proline-glutamate/proline-proline-glutamate (protein family)
PRR pattern recognition receptor
RIPK1/3 receptor-interacting protein kinase 1/3
RNS reactive nitrogen species
ROS reactive oxygen species
TB tuberculosis
TLR toll-like receptor
TNF tumor necrosis factor
TNFR tumor necrosis factor receptor
ZMP-1 zinc metalloprotease-1

1. Introduction

Mycobacterium tuberculosis (Mtb) causes tuberculosis (TB) and is responsible for the highest mortality rate among bacterial infections. Despite several decades of research on TB, disease management has several limitations, including imprecise diagnostics, long treatment duration leading to noncompliance, disease relapse, (multiple) drug resistance, and synergism with other infections and metabolic diseases. To identify novel targets that overcome the present limitations of disease management, we must look at novel host-pathogen interactions for host-directed therapy that could target tuberculosis and several other intracellular infections. The success of *Mtb* in persisting and causing disease is due to its ability to manipulate host innate defenses. Programmed cell death is critical for host-pathogen interactions in innate defenses. Programmed cell death pathways, such as pyroptosis and necroptosis, are crucial for bacterial clearance, the activation of immune responses, and the maintenance of tissue integrity.

Multiple cell death modalities are activated upon mycobacterial infection, resulting in complex interactions between the host and pathogen [1]. This process is influenced by factors from both the host and pathogen, which shape the immune response and determine the outcome of the infection. Understanding the role of host cell death modalities during *Mtb* infection in human macrophages is essential, as it influences the outcome of host-pathogen interactions [2]. As only 10% of infections lead to the successful establishment of TB, the remainder are cleared by autophagy and various cell death modalities, including apoptosis, pyroptosis [3], and necroptosis. New forms of cell death have been described, including ferroptosis [4], NETosis [5], disulfidptosis [6], and PANoptosis [7]. These forms of cell death differ mechanistically in terms of inflammatory response and outcome of infection and have been extensively reviewed [2,7–9] to name a few). Some forms of cell death play dual roles in infection by eliminating infected cells and facilitating bacterial dissemination under specific conditions. The influence of cell death on infection progression is multifaceted. Some forms of cell death can facilitate pathogen survival, whereas others can enhance host defense mechanisms. In certain instances, the timing of cell death can alter the infection results. Therefore, we need to systematically understand the role of cell death mechanisms over time and space involving multiple cell types, which directly or indirectly influence the outcome of infection. This chapter examines two forms of programmed cell death other than apoptosis that significantly affect the outcome of *Mtb* infection in human host macrophages. We also underscored their dual role as protective mechanisms and contributors to disease pathology, which has not been sufficiently addressed.

2. Programmed cell death modalities as host defense

Apoptosis is one of the innate defense mechanisms that is well-characterized for eliminating intracellular infections across evolution, both bacterial and viral. This classical programmed cell death maintains cellular homeostasis. Apoptosis involves a cascade of biochemical processes and enzymatic reactions mediated by the hallmark enzymes, caspases, which act on their specific substrates to facilitate cell morphological changes, including DNA fragmentation and packaging of the disintegrating cell into tiny apoptotic bodies. The induction of apoptosis clears the infected cells, triggering minimal or no inflammatory response, as the apoptotic bodies are efferocytosed (phagocytosis of apoptotic bodies by macrophages) before they become lytic. The role of apoptosis in *Mtb* infection has been widely studied concerning various *Mtb* factors and has been reviewed extensively [9–13], and is not within the scope of the present chapter. During *Mtb* infection, different pathogen-associated molecular patterns (PAMPs), including peptidoglycans, lipoarabinomannans (LAMs), surface proteins of the proline-glutamic acid-proline-proline-glutamic acid (PE/PPE) protein family, and secreted proteins such as ESAT-6, LpqH, and many others, are recognized by host receptors such as toll-like receptors 1/2 (TLR1/2), Nod-like receptors (NLRs), Dectin-1, and tumor necrosis factor receptor (TNFR), which activate the apoptosis pathway leading to activation of the apoptosome that further triggers the cascade of caspases-3, 6, and 9, culminating in apoptosis (reviewed in Refs. [14,15]. However, *Mtb* trapped in apoptotic bodies can enter uninfected macrophages during efferocytosis and persist because it inhibits phagolysosome fusion [16,17]. The timing of apoptosis and mycobacterial load can alter disease progression and outcomes [18].

Although apoptosis is the classical programmed cell death pathway, variations in the players involved in the mechanism confer different immune responses and infection outcomes [19]. Although several variations of classical programmed cell death have been reported, we focus on two significant variations involving inflammation, namely pyroptosis and necroptosis, which are described below.

2.1 Pyroptosis is a regulated form of cell death that triggers an inflammatory response

Pyroptosis is a regulated form of cell death that is accompanied by lysis and inflammation [20]. Unlike apoptosis, it has an intact nucleus and causes inflammation from membrane swelling and subsequent lysis. Biochemical markers include caspase-1 and gasdermin cleavage, and immunological markers include IL-1β and IL-18 secretions. In canonical pyroptosis, host pattern recognition receptors (PRRs) sense pathogen-associated molecular patterns (PAMPs) or damage-associated molecular patterns

(DAMPs) (such as reactive oxygen species (ROS), Ca^{2+} flux, or mitochondrial DNA (mtDNA)) and activate NLR family pyrin domain containing 3 (NLRP3), which undergoes oligomerization. This process recruits Apoptosis-associated speck-like protein containing a CARD-1 (ASC-1) adaptor molecules that polymerize and recruit caspase-1 to form the inflammasome. The inflammasome converts pro-caspase-1 and pro-IL-18 into their active forms, caspase-1 and IL-18 cytokines. Caspase-1 cleaves pro-gasdermin, generating fragments that oligomerize and form pores in the plasma membrane, releasing cytokines, alarmins, and other fluids. This causes osmotic lysis and inflammation [21,22]. In non-canonical signaling, upon sensing LPS from Gram-negative bacteria, caspase-4/5 is activated in humans or caspase-11 in mice, which cleaves gasdermin, followed by pyroptotic cell death. Immune cells recruited by cytokines are crucial for clearing debris, eliminating pathogens, and maintaining homeostasis [21]. Signaling is mainly intrinsic but can also be triggered by receptors such as TNFR [23].

2.2 Necroptosis is a programmed inflammatory cell-death pathway

Necrosis was previously believed to be accidental cell death and that it is not regulated. Emerging evidence from cell death research has indicated that necrosis can be regulated under certain conditions, which is now termed "necroptosis." Necroptosis is a distinct pathway of programmed cell death (PCD), characterized by its inflammatory nature and independence from caspase activation, differentiating it from pyroptosis and apoptosis. This form of cell death is similar to classical necrosis. Morphologically, it is characterized by the swelling of cells and organelles such as mitochondria, condensation of chromatin without fragmentation, and subsequent cell lysis, releasing intracellular contents. It is signaled by the receptor-interacting protein kinases RIPK1 and RIPK3 in conjunction with the pseudo kinase MLKL. Necroptosis may be activated through the TNF receptor signaling pathway or by pattern recognition receptors (PRRs). Death receptors, including TNFR1, FAS, DR4, and DR5, recognize cognate ligands but fail to undergo apoptosis, resulting in necroptosis [19]. Necrostatin-1 inhibits RIPK1 activity, which typically signals apoptosis; consequently, inhibition of apoptosis results in necroptosis [24,25]. The genetic pathways of necroptosis are believed to have evolved in higher vertebrates and seem to play a crucial role as a checkpoint to block disease progression [19].

In summary, although programmed, these three types of cell death have fundamental differences. Morphologically, apoptosis is characterized by cell shrinkage, nuclear condensation, and fragmentation into vesicles known as apoptotic bodies, while the cell membrane remains intact. In contrast, pyroptosis involves lytic cell death, marked by plasma membrane rupture, release of intracellular contents, and cell swelling. Necroptosis resembles necrosis, showing similar signs of cell swelling, organelle damage, and plasma membrane rupture; however, it is a regulated form of cell death. Mechanistically, apoptosis depends on caspase-2, -8, -9, and -10, which initiate the process, whereas caspase-3, -6, and -7 cleave cellular proteins to execute it. Pyroptosis is dependent on inflammasomes and mediated by caspase-1. It is associated with the cleavage of gasdermin proteins, which leads to the release of inflammatory cytokines such as IL-1β and IL-18. Unlike these two types of cell death, necroptosis is caspase-independent and involves a complex of proteins known as the necrosome, which includes RIP1 and RIP3. This complex activates mixed lineage kinase domain-like (MLKL), a pseudokinase, which results in plasma membrane permeabilization. Among these, apoptosis is relatively immunologically silent, pyroptosis is highly inflammatory, and necroptosis signals the immune system against cellular damage and induces an inflammatory response.

3. Mycobacteria modulate host-regulated cell death modalities

3.1 Pyroptosis is modulated by mycobacterial infection

Host sensors and several *Mtb* factors (Table 9.1) induce inflammasome activation and pyroptotic cell death during TB infection, leading to bacterial clearance and host tissue inflammation (Fig. 9.1). The interaction of mycobacterial PAMPs with host PRRs, such as TLR2/4, triggers inflammatory signaling that activates host defenses, such as macrophage effector functions, including phagocytosis, ROS/reactive nitrogen species (RNS) production, phagolysosome fusion, and autophagy toward bacterial clearance. Macrophage effector functions eliminate most bacteria; however, some bacteria persist in the phagosome owing to their adaptive mechanisms and resist fusion with lysosomes. These bacteria secrete factors that help evade innate defenses, primarily through cell death mechanisms. However, it should be noted that some of the secreted proteins and lipids also promote host cell death pathways. The primary system that aids in the interaction of *Mtb* proteins with the host signaling partners is the ESX-1 secretion system, which not only secretes a specific set of *Mtb* proteins from its cell wall but also creates pores in the host phagosome, allowing *Mtb* proteins to enter the host cytoplasm. The pore-forming ability of ESAT-6 secretion system 1 (ESX-1) (with other *Mtb* factors) compromises the membrane integrity of the host, leading to leakage of ions such as Ca^{2+}, Na^{+}, and K^{+}, and signaling toward the inflammatory cell death pathway, pyroptosis. Another *Mtb*-secreted protein, Rv1579c, enters the host cytoplasm, interacts with the receptor for activated C kinase 1 (RACK1), and triggers NLRP3-mediated gasdermin D pyroptosis [26].

Several other secreted proteins have been reported to activate the NLRP3-based inflammasome, such as LpqH [35] and PPE13 [36], which may not directly result in pyroptosis but could accentuate inflammatory cell death mediated by other agents. The PPE60 [29], ESAT-6 [37], and Rv1579c [26] have been shown to trigger canonical pyroptosis. One *Mtb* protein that was shown to induce non-canonical pyroptosis (caspase-1 independent pyroptosis) is PE_PGRS19, which promotes pyroptosis through the caspase-11/gasdermin pathway in mouse macrophages, showing cell membrane swelling that subsequently ruptured, consistent with canonical pyroptosis, along with biochemical markers, including caspase-11 and gasdermin cleavage [31].

Mtb proteins that inhibit host pyroptosis can aid the persistence of host macrophages. These proteins can inhibit inflammasome formation at various levels. For example, EST12 (Rv3364c) protein binds to cathepsin G on the membrane, limiting its enzymatic activity and preventing pyroptosis in human U937 monocyte-derived macrophages [38]. However, the deletion mutant failed to inhibit cell death. Similarly, PknF, a serine threonine kinase, inhibits pyroptosis in bone marrow-derived macrophages by inhibiting NLRP3 activation and reducing IL-1β, which was not so in the ΔPknf mutant [33]. However, the levels of IL-6 remained unchanged compared to *Mtb*-infected cells, indicating that the mutant did not affect the priming step of inflammasome activation.

However, the activation stage was altered because K^{+} efflux, Cl^{-} efflux, and the production of ROS led to pyroptosis mediated by the *Mtb PknF* mutant strain. ZMP-1 is another secreted *Mtb* protein known to inhibit host pyroptosis [39], the mechanism of which was recently explored. GRIM-19 is a substrate of ZMP-1, which is essential for activating the NLRP3 inflammasome [40]. A comprehensive

Table 9.1 List of Mycobacterial proteins influencing host pyroptosis and necroptosis.

Pathway	*Mtb* protein/factor	Host signaling factors/ pathways affected	Remarks	Refs.
Pyroptosis	ESAT-6	NLRP3/ASC/Casp-1/IL-β	Activates pyroptosis	[8,27]
	Rv1579c	RACK1/UCHL5/NLRP-3/Casp-1/GSDMD/IL-1β	Activates pyroptosis	[2]
	LpqH	NLRP-3/ASC-1/Casp-1/IL-1β	Activates inflammasome without cell death	[7]
	PPE13	NLRP-3/ASC-1/Casp-1/IL-1β	Activates inflammasome	[28]
	PPE60	NLRP3/Casp-1/IL-1β/GSDMD	Promotes pyroptosis	[29]
	PE_PGRS19	Casp-11/GSDMD (non-canonical pathway)	Promotes pyroptosis	[30]
	Rv3364c	Prevents Casp-1-dependent apoptosis by inhibiting Cathepsin G activity	Inhibits pyroptosis	[31]
	ZMP-1	Cleaves GRIM-19, necessary for NLRP3 activation and IL-1β secretion	Inhibits pyroptosis	[32]
	PknF	Inhibits NLRP3 inflammasome	Inhibits pyroptosis	[26]
Necroptosis	TNT	Depletes NAD+, activating RIPK3 and MLKL	Activates necroptosis	[33]
	PstS1 (38 kDa lipoprotein)	Induces TNF	Activates necroptosis	[34]

ASC, *apoptosis-associated speck-like protein containing a CARD;* CASP, *caspase;* FAK, *focal adhesion kinase;* GSDMD, *gasdermin D;* MLKL, *mixed lineage kinase domain-like pseudokinase; (*NAD+*), nucleotide adenine dinucleotide+;* NLRP3, *NLR family pyrin domain containing 3;* PDIM, *phthiocerol dimycocerosate;* RACK1, *receptor for activated C kinase 1;* RIPK3, *receptor-interacting serine/threonine kinase 3;* ROS, *reactive oxygen species;* TNF, *tumor necrosis factor;* TNT, *tuberculosis necrotizing toxin;* UCHL5, *ubiquitin C-terminal hydrolase L5.*

screening of all factors secreted by *Mtb* during infection will likely identify novel candidates influencing pyroptosis, highlighting its crucial role in TB progression.

3.2 Mycobacteria promote necroptosis for dissemination

Macrophages infected with *Mtb* at high MOI leads to necrosis. Necrosis is believed to occur in the later stages of TB and helps in dissemination. Specific *Mtb* secreted proteins induce host necroptosis, facilitating the spread of *Mtb* to other macrophages or their escape from the host to infect a new one. *Mtb* ESAT-6, apart from causing different forms of cell death such as apoptosis and pyroptosis, also induces necroptosis. This typically occurs when the bacterial burden is high. Phthiocerol dimycocerosates (PDIM), an *Mtb* cell wall lipid associated with virulence, also aids in the phagosomal escape of *Mtb* factors that subsequently trigger host necroptosis.

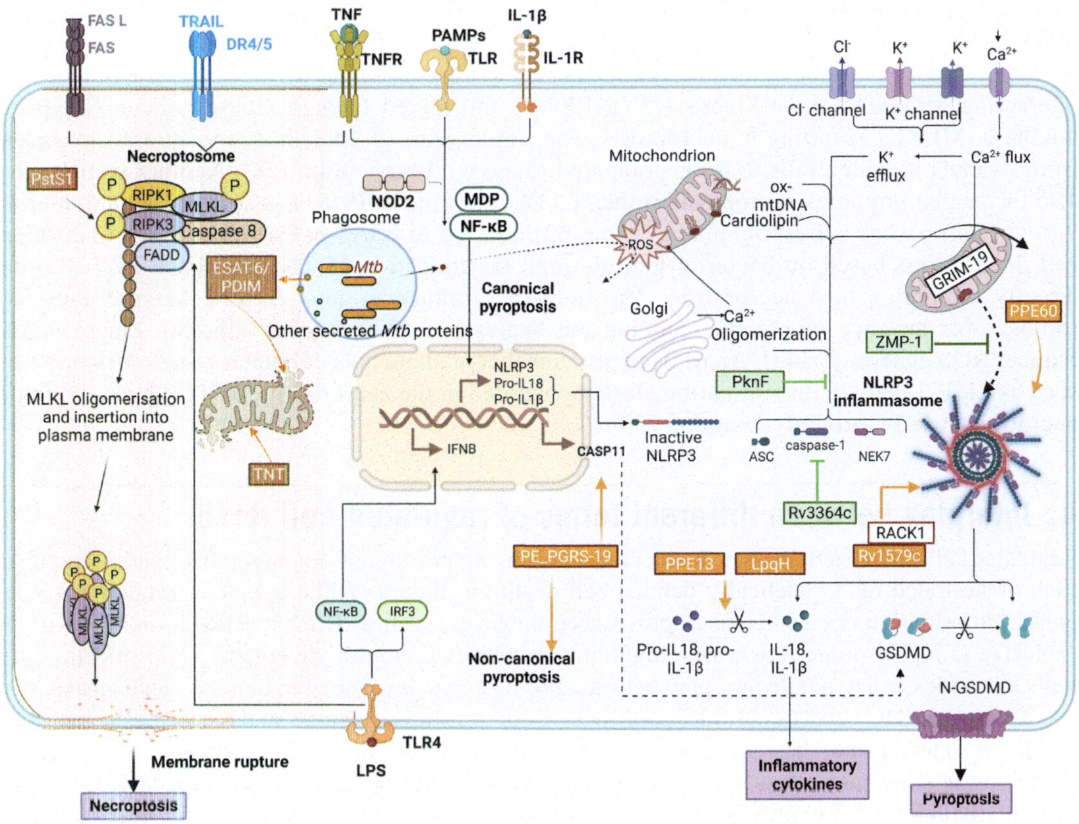

FIGURE 9.1 Schematic of *Mtb* Interference With Host Programmed Pyroptosis and Necroptosis.

Several surface receptors of the host, including pattern recognition receptors (PRRs) and death receptors, recognize cognate ligands from *Mycobacterium tuberculosis* (*Mtb*) and trigger a cascade of signaling reactions with distinct adaptor proteins. Receptors, such as tumor necrosis factor receptor (TNFR) and toll-like receptor 4 (TLR4), mediate programmed inflammatory cell death by interacting with adaptor proteins. Receptor-interacting serine/threonine kinase 1/3 (RIPK1/RIPK3) and mixed lineage kinase domain-like pseudokinase (MLKL) mediate the formation of necroptosomes and subsequent necroptosis. In contrast, NLR family pyrin domain containing 3 (NLRP3), apoptosis-associated speck-like protein containing a CARD (ASC), and caspase-1 mediate the formation of inflammasomes and subsequent pyroptosis. Both forms of death cause membrane swelling and release of inflammatory cytokines. However, pore size, structure, and type of molecules released differ. The loss of membrane integrity in pyroptosis is due to pores caused by oligomerization of gasdermin D fragments, and membrane rupture in necroptosis is due to large pores caused by oligomerization of MLKL proteins. The pores caused by gasdermin D (GSDMD) are relatively small and allow the release of inflammatory cytokines such as IL-1β and IL-18, ions, and water. In contrast, MLKL forms larger pores than gasdermin D. They disrupt membrane integrity more extensively, leading to cell lysis, the release of damage-associated molecular patterns (DAMPs), and excessive immune cell infiltration. Several *Mtb* proteins modulate pyroptosis in host cells. While some proteins accentuate host proptosis (marked in orange boxes and arrows, namely PE_PGRS-19, PPE13, Rv1579c, LpqH, and PPE60), other proteins inhibit host inflammasome signaling (marked in green boxes, namely Rv3364c, ZMP-1, and PknF). However, in the case of necroptosis, *Mtb* proteins have only been demonstrated to activate necroptosis (marked in red boxes in the left panel with peach background, namely PstS1, ESAT-6/PDIM, and TNT).

Created using Biorender.com.

However, the underlying molecular mechanisms remain unclear [41]. Tuberculosis necrotizing toxin (TNT) is a nicotinamide adenine dinucleotide+ (NAD+) glycohydrolase that depletes NAD+, depolarizes mitochondria, and impairs ATP synthesis, signaling the activation of receptor-interacting serine/threonine kinase 1/3 (RIPK1/3) and mixed lineage kinase domain-like pseudokinase (MLKL), resulting in necroptosis. The significance of TNT-induced necroptosis is clear from its mutant, which fails to grow compared to the wild-type strain [28]. Another well-studied *Mtb* factor that promotes necroptosis is the 38 kDa lipoprotein (PstS1), associated with numerous virulence properties in host immune evasion [32]. Similar to ESAT-6, PstS1 also leads to different cell death forms based on bacterial load through the induction of TNF and FasL [42]. Besides directly promoting host necroptosis, *Mtb* indirectly influences host factors that promote necroptosis; for instance, gradually reducing the expression of host focal adhesion kinase (FAK) reduces ROS generation [43]. Apart from proteins that mediate host cell necroptosis, *Mtb* proteins, such as ZMP-1, help in dissemination, as demonstrated in the zebrafish model [44] by clearing the necrotic matrix by proteolytic action [30].

4. Interplay between different forms of regulated cell death

Regulated cell death encompasses distinct mechanisms, such as apoptosis, pyroptosis, and necroptosis, each orchestrated by a genetically defined cell death machinery. While apoptosis-related genes are present in primitive organisms, necroptosis-specific genes, such as RIPK1, RIPK3, and MLKL, are exclusive to higher-order vertebrates, highlighting their evolutionary divergence. *Mtb* infection presents a complex and dynamic interplay between pro-host and pro-pathogen defense mechanisms, with extensive crosstalk across multiple pathways. Both pro-inflammatory and necroptosis pathways frequently interact and share molecular players, such as PRRs and TNF signaling, synergistically enhancing host defense responses. Caspase-3-mediated apoptosis can lead to secondary pyroptosis, particularly in DFNA-5-expressing cells. Caspase-3 cleaves GSMD-related DFNA-5, generating pore-forming fragments that induce pyroptotic cell death. Conversely, DFNA-5-deficient cells undergo apoptotic body disassembly [34]. Moreover, uncleared apoptotic bodies can undergo lysis, triggering secondary necrosis [45].

Adding to this complexity, *Mtb* factors actively suppress multiple cell death pathways, modulate inflammatory responses, and influence bacterial persistence. Additionally, the type of inflammatory cell death is often dictated by bacterial load, which affects the trajectory of disease progression. Dysregulation of these interconnected pathways can lead to excessive inflammation and tissue damage, exacerbating the severity of infection.

5. Cell death modulators as targets for therapeutic intervention

Given the significance of host cell pyroptosis and necroptosis in the progression of TB, intervening in pyroptosis and necroptosis pathways holds great promise as a host-directed therapy. These interventions may assist in managing both disease pathology and bacterial clearance. Therefore, modulating inflammasome activation, RIPK signaling, or MLKL activity may enhance protective immune responses while reducing excessive inflammation. Recent progress in understanding cell death processes has significant potential for therapeutic development. RIPK1 has been proposed as a therapeutic target for autoimmune, inflammatory, neurodegenerative, ischemic, and acute conditions, including sepsis, for which several small-molecule inhibitors are in different phases of clinical trials [46]. Notably, the RIPK1 inhibitor SAR443820 (DNL788) was used in a phase IIb human clinical study for the treatment of amyotrophic lateral sclerosis (neurodegenerative disease) and multiple sclerosis (autoimmune disease of the CNS). These inhibitors may hold great promise for managing TB, wherein excessive inflammation-induced pathology and dissemination can be prevented. Similarly, pyroptosis modulators are emerging as promising therapeutic targets for various inflammatory diseases and cancers that could be explored for TB, in which pyroptosis plays a crucial role. Targeting gasdermin D, the primary executor of pyroptosis, can help to minimize excessive inflammation. Small-molecule inhibitors such as GI-Y1 [47] and disulfiram [48], among others, are being investigated to prevent GSDMD activation in early drug development and clinical trials [49].

6. Concluding remarks

Although apoptosis and necrosis, the classic forms of cell death, are well studied in TB progression, other cell death modalities observed during mycobacterial infection of host macrophages remain underexplored regarding their impact on infection dynamics and disease outcome. The interplay between regulated and unregulated cell death mechanisms involves a complex network of overlapping molecular players that shape the balance between the host defense and disease pathology. Although these pathways are protective, they can also contribute to tissue damage and bacterial persistence, thereby prolonging the infection. Specific mycobacterial proteins are involved in multiple pathways that facilitate pathogen survival. Targeting these proteins presents an opportunity to restore the host's ability to regulate innate immune mechanisms and eliminate *Mtb*. Systematic research is essential to unravel the intricate nature of programmed cell death (PCD) in TB and to translate these insights into effective therapeutic strategies. Under conditions of bacterial persistence, certain mycobacterial factors, such as ZMP-1, Rv3364c, and PknF, inhibit inflammasome activation, thereby preventing pyroptosis and further complicating disease progression.

Acknowledgments

SERB SUPRA (SPR/2021/000137) and ICMR (5/8/5/21/ITRC/Diag/2021/ECD-1) support to SB. Furthermore, the DST-FIST to the Department of Biochemistry (SR/FST/LS-II/2023/1172), the DBT-BUILDER grant to the School of Life Sciences (BT/INF/22/SP41176/2020), and the Institution of Eminence-supported projects RC1-20-017 to S.B. The IoE to the University of Hyderabad, MHRD (F11/9/2019- U3(A)), infrastructure support to the Department of Biochemistry, School of Life Sciences, and shared facilities of the University of Hyderabad are acknowledged. We acknowledge the assistance of Microsoft Copilot in generating text suggestions for grammar and clarity. Fig. 1 was created using BioRender (F60, M., 2025) https://BioRender.com/722lfdd).

Conflict of interest

The authors declare no conflicts of interest.

References

[1] Ramon-Luing LA, Olvera Y, Flores-Gonzalez J, Palacios Y, Carranza C, Aguilar-Duran Y, Vargas MA, Gutierrez N, Medina-Quero K, Chavez-Galan L. Diverse cell death mechanisms are simultaneously activated in macrophages infected by virulent *Mycobacterium tuberculosis*. Pathogens 2022;11(5):492. https://doi.org/10.3390/pathogens11050492.

[2] Mohareer K, Asalla S, Banerjee S. Cell death at the cross roads of host-pathogen interaction in *Mycobacterium tuberculosis* infection. Tuberculosis 2018;113:99–121. https://doi.org/10.1016/j.tube.2018.09.007.

[3] Li L, Dickinson MS, Coers J, Miao EA. Pyroptosis in defense against intracellular bacteria. Semin Immunol 2023;69:101805. https://doi.org/10.1016/j.smim.2023.101805.

[4] Gan B. Ferroptosis hijacking by *Mycobacterium tuberculosis*. Nat Commun 2023;14(1). https://doi.org/10.1038/s41467-023-37149-w.

[5] Cavalcante-Silva LHA, Almeida FS, Andrade AGD, Comberlang FC, Cardoso LL, Vanderley SER, Keesen TSL. *Mycobacterium tuberculosis* in a trap: the role of neutrophil extracellular traps in tuberculosis. Int J Mol Sci 2023;24(14). https://doi.org/10.3390/ijms241411385.

[6] Xiao F, Li H-L, Yang B, Che H, Xu F, Li G, Zhou C-H, Wang S. Disulfidptosis: a new type of cell death. Apoptosis 2024;29(9–10):1309–29. https://doi.org/10.1007/s10495-024-01989-8.

[7] Nisa A, Kipper FC, Panigrahy D, Tiwari S, Kupz A, Subbian S. Different modalities of host cell death and their impact on *Mycobacterium tuberculosis* infection. Am J Physiol Cell Physiol 2022;323(5):C1444. https://doi.org/10.1152/AJPCELL.00246.2022.

[8] Parandhaman DK, Narayanan S. Cell death paradigms in the pathogenesis of *Mycobacterium tuberculosis* infection. Front Cell Infect Microbiol 2014;4. https://doi.org/10.3389/fcimb.2014.00031.

[9] Ruan H, Lyu M, Lai H, Niu L, Zhao Z, Liu T, Lei S, Ying B. Host-pathogen dialogues in different cell death modes during *Mycobacterium tuberculosis* infection. Interdiscip Med 2024;2(2). https://doi.org/10.1002/INMD.20230044.

[10] Behar SM, Martin CJ, Booty MG, Nishimura T, Zhao X, Gan HX, Divangahi M, Remold HG. Apoptosis is an innate defense function of macrophages against *Mycobacterium tuberculosis*. Mucosal Immunol 2011;4(3):279–87. https://doi.org/10.1038/mi.2011.3.

[11] Koul A, Herget T, Klebl B, Ullrich A. Interplay between mycobacteria and host signalling pathways. Nat Rev Microbiol 2004;2(3):189–202. https://doi.org/10.1038/nrmicro840.

[12] Lee J, Hartman M, Kornfeld H. Macrophage apoptosis in tuberculosis. Yonsei Med J 2009;50(1):1–11. https://doi.org/10.3349/ymj.2009.50.1.1.

[13] Liu CH, Liu H, Ge B. Innate immunity in tuberculosis: host defense vs pathogen evasion. Cell Mol Immunol 2017;14(12):963–75. https://doi.org/10.1038/cmi.2017.88.

[14] Brightbill HD, Libraty DH, Krutzik SR, Yang RB, Belisle JT, Bleharski JR, Maitland M, Norgard MV, Plevy SE, Smale ST, Brennan PJ, Bloom BR, Godowski PJ, Modlin RL. Host defense mechanisms triggered by microbial lipoproteins through toll-like receptors. Science 1999;285(5428):732–6. https://doi.org/10.1126/science.285.5428.732.

[15] Wang Y, Shi Q, Qi C, Zhou X, Yuan H, Jia X, Liu S, Li Q, Ge L. Emerging advances in identifying signal transmission molecules involved in the interaction between *Mycobacterium tuberculosis* and the host. Front Cell Infect Microbiol 2022;12. https://doi.org/10.3389/fcimb.2022.956311.

[16] Krishnan V, Nath S, Nair P, Das B. *Mycobacterium tuberculosis* and its clever approaches to escape the deadly macrophage. World J Microbiol Biotechnol 2023;39(11). https://doi.org/10.1007/s11274-023-03735-9.

[17] Vu A, Glassman I, Campbell G, Yeganyan S, Nguyen J, Shin A, Venketaraman V. Host cell death and modulation of immune response against *Mycobacterium tuberculosis* infection. Int J Mol Sci 2024;25(11):6255. https://doi.org/10.3390/ijms25116255.

[18] Mittal E, Deore S, Kumar M, Krishnasastry MV. When to die is the question: need and manipulation of cell death by mycobacterium. Curr Mol Biol Rep 2020;6(3):103–15. https://doi.org/10.1007/s40610-020-00137-w.

[19] Yuan J, Ofengeim D. A guide to cell death pathways. Nat Rev Mol Cell Biol 2024;25(5):379–95. https://doi.org/10.1038/s41580-023-00689-6.

[20] Cookson BT, Brennan MA. Pro-inflammatory programmed cell death [2]. Trends Microbiol 2001;9(3): 113–4. https://doi.org/10.1016/s0966-842x(00)01936-3.

[21] Boucher D, Monteleone M, Coll RC, Chen KW, Ross CM, Teo JL, Gomez GA, Holley CL, Bierschenk D, Stacey KJ, Yap AS, Bezbradica JS, Schroder K. Caspase-1 self-cleavage is an intrinsic mechanism to terminate inflammasome activity. J Exp Med 2018;215(3):827–40. https://doi.org/10.1084/jem.20172222.

[22] Man SM, Karki R, Kanneganti TD. Molecular mechanisms and functions of pyroptosis, inflammatory caspases and inflammasomes in infectious diseases. Immunol Rev 2017;277(1):61–75. https://doi.org/10.1111/imr.12534.

[23] Gao K, Liu Y, Sun C, Wang Y, Bao H, Liu G, Ou J, Sun P. TNF-α induces mitochondrial dysfunction to drive NLRP3/Caspase-1/GSDMD-mediated pyroptosis in MCF-7 cells. Sci Rep 2024;14(1). https://doi.org/10.1038/s41598-024-76997-4.

[24] Christofferson DE, Yuan J. Necroptosis as an alternative form of programmed cell death. Curr Opin Cell Biol 2010;22(2):263–8. https://doi.org/10.1016/j.ceb.2009.12.003.

[25] Takahashi N, Duprez L, Grootjans S, Cauwels A, Nerinckx W, Duhadaway JB, Goossens V, Roelandt R, Van Hauwermeiren F, Libert C, Declercq W, Callewaert N, Prendergast GC, Degterev A, Yuan J, Vandenabeele P. Necrostatin-1 analogues: critical issues on the specificity, activity and in vivo use in experimental disease models. Cell Death Dis 2012;3(11). https://doi.org/10.1038/cddis.2012.176.

[26] Qu Z, Zhou J, Zhou Y, Xie Y, Jiang Y, Wu J, Luo Z, Liu G, Yin L, Zhang X-L. Mycobacterial EST12 activates a RACK1–NLRP3–gasdermin D pyroptosis–IL-1β immune pathway. Sci Adv 2020;6(43). https://doi.org/10.1126/sciadv.aba4733.

[27] Mishra B, Moura-Alves P, Sonawane A, et al. Mycobacterium tuberculosis protein ESAT-6 is a potent activator of the NLRP3/ASC inflammasome. Cell Microbiol 2010;114:219–24. https://doi.org/10.1111/j.1462-5822.2010.01450.x. In this issue.

[28] Pajuelo D, Gonzalez-Juarbe N, Tak U, Sun J, Orihuela CJ, Niederweis M. NAD+ depletion triggers macrophage necroptosis, a cell death pathway exploited by *Mycobacterium tuberculosis*. Cell Rep 2018;24(2):429–40. https://doi.org/10.1016/j.celrep.2018.06.042.

[29] Gong Z, Kuang Z, Li H, Li C, et al. Regulation of host cell pyroptosis and cytokines production by Mycobacterium tuberculosis effector PPE60 requires LUBAC mediated NF-κB signaling. Cell Immunol 2019;335:41–50. https://doi.org/10.1016/j.cellimm.2018.10.009. In this issue.

[30] Petrera A, Amstutz B, Gioia M, Hähnlein J, Baici A, Selchow P, Ferraris DM, Rizzi M, Sbardella D, Marini S, Coletta M, Sander P. Functional characterization of the *Mycobacterium tuberculosis* zinc metallopeptidase Zmp1 and identification of potential substrates. Bchm 2012;393(7):631–40. https://doi.org/10.1515/hsz-2012-0106.

[31] Qian J, Hu Y, Zhang X, Chi M, Xu S, Wang H, Zhang X. *Mycobacterium tuberculosis* PE_PGRS19 induces pyroptosis through a non-classical caspase-11/GSDMD pathway in macrophages. Microorganisms 2022;10(12). https://doi.org/10.3390/microorganisms10122473.

[32] Ramon-Luing L, Palacios Y, Ruiz A, Téllez-Navarrete N, Chavez-Galan L. Virulence factors of *Mycobacterium tuberculosis* as modulators of cell death mechanisms. Pathogens 2023;12(6):839. https://doi.org/10.3390/pathogens12060839.

[33] Rastogi S, Ellinwood S, Augenstreich J, Mayer-Barber KD, Briken V, Behr MA. *Mycobacterium tuberculosis* inhibits the NLRP3 inflammasome activation via its phosphokinase PknF. PLoS Pathog 2021;17(7): e1009712. https://doi.org/10.1371/journal.ppat.1009712.

[34] Rogers C, Fernandes-Alnemri T, Mayes L, Alnemri D, Cingolani G, Alnemri ES. Cleavage of DFNA5 by caspase-3 during apoptosis mediates progression to secondary necrotic/pyroptotic cell death. Nat Commun 2017;8(1). https://doi.org/10.1038/ncomms14128.

[35] Liu L, Zhai K, Chen Y, Chen X, Wang G, Wu L, Ashwood P. Effect and mechanism of *Mycobacterium tuberculosis* lipoprotein LpqH in NLRP3 inflammasome activation in mouse ana-1 macrophage. BioMed Res Int 2021;2021(1). https://doi.org/10.1155/2021/8239135.

[36] Yang Y, Xu P, He P, Shi F, Tang Y, Guan C, Zeng H, Zhou Y, Song Q, Zhou B, Jiang S, Shao C, Sun J, Yang Y, Wang X, Song H. Mycobacterial PPE13 activates inflammasome by interacting with the NATCH and LRR domains of NLRP3. FASEB J 2020;34(9):12820–33. https://doi.org/10.1096/fj.202000200RR.

[37] Basu S, Fowler BJ, Kerur N, Arnvig KB, Rao NA. NLRP3 inflammasome activation by mycobacterial ESAT-6 and dsRNA in intraocular tuberculosis. Microb Pathog 2018;114:219–24. https://doi.org/10.1016/j.micpath.2017.11.044.

[38] Danelishvili L, Everman JL, McNamara MJ, Bermudez LE. Inhibition of the plasma-membrane-associated serine protease cathepsin G by *Mycobacterium tuberculosis* Rv3364c suppresses caspase-1 and pyroptosis in macrophages. Front Microbiol 2012;2. https://doi.org/10.3389/fmicb.2011.00281.

[39] Master SS, Rampini SK, Davis AS, Keller C, Ehlers S, Springer B, Timmins GS, Sander P, Deretic V. *Mycobacterium tuberculosis* prevents inflammasome activation. Cell Host Microbe 2008;3(4):224–32. https://doi.org/10.1016/j.chom.2008.03.003.

[40] Kurane T, Matsunaga T, Ida T, Sawada K, Nishimura A, Fukui M, Umemura M, Nakayama M, Ohara N, Matsumoto S, Akaike T, Matsuzaki G, Takaesu G. GRIM-19 is a target of mycobacterial Zn2+ metalloprotease 1 and indispensable for NLRP3 inflammasome activation. FASEB (Fed Am Soc Exp Biol) J 2022; 36(1). https://doi.org/10.1096/fj.202101074RR.

[41] Quigley J, Hughitt VK, Velikovsky CA, Mariuzza RA, El-Sayed NM, Briken V. The cell wall lipid PDIM contributes to phagosomal escape and host cell exit of *Mycobacterium tuberculosis*. mBio 2017;8(2). https://doi.org/10.1128/mBio.00148-17.

[42] Sanchez A, Espinosa P, Esparza MA, Colon M, Bernal G, Mancilla R. *Mycobacterium tuberculosis* 38-kDa lipoprotein is apoptogenic for human monocyte-derived macrophages. Scand J Immunol 2009;69(1):20–8. https://doi.org/10.1111/j.1365-3083.2008.02193.x.

[43] Afriyie-Asante A, Dabla A, Dagenais A, Berton S, Smyth R, Sun J. *Mycobacterium tuberculosis* exploits focal adhesion kinase to induce necrotic cell death and inhibit reactive oxygen species production. Front Immunol 2021;12. https://doi.org/10.3389/fimmu.2021.742370.

[44] Vemula MH, Medisetti R, Ganji R, Jakkala K, Sankati S, Chatti K, Banerjee S. *Mycobacterium tuberculosis* zinc metalloprotease-1 assists mycobacterial dissemination in zebrafish. Front Microbiol 2016;7. https://doi.org/10.3389/fmicb.2016.01347.

[45] Jorgensen I, Zhang Y, Krantz BA, Miao EA. Pyroptosis triggers pore-induced intracellular traps (PITs) that capture bacteria and lead to their clearance by efferocytosis. J Exp Med 2016;213(10):2113–28. https://doi.org/10.1084/jem.20151613.

[46] Mifflin L, Ofengeim D, Yuan J. Receptor-interacting protein kinase 1 (RIPK1) as a therapeutic target. Nat Rev Drug Discov 2020;19(8):553–71. https://doi.org/10.1038/s41573-020-0071-y.

[47] Zhong L, Han J, Fan X, Huang Z, Su L, Cai X, Lin S, Chen X, Huang W, Dai S, Ye B. Novel GSDMD inhibitor GI-Y1 protects heart against pyroptosis and ischemia/reperfusion injury by blocking pyroptotic pore formation. Basic Res Cardiol 2023;118(1). https://doi.org/10.1007/s00395-023-01010-4.

[48] Guo W, Chen S, Li C, Xu J, Wang L. Application of disulfiram and its metabolites in treatment of inflammatory disorders. Front Pharmacol 2022;12. https://doi.org/10.3389/fphar.2021.795078.

[49] Dai Z, Liu WC, Chen XY, Wang X, Li JL, Zhang X. Gasdermin D-mediated pyroptosis: mechanisms, diseases, and inhibitors. Front Immunol 2023;14. https://doi.org/10.3389/fimmu.2023.1178662.

CHAPTER

10 Unraveling the cellular battlefield: *Mycobacterium tuberculosis* and its interplay with host cell death pathways

Korak Chakraborty[1, 2] and Tirumalai R. Raghunand[1, 2]

[1]*CSIR—Centre for Cellular and Molecular Biology, Hyderabad, Telangana, India;* [2]*Academy of Scientific and Innovative Research (AcSIR), Ghaziabad, Uttar Pradesh, India*

Highlights

1. *Mycobacterium tuberculosis* (*Mtb*) modulates host cell death pathways to establish infection.
2. Apoptosis and autophagy are pro-host mechanisms that aid in *Mtb* clearance.
3. *Mtb* proteins like ESAT-6, NuoG, and PknE inhibit apoptosis to promote survival.
4. Necrosis is a pro-*Mtb* mechanism that helps in bacterial dissemination.
5. *Mtb* induces necroptosis, pyroptosis, and ferroptosis to cause pathological damage.
6. Targeting host cell death pathways is a potential host-directed therapeutic strategy.
7. Understanding *Mtb*-host cell death interplay is crucial for developing novel tuberculosis (TB) treatments.

Abbreviations

ABC ATP-binding cassette
AIM2 absent in melanoma 2
Akt Akt serine/threonine kinase
Apaf1 apoptotic protease activating factor 1
APCs antigen-presenting cells
ASC apoptosis-associated speck-like protein containing CARD
ASK-1 apoptosis signal-regulating kinase 1
ATGs autophagy-related genes
Bak Bcl2 antagonist/killer
Bax Bcl2-associated X protein
Bcl-2 B-cell lymphoma 2
BECN1 beclin-1
Bid BH3 interacting domain death agonist
cFLIP cellular FLICE-like inhibitory protein
CR1/CR3 complement receptors 1 and 3
DD death domains

Navigating Tuberculosis. https://doi.org/10.1016/B978-0-443-27735-1.00003-4

DISC death- inducing signal complex
DKO double knockout
Eis enhanced intracellular survival
ER endoplasmic reticulum
ESAT-6 early secreted antigen 6
FADD fas- associated death domain
FAK focal adhesion kinase
Fas fas cell surface death receptor
FasL fas ligand
Fer-1 ferrostatin-1
GPX4 glutathione peroxidase 4
GSDMD gasdermin D
GSH glutathione
HBHA heparin-binding hemagglutinin adhesin
HDAC histone deacetylases
HDT host-directed therapeutic
HIV human immunodeficiency virus
Hly listeriolysin
H~2~O~2~ hydrogen peroxide
IFNG interferon gamma
IL-10 interleukin 10
IL-17 interleukin 17
IRF-3 interferon-regulatory factor 3
JNK janus kinase
LAM lipoarabinomannan
LAP LC3-associated phagocytosis
LC3 microtubule-associated protein 1 light chain 3
M. bovis *mycobacterium bovis*
M. smegmatis *mycobacterium smegmatis*
MAMPs microbe-associated molecular patterns
MAPK mitogen-activated protein kinase
MAPK1/ERK1 mitogen-activated protein kinase 1
MAPK14/p38 mitogen-activated protein kinase 14
Mcl-1 MCL1 apoptosis regulator, BCL2 family member
MDM monocyte-derived macrophages
MG methylglyoxal
MHC major histocompatibility complex
MLKL mixed-lineage kinase domain-like protein
Mtb *mycobacterium tuberculosis*
MTOR mechanistic target of rapamycin kinase
NAD^+^ nicotinamide adenine dinucleotide
NADH nicotinamide adenine dinucleotide hydrogen
NFKB nuclear factor kappa B
NLR nucleotide-binding and leucine-rich repeat receptors
NLRP3 NLR family pyrin domain-containing 3
NO nitric oxide
NuoG NADH dehydrogenase I subunit G
PAMPs pathogen-associated molecular patterns

PARP Poly(ADP-ribose) polymerase
PDIMs phthiocerol dimycocerosates
PE/PPE proline-glutamic acid/proline-proline-glutamic acid (protein family)
PGRS polymorphic GC-rich sequences
PI3K phosphoinositide- 3- kinase
PIH pyridoxal isonicotinoyl hydrazone
PIK3C3 phosphatidylinositol 3-kinase catalytic subunit type
PknE protein kinase E
PknG protein kinase G
PMNs polymorphonuclear neutrophilic granulocytes
PRKAA2/AMPK protein kinase AMP-activated catalytic subunit alpha 2
PRRs pattern recognition receptors
PSF protein-associated splicing factor
RAB14 RAB14, member RAS oncogene family
RAB5A RAB5A, member RAS oncogene family
RALBP1/ERK ralA binding protein 1/extracellular regulated kinase
rBCG recombinant BCG
RIPKs receptor-interacting protein kinases
RNIs reactive nitrogen intermediates
ROIs reactive oxygen intermediates
siRNA small interfering/silencing RNA
SLs sulfolipids
SodA superoxide dismutase A
TB tuberculosis
TGF-B tumor growth factor beta
Th1 T-helper 1
TLRs toll-like receptors
TNF tumor necrosis factor
TNFR TNF receptor
TNT tuberculosis necrotizing toxin
WT wild-type

1. Introduction

The extraordinary success of *Mtb*, the causative agent of human TB, has been attributed to its versatility in the subversion of host immune responses, facilitating long-term persistence. The probability of a healthy individual being infected with *Mtb* depends on the infection strain, infection dose, and immunocompetence of the individual. In most cases, the infecting pathogen is cleared from the host by alveolar macrophages, the primary site of *Mtb;* however, in some cases, the bacteria can establish a successful infection. To prevent the spread of infection, the host immune system recruits different types of immune cells, such as dendritic cells, Neutrophils, B cells, and T cells, to the site of infection. To contain the infection, these immune cells organize in concentric layers around the bacteria, which are trapped at the core, leading to granuloma formation, a hallmark feature of TB.

2. Host immune responses to *Mtb* infection

Apart from the formation of granulomas, the host immune system elicits various defense responses that are considered checkpoints during the progression of TB infection. After inhalation of contaminated droplets, only a tiny fraction of *Mtb* reaches the alveoli of the respiratory tract, where it comes into contact with phagocytes such as macrophages, dendritic cells, and polymorpho-nuclear cells (PMNs), which represent the first line of defense against invading pathogens. Once encountered, macrophages recognize pathogens via PRRs, which bind to their surfaces (PAMPs). In the case of mycobacterial infection, phagocytic cells interact with *Mtb* through both surface-associated (TLRs, C-type lectin receptors, and mannose receptors) and internal receptors (NOD-like receptors). Additionally, phagocytosis of *Mtb* has been reported to be mediated by human monocyte CR1 and CR3. The mannose receptor expressed on mature macrophages, but not on fresh monocytes, facilitates the internalization of virulent *Mtb*. This receptor binds to LAM, which is surface-exposed and contains terminal mannose residues. The interaction of *Mtb* proteins with TLRs leads to the activation of a downstream cascade of reactions that trigger various pathways, such as the NFKB and MAPK pathways, resulting in the expression of multiple inflammatory mediators. ESAT 6 of *Mtb* is a TLR2 ligand that promotes apoptosis of infected macrophages via NFKB activation [1]. *Mtb* proteins belonging to the PE/PPE family have also been reported to bind to different TLRs and perform immunomodulation, contributing to the establishment and maintenance of *Mtb* infection. PE35 and PPE68 have been reported to interact with each other, and the heterodimeric complex binds to TLR2 and induces secretion of the anti-inflammatory cytokine IL-10 via phosphorylation of MAPK14/p38 and MAPK1/ERK1. Similarly, another complex formed by PE9-PE10 binds to TLR4, resulting in immunomodulation and induction of interferon (IFN) expression by IRF-3 [2].

The expression of various cytokines during *Mtb* infection is an early event in the innate immune response. Some major cytokines that play a role in host defense responses against *Mtb* infection are TNF and IFNs. TNF has been reported to be crucial in activating macrophages, which kill or inhibit the growth of *Mtb* by producing high levels of reactive nitrogen and oxygen intermediates. It is also involved in initiating the extrinsic pathway of apoptosis, which helps clear the infection. It is also a proinflammatory cytokine that facilitates the recruitment of different immune cells to the site of infection. Type II IFN, IFNG also plays a critical role in macrophage activation, enhancing antigen presentation by increasing the expression of MHC molecules on macrophages and triggering the Th1 immune response.

In addition to innate immunity, host adaptive immune responses are crucial in *Mtb* infection. The onset of adaptive immunity against *Mtb* occurs in a delayed manner compared to other infections owing to the delay in activating $CD4^+$ T cells in the lymph nodes. It has also been reported that the activation of Ag85 B-specific T-cells depends on the production of antigens by *Mtb* in the lymph nodes but not in the lungs. In another study, both $CD4^+$ and $CD8^+$ T cells were reported to contribute to immunity against *Mtb* in mice, whereas, in humans, an analysis of patients co-infected with TB and HIV showed that the mortality rate was correlated with the $CD4^+$ cell count, representing the degree of immune deficiency. T cells also produce IFNG, which is widely believed to be important in providing immunity against TB. However, specific IFNG-independent mechanisms have been reported in which $CD4^+$ T cells contribute to their protective function. *Mtb*-specific $CD4^+$ T cells produce IL-17, a major proinflammatory cytokine involved in recruiting neutrophils at the site of infection. IL-23 is essential for IL-17A response during *Mtb* infection since its transient expression in the lungs of *Mtb*-infected mice upregulates the production of IFNG and IL-17A by T cells, which reduces the bacterial burden in the lungs. *Mtb* infection in mice with a genetically inactivated IL-17 receptor. A subunit resulted in a defect in eliciting long-term control, with a progressive increase in lung bacteria and impairment in neutrophil recruitment.

Host cell death is fundamental to the innate immune response against infections by invading pathogens. Various host cell death modalities are involved in *Mtb* infection, some of which are advantageous to the bacteria that aid in pathogenesis. In contrast, others are favorable to the host, using which pathogen clearance is achieved [3]. Host cell death is classified into three major types: apoptosis, autophagy, and necrosis [4]. Recent research in this field has demonstrated the involvement of new modes and pathways of nonapoptotic regulated cell death, such as pyroptosis, necroptosis, and ferroptosis. *Mtb* infection in macrophages leads to the initiation of multiple cell death outcomes depending on whether the pathogen requires infection progression or survival within the host system. The onset of programmed cell death or apoptosis is a pro-host mechanism in which the plasma membrane of host cells remains intact, and the bacterial burden is reduced in the system. In contrast, necrosis involves cell lysis, allowing *Mtb* to exit the macrophage and infect neighboring healthy cells. Therefore, in addition to using several other strategies, *Mtb* evades apoptosis and promotes necrosis, which aids its survival and dissemination within the host [4].

3. Diverse cell death mechanisms and evasion strategies of *Mtb*

When a host inhales an aerosol contaminated with *Mtb*, the pathogen enters the lung airways. It is internalized into alveolar macrophages and other mononuclear phagocytes, such as dendritic cells, via entry receptors. Once phagocytosed, the host and bacteria are involved in a conflict in which the host elicits different immune responses to clear the infection, and the pathogen adapts various measures to evade host defense mechanisms. One such strategy is the reprogramming of host cell death pathways. Therefore, it is of utmost importance to recognize the different cell death modalities that the host utilizes to prevent the spread of the infection and how *Mtb* virulence factors play a crucial role in withstanding these mechanisms and determining the outcome of the infection.

3.1 Apoptosis: A pro-host mechanism of death

Once *Mtb* infects a host cell, the immune system signals various immune cells to kill the infected cell and clear the invading pathogen through programmed cell death, a process known as apoptosis. It is a tightly regulated energy-driven form of cell death in which plasma membrane integrity is maintained, helping to contain the nuclear and cytoplasmic contents of cells undergoing death. Apoptotic cells disintegrate into membrane-bound vesicles called apoptotic bodies, including the contents of dying cells and foreign pathogens. These apoptotic bodies produce "find-me" and "eat-me" signals that lead to their engulfment and clearance by phagocytes, with the production of specific anti-inflammatory cytokines such as TGF-B and IL-10 that ensure immunologically silent elimination of the dying cells. Apoptosis is the preferred mode of cell death because it is considered beneficial for the host in controlling mycobacterial infections. By containing bacteria within dead macrophages and facilitating their phagocytosis by uninfected cells, it prevents bacterial growth and the spread of disease. Infected cells are then processed and presented to $CD4^+$ and $CD8^+$ T cells by APCs, which trigger the host's adaptive immune responses.

Host apoptosis during *Mtb* infection can proceed through one of two pathways: extrinsic or ligand-mediated, or intrinsic or mitochondria-mediated, both of which culminate in the execution of cell death by caspases, a family of cysteine-aspartate proteases. The extrinsic pathway is initiated by the interaction of death receptors (Fas and TNFR) with their respective ligands (FasL and TNF). The binding of FasL to its transmembrane Fas receptor leads to the interaction of DD and recruitment of the adapter protein, FADD, along with several other proteins, forming a multimeric DISC. This leads to proteolytic cleavage of the initiator caspase, caspase 8, which activates caspase 3, the executioner caspase that carries out the apoptotic process in the cell.

In contrast, different cellular stress conditions, such as starvation and DNA damage, trigger the intrinsic or mitochondria-mediated apoptotic pathway. These diverse factors lead to the involvement of various groups of the Bcl-2 protein family that activate apoptosis [5]. The proapoptotic members of this family activate proteins such as Bcl2 antagonist/killer (Bak) and Bcl2-associated X, apoptosis regulator (Bax), which translocate from the cytosol to the outer mitochondrial membrane and induce mitochondrial membrane permeability. Due to the disruption of mitochondrial membrane integrity, several proteins, such as cytochrome C, are predominantly present in the inter-mitochondrial membrane space and released into the cytoplasm. Upon its release, cytochrome C binds to Apaf1, which recruits the initiator caspase, caspase 9, and other molecules to form a multimeric complex called the apoptosome, which activates caspase 3, leading to apoptosis. Macrophages can readily detect infections by non-pathogenic mycobacteria, such as *M. smegmatis*, or attenuated strains, such as *M. bovis* BCG and *Mtb* H37Ra; however, virulent strains of *Mtb* have coevolved with humans and have developed several strategies to evade apoptosis-related bactericidal effects.

3.1.1 Mtb *proteins involved in the inhibition of apoptosis*

The first report of an *Mtb* gene involved in anti-apoptotic activity was SodA, encoded by *SodA* [6], which catalyzes the conversion of a superoxide anion to H_2O_2. *Mtb* has been observed to produce and secrete iron cofactor SOD at much higher levels than non-pathogenic mycobacteria. The SOD-attenuated *Mtb* H37Rv strain was shown to have reduced bacterial burden in mice and increased interstitial infiltration with mononuclear cells, compared to minimal infiltration in cases of infection with WT *Mtb*. Moreover, the SOD-diminished strain was reported to exhibit increased apoptosis. In contrast, the WT strain showed minimal apoptotic cells even after 14 days of infection, providing evidence of the involvement of *SodA* in apoptosis inhibition. Since the SOD-attenuated strain showed increased susceptibility to ROIs, it was hypothesized that SOD might be helpful in *Mtb* survival and apoptosis inhibition by neutralizing phagosomal ROIs (Table 10.1). In another study, *SodA* overexpression in *M. bovis* BCG had multiple effects on innate and adaptive immune responses, with reduced protective ability of the vaccine against *Mtb* infection in guinea pigs. Macrophage infection with the *SodA*-overexpressing strain decreased apoptosis, disrupted cytokine balance, and skewed $CD4^+$ T cell responses in mice. The modulation of innate immune responses by SOD leads to weakened and reduced antigen presentation to T cells, thereby disrupting adaptive immunity.

In an attempt to delineate *Mtb* genes involved in the inhibition of host cell apoptosis, a gain-of-function screen was performed, which showed that multiple genetic loci in *Mtb* controlled the

Table 10.1 *Mtb* proteins suppress or induce host macrophage apoptosis.

Protein	Function	Regulation	Predicted mechanism
SodA	Superoxide dismutase	Inhibition	Neutralization of phagosomal ROS (predicted)
NuoG	Type I NADH dehydrogenase	Inhibition	Neutralizes NOX2 derived ROS
PknE	Serine-threonine kinase	Inhibition	Inhibits NO-stress mediated apoptosis
Rv3654c	Part of 7-gene operon	Inhibition	Cleavage of host PSF leading to less caspase-8 levels in host cells
Rv3033	Outer membrane protein (predicted)	Inhibition	Inhibition of caspase-9 mediated intrinsic apoptosis
ESAT 6	Component of ESX-1 type VII secretion system	Promotion	Promotion of extrinsic pathway in TNF dependent manner
PE_PGRS33	PE family protein	Promotion	TLR-2 mediated ASK-1 activation & activation of caspase proteins
HBHA	Hemagglutination and adhesion activity	Promotion	Inducing intrinsic mode of apoptosis
OppABCD	Oligopeptide transporter protein	Promotion	OppA uptakes GSH leading to increased MG resulting in TNF mediated apoptosis
PstS	Phosphate binding lipoprotein	Promotion	Induces apoptosis via TLR2 and activation of TNF and Fas receptors

Superoxide dismutase A; SodA, Reactive oxygen species; ROS, Nicotinamide adenine dinucleotide hydrogen; NADH, NADH oxidase 2; NOX2, Nitric oxide; NO, Protein slicing factor; PSF, ESX1-associated antigen 6; ESAT6, Type VII secretion system; T7SS, Tumor necrosis factor, TNF, Proline-glutamic acid; PE, Polymorphic GC-rich sequences; PGRS, Apoptosis signal regulating kinase 1; ASK-1, Heparin-binding hemagglutinin; HBHA, Glutathione; GSH, Methyl glyoxal; MG, Toll-like receptor 2; TLR2, Fas cell surface death receptors; FasRs.

regulation of infection-induced macrophage apoptosis. One such locus that showed increased virulence in a severe combined immunodeficiency (SCID) mouse infection model was characterized and reported to possess a large operon that encodes 14 subunits of the *Mtb* type I NADH dehydrogenase complex. *Mtb Rv3151,* encoding *NuoG,* was found to be a virulence gene essential for its pathogenesis since infection of SCID mice with *MtbΔNuoG* showed increased apoptosis of host cells. The expression of *Mtb NuoG* in *Mycobacterium kansasii(M. kansasii),* a non-pathogenic mycobacterium, induced its ability to inhibit apoptosis. In contrast, the deletion of *NuoG* from virulent *Mtb* reduced virulence in mice and abrogated the ability to inhibit apoptosis. In another study by the same group, phagosomes containing *MtbΔNuoG* showed increased ROS levels compared with WT *Mtb* phagosomes. Moreover, NADPH oxidase (NOX2)-deficient macrophages showed increased ROS, reducing apoptosis when challenged with a *NuoG*-deleted *Mtb* strain. These observations imply that *NuoG,* which encodes a functional NADH dehydrogenase I, is essential for neutralizing phagosomal ROS produced by NOX2 and that *Mtb* evades TNF-mediated host cell apoptosis via a NuoG-dependent mechanism (Table 10.1).

In addition to neutralizing ROIs produced during infection, *Mtb* also produces proteins to clear RNIs, which play a role in eliminating invading pathogens. *Mtb* PknE, a eukaryotic serine-threonine kinase, responds to NO stress. The *PknE* promoter showed increased activity under the influence of NO donors. However, its activity was diminished under oxidative stress conditions (H_2O_2 treatment). The *PknE* deletion mutant in *Mtb* H37Rv was reported to enhance apoptosis, simultaneously inhibit necrosis and proliferation in a human macrophage model, and reduce proinflammatory cytokine levels. In a follow-up investigation to understand the role of *PknE* in regulating multiple facets of apoptosis and aiding bacillary survival, the group performed a comparative transcriptomic analysis of macrophages infected with *MtbΔPknE* versus those infected with WT and *PknE*-complemented *Mtb* strains. *MtbΔPknE*-infected macrophages showed reduced inflammatory responses, increased expression of proapoptotic genes such as *Bax* and *Bid*, and reduced levels of the anti-apoptotic gene *Mcl-1*. Moreover, in *PknE*-deleted *Mtb*-infected macrophages, Akt activation, which activates anti-apoptotic genes, was suppressed. The expression of the β-chemokine family has been previously shown to have a protective role against *Mtb*, and infection with *MtbΔPknE* increased β-chemokine expression, indicating the various ways in which *Mtb* PknE facilitates *Mtb* survival by abolishing host cell apoptosis (Fig. 10.1).

Rv3654c and Rv3655c, two *Mtb* proteins that are part of a 7-gene operon (*Rv3654c-Rv3660c*), were found to inhibit host cell apoptosis during *Mtb* infection. Both of these proteins were found to be secreted out of the phagosome by a putative secretory apparatus, and their expression was increased following *Mtb* infection. This study showed that *Mtb* Rv3654c interacts with and cleaves the host PSF, inhibiting the splicing activity that converts caspase 8 pre-mRNA into its active form. Consistent with these results, siRNA knockdown of PSF diminished the levels of active caspase 8, thereby inhibiting the extrinsic apoptosis pathway. Similarly, infection of macrophages with a strain of *M. smegmatis* overexpressing *Rv3033* led to reduced apoptosis, with a concomitant increase in intracellular bacillary survival, indicative of its anti-apoptotic activity [7]. Its anti-apoptotic activity was observed to occur through inhibition of the caspase 9-mediated intrinsic apoptosis pathway. This inhibited mitochondrial cytochrome C release in the cytosol and blocked the endoplasmic reticulum-stress-mediated apoptosis [7].

3.1.2 **Mtb** *proteins responsible for promoting host cell apoptosis*

Although the inhibition of host cell apoptosis during *Mtb* infection is one of the prime mechanisms that facilitates bacillary survival, certain *Mtb* virulence factors are also involved in the induction of apoptosis in host macrophages. There is a delicate balance between apoptosis-inducing and apoptosis-inhibiting *Mtb* factors, and maintaining this balance results in pathogen survival in host macrophages. *M.tb.* ESAT-6 induces apoptosis in the THP-1 macrophages in a dose-dependent manner. The expression of various caspase genes (caspase-1, -3, -5, -7, and -8) was also upregulated in ESAT-6-treated cells, indicating the induction of receptor-mediated or extrinsic apoptosis pathways. Consistent with these results, ESAT-6-mediated apoptosis was abrogated by pan-caspase inhibitor treatment. PE_PGRS33, a member of the PE_PPE protein family, induces apoptosis in a TNF-dependent manner. This protein triggered the release of TNF from infected macrophages, activating downstream MAPKs, such as MAPK14/p38 and JNK. Deletions in this protein's PGRS domain abrogated the TNF release event. Increased apoptosis in PE_PGRS33-challenged macrophages depended on its interaction with TLR2, activating ASK-1, and apoptosis-associated caspase proteins (caspase 3 and caspase 8). *Mtb* HBHA, when ectopically expressed in *M. smegmatis*, showed evidence of apoptosis-mediated death in murine macrophages, such as increased DNA fragmentation, activation of caspases, and PARP cleavage. HBHA-mediated apoptotic cells exhibited increased ROS production and Bax activation,

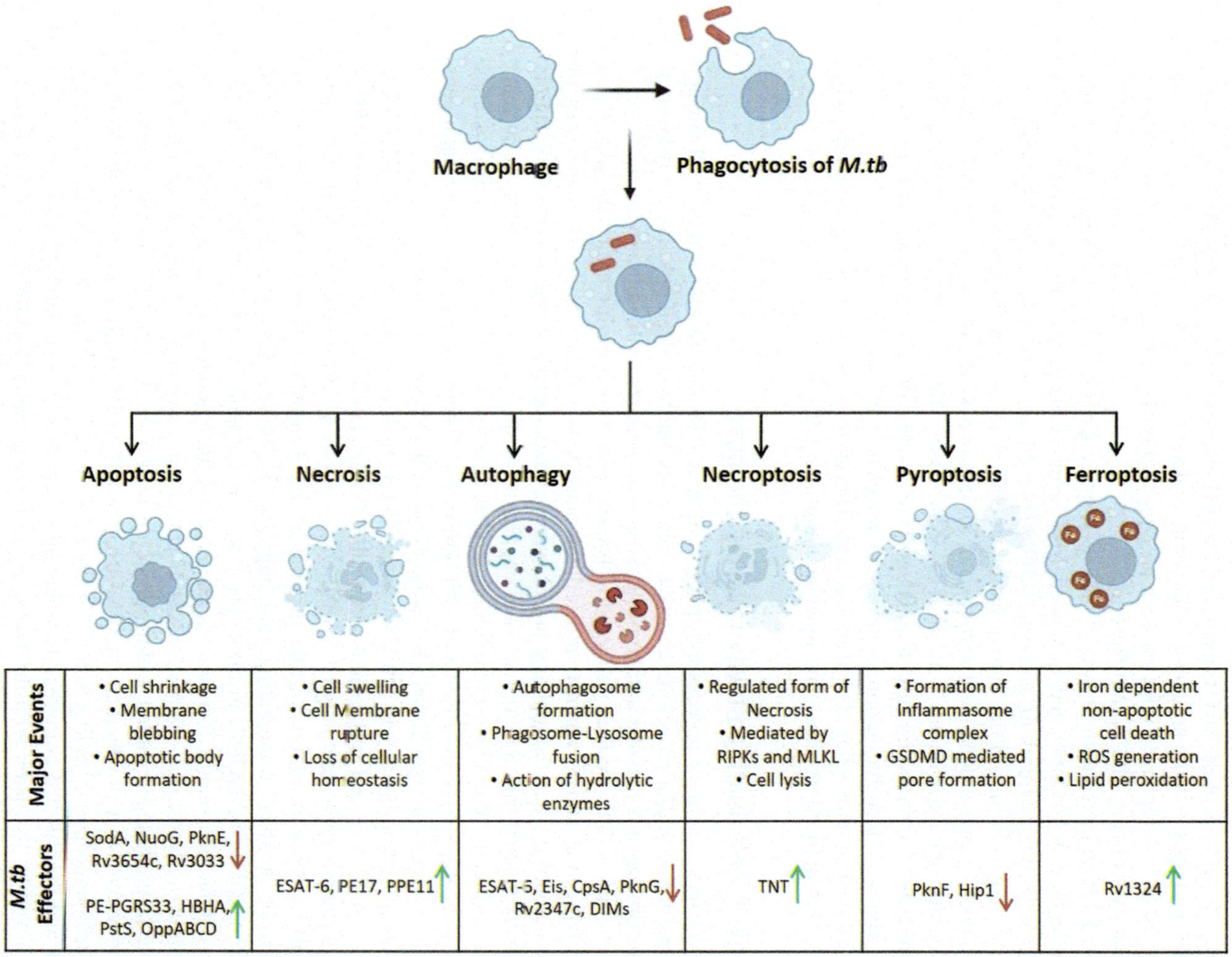

FIGURE 10.1 Schematic Representation of Various Host Cell Death Mechanisms During *Mycobacterium tuberculosis* (*Mtb*) Infection and the Role of different *Mtb* Effector Proteins in Regulating These Processes.

Mtb encounters macrophages in the alveoli of the host lung, which engulf the pathogen by phagocytosis and undergo diverse modes of host cell death, as depicted in the figure. *Mtb* effector proteins modulate host cell death mechanisms by interfering with host signaling pathways, immune responses, and cellular integrity. The interplay between these effector proteins and host cell death pathways contributes to pathogenesis as well as to immune evasion strategies adopted by *Mtb*.

Figure created using BioRender.

followed by translocation to the mitochondrial membrane. HBHA-treated macrophages promote mitochondrial membrane destabilization, leading to cytochrome C release and activating the intrinsic apoptosis pathway. An oligopeptide ABC transporter, OppABCD, encoded by *Mtb oppABCD*, binds to glutathione (GSH) and regulates the apoptosis of host macrophages. An increase in intracellular GSH levels was observed in macrophages infected with an *OppDA* knockout strain of *M. smegmatis* (*M. smΔOppDA*), thereby reducing MG accumulation.

In contrast, WT *Mtb*-infected cells showed a reduction in the GSH content of macrophages owing to uptake by functional *Mtb* OppD and enhanced MG levels. Consistent with the above results, cytokine levels in cells infected with *M. smegmatisΔOppDA* were reported to be reduced, along with decreased cellular apoptosis due to the detoxification of MG. The 38 kDa lipoprotein of *Mtb*, also known as PstS, induced caspase-dependent apoptosis in MDMs in a TLR-2-dependent manner, which was confirmed by blocking experiments with the TLR-2 antibody. The observed increase in host cell apoptosis was mediated by increased time-dependent TNF production and increased expression of TNF receptors, which reduced the apoptosis levels. A novel role for the Fas receptor and FasL in triggering PstS-induced apoptosis has also been reported. Therefore, *Mtb* PstS proved apoptogenic for human MDMs, which induced extrinsic and intrinsic apoptotic pathways.

3.2 Autophagy: A cell death pathway favoring the host

Autophagy is a conserved cellular process that plays a fundamental role in maintaining cellular homeostasis by sequestering and/or recycling defective organelles and misfolded proteins via lysosomal degradation. This process involves the uptake of damaged components by membrane-bound vesicles called autophagosomes, which are then delivered to lysosomes for their elimination. Autophagy can also act as a defense mechanism against intracellular pathogens. Of the three common types of autophagy, microautophagy, macroautophagy, and chaperone-mediated autophagy, macroautophagy is the most extensively studied. Microautophagy and chaperone-mediated autophagy transport proteins from the cytoplasm directly to the lysosomal lumen, whereas macroautophagy results in phagosomes that engulf damaged components and fuse them with lysosomes. Autophagy also eliminates pathogens from the host and is synonymous with selective autophagy, called xenophagy [8].

During *Mtb* infection, autophagy is activated when PRRs, such as TLRs, NLRs, and RIG-I-like receptors (RLRs) on the host cell surface interact with *Mtb*-associated molecular patterns (MAMPs) [9]. These events trigger downstream signaling pathways such as MAPK14/p38 and RALBP1/ERK, leading to the subsequent activation of various proteins encoded by ATGs. A group of at least 32 *Atg* genes is known to be involved in mediating autophagy [6]. TLR4 was the first PRR to be linked to autophagy. NLRs (NOD1 and NOD2) in the cytosol activate autophagy by binding to bacterial cell walls and triggering autophagosome formation at bacterial entrance points, facilitating the elimination of invading bacteria via lysosomal destruction. After activation, autophagy eliminates pathogens by forming a double-layered vesicle called a phagophore derived from the ER. The initiation of phagophore formation occurs when downstream MAPK14/p38 or RIP1/ERK pathways activate the binding of MAMPs to specific receptors, leading to the phosphorylation of PI3K, which signals other proteins, such as BECN1 and the PIK3C3 complex. The phagophore matures and extends, forming the autophagosome, followed by its fusion with the lysosome to form an autolysosome, within which pathogens are killed because of the highly acidic environment and presence of hydrolytic enzymes [8,10].

3.2.1 *Role of* Mtb *proteins in regulating autophagy*

Although autophagy induction is a host defense mechanism against mycobacterial infection, *Mtb* has developed several strategies to subvert autophagy-related immune clearance [11]. Many studies have reported the role of ESAT-6 in *Mtb* pathogenesis. In cell death, ESAT-6 plays a significant role in inhibiting autophagy by preventing autophagosome-lysosome fusion. Moreover, ESAT-6 causes lysosomal dysfunction by dampening the activity of lysosomal enzymes and has been reported to induce the proliferation of *M. bovis* BCG within macrophages [4,6,12]. Another *Mtb* protein, Eis, a GCN-5-related family of N-acetyltransferase members, has also been reported to regulate host autophagy (Fig. 10.1). Macrophages infected with the *Eis*-deletion mutant of *Mtb* H37Rv strain (*MtbΔEis*) accumulated autophagic vesicles, promoting autophagy. This was confirmed by the increased recruitment of MAP1LC3/LC3 to punctate structures and the cleavage of LC3 to LC3-II, which is a marker for the extent of autophagosome formation. An increase in ROS production was also observed upon infection with *MtbΔEis*, making cells sensitive to autophagy, which was reversed by Eis overexpression.

Mtb inhibits the non-canonical autophagy pathway known as LC3-associated phagocytosis (LAP). In contrast to autophagy, this process is characterized by forming a single membrane phagosome that rapidly matures with the phagosomal membrane conjugated to the LC3 family proteins, ultimately leading to lysosomal fusion. LAP requires NADPH oxidase, which generates ROS during phagosomal maturation and kills *Mtb*. An exported *Mtb* protein, CpsA, is reported to prevent its lysosomal clearance by inhibiting LC3-associated lysosomal trafficking and preventing the recruitment of NADPH oxidase to mycobacterial phagosomes, thereby ceasing ROS production [13]. *Mtb* PknG enhances the intracellular survival of the pathogen by inhibiting autophagosome maturation, thereby blocking autophagic flux. PknG targets RAB14, a small host GTPase, and its interaction inhibits RAB14-GTP hydrolysis, which in turn blocks phagosome maturation, helping *Mtb* escape from the host's autophagic responses [14].

Rv2347c, an effector molecule of the *Mtb* T7SS, has been reported to prevent phagosome maturation by inhibiting the expression of the early phagosome marker RAB5A, thereby affecting its fusion with lysosomes. Lipid virulence factors of *Mtb,* such as SLs and PDIMs, modulate host autophagy at various levels in infected human macrophages. The *Mtb* strains mutant for the above-mentioned lipids showed activation of autophagy pathways in a MyD88-dependent manner, where SLs acted as TLR2 antagonists and PDIMs prevented acidification of autophagic compartments [15].

3.3 Necrosis: A process favoring *Mtb* survival

Necrosis is an unregulated and unprogrammed mode of cell death initiated by cytotoxic signals from physical or chemical damage. It leads to the swelling of cell organelles, resulting in the rupture of the plasma membrane and cell lysis. These events lead to the release of intracellular contents into the surrounding environment. Spilled components contribute to inflammatory responses by activating inflammasomes and secreting proinflammatory cytokines. Unlike apoptosis, which is a programmed means of cell death, necrosis does not involve events such as pyknosis (chromatin condensation) or karyorrhexis (nuclear fragmentation), nor does it affect the formation of autophagic vesicles. During infection, the onset of necrosis provides an advantage to bacteria, as it is released into the extracellular milieu and infects macrophages in the neighborhood. Virulent *Mtb* strains have been reported to interfere with the apoptosis of infected cells, which is a host-favorable process; instead, they are reprogrammed to undergo necrosis. In doing so, *Mtb* maintains a favorable host cell environment conducive to its persistence. It allows it to evade various apoptosis-related bactericidal effects and avoid the priming of naive T cells, leading to a delay in initiating cell-mediated immunity. In addition to helping in dissemination, necrosis has also been reported to provide a niche for *Mtb* replication, as inhibiting necrosis using specific inhibitors was observed to limit *Mtb* replication [16]. *Mtb* is also known to induce necrosis of human neutrophils in an ESX-1-dependent manner, thereby affecting the efficient removal of infected neutrophils from the lungs of patients with active TB. ROS generated by necrotic neutrophils were found to drive this process, as inhibition of ROS production promoted apoptosis, aiding in eliminating dying neutrophils and controlling *Mtb* infection [17].

3.3.1 Mtb *proteins influencing necrosis in host cells*

In human macrophages, FAK plays a role in restricting the intracellular survival of *Mtb,* predominantly by blocking macrophage cell death and producing ROS. In response, *Mtb* exploits and downregulates FAK in infected macrophages, induces necrotic cell death, and modulates the immunological profile of macrophages [18]. Some *Mtb* proteins also activate necrosis in the host macrophages. *Mtb* ESAT-6 has been reported to increase Ca^{2+} influx in infected neutrophils, activating calpains, a group of calcium activated cysteine proteases, resulting in necrotic cell death. *Mtb* PE17 (Rv1646) increased the intracellular survival of non-pathogenic *M. smegmatis* in infected macrophages and mice, along with downregulation of proinflammatory cytokines and induction of macrophage necrosis [19]. Similarly, the expression of PPE11 in *M. smegmatis* enhanced its survival within macrophages and increased the bacterial burden in a mouse infection model. Recombinant *M. smegmatis* expressing PPE11 also showed higher levels of necrosis in infected macrophages, as evidenced by increased lactate dehydrogenase release at different time intervals post-infection [20]. Induction of necrosis is one such strategy evolved by *Mtb* to evade cell death-mediated immune responses, helping the pathogen survive and disseminate throughout the host. Although necrosis is considered a passive and unprogrammed form of cell death, recent studies have suggested that specific signaling pathways can also regulate necrosis. This form of programmed necrosis is known as necroptosis [21]. Apart from necroptosis, *Mtb*-infected macrophages induce necrosis through various pathways, such as pyroptosis and ferroptosis, which are discussed below.

3.4 Necroptosis

Necroptosis is a form of regulated caspase-dependent necrosis that displays features similar to necrotic cell death. Necroptosis is primarily initiated during pathological conditions, such as infections, by the action of various death receptors, such as TNFR and Fas receptor [22]. Necroptosis is mediated by RIPKs and MLKL. Once the death receptors are engaged, RIPK1, a key player in the necroptosis pathway, is recruited to the death receptor signaling complexes and mediates death receptor signaling by activating the NFKB and MAPK pathways. NFKB induces the cFLIP, preventing death receptor-induced apoptosis [23]. RIPK1 undergoes a conformational change via autophosphorylation and recruits RIPK3, another vital member of the RIPK family. RIPK3 phosphorylates MLKL, which oligomerizes and ruptures the plasma membrane [22,24], leading to ion influx, cell swelling, and lysis, resulting in necrotic cell death.

Some *Mtb* proteins have been reported to be involved in activating the necroptotic pathway because of their beneficial roles in pathogen survival. In one study, it was demonstrated that the tuberculosis necrotizing toxin (TNT), produced by *Mtb* following phagocytosis, translocates into the host cell cytosol, where it depletes NAD^+. The resulting reduction in cytosolic NAD^+ levels is sensed by RIPK3, which subsequently phosphorylates MLKL. Phosphorylated MLKL oligomerizes and inserts into the plasma membrane to form pores, ultimately inducing necroptosis [25].

3.5 Pyroptosis

Pyroptosis is a caspase-dependent, unique mode of cell death that involves inflammation. It is mediated by the formation of an inflammasome complex induced by the activation of inflammasome sensors, such as the NLR family, cytosolic DNA sensor AIM2, and pyrin receptor, which bind to a variety of PAMPs and DAMPs released by pathogens [26]. NLRP3 is an inflammasome sensor that is vital in triggering inflammation. When these sensors and PAMPs interact, NLRs or AIM2 oligomerize and recruit ASC. This activates caspase 1, which cleaves the effector molecule, GSDMD, thereby promoting its oligomerization to form large pores in the plasma membrane, leading to cell death. GSDMD can be cleaved by inflammatory caspases other than caspase 1, such as caspase 4, caspase 5, and caspase 11, of which caspase 11 is present in rodents that lack caspase 4 and caspase 5. Caspase 1 cleaves GSDMD and activates the inflammatory cytokines IL-1B and IL-18 from their precursor forms. The production of these cytokines contributes to the host inflammatory responses produced by cells undergoing pyroptosis. The secretion of IL-1B and IL-18 is believed to occur through pores formed in the plasma membrane by caspase 1-mediated cleavage of GSDMD.

Mtb expressing a functional ESX-1 secretion system producing ESAT-6 was reported to be a prerequisite for both caspase 1 activation and IL-1B secretion and is therefore vital to inflammasome complex formation. *Mtb* mutants lacking ESAT-6 were observed to activate low levels of caspase 1 and secrete reduced levels of IL-1B compared with wild-type *Mtb*, which was restored upon complementation. *Mtb* can also inhibit the host inflammatory responses generated through IL-1B secretion by preventing NLRP3 inflammasome activation and subsequently inhibiting pyroptosis. The serine-threonine kinase PknF is associated with NLRP3 inflammasome inhibition. Infection of macrophages with an *MtbΔPknF* strain induced IL-1B secretion by increased caspase 1 activation, enhanced K^+ and Cl^- efflux, and increased production of ROS, thereby inducing pyroptosis (Fig. 10.1) [27,28]. The *Mtb* Hip1 also inhibits NLRP3 inflammation-mediated pyroptosis by dampening TLR2-mediated cell signaling and affecting the secretion of inflammatory cytokines involved in host defense responses via inflammasome activation [27].

3.6 Ferroptosis

Ferroptosis is an iron-dependent, nonapoptotic form of cell death distinct from other host cell death modalities. The trigger for ferroptosis is based on intracellular concentrations of labile iron and ROS generated by iron overload. ROS has a dual nature: limited ROS production causes oxidative stress. It helps in host defense against foreign pathogens, whereas excessive ROS production causes oxidative distress, which is harmful to the host [29]. The host antioxidant enzymes such as SOD, catalase, and glutathione peroxidase are responsible for neutralizing excess ROS. ROS levels exceed antioxidant capacity when cells experience oxidative distress due to iron overload, resulting in membrane damage and cell death [30]. Ferroptosis is also known to be induced by erastin, a small molecule that inhibits cystine import, thereby depleting glutathione levels within the system. Host GPX4 uses glutathione to convert toxic hydroperoxides into the corresponding alcohols, thereby reducing cell toxicity levels. Under glutathione-depleted conditions, GPX4 is inactivated, leading to increased lipid peroxidation. Lipid peroxidation, together with reduced GPX4 activity, causes damage to various cell components, leading to reduced cell volume and plasma membrane destabilization, resulting in cell death [31]. The use of iron chelators, such as PIH (pyridoxal isonicotinoyl hydrazone), can suppress iron-dependent ferroptosis in *Mtb*-infected macrophages. Similarly, lipid peroxidation inhibitors like ferrostatin-1 (Fer-1) have been shown to mitigate ferroptotic cell death. Treatment of *Mtb*-infected animals with Fer-1 significantly reduced both bacterial burden and granulomatous inflammation. Furthermore, expression of *Mtb* Rv1324 in *M. smegmatis* (Rv1324/Msm) has been reported to enhance ROS production and trigger oxidative stress responses. Mice infected with Rv1324/Msm exhibited increased lipid peroxidation, reduced GSH levels, and diminished GPX4 activity, hallmarks of ferroptosis. These effects were attenuated by Fer-1 treatment, indicating that Rv1324 promotes ferroptosis in a lipid peroxidation-dependent manner [31,32].

Some studies have also reported the possibility of overlap between ferroptosis and other modes of cell death. Lipid peroxidation, a triggering factor for ferroptosis in cells, was found to alleviate pyroptotic cell death. GPX4, which inhibits lipid peroxidation, negatively regulates pyroptosis in macrophages and prevents sepsis in mice. GPX4-deficient mice were observed to have increased lipid peroxidation and increased caspase 11-mediated GSDMD cleavage, leading to pyroptosis [33]. In another study, ferroptosis was shown to be induced by autophagy through the degradation of ferritin [34], an intracellular iron-storage protein complex. Moreover, since ferritin expression in cells prevents ferroptosis, its degradation by autophagy increases intracellular iron levels, leading to lipid peroxidation and induction of ferroptosis [31]. Hence, *Mtb*-induced autophagy under appropriate conditions can trigger ferroptotic macrophage death.

4. Therapeutic strategies targeting host cell death pathways in *Mtb* infection

Since *Mtb* manipulates host cell death mechanisms to establish infection, and the outcome of this host-pathogen battle determines the fate of the disease, targeting host cell death pathways can be considered a possible HDT strategy for TB treatment. The role of different host cell death modalities discussed in this chapter establishes that, while the host possesses the ability to induce specific modes of cell death to clear the pathogen, *Mtb* has evolved multiple ways to reprogram these pathways to favor its own survival. The development of small-molecule activators of proapoptotic proteins or inhibition of anti-apoptotic factors upregulated by *Mtb*, targeting effector molecules interfering with autophagy pathways, developing inhibitors of *Mtb*-induced necrotic cell death, and fine-tuning different processes can be considered possible therapeutic options to treat the disease.

Certain modifications to the existing TB vaccine, BCG, have been shown to confer increased protection against infection. A rBCG strain expressing Hly from *Listeria monocytogenes* was constructed, which additionally lacked mycobacterial UreC, a protein that neutralizes the phagosomal pH and inhibits phagosome maturation. Deletion of *UreC* maintains an acidic environment in phagosomes, enhancing the activity of Hly. Administration of the rBCG strain secreting Hly showed improved and prolonged protection against virulent *Mtb* H37Rv and Beijing/W-infected mice compared to parental BCG vaccination. The enhanced efficacy of *Hly*$^{+}$rBCG vaccination was attributed to increased mycobacterial antigen cross-presentation due to perforation of the phagosomal membrane by Hly and induced apoptosis, which led to priming of $CD8^+$ T cells, resulting in induced T-cell-mediated immunity. Moreover, the rBCGΔ*UreC hly*$^+$ vaccine candidate enhanced AIM2 inflammasome activation and autophagy in THP-1 macrophages [35]. Based on this approach, VPM1002, the vaccine candidate, has successfully completed phase I and phase IIa clinical trials, and phase III trials are in progress [36,37]. The same group deleted the anti-apoptotic gene *NuoG*, which encodes the NADH dehydrogenase I subunit, in a separate study from rBCGΔ*UreC hly*$^+$. This strain showed improved protection over the parental rBCG vaccine candidate in terms of reduced bacillary load in mice, improved lung pathology, and stronger elicitation of immune responses [38].

Along with apoptosis, autophagy is also the preferred response of the host against invading pathogens; therefore, targeting autophagy has emerged as a potential therapeutic strategy. Using rapamycin, an inhibitor of MTOR, autophagy was induced in macrophages, which upregulated Ag85B expression, leading to enhanced Ag85B-specific T-cell responses in vivo. Rapamycin-induced autophagy has also been reported to colocalize *Mtb* with autophagosomes, followed by their targeting to lysosomes. Administration of a recombinant BCG strain overexpressing Ag85B in mice led to a higher level of immunization compared to parental BCG in terms of lowered lung bacterial load up to 90 days post-infection and higher levels of effector memory immune response when challenged with the virulent *Mtb* Erdman strain. Another vaccine candidate, *Mtb* Δ*FbpA*Δ*SapM*, lacks both *FbpA* and *SapM* encoding Ag85B and an acid phosphatase that dephosphorylates phosphatidyl-inositol-3-phosphate on the phagosome membrane, respectively, and has been shown to be highly immunogenic in vitro and in vivo. The DKO candidate expressed lysosomal markers, proving its localization to lysosomes, and showed enhanced expression of autophagy-inducing genes compared to WT *Mtb* H37Rv. The enhanced autophagy in infected macrophages and dendritic cells led to increased antigen presentation to T cells, and in comparison to BCG, the DKO candidate elicited stronger and long-lasting protection against *Mtb* Erdman infection, demonstrating a new approach for TB immunization [39].

Apart from the above-described approaches, various molecules and pharmacological agents have been shown to specifically regulate cell death mechanisms and elicit antimicrobial defense responses against *Mtb* infection. The antidiabetic drug metformin abrogated the intracellular growth of *Mtb* H37Rv in human monocyte-derived macrophages (MDMs) by activating the PRKAA2/AMPK pathway. Along with the induction of autophagy and intrinsic apoptotic pathways, the drug also enhanced mitochondrial ROS production and phagosome-lysosome fusion. In addition, metformin was reported to increase the efficacy of the anti-TB drug isoniazid. The HDAC inhibition using specific inhibitors reduces intracellular *Mtb* survival in both M1 and M2 macrophages. In addition, macrophages treated with HDAC inhibitors during differentiation from monocytes showed an increased intrinsic capacity to control *Mtb* survival [40].

The tumor suppressor gene p53 has been reported to modulate different apoptotic pathways by regulating the transcription of its target genes. Macrophages lacking p53 show increased intracellular *Mtb* survival and suppressed activation of caspase proteins. Nutlin-3, a p53 activator molecule, enhances apoptosis by increasing ROS and NO production and enhancing proinflammatory cytokine expression [41]. Baicalein from *Scutellaria radix* inhibits pyroptosis in *Mtb*-infected macrophages by disrupting the assembly of AIM2 and NLRP3 inflammasomes. The expression levels of different proteins involved in inflammasome-mediated cell death, such as NLRP3, AIM2, and ASC, and their interactions were reduced in baicalein-treated cells. Baicalein was also reported to induce autophagy in

Mtb-infected macrophages by inactivating the Akt-MTOR pathway, a negative regulator of autophagy, thereby providing a potential new adjunct HDT drug [42].

5. Conclusions and perspectives

Although there have been advancements in HDTs targeting both host effector molecules and *Mtb* virulence factors regulating cell death mechanisms, many technical challenges still exist in exploiting host cell death modalities to treat *Mtb* infection. The primary hurdle is to capture the precise events that characterize the non-classical modes of cell death discovered in recent years, such as pyroptosis, ferroptosis, and necroptosis. Moreover, some cell death pathways involve severe tissue damage; therefore, it is essential to balance the extent of induced host cell death to promote bacterial clearance from the system while employing novel HDTs. To target host cell death mechanisms, it is critical to ensure specificity and avoid the disruption of normal cellular processes. HDTs can also be combined with conventional anti-TB drugs to yield synergistic effects in TB treatment. The PE/PPE multigene family in mycobacteria can also be used as potential vaccine targets because the link between these proteins and *Mtb* pathogenesis has been well documented, and they contribute widely to regulating immune responses and modulation of cell death. Since reprogramming host cell death mechanisms has proven to be a potent strategy for *Mtb* to subvert host defense responses, there is an urgent need to gain a deeper understanding of *Mtb* pathogenesis in relation to host cell death. Doing so will enable the development of a new class of treatments against TB, adding to the arsenal that we desperately need to eradicate this scourge, which has plagued humankind for millennia.

Declaration

We declare that the table and figure used in the manuscript are original (created for this book chapter).

Acknowledgments

We acknowledge funding support from the Council of Scientific and Industrial Research, Government of India (CSIR) (to T. R. R.) (RDSF-IHP240001). K. C. is supported by a Senior Research Fellowship from the CSIR.

References

[1] Yang S, Li F, Jia S, Zhang K, Jiang W, Shang Y, Chang K, Deng S, Chen M. Early secreted antigen ESAT-6 of *Mycobacterium tuberculosis* promotes apoptosis of macrophages via targeting the MicroRNA155-SOCS1 interaction. Cell Physiol Biochem 2015;35(4):1276–88. https://doi.org/10.1159/000373950.

[2] Tiwari B, Ramakrishnan UM, Raghunand TR. The *Mycobacterium tuberculosis* protein pair PE9 (Rv1088)-PE10 (Rv1089) forms heterodimers and induces macrophage apoptosis through toll-like receptor 4. Cell Microbiol 2015;17(11):1653–69. https://doi.org/10.1111/cmi.12462.

[3] Mohareer K, Asalla S, Banerjee S. Cell death at the cross roads of host-pathogen interaction in *Mycobacterium tuberculosis* infection. Tuberculosis 2018;113:99–121. https://doi.org/10.1016/j.tube.2018.09.007.

[4] Vu A, Glassman I, Campbell G, Yeganyan S, Nguyen J, Shin A, Venketaraman V. Host cell death and modulation of immune response against *Mycobacterium tuberculosis* infection. Int J Mol Sci 2024;25(11):6255. https://doi.org/10.3390/ijms25116255.

[5] Kale J, Osterlund EJ, Andrews DW. BCL-2 family proteins: changing partners in the dance towards death. Cell Death Differ 2018;25(1):65–80. https://doi.org/10.1038/cdd.2017.186.

[6] Nisa A, Kipper FC, Panigrahy D, Tiwari S, Kupz A, Subbian S. Different modalities of host cell death and their impact on *Mycobacterium tuberculosis* infection. Am J Physiol Cell Physiol 2022;323(5):C1444. https://doi.org/10.1152/AJPCELL.00246.2022.

[7] Zhang W, Lu Q, Dong Y, Yue Y, Xiong S. Rv3033, as an emerging anti-apoptosis factor, facilitates mycobacteria survival via inhibiting macrophage intrinsic apoptosis. Front Immunol 2018;9. https://doi.org/10.3389/fimmu.2018.02136.

[8] Castrejón-Jiménez NS, Leyva-Paredes K, Hernández-González JC, Luna-Herrera J, García-Pérez BE. The role of autophagy in bacterial infections. BioScience Trends 2015;9(3):149–59. https://doi.org/10.5582/bst.2015.01035.

[9] Zihad SMNK, Sifat N, Islam MA, Monjur-Al-Hossain ASM, Yasif Kayes Sikdar KM, Sarker Md MR, Shilpi JA, Uddin SJ. Role of pattern recognition receptors in sensing *Mycobacterium tuberculosis*. Heliyon 2023;9(10):e20636. https://doi.org/10.1016/j.heliyon.2023.e20636.

[10] Paulus GLC, Xavier RJ. Autophagy and checkpoints for intracellular pathogen defense. Curr Opin Gastroenterol 2015;31(1):14–23. https://doi.org/10.1097/MOG.0000000000000134.

[11] Chai Q, Wang L, Liu CH, Ge B. New insights into the evasion of host innate immunity by *Mycobacterium tuberculosis*. Cell Mol Immunol 2020;17(9):901–13. https://doi.org/10.1038/s41423-020-0502-z.

[12] Dong H, Wu J, Runpeng Z, Xuewei X, Mu M, Ru C, Xing Y, Ni S, Zhang R. ESAT6 inhibits autophagy flux and promotes BCG proliferation through MTOR. Biochem Biophys Res Commun 2016;477(2):195–201. https://doi.org/10.1016/j.bbrc.2016.06.042.

[13] Köster S, Upadhyay S, Chandra P, Papavinasasundaram K, Yang G, Hassan A, Grigsby SJ, Mittal E, Park HS, Jones V, Hsu F-F, Jackson M, Sassetti CM, Philips JA. *Mycobacterium tuberculosis* is protected from NADPH oxidase and LC3-associated phagocytosis by the LCP protein CpsA. Proc Natl Acad Sci 2017;114(41). https://doi.org/10.1073/pnas.1707792114.

[14] Ge P, Lei Z, Yang Y, Lu Z, Qiang L, Chai Q, Zhang Y, Zhao D, Li B, Pang Y, Liu CH, Jing Wang M. Tuberculosis PknG manipulates host autophagy flux to promote pathogen intracellular survival. Autophagy 2022;18(3):576–94. https://doi.org/10.1080/15548627.2021.1938912.

[15] Bah A, Sanicas M, Nigou J, Guilhot C, Astarie-Dequeker C, Vergne I. The lipid virulence factors of *Mycobacterium tuberculosis* exert multilayered control over autophagy-related pathways in infected human macrophages. Cells 2020;9(3):666. https://doi.org/10.3390/cells9030666.

[16] Lerner TR, Borel S, Greenwood DJ, Repnik U, Russell MRG, Herbst S, Jones ML, Collinson LM, Griffiths G, Gutierrez MG. *Mycobacterium tuberculosis* replicates within necrotic human macrophages. J Cell Biol 2017;216(3):583–94. https://doi.org/10.1083/jcb.201603040.

[17] Dallenga T, Repnik U, Corleis B, Eich J, Reimer R, Griffiths GW, Ulrich E, Schaible M. Tuberculosis-induced necrosis of infected neutrophils promotes bacterial growth following phagocytosis by macrophages. Cell Host Microbe 2017;22(4):519. https://doi.org/10.1016/j.chom.2017.09.003.

[18] Afriyie-Asante A, Dabla A, Dagenais A, Berton S, Smyth R, Sun J. *Mycobacterium tuberculosis* exploits focal adhesion kinase to induce necrotic cell death and inhibit reactive oxygen species production. Front Immunol 2021;12. https://doi.org/10.3389/fimmu.2021.742370.

[19] Li Z, Liu H, Li H, Dang G, Cui Z, Song N, Wang Q, Liu S, Chen L. PE17 protein from *Mycobacterium tuberculosis* enhances Mycobacterium smegmatis survival in macrophages and pathogenicity in mice. Microb Pathog 2019;126:63–73. https://doi.org/10.1016/j.micpath.2018.10.030.

[20] Peng X, Luo T, Zhai X, Zhang C, Suo J, Ma P, Wang C, Bao L. PPE11 of *Mycobacterium tuberculosis* can alter host inflammatory response and trigger cell death. Microb Pathog 2019;126:45–55. https://doi.org/10.1016/j.micpath.2018.10.031.

[21] Chan FKM, Luz NF, Moriwaki K. Programmed necrosis in the cross talk of cell death and inflammation. Annu Rev Immunol 2015;33:79–106. https://doi.org/10.1146/annurev-immunol-032414-112248.

[22] Liu XQ, Jiao Z, Gao YC, Zheng XY, Huang H. Progress in development and application of decision-making systems for individualized dosing. Chin Pharmaceut J 2019;54(1):1–8. https://doi.org/10.11669/cpj.2019.01.001.

[23] Tsuchiya Y, Nakabayashi O, Nakano H. FLIP the switch: regulation of apoptosis and necroptosis by cFLIP. Int J Mol Sci 2015;16(12):30321–41. https://doi.org/10.3390/ijms161226232.

[24] Green DR. The coming decade of cell death research: five riddles. Cell 2019;177(5):1094–107. https://doi.org/10.1016/j.cell.2019.04.024.

[25] Pajuelo D, Gonzalez-Juarbe N, Tak U, Sun J, Orihuela CJ, Niederweis M. NAD(+) depletion triggers macrophage necroptosis, a cell death pathway exploited by *Mycobacterium tuberculosis*. Cell Rep 2018;24(2).

[26] Bertheloot D, Latz E, Franklin BS. Necroptosis, pyroptosis and apoptosis: an intricate game of cell death. Cell Mol Immunol 2021;18(5):1106–21. https://doi.org/10.1038/s41423-020-00630-3.

[27] Rastogi S, Ellinwood S, Augenstreich J, Mayer-Barber KD, Briken V, Behr MA. *Mycobacterium tuberculosis* inhibits the NLRP3 inflammasome activation via its phosphokinase PknF. PLoS Pathog 2021;17(7):e1009712. https://doi.org/10.1371/journal.ppat.1009712.

[28] Rastogi S, Briken V. Interaction of mycobacteria with host cell inflammasomes. Front Immunol 2022;13. https://doi.org/10.3389/fimmu.2022.791136.

[29] Herb M, Schramm M. Functions of ROS in macrophages and antimicrobial immunity. Antioxidants 2021;10(2):313. https://doi.org/10.3390/antiox10020313.

[30] Stockwell BR, Friedmann Angeli JP, Bayir H, Bush AI, Conrad M, Dixon SJ, Fulda S, Gascón S, Hatzios SK, Kagan VE, Noel K, Jiang X, Linkermann A, Murphy ME, Overholtzer M, Oyagi A, Pagnussat GC, Park J, Ran Q, Rosenfeld CS, Salnikow K, Tang D, Torti FM, Torti SV, Toyokuni S, Woerpel KA, Zhang DD. Ferroptosis: a regulated cell death nexus linking metabolism, redox biology, and disease. Cell 2017;171(2):273–85. https://doi.org/10.1016/j.cell.2017.09.021.

[31] Amaral EP, Costa DL, Namasivayam S, Riteau N, Kamenyeva O, Mittereder L, Mayer-Barber KD, Andrade BB, Sher A. A major role for ferroptosis in *Mycobacterium tuberculosis*–induced cell death and tissue necrosis. J Exp Med 2019;216(3):556–70. https://doi.org/10.1084/jem.20181776.

[32] Shi X, Li C, Lin C, Ullah H, Sha S, Kang J, Ma X, Ma Y, Singh A. *Mycobacterium tuberculosis* Rv1324 protein contributes to mycobacterial persistence and causes pathological lung injury in mice by inducing ferroptosis. Microbiol Spectr 2023;11(1). https://doi.org/10.1128/spectrum.02526-22.

[33] Kang R, Zeng L, Zhu S, Xie Y, Liu J, Wen Q, Cao L, Xie M, Ran Q, Kroemer G, Wang H, Billiar TR, Jiang J, Tang D. Lipid peroxidation drives gasdermin D-mediated pyroptosis in lethal polymicrobial sepsis. Cell Host Microbe 2018;24(1):97. https://doi.org/10.1016/j.chom.2018.05.009.

[34] Hou W, Xie Y, Song X, Sun X, Lotze MT, Zeh HJ, Kang R, Tang D. Autophagy promotes ferroptosis by degradation of ferritin. Autophagy 2016;12(8):1425–8. https://doi.org/10.1080/15548627.2016.1187366.

[35] Saiga H, Nieuwenhuizen N, Gengenbacher M, Koehler AB, Schuerer S, Moura-Alves P, et al. The recombinant BCG DeltaureC::hly vaccine targets the AIM2 inflammasome to induce autophagy and inflammation. J Infect Dis 2015;211(11).

[36] Kaufmann SHE. Vaccination against tuberculosis: revamping BCG by molecular genetics guided by immunology. Front Immunol 2020;11. https://doi.org/10.3389/fimmu.2020.00316.

[37] Loxton AG, Knaul JK, Grode L, Gutschmidt A, Meller C, Eisele B, Johnstone H, Van Der Spuy G, Maertzdorf J, Kaufmann SHE, Hesseling AC, Walzl G, Cotton MF. Safety and immunogenicity of the

recombinant mycobacterium bovis BCG vaccine VPM1002 in HIV-unexposed newborn infants in South Africa. Clin Vaccine Immunol 2017;24(2). https://doi.org/10.1128/CVI.00439-16.

[38] Gengenbacher M, Nieuwenhuizen N, Vogelzang A, Liu H, Kaiser P, Schuerer S, Lazar D, Wagner I, Mollenkopf H-J, Kaufmann SHE, Rappuoli R, Sher A, Goldfeld A, Stenger S, Manganelli R. Deletion of nuoG from the vaccine candidate Mycobacterium bovis BCG Δ ureC :: hly improves protection against tuberculosis. mBio 2016;7(3). https://doi.org/10.1128/mbio.00679-16.

[39] Mishra A, Khan A, Singh VK, Glyde E, Saikolappan S, Garnica O, et al. The DeltafbpADeltasapM candidate vaccine derived from *Mycobacterium tuberculosis* H37Rv is markedly immunogenic in macrophages and induces robust immunity to tuberculosis in mice. Front Immunol 2024;15.

[40] Moreira JD, Koch BEV, van Veen S, Walburg KV, Vrieling F, Mara Pinto Dabés Guimarães T, Meijer AH, Spaink HP, Ottenhoff THM, Haks MC, Heemskerk MT. Functional inhibition of host histone deacetylases (HDACs) enhances in vitro and in vivo anti-mycobacterial activity in human macrophages and in zebrafish. Front Immunol 2020;11. https://doi.org/10.3389/fimmu.2020.00036.

[41] Lim Y-J, Lee J, Choi J-A, Cho S-N, Son S-H, Kwon S-J, Son J-W, Song C-H. M1 macrophage dependent-p53 regulates the intracellular survival of mycobacteria. Apoptosis 2020;25(1–2):42–55. https://doi.org/10.1007/s10495-019-01578-0.

[42] Ning B, Shen J, Liu F, Zhang H, Jiang X, Prisic S. Baicalein suppresses NLRP3 and AIM2 inflammasome-mediated pyroptosis in macrophages infected by *Mycobacterium tuberculosis* via induced autophagy. Microbiol Spectr 2023;11(3). https://doi.org/10.1128/spectrum.04711-22.

SECTION

Therapeutic frontiers and drug repurposing

III

CHAPTER

Repurposing potential of cell wall modulating drugs: A new frontier in TB management

11

Sana Jamily[1], Sarah Hamid[1], Faizan Ansari[1], Vikas Kumar[2] and Sonam Grover[1]

[1]*Department of Molecular Medicine, School of Interdisciplinary Sciences and Technology, Jamia Hamdard, New Delhi, India;* [2]*Department of Biotechnology, Indira Gandhi National Tribal University, Amarkantak, Madhya Pradesh, India*

Highlights

1. Tuberculosis (TB) remains a major global health concern due to the rise of drug-resistant strains.
2. The complex cell wall of *Mycobacterium tuberculosis* (*Mtb*) contributes to its virulence and antibiotic resistance.
3. Repurposing cell wall-modulating drugs offers a novel approach to treating TB.
4. β-lactams, glycopeptides, and sulfonamides demonstrate significant antimycobacterial activity by targeting cell wall synthesis.
5. Drug repurposing accelerates the development of new TB treatments by utilizing existing safety data.
6. Challenges in repurposing cell wall-targeting drugs include potential resistance, pharmacokinetic limitations, and toxicity.
7. Combining repurposed drugs with existing TB therapies may enhance treatment effectiveness and reduce duration.

1. Introduction

TB, caused by the bacterium *Mtb*, remains one of the most lethal infectious diseases globally. Despite continuous advancements in medical research and treatment modalities, TB persists as a global health crisis, accounting for millions of deaths annually [1]. The bacterium's ability to adapt to its environment and persist within human hosts is largely attributed to its complex and highly protective cell wall [2]. This distinctive structure renders the bacterium inherently resistant to a wide range of environmental stresses and challenges the efficacy of conventional antibiotics, thereby complicating treatment. The standard TB treatment regimen, known as Directly Observed Treatment Short-course (DOTS), comprises a combination of several drugs, which must be administered over a 6-month period [3]. DOTS is effective in treating drug-susceptible TB. However, the prolonged treatment duration, severe side effects, and risk of noncompliance have led to the emergence of drug-resistant strains. Multidrug-resistant (MDR) and extensively drug-resistant (XDR) TB have become significant public

Navigating Tuberculosis. https://doi.org/10.1016/B978-0-443-27735-1.00002-2

health concerns, underscoring the necessity for novel treatment strategies [4]. The rise of MDR-TB presents an urgent challenge to global TB control, as standard therapies become ineffective [5]. The development of resistance to first-line drugs such as rifampicin (RIF) and isoniazid (INH) highlights the need for more effective, targeted therapeutic options [4].

The development of new drugs for TB treatment is a protracted and costly endeavor. New drug candidates, such as bedaquiline and delamanid, have only recently been authorized following years of research and clinical trials [6]. The lengthy development timelines and high costs associated with new drug discovery pose significant barriers to timely TB treatment solutions. This slow progress underscores the need for alternative strategies, one of which is drug repurposing. Drug repurposing involves identifying existing drugs that are already approved for other diseases and assessing their potential efficacy against TB [7]. The advantage of this approach is that it circumvents the lengthy and costly drug discovery process, facilitating faster implementation of treatment options. Repurposing drugs offers a valuable opportunity to address the complex challenges associated with TB treatment. This aspect is particularly crucial in TB treatment, where the emergence of resistance to single-drug treatments is increasingly problematic.

The cell wall is essential for bacterial survival, virulence, and resistance to antibiotics. The cell wall is composed of complex molecules, including mycolic acids, peptidoglycans (PGs), and arabinogalactans (AGs), which collectively form a robust, waxy barrier that protects the bacterium from host immune responses and numerous antibiotics. Targeting the enzymes responsible for synthesizing these components presents a promising therapeutic strategy. Several existing drugs, initially developed for other infections, have shown potential in disrupting the mycobacterial cell wall. Glycopeptides such as vancomycin, traditionally used to treat Gram-positive bacterial infections, have demonstrated promise in modulating the cell wall and enhancing the efficacy of TB drugs [8]. The mycobacterial cell wall remains a critical focus for drug development, with numerous candidates targeting its biosynthesis currently under investigation. Recent advancements in our understanding of the molecular processes underlying cell wall production have facilitated novel drug design. For instance, enzymes such as DprE1, InhA, and MmpL3 have emerged as key drug targets, with inhibitors of these enzymes showing considerable promise in preclinical studies [9]. In this chapter, we explore the potential of cell wall-modulating drugs in TB treatment, emphasizing repurposed drugs and their mechanisms of action. Additionally, we examine advancements in drug delivery technologies, such as nano-formulations, which enhance the bioavailability and effectiveness of these treatments.

2. The mycobacterial fortress: Structural and functional insights into the *Mycobacterium tuberculosis* cell wall

The cell wall of *Mtb* is a critical determinant of its pathogenicity and resistance, characterized by a complex structure comprising three primary layers: the plasma membrane, the cell wall core, and the outermost capsule [10]. First comprehensively described in 1995 [11], the cell wall features a thick peptidoglycan (PG) layer covalently linked to AG, which anchors long-chain mycolic acids, forming a rigid, waxy barrier that constitutes approximately 60% of the cell's mass and imparts high hydrophobicity [12]. The AG layer significantly influences the shape and impermeability of the cell wall [13]; while the outer membrane incorporates solvent-extractable lipids, such as waxes and non-covalently linked glycolipids, intercalated into the mycolate layer [14]. The outermost capsule is primarily composed of proteins and carbohydrates [15]. This multilayered envelope functions not only as a structural defense but also as a dynamic immunomodulatory interface. Key lipids, including sulfoglycolipids and phthiocerol dimycocerosates, inhibit host immune detection by targeting toll-like receptor 2 and concealing pathogen-associated molecular patterns [12]. Additionally, various cell wall components interfere with antigen processing and presentation via major histocompatibility complex class I and II pathways and can suppress innate immune mechanisms such as autophagy [16]. Collectively, the lipid- and carbohydrate-rich composition of the *Mtb* cell wall creates a formidable permeability barrier and supports immune evasion, rendering its biosynthesis an attractive target for therapeutic intervention [17].

The *Mtb* cell wall derives its structural integrity and pathogenicity from three essential components: PG, AG, and mycolic acids. PG forms the innermost layer, consisting of long glycan chains composed of alternating sugars cross-linked by short peptide side chains, and is synthesized outside the cell membrane by enzymes such as the bifunctional penicillin-binding proteins (PBPs) PonA1 and PonA2, which both polymerize glycan strands and cross-link peptides [18,19]. A hallmark of mycobacterial PG is the prevalence of 3 → 3 cross-links between the third amino acids of stem peptides, formed by L,D-transpeptidases like LdtB, in addition to the conventional 4 → 3 cross-links catalyzed by PBPs [20,21]. AG, a linear galactan made of alternating β-(1 → 5)- and β-(1 → 6)-linked D-galactofuranose residues, is appended with branched arabinan chains containing up to 31 D-arabinofuranose units. This component plays a critical role in cell wall integrity, transmission, and is a valuable target for anti-TB drug development [11,22]. Mycolic acids, the hallmark outermost lipids, are long-chain β-hydroxy fatty acids essential for *Mtb* survival and are classified into α-, methoxy-, and keto-mycolic acids. α-mycolic acids are the most abundant (>70%) and include cis,cis-dicyclopropyl structures, while methoxy- and keto-mycolates form minor components (10%–15%) [23–26]. These lipid layers contribute to the bacterium's impermeability, virulence, and resistance, highlighting their importance in TB pathogenesis and drug targeting (Fig. 11.1).

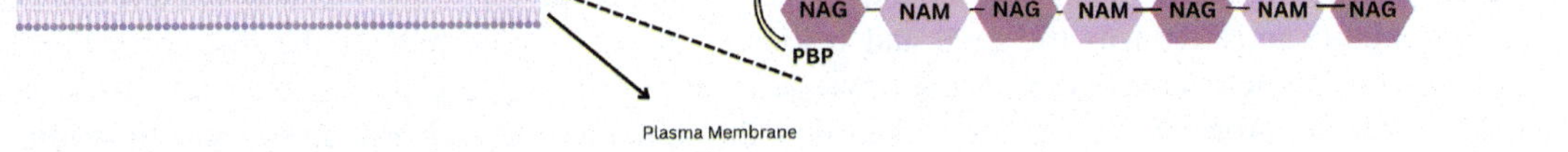

FIGURE 11.1 Tripartite Architecture of the *Mtb* Cell Wall.

The figure illustrates the structure of *Mtb* cell wall has three layers: the plasma membrane, peptidoglycan connected to arabinogalactan, and an outer layer of mycolic acids called the mycomembrane. Key components include lipoarabinomannan (LAM), phospholipids, and capsular polysaccharides. This intricate structure helps the bacteria withstand drugs and evade the immune system.

The *Mtb* cell wall is a structurally intricate and biochemically unique barrier that plays a central role in significantly limiting the entry of hydrophilic antibiotics and enhancing the bacterium's survival within host macrophages [13,23]. Embedded within this cell wall are several critical glycolipids, phosphatidyl-myo-inositol mannosides (PIMs), lipomannan (LM), and lipoarabinomannan (LAM), which are noncovalently attached to the inner membrane and potentially the outer membrane, and function as major virulence factors [27]. In PIMs, the acylated phosphatidylinositol anchor is decorated with up to six α-D-mannopyranosyl residues, while LM and LAM extend this mannose core with a branched arabinan domain, contributing to immune modulation [27,28]. These molecules, particularly LAM, interfere with macrophage activation and cytokine production, thereby dampening the host immune response [29]. Moreover, the cell wall houses enzymes such as DprE1 and InhA, essential for cell wall biosynthesis and validated targets for anti-TB drug development [23]. Altogether, the *Mtb* cell envelope not only fortifies the pathogen against environmental and host-derived threats but also actively facilitates immune evasion, establishing it as a pivotal focus in TB research and therapy.

3. Breaking the resistance: Challenges and strategies in combating drug-resistant tuberculosis

One of the primary challenges in the treatment of TB is the increasing prevalence of drug-resistant *Mtb*. Multidrug-resistant TB (MDR-TB) arises when the bacteria develop resistance to at least INH and RIF, the two most potent first-line medications [11]. The subsequent stage, XDR-TB, involves resistance to fluoroquinolones and at least one of the injectable second-line drugs. These forms of TB necessitate prolonged treatment durations and the use of more toxic drugs, rendering them both costly and difficult to manage [10]. The emergence of antibiotic resistance is linked to factors such as inadequate healthcare systems and suboptimal treatment plans. Antibiotics targeting actively dividing cells are less effective against *Mtb* due to its ability to enter a latent state [12]. These challenges demonstrate the urgent need for novel therapeutic strategies that can overcome drug resistance, reduce treatment durations, and enhance patient adherence.

4. Old drugs, new tricks: The promise of drug repurposing in tuberculosis therapy

The development of a novel therapeutic agent for TB is an extensive and costly endeavor, typically requiring 15–20 years before a new drug becomes commercially available. Drug repurposing has emerged as a promising alternative, addressing the urgent demand for effective therapies for this disease. By utilizing drugs that have already received approval for other medical conditions, this approach significantly reduces the time and resources typically required for the discovery of new treatments [13]. These preexisting drugs possess an established safety profile. Several drugs with potential anti-TB properties have been identified through this method. For instance, clofazimine, initially developed for leprosy, has demonstrated considerable efficacy against MDR-TB due to its ability to interfere with the synthesis of the *Mtb* cell wall [14]. Similarly, linezolid, an antibiotic originally designed for gram-positive bacterial infections, has been repurposed for the treatment of XDR-TB.

The significance of drug repurposing also extends to indirectly targeting the bacterium. Other repurposed drugs, such as metformin and statins, are known to modulate the host's immunological response, thereby enhancing the body's ability to combat TB [13]. This dual approach, which combines host-directed and pathogen-targeted therapies, provides a comprehensive strategy for TB treatment. In summary, while TB remains a significant global health concern, advancements in understanding *Mtb* biology and the potential of drug repurposing may lead to improved treatment outcomes. Focusing on the bacterium's cell wall and employing existing drugs could open new avenues for effective and accessible TB treatments [15].

5. Breaking the barrier: Repurposing cell wall inhibitors to combat drug-resistant tuberculosis

Mtb is the most common infectious agent that causes death worldwide. It additionally acts as a hydrophobic barrier to protect the host defenses and the drug. The bacteria can enter new host macrophages through sputum [17]. By regulating their immune response, it harms the host cells. For this reason, the *Mtb* cell envelope is a desirable target for vaccine and drug development. Effective treatment of the disease was affected by co-infection with human immune deficiency virus and the emergence of MDR-TB. Thus, the study and creation of new drugs is one of the primary foundations of TB treatment [16]. Gene products in *Mtb* that control vital aspects of mycobacterial structure and metabolism, including the formation of cell wall, DNA replication, RNA synthesis, protein synthesis, energy, and folate metabolism, constitute promising targets for treatment. The low permeability of the cell wall is essential for mycobacteria survival within the host because mycobacterium porin proteins are ineffective at permitting hydrophilic solutes and antimicrobials to enter the cytoplasm via the cell wall [18]. To reduce the length of therapy for the increasing number of XDR and MDR-TB patients worldwide, more effective drugs must be developed. Promising treatment targets that are essential for *Mtb* development and do not have analogues in mammalian systems are provided by the production pathways of the cell wall components [19]. These targets provide the basis of existing drugs the potential to create new anti-TB drugs [20].

Repurposing existing pharmacological agents to target the *Mtb* cell wall has emerged as a promising strategy to address drug-resistant TB. Enzymes such as DprE1, involved in AG biosynthesis, and InhA, which catalyzes the formation of mycolic acids, are crucial for maintaining cell wall integrity and have been validated as drug targets [21]. Inhibitors of DprE1 disrupt the synthesis of key carbohydrate structures, while agents targeting InhA, such as INH, prevent the formation of the protective lipid layer. Additionally, MmpL3, a transporter responsible for exporting mycolic acid precursors to the outer membrane, is essential for cell wall assembly and has been effectively targeted by new drug candidates [21] (Table 11.1). Several repurposed drugs have demonstrated efficacy in this context. β-lactams, including carbapenems and amoxicillin when combined with clavulanic acid, overcome β-lactamase-mediated resistance and inhibit PG synthesis [30]. Glycopeptides like vancomycin, although primarily effective against Gram-positive bacteria, have shown limited but notable activity against *Mtb* by targeting PG. Sulfonamides such as sulfadiazine inhibit folate biosynthesis, thereby indirectly impairing cell wall precursor formation [22]. However, despite these therapeutic advances, the impermeability of the *Mtb* cell wall, attributable to the AG layer, and the action of efflux pumps further complicate drug delivery [23]. The bacterium's slow replication rate and ability to

Table 11.1 Enzymes involved in cell wall biosynthesis as druggable targets.

Drugs class/compound	Target in cell wall biosynthesis	Mechanism of action	Current status
Isoniazid (INH)	INH (fatty acid synthase II)	Inhibits mycolic acid synthesis, essential for cell wall integrity.	Frontline TB-drug
Ethambutol (Emb)	Arabinosyltransferase (EmbA, EmbB)	Blocks arabinogalactan biosynthesis.	Frontline TB-drug
Rifampicin	RNA polymerase	Inhibits bacterial RNA synthesis, indirectly affecting cell wall processes.	Frontline TB-drug

enzymatically inactivate drugs further diminish treatment efficacy [17]. Consequently, repurposed drugs that interfere with cell wall synthesis not only disrupt the physical barrier protecting *Mtb* but also enhance the pathogen's susceptibility to immune clearance and concurrent therapies, offering a vital route toward shortening treatment duration and overcoming resistance [7].

There are two primary classes of antibiotics that inhibit bacterial cell wall synthesis: β-lactams and glycopeptide antibiotics. Both classes function by obstructing the production of PG, a crucial component of the bacterial cell wall [24]. Despite their shared target, they employ distinct mechanisms, β-lactams act directly, whereas glycopeptides exert their effects indirectly. All β-lactam antibiotics, including penicillin, cephalosporins, carbapenems, and monobactams, possess a characteristic structure known as a β-lactam ring, which consists of three carbon atoms and one nitrogen atom. The term "β-lactam" is derived from the bond between the nitrogen atom and the β-carbon atom within the ring. Generally, β-lactam antibiotics function by binding to specific bacterial proteins termed PBPs. These proteins, which include enzymes such as D, D-transpeptidases and D, D-carboxypeptidases, facilitate the cross-linking of PG chains, thereby fortifying the bacterial cell wall. By irreversibly binding to these PBPs, β-lactams inhibit the formation of these cross-links, ultimately impairing the proper construction of the bacterial cell wall [25].

The chemical structure of β-lactam antibiotics exhibits variations that influence their binding affinity to different PBPs. This structural diversity impacts their antimicrobial efficacy and the spectrum of bacterial targets. As previously noted, β-lactam antibiotics function by inhibiting PBPs, which are enzymes responsible for cross-linking peptide chains during the terminal phase of PG synthesis [26]. This process is crucial for maintaining the integrity and stability of the bacterial cell wall. The bactericidal action of β-lactams is attributed to their structural resemblance to the PG precursor, particularly the D-alanyl-D-alanine group (D-Ala-D-Ala). At the molecular level, β-lactams target PBPs by covalently binding to a critical serine residue via their β-lactam ring, forming a stable acyl-enzyme complex. This interaction results in the opening of the β-lactam ring, leading to irreversible enzyme inactivation [27] Consequently, PG synthesis is inhibited, compromising the bacterial cell wall and ultimately causing cell death. Recent developments in β-lactam antibiotics, such as cefiderocol and ceftobiprole, offer promising strategies against resistance to conventional β-lactams. Cefiderocol, a member of the novel siderophore cephalosporins, targets Gram-negative bacteria by primarily binding to PBP-3. Its unique structure, featuring a catechol group, facilitates iron binding and enables the antibiotic to penetrate the bacterial cell via active transport [28]. This mechanism circumvents common resistance mechanisms, such as reduced porin expression or increased efflux pump activity. Ceftobiprole, another novel β-lactam, exhibits potent bactericidal activity by binding to multiple PBPs in both Gram-positive and Gram-negative bacteria. It forms a highly stable inhibitory complex with resistant PBPs, such as PBP 2′ (PBP 2a) and PBP 2x, which are prevalent in beta-lactam-resistant staphylococci and streptococci. Its stability and extended side chain enhance its efficacy in inhibiting bacterial cell wall synthesis [29].

Glycopeptide antibiotics, including vancomycin, teicoplanin, oritavancin, and telavancin, represent another critical class of cell wall-targeting agents [31]. Unlike β-lactams, glycopeptides exert their effect by binding to the D-Ala-D-Ala terminus of PG precursors in Gram-positive bacteria. For instance, vancomycin binds to Lipid II, a precursor in PG synthesis, through a specific binding pocket in its molecular structure [32]. This interaction prevents PBPs from incorporating Lipid II into the mature PG layer, thereby disrupting the formation of the bacterial cell wall. The strong binding is stabilized by five hydrogen bonds, ensuring high affinity between vancomycin and its target. Consequently, the bacterial cell wall is compromised, leading to osmotic shock and cell death [33] (Fig. 11.2).

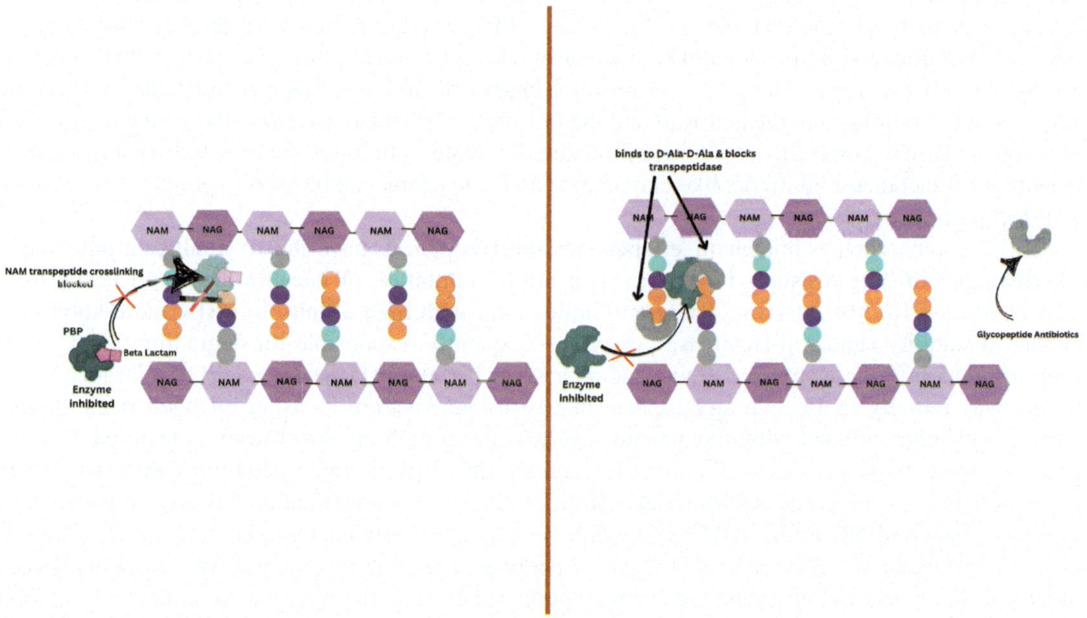

FIGURE 11.2 Mechanism of Action of β-Lactam and Glycopeptide Antibiotics.

Bactericidal mechanism of (1) β-lactams and (2) glycopeptide antibiotics; NAG—N-acetylglucosamine, NAM—N-acetylmuramic acid, β-lactam antibiotics inhibit bacterial cell wall synthesis by binding and inactivating transpeptidase enzymes, which cross-link peptidoglycan strands. This inhibits the formation of peptide bridges between N-acetylglucosamine (NAG) and N-acetylmuramic acid (NAM) chains, weakening the bacterial cell wall and leading to cell lysis.

5.1 Reviving the old guard: β-lactams, glycopeptides, and sulfonamides in the fight against drug-resistant TB

PBPs have been recognized as critical components of bacterial physiology, particularly in the synthesis of the cell wall PG layer. They serve as the primary targets for β-lactam antibiotics, which constitute the most widely used class of antibacterial agents [10]. PBPs facilitate the final stages of PG synthesis, while β-lactams inhibit these essential enzymatic functions. The development of bacterial resistance is largely attributed to the evolution of β-lactamases, enzymes that hydrolyze and deactivate β-lactams. This traditional understanding has informed treatment strategies for bacterial infections using β-lactams for decades. In *Mtb*, PonA1 and PonA2, two prominent PBPs, have demonstrated greater efficacy than BlaC in inactivating penicillins and carbapenems. Specific PBPs in *Mtb* can actively neutralize β-lactam drugs, rather than merely serving as passive targets. Lesser-known PBPs, such as DacB, DacB1, DacB2, and Rv2864c, exhibited increased resistance to all tested β-lactam inhibitors. Rv1730c, a member of the PBP family, displayed minimal affinity for β-lactams, indicating a diverse range of susceptibilities [10,34]. The conventional β-lactam-PBP-blaC model in *Mtb* does not sufficiently explain the dynamic interactions between β-lactams and the enzymes responsible for bacterial cell wall synthesis. Understanding the mechanisms of PBP-mediated β-lactam resistance is crucial for developing effective treatments and designing new β-lactam-based therapeutics for *Mtb* [12]. Certain PBPs in the bacterial cell wall are essential for cell wall synthesis, making them key targets for β-lactam antibiotics in the treatment of TB [12,24]. PBPs bind to β-lactam antibiotics such as ampicillin, amoxicillin, and imipenem at therapeutically relevant concentrations, suggesting their potential efficacy in TB treatment. However, the primary causes of *Mtb* resistance include the delayed absorption of these drugs into the cell wall and the presence of β-lactamase enzymes, which degrade the antibiotics. To overcome this resistance, β-lactamase-stable antibiotics such as imipenem (a carbapenem), or β-lactamase inhibitors like clavulanate and sulbactam, can be used to preserve the efficacy of β-lactam therapy [26].

Carbapenems, such as imipenem, ertapenem, and meropenem, have shown significant potential in the treatment of drug-resistant TB due to their strong inhibition of bacterial cell wall synthesis, a critical process for *Mtb* survival. These antibiotics bind with high affinity to PBPs and demonstrate enhanced stability against β-lactamases, enzymes frequently responsible for degrading β-lactam antibiotics, thus offering a strategic advantage over other β-lactams [14]. Amoxicillin, a β-lactam antibiotic, was initially ineffective against *Mtb* due to its inactivation by BlaC-encoded penicillinase. However, when combined with clavulanate, a β-lactamase inhibitor, its efficacy is restored, as clavulanate inhibits the degradation of amoxicillin and facilitates its binding to DD-transpeptidases, PBPs crucial for PG cross-linking and bacterial viability [30,35]. This combination has proven particularly effective against MDR strains of *Mtb* and exhibits significant early bactericidal activity, especially in cases of extracellular TB infections [11]. Glycopeptides, particularly vancomycin, traditionally used against Gram-positive pathogens, are being repurposed for TB therapy due to their mechanism of inhibiting PG biosynthesis by binding to the D-Ala-D-Ala termini of PG precursors, thus blocking transglycosylation and transpeptidation steps essential for cell wall cross-linking. In *Mtb*, vancomycin interferes with PG synthesis by interacting with enzymes such as PonA2 and LdtB; however, the bacterium's unique lipid-rich cell wall poses a barrier to vancomycin penetration, conferring intrinsic resistance. Nonetheless, disruption of these lipid components has been shown to significantly enhance vancomycin's efficacy against *Mtb* [32,36].

Sulfonamides, also known as sulfa medicines, are broad-spectrum antibiotics that work against bacteria that are Gram-positive or Gram-negative. Their initial use as a monotherapy for TB discontinued because of their toxic effects and decreased efficacy when compared to other drugs such as streptomycin and INH. But another sulfonamide known as sulfadiazine has been used to treat MDR-TB with improved efficacy and safety [22]. Sulfadiazine shows the potential as a treatment for drug-resistant TB, including strains that are MDR and XDR. By blocking the enzyme dihydropteroate synthase, it stops the production of folic acid, which is needed for *Mtb* to grow. Sulfadiazine efficiently kills MTC strains at patient-achievable concentrations without causing any observable resistance. This makes sulfadiazine a useful second-line treatment for TB that is resistant, particularly when combined with other effective antibiotics. It might provide a useful supplement to the small number of drugs now available to treat MDR TB [10].

5.2 Molecular mechanisms of action and resistance to first-line antituberculosis drugs

The standard first-line treatment regimen for TB comprises four drugs: INH, RIF, ethambutol (EMB), and pyrazinamide (PZA). These are administered during the initial 2-month intensive phase, followed by INH and RIF during the subsequent 4-month continuation phase (TB) Treatment & Management: Approach [37]. INH functions by inhibiting the synthesis of mycolic acids, which are essential components of the mycobacterial cell wall. This inhibition occurs through the targeting of the InhA enzyme, following activation by the bacterial catalase-peroxidase enzyme KatG, in conjunction with NAD+ [38]. Resistance to INH typically results from mutations in katG, inhA, ahpC, kasA, and ndh; however, approximately 10% of resistant strains do not exhibit mutations in these genes, indicating the presence of alternative resistance mechanisms [39]. RIF exerts its effect by binding to the β-subunit of DNA-dependent RNA polymerase, encoded by the rpoB gene, thereby inhibiting the initiation of RNA synthesis [40]. More than 95% of RIF-resistant mutations are located within the 81-bp RIF resistance-determining region of the rpoB gene, predominantly as nonsynonymous substitutions [41,42]. EMB disrupts the biosynthesis of the AG component of the mycobacterial cell wall by inhibiting arabinosyl transferases encoded by the embCAB operon, primarily affecting embB. This inhibition increases cell wall permeability, thereby enhancing the efficacy of companion drugs [43,44]. PZA, a prodrug converted into the active molecule pyrazinoic acid by pyrazinamidase (encoded by pncA), impairs Coenzyme A biosynthesis by targeting aspartate decarboxylase (PanD) and exhibits sterilizing activity against nonreplicating *Mtb* in necrotic granulomas [45]. Resistance to PZA is predominantly associated with mutations in pncA and, to a lesser extent, in the ribosomal protein S1 gene (rpsA) [46].

5.3 Navigating the complex landscape of tuberculosis therapy: challenges in drug resistance, toxicity, and pharmacokinetics

Despite their proven efficacy, first-line anti-TB drugs face several critical challenges that compromise treatment success. Drug resistance, primarily resulting from genetic mutations in the enzymes or pathways targeted by these drugs, significantly hinders therapeutic efficacy. Common mutations in genes such as katG, inhA, and rpoB confer resistance to INH and RIF, respectively [39,41]. In addition to resistance, adverse effects, such as hepatotoxicity from INH, optic neuritis from EMB, and peripheral neuropathy, contribute to poor patient adherence. The prolonged treatment duration, typically

6 months or more, is necessitated by the slow-growing nature of *Mtb* and the primarily bacteriostatic action of some drugs, further increasing the likelihood of noncompliance. Accessibility issues compound these challenges, especially in low-resource settings where second-line drugs are often unavailable or prohibitively expensive (TB) Treatment & Management: Approach [37]. While drug repurposing strategies, particularly those targeting the bacterial cell wall, offer promising avenues for novel anti-TB therapies, they are not without limitations. Barriers include restricted public access to comprehensive drug data, the complexity of managing diverse computational outputs, and the resource-intensive nature of data integration and analysis, all of which can delay drug development timelines [13]. Pharmacokinetic (PK) variability, influenced by host genetic differences, comorbidities, diet, and drug–drug interactions, can result in subtherapeutic drug levels despite standardized dosing regimens, thereby diminishing efficacy and promoting resistance [47,48]. Additionally, the cytotoxic and mutagenic effects of anti-TB drugs warrant concern: INH metabolites exhibit genotoxicity; EMB may cause irreversible optic neuropathy; and PZA, while effective, has been associated with chromosomal and sperm morphological abnormalities in animal models. Notably, the combined use of these drugs may produce synergistic mutagenic effects, emphasizing the need for thorough toxicological assessments to safeguard patient health during TB therapy [49].

6. Conclusions and perspectives

Novel therapeutic strategies are imperative to address the persistent challenge of TB, exacerbated by the emergence of *Mtb* strains that are XDR and MDR. A significant obstacle in combating *Mtb* is its intricate and essential cell wall, composed of PG, AG, and mycolic acid [50]. There exists substantial potential for the development of innovative anti-TB drugs by targeting the synthesis and remodeling of this protective barrier. Various families of repurposed drugs exhibit potent antimycobacterial activity by targeting the *Mtb* cell wall. β-lactam antibiotics, particularly carbapenems, in combination with β-lactamase inhibitors such as clavulanate, have been demonstrated to effectively inhibit the production of PGs, which are crucial for maintaining the integrity of bacterial cell wall [25]. Although glycopeptides encounter challenges in penetrating the mycobacterial cell wall, they nonetheless target the same critical PG synthesis process and warrant further investigation with enhanced delivery methods. Additionally, sulfonamides offer a unique mechanism of action that can be effective against drug-resistant bacteria by indirectly affecting cell wall formation pathways through the inhibition of folate synthesis. Targeting the cell wall presents a strategic advantage, as it is vital to bacterial survival, pathogenicity, and antibiotic resistance [12]. These repurposed drugs may compromise the structural integrity of *Mtb* by inhibiting the synthesis of essential cell wall components, potentially rendering it more susceptible to the host immune system and other co-administered treatments. This approach holds promise for developing shorter and more effective TB treatment regimens, in addition to combating drug resistance. In conclusion, repurposing drugs that target the cell wall represents a novel and effective strategy in the ongoing battle against TB. By leveraging existing knowledge and safety data on these drugs, we can expedite the development of urgently needed therapeutic options to address the escalating threat of drug-resistant TB and ultimately enhance patient outcomes. Further research is necessary to optimize the use of these repurposed drugs, explore synergistic combinations, and conduct comprehensive clinical trials to realize their potential in addressing this global health challenge.

References

[1] Bacterial pathogens of public health importance to guide research, development and strategies to prevent and control antimicrobial resistance. WHO bacterial priority pathogens list (2024).

[2] Silhavy TJ, Kahne D, Walker S. The bacterial cell envelope. Cold Spring Harbor Perspect Biol 2010;2(5): a000414. https://doi.org/10.1101/cshperspect.a000414.

[3] Fatima S, Bhaskar A, Dwivedi VP. Repurposing immunomodulatory drugs to combat tuberculosis. Front Immunol 2021;12. https://doi.org/10.3389/fimmu.2021.645485.

[4] Implementing the WHO stop TB strategy: a handbook for national tuberculosis control programmes (Internet). Geneva: World Health Organization (2008).

[5] Liebenberg D, Gordhan BG, Kana BD. Drug resistant tuberculosis: implications for transmission, diagnosis, and disease management. Front Cell Infect Microbiol 2022;12. https://doi.org/10.3389/fcimb.2022.943545.

[6] Zhu H, Zhou X, Zhuang Z, Li L, Bi J, Mi K. Advances of new drugs bedaquiline and delamanid in the treatment of multi-drug resistant tuberculosis in children. Front Cell Infect Microbiol 2023;13. https://doi.org/10.3389/fcimb.2023.1183597.

[7] Zhan P, Yu B, Ouyang L. Drug repurposing: an effective strategy to accelerate contemporary drug discovery. Drug Discov Today 2022;27(7):1785–8. https://doi.org/10.1016/j.drudis.2022.05.026.

[8] Sarkar P, Xu W, Vázquez-Hernández M, Dhanda G, Tripathi S, Basak D, et al. Enhancing the antibacterial efficacy of vancomycin analogues: targeting metallo-β-lactamases and cell wall biosynthesis. Chem Sci 2024;15(39):16307–20. https://doi.org/10.1039/d4sc03577a. PubMed PMID: 39309102; PubMed Central PMCID: PMC11409854.

[9] Amado PSM, Woodley C, Cristiano MLS, O'Neill PM. Recent advances of DprE1 inhibitors against *Mycobacterium tuberculosis*: computational analysis of physicochemical and ADMET properties. ACS Omega 2022;7(45):40659–81. https://doi.org/10.1021/acsomega.2c05307.

[10] Seung KJ, Keshavjee S, Rich ML. Multidrug-resistant tuberculosis and extensively drug-resistant tuberculosis. Cold Spring Harb Perspect Med 2015;5(9). https://doi.org/10.1101/cshperspect.a017863.

[11] Vishwakarma D, Gaidhane A, Sahu ASR. Multi-drug resistance tuberculosis (MDR-TB) challenges in India: a review. Cureus 2023;15(12).

[12] Laxminarayan R, Duse A, Wattal C, Zaidi AKM, Wertheim HFL, Sumpradit N, Vlieghe E, Hara GL, Gould IM, Goossens H, Greko C, So AD, Bigdeli M, Tomson G, Woodhouse W, Ombaka E, Peralta AQ, Qamar FN, Mir F, Kariuki S, Bhutta ZA, Coates A, Bergstrom R, Wright GD, Brown ED, Cars O. Antibiotic resistance-the need for global solutions. Lancet Infect Dis 2013;13(12):1057–98. https://doi.org/10.1016/S1473-3099(13)70318-9.

[13] Sharma K, Ahmed F, Sharma T, Grover A, Agarwal M, Grover S. Potential repurposed drug candidates for tuberculosis treatment: progress and update of drugs identified in over a decade. ACS Omega 2023;8(20): 17362–80. https://doi.org/10.1021/acsomega.2c05511.

[14] Stadler JAM, Maartens G, Meintjes G, Wasserman S. Clofazimine for the treatment of tuberculosis. Front Pharmacol 2023;14. https://doi.org/10.3389/fphar.2023.1100488.

[15] Nair A, Greeny A, Nandan A, Sah RK, Jose A, Dyawanapelly S, Junnuthula V, Athira KV, Sadanandan P. Advanced drug delivery and therapeutic strategies for tuberculosis treatment. J Nanobiotechnol 2023;21(1). https://doi.org/10.1186/s12951-023-02156-y.

[16] Xu X, Dong B, Peng L, Gao C, He Z, Wang C, Zeng J. Anti-tuberculosis drug development via targeting the cell envelope of *Mycobacterium tuberculosis*. Front Microbiol 2022;13. https://doi.org/10.3389/fmicb.2022.1056608.

[17] Chai Q, Lu Z, Liu CH. Host defense mechanisms against *Mycobacterium tuberculosis*. Cell Mol Life Sci 2020;77(10):1859–78. https://doi.org/10.1007/s00018-019-03353-5.

[18] Bhat ZS, Rather MA, Maqbool M, Ahmad Z. Drug targets exploited in *Mycobacterium tuberculosis*: pitfalls and promises on the horizon. Biomed Pharmacother 2018;103:1733–47. https://doi.org/10.1016/j.biopha.2018.04.176.
[19] Palomino JC, Martin A. Drug resistance mechanisms in *Mycobacterium tuberculosis*. Antibiotics 2014;3(3):317–40. https://doi.org/10.3390/antibiotics3030317.
[20] Schneider T, Sahl HG. An oldie but a goodie—cell wall biosynthesis as antibiotic target pathway. Int J Med Microbiol 2010;300(2–3):161–9. https://doi.org/10.1016/j.ijmm.2009.10.005.
[21] Fil M, Alibert S. Unveiling DprE1 as a key target in the fight against tuberculosis: insights and Perspectives on developing novel antimicrobial agents. BioMed 2024;4(3):220–36. https://doi.org/10.3390/biomed4030018.
[22] Ovung A, Bhattacharyya J. Sulfonamide drugs: structure, antibacterial property, toxicity, and biophysical interactions. Biophys Rev 2021;13(2):259–72. https://doi.org/10.1007/s12551-021-00795-9.
[23] Sander NG, Sampson SL, Van Rie A. Mechanisms of drug-induced tolerance in *Mycobacterium tuberculosis*. Clin Microbiol Rev 2020;34(1). https://doi.org/10.1128/cmr.00141-20.
[24] Uddin TM, Chakraborty AJ, Khusro A, Zidan BRM, Mitra S, Emran TB, Dhama K, Ripon MKH, Gajdács M, Sahibzada MUK, Hossain MJ, Koirala N. Antibiotic resistance in microbes: history, mechanisms, therapeutic strategies and future prospects. J Infect Public Health 2021;14(12):1750–66. https://doi.org/10.1016/j.jiph.2021.10.020.
[25] Cho H, Uehara T, Bernhardt TG. Beta-lactam antibiotics induce a lethal malfunctioning of the bacterial cell wall synthesis machinery. Cell 2014;159(6):1300–11. https://doi.org/10.1016/j.cell.2014.11.017.
[26] Bush K, Bradford PA. β-lactams and β-lactamase inhibitors: an overview. Cold Spring Harb Perspect Med 2016;6(8):a025247. https://doi.org/10.1101/cshperspect.a025247.
[27] Peptidoglycan, structure, synthesis, and regulation—PMC (2025).
[28] Bahr G, González LJ, Vila AJ. Metallo-β-lactamases in the age of multidrug resistance: from structure and mechanism to evolution, dissemination, and inhibitor design. Chem Rev 2021;121(13):7957–8094. https://doi.org/10.1021/acs.chemrev.1c00138.
[29] Nauta KM, Ho TD, Ellermeier CD. The penicillin-binding protein pbpp is a sensor of β-lactams and is required for activation of the extracytoplasmic function σ factor σp in bacillus thuringiensis. mBio 2021;12(2). https://doi.org/10.1128/mBio.00179-21.
[30] Evans J, Hanoodi, Wittler M. Amoxicillin clavulanate, treasure Island (FL). StatPearls Publishing; 2025.
[31] Kakoullis L, Papachristodoulou E, Chra P, Panos G. Mechanisms of antibiotic resistance in important gram-positive and gram-negative pathogens and novel antibiotic solutions. Antibiotics 2021;10(4):415. https://doi.org/10.3390/antibiotics10040415.
[32] Stogios PJ, Savchenko A. Molecular mechanisms of vancomycin resistance. Protein Sci 2020;29(3):654–69. https://doi.org/10.1002/pro.3819.
[33] Münch D, Sahl HG. Structural variations of the cell wall precursor lipid II in Gram-positive bacteria—impact on binding and efficacy of antimicrobial peptides. Biochim Biophys Acta BBA—Biomembr 2015;1848(11):3062–71.
[34] Kumar G, Galanis C, Batchelder HR, Townsend CA, Lamichhane G. Penicillin binding proteins and β-lactamases of *Mycobacterium tuberculosis*: reexamination of the historical paradigm. mSphere 2022;7(1). https://doi.org/10.1128/msphere.00039-22.
[35] Kurz SG, Bonomo RA. Reappraising the use of β-lactams to treat tuberculosis. Expert Rev Anti Infect Ther 2012;10(9):999–1006. https://doi.org/10.1586/eri.12.96.
[36] Reynolds B. DoseMeRx, vancomycin mechanism of action | resistance and more. 2020.
[37] (TB) Treatment & management: approach considerations, treatment during pregnancy. Tuberculosis (2025).
[38] Favrot L, Ronning DR. Targeting the mycobacterial envelope for tuberculosis drug development. Expert Rev Anti Infect Ther 2012;10(9):1023–36. https://doi.org/10.1586/eri.12.91.

[39] Vilchèze C, Jacobs JR WR, Hatfull GF, Jacobs Jr WR. Resistance to isoniazid and ethionamide in *Mycobacterium tuberculosis*: genes, mutations, and causalities. Microbiol Spectr 2014;2(4). https://doi.org/10.1128/microbiolspec.mgm2-0014-2013.

[40] Lei T, Qiang T, Liang C, Ren X, Jia M, Zhang J, Li J, Wan M, YuWen X, Han L, Cao W, Liu H. RNA-dependent RNA polymerase (RdRp) inhibitors: the current landscape and repurposing for the COVID-19 pandemic. Eur J Med Chem 2021;213:113201. https://doi.org/10.1016/j.ejmech.2021.113201.

[41] Zaw MT, Emran NA, Lin Z. Mutations inside rifampicin-resistance determining region of rpoB gene associated with rifampicin-resistance in *Mycobacterium tuberculosis*. J Infect Public Health 2018;11(5):605–10. https://doi.org/10.1016/j.jiph.2018.04.005.

[42] Mosaei H, Zenkin N. Inhibition of RNA polymerase by rifampicin and rifamycin-like molecules. EcoSal Plus 2020;9(1). https://doi.org/10.1128/ecosalplus.ESP-0017-2019.

[43] Goude R, Amin AG, Chatterjee D, Parish T. The arabinosyltransferase EmbC is inhibited by ethambutol in *Mycobacterium tuberculosis*. Antimicrob Agents Chemother 2009;53(10):4138–46. https://doi.org/10.1128/AAC.00162-09.

[44] Vilchèze C, Cell Wall M. A source of successful targets for old and new drugs. Appl Sci 2020;10(7):2278. https://doi.org/10.3390/app10072278.

[45] Gopal P, Nartey W, Ragunathan P, Sarathy J, Kaya F, Yee M, Setzer C, Sony Subramanian Manimekalai M, Dartois V, Grüber G, Dick T. Pyrazinoic acid inhibits mycobacterial Coenzyme A biosynthesis by binding to aspartate decarboxylase PanD. ACS Infect Dis 2017;3(11):807–19. https://doi.org/10.1021/acsinfecdis.7b00079.

[46] Khan MT, Malik SI, Ali S, Masood N, Nadeem T, Khan AS, Afzal MT. Pyrazinamide resistance and mutations in pncA among isolates of *Mycobacterium tuberculosis* from Khyber Pakhtunkhwa, Pakistan. BMC Infect Dis 2019;19(1). https://doi.org/10.1186/s12879-019-3764-2.

[47] Levison ME, Levison JH. Pharmacokinetics and pharmacodynamics of antibacterial agents. Infect Dis Clin 2009;23(4):791–815. https://doi.org/10.1016/j.idc.2009.06.008.

[48] Koziolek M, Alcaro S, Augustijns P, Basit AW, Grimm M, Hens B, Hoad CL, Jedamzik P, Madla CM, Maliepaard M, Marciani L, Maruca A, Parrott N, Pávek P, Porter CJH, Reppas C, van Riet-Nales D, Rubbens J, Statelova M, Trevaskis NL, Valentová K, Vertzoni M, Čepo DV, Corsetti M. The mechanisms of pharmacokinetic food-drug interactions—a perspective from the UNGAP group. Eur J Pharmaceut Sci 2019;134:31–59. https://doi.org/10.1016/j.ejps.2019.04.003.

[49] Researchgate, In vitro toxic action potential of anti tuberculosis drugs and their combinations | Request PDF (2025).

[50] Chai Q, Zhang Y, Liu CH. *Mycobacterium tuberculosis*: an adaptable pathogen associated with multiple human diseases. Front Cell Infect Microbiol 2018;8(May). https://doi.org/10.3389/fcimb.2018.00158.

CHAPTER

12

Metabolic repurposing in Mycobacterium-infected macrophages: A strategy for enhanced virulence and survival

Kashish Choudhary[1,a], Kunal Malik[1,a], Ishan[1], Vandana Pawar[2] and Sonam Grover[1]

[1]*Department of Molecular Medicine, School of Interdisciplinary Sciences and Technology, Jamia Hamdard, New Delhi, India;* [2]*Department of Biotechnology, Indira Gandhi National Tribal University, Amarkantak, Madhya Pradesh, India*

Highlights

1. *Mycobacterium tuberculosis* (*Mtb*) rewires host macrophage metabolism to enhance its virulence and survival.
2. *Mtb* induces the Warburg effect in macrophages, promoting aerobic glycolysis and pro-inflammatory responses.
3. *Mtb* reprograms its own metabolism to utilize host-derived lipids and cholesterol for persistence.
4. Metabolic adaptation allows *Mtb* to enter a dormant, non-replicating state within granulomas.
5. Slow growth, efflux pump overexpression, and ROS detoxification contribute to *Mtb*'s drug resistance.
6. Targeting the metabolic interplay between *Mtb* and host macrophages may provide novel therapeutic strategies.
7. Understanding the link between metabolic rewiring and drug resistance is crucial for combating tuberculosis.

Abbreviations

ABC ATP-binding cassette
ATP adenosine triphosphate
BDQ bedaquiline
CFZ clofazimine
DosR dormancy survival regulator
ESAT-6 early secreted antigenic target 6 kDa
FAO fatty acid oxidation
HDTs host-directed therapies
HIF-1α hypoxia-inducible factor-1 alpha
HO-1 heme oxygenase-1
ICL isocitrate lyase

[a]Authors contributed equally.

Navigating Tuberculosis. https://doi.org/10.1016/B978-0-443-27735-1.00004-6

IFN-γ interferon-gamma
IL-10 interleukin-10
IL-1β interleukin-1 beta
InhA enoyl-ACP reductase gene
KatG catalase-peroxidase
LAM lipoarabinomannan
LTBI latent tuberculosis infection
M1 classically activated macrophages
M2 alternatively activated macrophages
Mce1/Mce4 mammalian cell entry 1 and 4 complexes
MDR-TB multidrug-resistant tuberculosis
MFS major facilitator superfamily
MHC II major histocompatibility complex class II
MIC minimum inhibitory concentration
miR-21 microRNA-21
Mtb mycobacterium tuberculosis
NAD+ nicotinamide adenine dinucleotide (oxidized)
NADH nicotinamide adenine dinucleotide (reduced)
NADPH nicotinamide adenine dinucleotide phosphate
NRP non-replicating persistent
PDIMs phthiocerol dimycocerosates
PFK phosphofructokinase
PRRs pattern recognition receptors
RNA ribonucleic acid
RNAP RNA polymerase
RND resistance-nodulation-division
ROS reactive oxygen species
rpoB RNA polymerase beta subunit gene
SigB/SigE sigma factors B and E
SOD superoxide dismutase
TB tuberculosis
TCA tricarboxylic acid cycle
TDM trehalose dimycolate
TG triacylglycerol
TLRs toll-like receptors
TMM trehalose monomycolate
TNF-α tumor necrosis factor-alpha
WHO World Health Organization
XDR-TB extensively drug-resistant tuberculosis

1. Introduction

Tuberculosis (TB), caused by the obligate intracellular pathogen *Mycobacterium tuberculosis* (*Mtb*), remains a leading cause of global morbidity and mortality, with an estimated. 8 million new cases and 2.5 million deaths reported in the 2024 WHO report [1]. This has escalated into a global health emergency due to the emergence of extensively drug-resistant (XDR-TB) and multidrug-resistant (MDR-TB) strains [2]. The slow-growing, acid-fast *Mtb* bacterium is specifically adapted to survive and proliferate within host macrophages. It evades immune defenses through various mechanisms, such as modifying host immunological responses and preventing phagosome-lysosome fusion [3]. TB is transmitted via airborne particles released when an infected individual coughs or sneezes. Upon inhalation, *Mtb* can establish a latent infection that remains asymptomatic for years until reactivation occurs when the immune system is compromised [4]. The World Health Organization (WHO) classifies drug-resistant TB as a critical priority, underscoring the urgent need to elucidate the molecular and metabolic mechanisms underlying its resilience. To create a replicative niche, *Mtb* alters host metabolic pathways, inhibits autophagy, and disrupts phagosomal acidification following phagocytosis [5]. A crucial adaptation mechanism that enhances bacterial survival and virulence in the harsh intracellular environment is the metabolic repurposing in macrophages infected with *Mtb* [6]. *Mtb* modifies the host cell's metabolism post-infection to establish a nutrient-rich niche that supports its persistence. This involves redirecting resources for bacterial benefit by altering macrophage metabolic pathways such as the tricarboxylic acid cycle (TCA), fatty acid oxidation, and glycolysis [5]. To induce the Warburg effect, which reduces the production of reactive oxygen species (ROS) and fosters a survival-conducive environment, the pathogen also shifts host cells from oxidative phosphorylation to aerobic glycolysis (Fig. 12.1). In an effort to promote long-term intracellular viability and immune evasion, *Mtb* simultaneously modifies its own metabolic processes to utilize cholesterol and host-derived lipids as carbon sources [6]. These metabolic interactions result in a paradox where the host's defense mechanisms inadvertently aid the pathogen's survival. Granuloma formation generates a range of micro-niches with varying levels of oxygen, nutrients, and antimicrobial compounds, although it is intended to contain the infection [7]. These microenvironments support bacterial

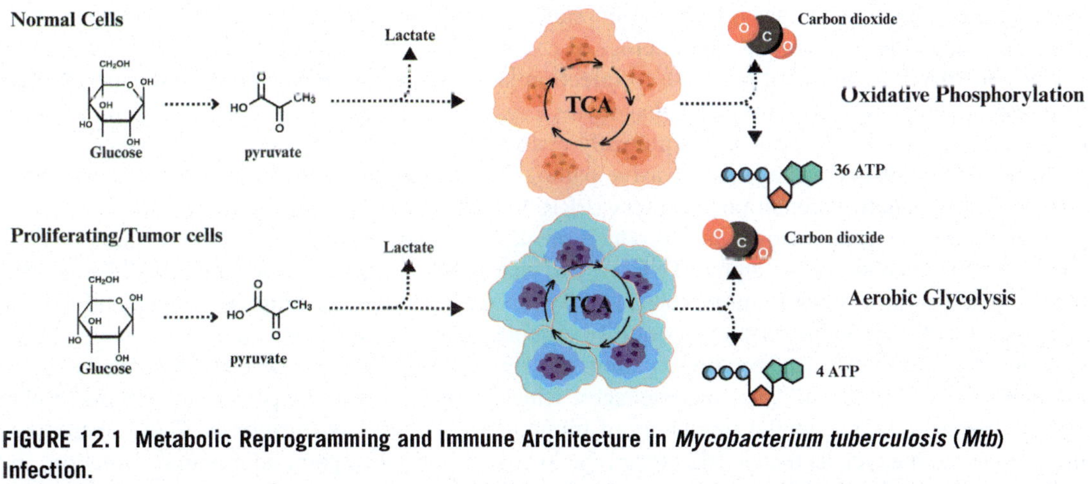

FIGURE 12.1 Metabolic Reprogramming and Immune Architecture in *Mycobacterium tuberculosis* (*Mtb*) Infection.

This schematic summarizes the formation of granulomas during *Mtb* infection, highlighting key immunological events such as immune cell recruitment, cytokine-mediated signaling, macrophage activation, and the development of a fibrotic encapsulated structure. It also contrasts cellular metabolic states, specifically the shift from oxidative phosphorylation to aerobic glycolysis (Warburg effect), in infected immune cells within the granulomatous environment.

subpopulations with distinct metabolic states and varying degrees of antibiotic susceptibility. For instance, NRP (Non-Replicating Persistent) *Mtb* with enhanced antioxidant systems (such as thioredoxin and mycothiol) that counteract ROS and improve antibiotic tolerance are observed in hypoxic granuloma cores [8].

Metabolic repurposing is intricately associated with drug resistance in *Mtb*, collectively enhancing the pathogen's adaptability to antibiotic stress [9]. The efflux pumps of the ATP-binding cassette (ABC) and major facilitator superfamily (MFS), which actively expel a broad spectrum of structurally diverse antibiotics, such as bedaquiline and clofazimine, are frequently upregulated in MDR and XDR strains [10]. This chapter provides insights into the metabolic repurposing in *Mtb* that enables survival under adverse conditions such as hypoxia, nutrient deprivation, and drug treatment. Furthermore, *Mtb*'s metabolic interdependencies with the host, including its reliance on scavenged lipids, redox homeostasis mediated by NADH/NAD+ shuttles, and adaptation to the diverse nutrient environment of granulomas, highlight how host-pathogen interactions metabolically codetermine bacterial persistence [11]. When non-replicating bacilli evade drugs targeting replicative processes and host-derived metabolites inadvertently extend bacterial survival, these mechanisms collectively impede therapeutic success. Novel strategies to combat resistance include pharmacological adjuvants such as efflux pump inhibitors (e.g., verapamil), which enhance intracellular antibiotic concentrations, and host-directed treatments (HDTs) that disrupt metabolic symbiosis.

2. Frontline defenders: Macrophage dynamics in the immune response to tuberculosis

Macrophages play an essential role in the immune system's defense against the TB pathogen, *Mtb*. They are integral to maintaining immunological control during the chronic stages of the disease and are also involved in the initial response to infection [12]. As key components of the innate immune response, macrophages are proficient phagocytes capable of identifying, engulfing, and eliminating pathogens and cellular debris [7]. Ubiquitously distributed throughout tissues, macrophages serve as the primary line of defense against invasive pathogens. Originating from blood monocytes, macrophages utilize pattern recognition receptors (PRRs), such as toll-like receptors (TLRs), to identify pathogens, initiate phagocytosis, and produce pro-inflammatory cytokines, including interleukin-1 (IL-1), tumor necrosis factor-alpha (TNF-α), and interleukin-6 (IL-6), which facilitate the recruitment and activation of other immune cells [13]. By processing microbial antigens and presenting them to helper T cells via major histocompatibility complex class II (MHC II) molecules, macrophages significantly contribute to antigen presentation, thereby bridging the gap between innate and adaptive immunity [14]. The versatility of macrophages, which allows them to adopt various functional phenotypes, renders them essential regulators of immune homeostasis and response. These phenotypes include the alternatively activated M2 phenotype, involved in tissue repair and immunoregulation, and the classically activated M1 phenotype, which is pro-inflammatory and microbicidal [15]. Upon infection, macrophages are the initial immune cells to encounter *Mtb* and employ diverse mechanisms to combat the pathogen.

3. Master of intracellular evasion: Mechanisms of *Mycobacterium tuberculosis* survival within macrophages

Mtb has developed a sophisticated array of defense mechanisms to evade the host immune system and persist within macrophages, which are cells specifically designed to eliminate infections [16]. Upon inhalation, alveolar macrophages engulf *Mtb* through phagocytosis, a process that typically destroys pathogens within acidified, enzyme-rich phagolysosomes [7] (Fig. 12.2). However, *Mtb* actively inhibits the maturation of the phagosome and its fusion with lysosomes, thereby avoiding exposure to acidic environments and degradative enzymes. The alteration of the phagosomal environment within macrophages is a primary strategy employed by *Mtb*. It prevents bacterial degradation by obstructing the fusion of phagosomes with lysosomes [17]. The early secreted antigenic target 6 kDa (ESAT-6), the serine/threonine protein kinase PknG, and cell wall lipids such as lipoarabinomannan (LAM) and phthiocerol dimycocerosates (PDIMs) collectively interfere with host signaling pathways and phagosomal membrane trafficking [18]. Furthermore, *Mtb* enhances its intracellular survival by counteracting oxidative and nitrosative stress through the production of detoxifying enzymes such as superoxide dismutase (SOD), catalase-peroxidase (KatG), and peroxynitrite reductase [19].

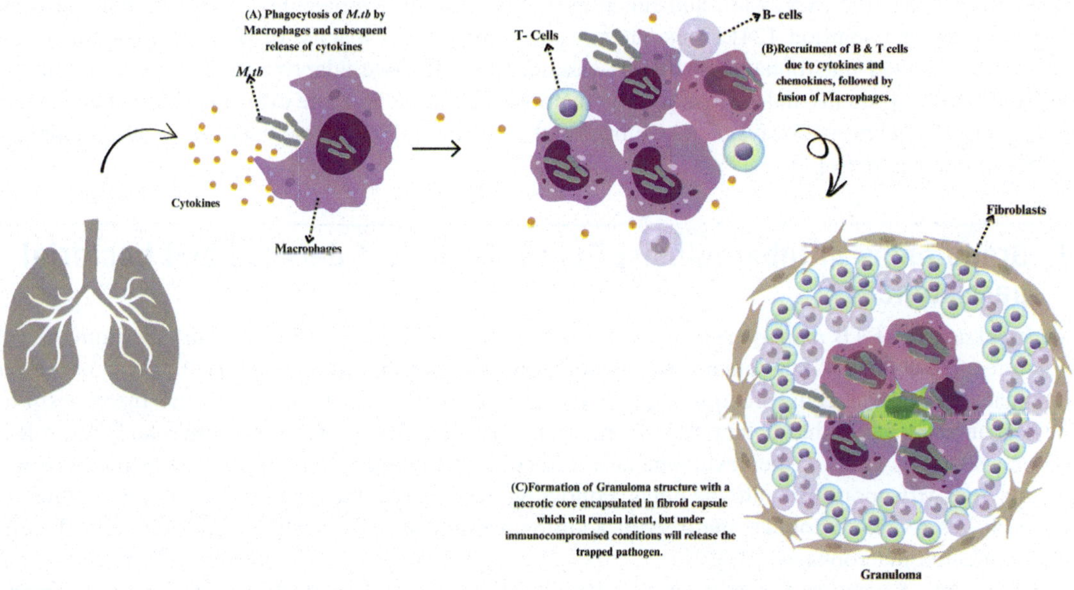

FIGURE 12.2 Metabolic Reprogramming and Immune Architecture in *Mycobacterium tuberculosis* (*Mtb*) Infection.

This schematic summarizes the formation of granulomas during *Mtb* infection, highlighting key immunological events such as immune cell recruitment, cytokine-mediated signaling, macrophage activation, and the development of a fibrotic encapsulated structure. It also contrasts cellular metabolic states, specifically the shift from oxidative phosphorylation to aerobic glycolysis (Warburg effect), in infected immune cells within the granulomatous environment.

Additionally, *Mtb* suppresses macrophage activation and diminishes the adaptive immune response by upregulating anti-inflammatory cytokines like interleukin-10 (IL-10) and downregulating pro-inflammatory cytokines such as interleukin-12 (IL-12) and TNF-α. To evade immune detection, the bacteria may enter a dormant, non-replicating state, reducing its metabolic activity and antigen presentation. Collectively, these strategies enable *Mtb* to establish a niche within macrophages, persist for extended periods, and contribute to the establishment of latent tuberculosis infection (LTBI) [20].

The type of cell death occurring in infected macrophages is influenced by *Mtb*. Virulent strains of *Mtb* have been shown to induce necrosis, or uncontrolled cell death, rather than apoptosis, or programmed cell death [21]. This results in cellular lysis and the release of bacteria into the extracellular space, facilitating further infection spread. Conversely, attenuated strains often induce apoptosis, which serves as a defense mechanism by reducing bacterial viability and exposing dendritic cells to antigens, thereby enhancing adaptive immunity [22]. Autophagy, an essential component of the innate immune response, targets intracellular pathogens for elimination. Nevertheless, *Mtb* has developed defenses against autophagic degradation [23]. Despite this defense mechanism, *Mtb* can persist within macrophages by inhibiting the formation of autophagosomes or interfering with autophagic flux. In addition to survival, *Mtb* avoids eliciting stronger adaptive immune responses by obstructing the autophagy process.

Mtb evades T cell detection by modifying antigen presentation pathways. The bacterium possesses the ability to alter the expression and function of MHC class II molecules, thereby reducing T-cell activation and recognition [24]. This interference limits the capacity of T cells to identify and respond to infected macrophages, which is essential for effective adaptive immunity. *Mtb* employs secreted effector proteins and various genetic factors to facilitate immune evasion. These elements can modify host signaling pathways, enhance bacterial survival under adverse conditions, and disrupt the body's innate defences [25].

4. Granulomas in tuberculosis: Dual role in host defense and bacterial persistence

A critical aspect of TB pathogenesis is the formation of granulomas, which are intricate immune cell structures that serve to both contain the infection and facilitate its persistence [17]. These organized aggregates of immune cells, predominantly macrophages, function as both a protective mechanism for the host and a niche for the bacteria [26]. Comprehending the formation of granulomas and their role in TB persistence is vital for the development of effective therapeutic strategies. Alveolar macrophages phagocytose *Mtb* upon entry into the lungs. Following migration to the lung interstitium, these infected macrophages secrete cytokines that recruit additional immune cells, such as dendritic cells, T cells, and uninfected macrophages [27].

Granuloma formation is dependent on key cytokines, including TNF-α and interferon-gamma (IFN-γ). The structural integrity of granulomas relies on these cytokines' ability to activate macrophages, promoting their differentiation into epithelioid cells [28]. Throughout this process, macrophages undergo significant reprogramming, resulting in morphological changes that facilitate close intercellular connections [29]. This transformation is characterized by increased glycolysis and altered metabolic pathways, which contribute to the persistence of *Mtb*. Within granulomas, *Mtb* can transition into a non-replicating state, allowing it to withstand immune responses. The granuloma's necrotic core, typically composed of infected macrophages and cellular debris (known as caseous necrosis), is surrounded by layers of epithelioid macrophages, multinucleated giant cells, activated T cells, and occasionally a fibrotic capsule formed by fibroblasts. This structure confines the infection, preventing bacterial dissemination throughout the body, while simultaneously providing a microenvironment that permits *Mtb* to remain dormant and non-replicating for extended periods [30]. The acidic, nutrient-deficient, and hypoxic conditions within the granuloma inhibit bacterial growth but do not eradicate the organism, enabling it to persist as a latent infection. The host immunity declines due to factors such as human immune deficiency virus (HIV) infection, malnutrition, or aging; therefore, granuloma breakdown and reactivation of active TB may ensue, disrupting this dormant state. Thus, while granulomas are a crucial defense mechanism intended to prevent TB infection, they paradoxically also serve as a reservoir for *Mtb*, significantly enhancing the pathogen's persistence by shielding the bacteria from antibiotics [31] (Fig. 12.2). Research into granuloma biology has paved the way for host-directed therapies (HDTs) that modulate immune responses within granulomas, potentially improving treatment outcomes [32].

5. The shift in host metabolism during *Mtb* infection

Mtb infection induces significant metabolic alterations in host cells, particularly in macrophages, which play a crucial role in the immune response. This metabolic reprogramming encompasses the activation of the Warburg effect, modifications in glycolysis, and changes in fatty acid metabolism. Understanding these shifts is essential for elucidating the mechanisms of *Mtb* persistence and developing novel therapeutic strategies [33].

5.1 Metabolic warfare: the Warburg effect in macrophage defense against *Mycobacterium tuberculosis*

The Warburg effect describes the phenomenon where cells preferentially utilize glycolysis for energy production even in the presence of sufficient oxygen, resulting in increased lactate production. Upon infection with *Mtb*, macrophages transition from oxidative phosphorylation to aerobic glycolysis [34] (Fig. 12.1). The generation of pro-inflammatory cytokines such as IL-1β, which are vital for regulating bacterial growth within macrophages, is dependent on this metabolic shift. It has been demonstrated that inhibiting this glycolytic shift enhances *Mtb* intracellular survival, indicating that glycolysis confers protection against the infection [35]. Increased production of pro-inflammatory cytokines is associated with the metabolic shift toward glycolysis. For instance, heightened glycolytic activity during *Mtb* infection leads to a substantial upregulation of IL-1β production, whereas such conditions inhibit anti-inflammatory cytokines like IL-10 [36].

5.2 Metabolic tug-of-war: HIF-1α and miR-21 in the host response to *Mycobacterium tuberculosis*

Significant alterations in the glycolytic pathway occur when *Mtb* infects a human host, indicating a change in cellular metabolism that is intimately related to the host's immunological response [37]. This change facilitates activated immune cells in meeting their increased bioenergetic and biosynthetic demands by promoting the rapid synthesis of ATP and the generation of metabolic intermediates necessary for cytokine production and antimicrobial action. Hypoxia-inducible factor-1 alpha (HIF-1α), a crucial regulator of glycolysis, is stabilized by *Mtb* infection, which enhances the production of glycolytic enzymes such as pyruvate kinase, phosphofructokinase (PFK), and hexokinase. By inhibiting PFK, the anti-inflammatory microRNA-21 (miR-21) has been linked to the suppression of host glycolysis during *Mtb* infection, leading to bacterial growth while diminishing pro-inflammatory responses [38]. This regulatory mechanism illustrates how *Mtb* can advantageously alter host metabolism.

5.3 Fueling persistence: the dual role of host lipid metabolism in tuberculosis pathogenesis and immune modulation

During infection, *Mtb* utilizes fatty acids as its primary carbon source, a characteristic that enhances its pathogenicity by facilitating adaptation to nutrient-limited conditions within the host [39]. In macrophages, the principal host cells harboring the pathogen, the host's fatty acid metabolism undergoes significant alterations. *M.tb* exploits the host's lipid metabolic pathways to establish a nutrient-rich niche conducive to its survival and persistence. A notable alteration is the formation of lipid droplets in infected macrophages, leading to the development of "foamy macrophages," which serve as reservoirs of cholesterol and fatty acids within the host [40]. Through β-oxidation and the glyoxylate shunt, *Mtb* metabolizes these lipids as a carbon source, enabling its survival in intracellular environments with limited nutrients [41].

The shift from oxidative phosphorylation to glycolysis may influence macrophage polarization and function. For instance, increased glycolysis promotes a pro-inflammatory M1 macrophage phenotype, and alterations in metabolic pathways can create an immunosuppressive environment that supports *Mtb* survival [42]. Concurrently, *Mtb* infection often induces the downregulation of fatty acid oxidation (FAO) and a shift toward increased lipogenesis in host cells by modulating the expression of enzymes such as acetyl-CoA carboxylase and fatty acid synthase [43] (Fig. 12.3). This metabolic reprogramming benefits the pathogen but also impacts the host immune response, as fatty acid-derived lipid mediators like prostaglandins and leukotrienes influence inflammation and the recruitment of immune cells.

5.4 Fat and fate: the role of host-derived lipids in the survival strategy of *Mycobacterium tuberculosis*

One of the adaptations of *Mtb* is the reprogramming of its metabolism to utilize host-derived lipids, particularly fatty acids and cholesterol, as essential carbon sources during infection [44]. This metabolic shift significantly influences the pathogen's virulence, persistence, and ability to evade immune clearance. Unlike many other pathogens that primarily focus on carbohydrates such as glucose, *Mtb* exhibits a distinct preference for lipids, especially during the intracellular phase of infection. Within macrophages and granulomas, where glucose is limited, lipids abundant in host cell membranes and lipid droplets become accessible [45]. To effectively catabolize fats and cholesterol, *Mtb* modifies its metabolic pathways to exploit this change in nutrient availability. It imports host fatty acids through specialized transporter systems, such as the Mce1 and Mce4 complexes. Inside the bacterial cell, acyl-CoA synthetases activate fatty acids, producing fatty acyl-CoA intermediates.

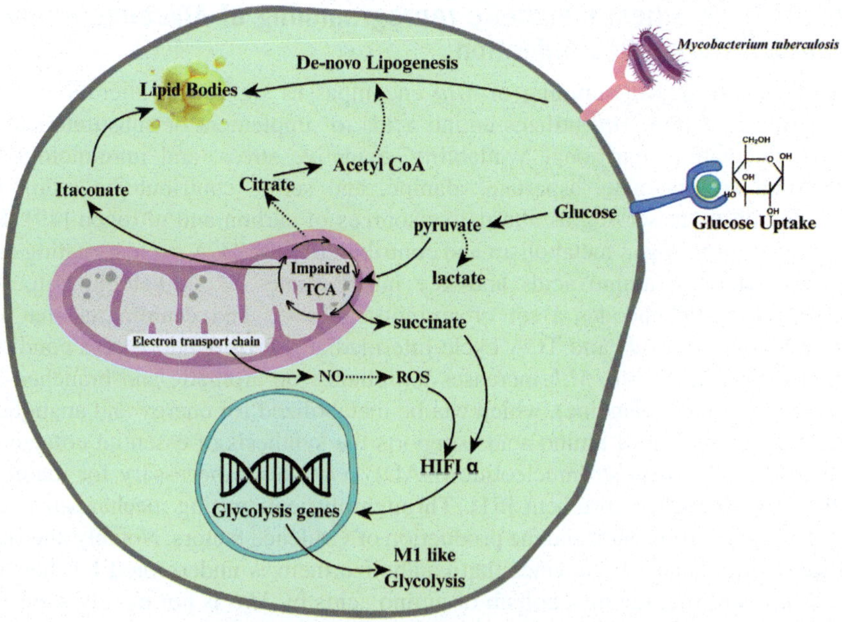

FIGURE 12.3 Immunometabolic Modulation of Macrophages During *Mycobacterium tuberculosis (Mtb)* Infection.
Mtb employs sophisticated strategies to inhibit host metabolic reprogramming by inducing type I interferon (IFN-β) and miR-21 expression. Infected M1-like macrophages display a disrupted tricarboxylic acid (TCA) cycle and accumulate succinate, which stabilizes hypoxia-inducible factor 1-alpha (HIF-1α), a key regulator of glycolysis and pro-inflammatory cytokine production, including interleukin-1β (IL-1β). Itaconate, a TCA cycle-derived metabolite, exerts immunoregulatory effects by inhibiting isocitrate lyases Icl1/2. *Mtb* infection also promotes reactive oxygen species (ROS) generation via nicotinamide adenine dinucleotide phosphate hydrogen (NADPH) oxidase activity. Concurrently, *Mtb* enhances intracellular lipid accumulation, leading to the formation of lipid droplets and the foamy macrophage phenotype. This altered phenotype may support bacterial survival by supplying cholesterol as a nutrient or hinder it through prostaglandin secretion.

These intermediates enter the β-oxidation pathway, yielding Acetyl-CoA, NADH, and $FADH_2$, which are crucial for energy production and biosynthesis. The glyoxylate shunt and β-oxidation are two catabolic processes specific to *Mtb* lipid metabolism [46]. This pathway circumvents the decarboxylation steps of the tricarboxylic acid (TCA) cycle, with enzymes such as malate synthase and isocitrate lyase (ICL) facilitating this adaptation, thereby enabling *Mtb* to preserve carbon skeletons during fatty acid catabolism [47]. This is particularly vital for survival in nutrient-restricted environments like the macrophage phagosome. The reliance on host lipids not only aids *Mtb* in avoiding immune clearance and adopting a dormant-like state but also in withstanding antibiotic treatment, thereby sustaining its persistence under stress. *Mtb*'s ability to metabolize cholesterol and interact with host lipid droplets also confers metabolic flexibility, which is essential for the long-term survival and reactivation of the mycobacterium [48].

5.5 Amino acids as allies: metabolic reprogramming of *Mycobacterium tuberculosis* during host infection

The dynamic metabolic reprogramming of *Mtb* encompasses significant alterations in amino acid processing. During infection, *Mtb* utilizes amino acids to supplement or substitute standard carbon sources due to fluctuating nutritional availability, oxidative stress, and immunological demands. Amino acids such as glutamate, aspartate, alanine, and serine contribute to redox balance and biosynthesis, in addition to serving as alternative sources of carbon and nitrogen [49]. For instance, glutamate is crucial for nitrogen metabolism and contributes to the TCA cycle by acting as a precursor for the synthesis of other amino acids and key intermediates like α-ketoglutarate. To maintain metabolic flexibility, *Mtb* encodes a set of aminotransferases and deaminases that facilitate interactions between amino acids and TCA cycle intermediates [50]. Under stress conditions such as hypoxia or nutritional limitation, *Mtb* increases its reliance on aromatic and branched-chain amino acids (valine, leucine, and isoleucine), which can be metabolized for energy and anabolic precursors. Furthermore, the metabolism of amino acids supports the synthesis of essential cofactors like glutathione and nicotinamide adenine dinucleotide $(NAD)^+$, which are necessary for maintaining redox balance in the phagosomal environment [51]. Through nutrient-sensing mechanisms, certain amino acids also regulate gene expression and the production of virulence factors. Notably, the significance of amino acid transporters and the enzymes that catabolize them is underscored by their upregulation during infection. Therefore, the metabolism of amino acids by *Mtb* is not merely supplementary but constitutes a fundamental component of its adaptive strategy to survive and proliferate within macrophages [52].

5.6 The silent survivor: metabolic reprogramming in dormant *Mycobacterium tuberculosis*

In *Mtb*, dormancy and persistence are critical survival strategies intricately linked to the bacterium's capacity for metabolic adaptation under extreme conditions [53]. Upon exposure to stressors such as hypoxia, nutrient deprivation, acidic pH, reactive oxygen and nitrogen species, or antibiotic pressure within the host environment, *Mtb* transitions into a non-replicating, metabolically altered state characterized by reduced growth and diminished metabolic activity. This latent state is associated with significant reprogramming of central metabolism, including a shift toward lipid and amino acid catabolism as alternative energy sources and the downregulation of energy-intensive processes [54]. Key metabolic pathways, such as the glyoxylate shunt and reductive branch of the TCA cycle, are activated, facilitating carbon conservation and survival in nutrient-limited conditions. Additionally, *Mtb* accumulates triacylglycerols in intracellular lipid inclusions to store energy during prolonged dormancy. The transcriptional response to hypoxia and other dormancy-inducing signals is primarily orchestrated by the dormancy survival regulator (DosR) regulon, which also enhances the expression of genes associated with stress adaptation, redox balance, and anaerobic respiration [55]. These metabolic adaptations also support persistence, a related mechanism wherein a subset of cells survives antibiotic treatment, enabling *Mtb* to evade host immunity and drug therapy. Consequently, these complex metabolic modifications ensure long-term survival and complicate the eradication of TB, particularly in latent and drug-tolerant infections [56].

6. Metabolic flexibility and drug resistance: The evasive tactics of *Mycobacterium tuberculosis*

Mtb is a well-known pathogen that has developed sophisticated strategies to resist anti-TB therapies over time. Several factors enable *Mtb* to survive in the hostile host environment, including the prevention of phagosome maturation, the recruitment of specialized immune cells to form antibiotic-resistant tubercles, and the ability to enter a dormant state by halting its metabolism (latent tuberculosis) [57]. These strategies indicate that *Mtb* has evolved multiple mechanisms to adapt to diverse hostile conditions, one of which is metabolic rewiring or metabolic flexibility [58]. Metabolic flexibility refers to the interaction between the pathogen and host, allowing the pathogen to successfully establish pathogenesis by evading the host's immune system. Understanding this complex mechanism is essential for developing novel therapeutics to overcome drug resistance. *Mtb* employs several methods of metabolic rewiring to develop drug resistance, such as cell wall remodeling, overexpression of efflux pumps, and ROS detoxification [59].

6.1 Metabolic plasticity under pressure: how *Mycobacterium tuberculosis* withstands antibiotic assault

Mtb has developed alternative strategies to modify its metabolic pathways under stress conditions. During antibiotic treatment, it alters its metabolism, exhibiting changes in its central carbon metabolism and shifting to multiple carbon sources, such as fatty acids, cholesterol, lactate, and pyruvate, to survive within macrophages. This metabolic flexibility enables it to colonize the host under stress condition [42] (Fig. 12.3).

6.1.1 Carbon control: adaptive rewiring of the TCA cycle and trehalose metabolism in Mycobacterium tuberculosis

Under stress conditions, *Mtb* modifies its carbon metabolism by altering the TCA cycle. It employs a bifurcated TCA cycle that prioritizes the production of metabolites such as succinate and malate, which are crucial for survival during hypoxic conditions. Additionally, *Mtb* possesses the Icl1 gene, which facilitates the glyoxylate shunt, thereby bypassing the oxidative branch of the TCA cycle and enabling the utilization of fatty acids as a carbon source under acidic conditions. Beyond rerouting carbon metabolism through the TCA cycle and utilizing alternative carbon sources for survival, *Mtb* regulates the flow of carbon metabolites via a trehalose catalytic shift (Fig. 12.3). Trehalose, a glucose derivative that constitutes cell wall sugars, surface trehalose monomycolate (TMM), and trehalose dimycolate (TDM) used in biofilm formation, is redirected to ensure adequate ATP synthesis [60].

6.1.2 Redox resilience: DosR-driven nitrate respiration in hypoxic Mycobacterium tuberculosis *survival*

Mtb is an obligate aerobe; however, under stress conditions such as hypoxia, it can temporarily engage in anaerobic respiration, indicating its capacity to adapt to varying oxygen levels. It utilizes nitrate and nitrite as electron acceptors in hypoxic conditions. During hypoxia, there is a significant expression of the DosR regulon. Proteins regulated by DosR facilitate several adaptations by reducing oxygen consumption and enhancing nitrate uptake for respiration [61]. Consequently, the bacterium can utilize nitrate as terminal electron acceptors in respiration instead of oxygen. This process, coupled with succinate production, enables the bacterium to maintain a proton motive force and generate ATP under anaerobic conditions, thereby supporting its survival even within granulomas [62].

6.1.3 Silencing the kill switch: metabolic dormancy and phenotypic resistance in Mycobacterium tuberculosis

Most antibiotics target metabolically active *Mtb*. In response to the stress imposed by antibiotic pressure, *Mtb* enters a slow metabolic state, thereby creating an antibiotic-tolerant condition that reduces susceptibility to first-line drugs such as Isoniazid, Rifampicin, Ethambutol, and Pyrazinamide [63]. This phenomenon is a form of phenotypic resistance. Isoniazid, a prodrug activated by the catalase-peroxidase enzyme KatG, targets mycolic acid biosynthesis by acting on the InhA gene. In slow-growing *Mtb*, resistance to Isoniazid is observed due to several factors, including reduced lipid synthesis, which decreases dependency, upregulation of the InhA gene, downregulation of the KatG gene, and upregulation of SigB and SigE, which induce resistance to Isoniazid [64]. Rifampicin, which disrupts RNA transcription by binding to RNA polymerase, involves mistranslation of the rpoB gene, reducing its binding affinity and thereby increasing tolerance. The mechanisms discussed above enable *Mtb* to develop resistance to antibiotic stress through metabolic rewiring and dormancy, allowing the bacterium to persist within the host. This metabolic flexibility permits *Mtb* to adapt to various host conditions, significantly contributing to its pathogenesis [65].

6.2 Latent lethality: non-replicating persistence and tuberculosis reactivation

The characteristic feature of Mycobacterium is its ability to persist within the host by remaining dormant and becoming active under favorable conditions, such as a weakened immune system [66]. It is hypothesized that *Mtb* enters a low metabolic state within granulomas through non-replicating persistent (NRP) cells, which, as the name suggests, do not replicate. This allows the bacterium to evade stressful conditions such as hypoxia, nutrient deprivation, and antibiotic treatment. The transformation into NRP cells is a crucial survival strategy that enables *Mtb* to endure the hostile host environment. NRP cells confer several advantages, the most significant being drug tolerance [67]. Antibiotics like rifampin target actively replicating *Mtb* cells by interfering with RNA synthesis through binding to RNA polymerase. However, since NRP cells exhibit reduced transcriptional activity, they demonstrate tolerance to rifampicin. It has been reported that rifampin's antibacterial activity is significantly diminished against NRP cells in in vitro models. As previously mentioned, by entering a latent state via NRP cells, *Mtb* evades detection by the immune system and increases drug tolerance, ultimately rendering therapy ineffective and leading to reactivation. Thus, NRP cells play a critical role in the establishment of LTBI. *Mtb* comprises 48 co-regulated genes known as the DosR regulon, which are upregulated under hypoxic conditions and are essential for the bacterium to enter a dormant or non-replicating state. During hypoxic conditions, the DosR regulon upregulates genes required for lipid metabolism, as discussed in carbon metabolism rewiring [68]. Consequently, NRP cells switch several carbon metabolism sources, such as fatty acids and cholesterol, accompanied by alternative ATP production necessary for the long-term persistence of NRP cells within the host. These metabolic flexibilities enable NRP cells to evade the host immune system and withstand rigorous antibiotic therapies, resulting in prolonged treatment [69].

6.3 Under pressure: efflux pump overexpression as a survival strategy in *Mycobacterium tuberculosis*

Efflux pumps are transport proteins that expel toxic substrates, such as antibiotics, from cells, thereby reducing the intracellular concentration of these drugs and preventing them from reaching their targets. This mechanism imparts tolerance to bacteria, which can be either intrinsic or acquired [70]. Under antibiotic stress, there is an overexpression of efflux pumps, which decreases the concentration of drugs below their minimum inhibitory concentration (MIC), rendering them ineffective. *Mtb* employs several efflux pump systems to expel antibiotics and toxic compounds, thereby contributing to drug resistance and persistence [71]. Among the key classes are ATP-Binding Cassette (ABC) transporters, which rely on ATP hydrolysis to actively transport substrates across membranes. Notable examples include Rv0194, Rv1819c (BacA), and the DrrABC system (Rv2936, Rv2937, and Rv2938), all of which have been associated with resistance to multiple drugs. The Major Facilitator Superfamily (MFS) transporters, such as Rv1258c (a Tap-like pump), utilize the proton motive force to function as drug-proton antiporters, conferring resistance to antibiotics like rifampin and fluoroquinolones [72]. Additionally, Resistance-Nodulation-Division (RND) transporters, such as the MmpS5/MmpL5 complex (Rv0676c/Rv0677c), form tripartite systems spanning the bacterial envelope and are linked to resistance against newer anti-TB drugs like bedaquiline (BDQ) and clofazimine (CFZ). These efflux systems exemplify the sophisticated mechanisms *Mtb* deploys to survive antibiotic pressure and evade therapeutic eradication [73].

6.3.1 Lipid remodeling in Mycobacterium tuberculosis*: a strategy for stress survival and drug resistance*

Under conditions of stress, *Mtb* undergoes a reconfiguration of its lipid metabolism, leading to modifications in its cell wall, such as thickening and changes in cell wall components, including the accumulation of triacylglycerol (TG) [74]. A significant alteration occurs in fatty acyls (FA), which *Mtb* utilizes as a carbon source and as building blocks for the formation of phospholipids, thereby imparting dormancy-like resistance traits. A study reported that arachidonic acid from the host can facilitate the resuscitation of dormant *Mtb*. During stress, *Mtb* is known to increase the production of mycolic acid, resulting in cell wall thickening and reduced permeability to lipophilic antibiotics such as rifampicin and isoniazid [75].

6.3.2 ROS detoxification enhances survival

Mtb effectively mitigates oxidative stress induced by antibiotics and host immune responses through its robust antioxidant systems, which are vital for its survival within host macrophages [76]. ROS, such as superoxide radicals and hydrogen peroxide, are detoxified by key enzymes including catalase and SOD, with *Mtb* utilizing multiple isoforms to withstand elevated ROS levels. Furthermore, the thioredoxin and glutathione systems sustain intracellular redox homeostasis, ensuring enzyme functionality and macromolecule repair under oxidative stress conditions. Heme oxygenase-1 (HO-1) plays a defensive role by degrading heme into carbon monoxide, a molecule possessing antioxidant and anti-inflammatory properties, and by alleviating nitrosative stress generated by host-derived reactive nitrogen species like nitric oxide [77]. The efficacy of these detoxification systems is critical for *Mtb*'s persistence during antibiotic treatment and immune challenges, as mutations in related genes can increase bacterial susceptibility to drugs, highlighting their significance in drug resistance mechanisms [78].

7. Conclusions and perspectives

Mtb survival and successful pathogenicity is a complex process involving intricate interactions between the pathogen and the host. In this chapter, we aim to elucidate the sophisticated mechanisms that enable *Mtb* to rewire its metabolic pathways and manipulate host immune cells, specifically macrophages, to enhance its virulence and persist despite anti-mycobacterial treatments. Understanding this intricate metabolic interplay is crucial in combating TB, particularly given the emergence of MDR and XDR strains.

During the initial infection, *Mtb* overcomes macrophage defenses through various mechanisms, such as arresting phagosome maturation and resisting oxidative and nitrosative stress, thereby establishing pathogenicity. The host's immune response leads to the formation of granulomas, complex structures composed of various immune cells, intended to contain the pathogen. However, these granulomas provide a protective niche for the bacterium, allowing it to survive and persist for extended periods. Within these structures, *Mtb* enters a dormant, non-replicating persistent (NRP) stage, exhibiting tolerance to antituberculosis therapies and ensuring its survival.

A critical strategy for *Mtb* survival is the reprogramming or rewiring of macrophage metabolism. This includes adaptations to varying oxygen concentrations (hypoxia), the Warburg effect, and the manipulation of host pro-inflammatory responses by altering cytokine production to support its survival. In addition to manipulating host metabolism, *Mtb* demonstrates remarkable metabolic plasticity, enabling it to utilize host lipids via pathways such as the glyoxylate shunt, adjust amino acid metabolism, and enter dormancy with reduced metabolic activity. These processes contribute to drug resistance through mechanisms including slow-growth tolerance, efflux pump overexpression, cell wall thickening, and enhanced ROS detoxification. The metabolic flexibility at the *Mtb*-macrophage interface is crucial for TB pathogenesis, persistence, and treatment outcomes, posing a significant therapeutic challenge due to the bacterium's resilience against intracellular stresses, immune evasion, and antibiotic assault. Therefore, there is a pressing need to develop novel therapeutic strategies that consider the importance of metabolic rewiring as potential targets.

References

[1] 9789240101531-eng.pdf (Internet). (cited 2024 Dec 9). Available from: https://iris.who.int/bitstream/handle/10665/379339/9789240101531-eng.pdf?sequence=1.

[2] Efflux as a mechanism for drug resistance in Mycobacterium tuberculosis | Pathogens and Disease | Oxford Academic (Internet). (cited 2024 Dec 9). Available from: https://academic.oup.com/femspd/article/63/1/1/548916.

[3] Pieters J. Evasion of host cell defense mechanisms by pathogenic bacteria. Curr Opin Immunol February 1, 2001;13(1):37–44.

[4] Magombedze G, Dowdy D, Mulder N. Latent tuberculosis: models, computational efforts and the Pathogen's regulatory mechanisms during dormancy. Front Bioeng Biotechnol August 27, 2013;1:4.

[5] Energy metabolism in health and diseases | Signal Transduction and Targeted Therapy (Internet). (cited 2025 May 1). Available from: https://www.nature.com/articles/s41392-025-02141-x.

[6] Ahmad F, Rani A, Alam A, Zarin S, Pandey S, Singh H, et al. Macrophage: a cell with many faces and functions in tuberculosis. Front Immunol May 6, 2022;13:747799.

[7] Hirayama D, Iida T, Nakase H. The phagocytic function of macrophage-enforcing innate immunity and tissue homeostasis. Int J Mol Sci December 29, 2017;19(1):92.

[8] Basheeruddin M, Qausain S. Hypoxia-inducible factor 1-alpha (HIF-1α): an essential regulator in cellular metabolic control. Cureus 2024;16(7):e63852.

[9] Kelly B, O'Neill LA. Metabolic reprogramming in macrophages and dendritic cells in innate immunity. Cell Res July 2015;25(7):771–84.

[10] Hasan Z, Razzak SA, Kanji A, Shakoor S, Hasan R. Efflux pump gene single-nucleotide variants associated with resistance in *Mycobacterium tuberculosis* isolates with discrepant drug genotypes. J Glob Antimicrob Resist September 1, 2024;38:128–39.

[11] Borah Slater K, Moraes L, Xu Y, Kim D. Metabolic flux reprogramming in Mycobacterium tuberculosis-infected human macrophages. cited 2025 Apr 15 Front Microbiol [Internet] November 17, 2023;14. Available from: https://www.frontiersin.orghttps://www.frontiersin.org/journals/microbiology/articles/10.3389/fmicb.2023.1289987/full.

[12] Liu CH, Liu H, Ge B. Innate immunity in tuberculosis: host defense vs pathogen evasion. Cell Mol Immunol December 2017;14(12):963–75.

[13] Wicherska-Pawłowska K, Wróbel T, Rybka J. Toll-like receptors (TLRs), NOD-like receptors (NLRs), and RIG-I-like receptors (RLRs) in innate immunity. TLRs, NLRs, and RLRs ligands as immunotherapeutic agents for hematopoietic diseases. Int J Mol Sci December 13, 2021;22(24):13397.

[14] Mantegazza AR, Magalhaes JG, Amigorena S, Marks MS. Presentation of phagocytosed antigens by MHC class I and II. Traffic Cph Den February 2013;14(2):135–52.

[15] Macrophages in immunoregulation and therapeutics | Signal Transduction and Targeted Therapy (Internet). (cited 2025 May 1). Available from: https://www.nature.com/articles/s41392-023-01452-1.

[16] Zhuang L, Yang L, Li L, Ye Z, Gong W. *Mycobacterium tuberculosis*: immune response, biomarkers, and therapeutic intervention. MedComm 2024;5(1):e419.

[17] Pagán AJ, Ramakrishnan L. Immunity and immunopathology in the tuberculous granuloma. Cold Spring Harb Perspect Med September 2015;5(9):a018499.

[18] Goldberg MF, Saini NK, Porcelli SA. Evasion of innate and adaptive immunity by*Mycobacterium tuberculosis*. Microbiol Spectr September 2014;2(5):MGM2.

[19] Kumar R, Singh P, Kolloli A, Shi L, Bushkin Y, Tyagi S, et al. Immunometabolism of phagocytes during *Mycobacterium tuberculosis* infection. cited 2025 Apr 15 Front Mol Biosci (Internet) October 14, 2019;6. Available from: https://www.frontiersin.orghttps://www.frontiersin.org/journals/molecular-biosciences/articles/10.3389/fmolb.2019.00105/full.

[20] Lin PL, Flynn JL. Understanding latent tuberculosis: a moving target. J Immunol Baltim Md 1950 2010 Jul 1;185(1):15–22.

[21] Host Cell Death and Modulation of Immune Response against Mycobacterium tuberculosis Infection (Internet). (cited 2025 May 2). Available from: https://www.mdpi.com/1422-0067/25/11/6255.

[22] Moraco AH, Kornfeld H. Cell death and autophagy in TB. Semin Immunol December 2014;26(6):497–511.

[23] Deretic V, Delgado M, Vergne I, Master S, De Haro S, Ponpuak M, et al. Autophagy in immunity against *Mycobacterium tuberculosis*: a model system to dissect immunological roles of autophagy. Curr Top Microbiol Immunol 2009;335:169–88.

[24] Harding CV, Boom WH. Regulation of antigen presentation by *Mycobacterium tuberculosis*: a role for Toll-like receptors. Nat Rev Microbiol April 2010;8(4):296–307.

[25] Sia JK, Rengarajan J. Immunology of *Mycobacterium tuberculosis* infections. Microbiol Spectr 2019;7(4). https://doi.org/10.1128/microbiolspec.gpp3-0022–2018.

[26] Rajaram MVS, Ni B, Dodd CE, Schlesinger LS. Macrophage immunoregulatory pathways in tuberculosis. Semin Immunol December 2014;26(6):471–85.

[27] Macrophage polarization in lymph node granulomas from cattle and pigs naturally infected with Mycobacterium tuberculosis complex—Fernanda Larenas-Muñoz, Mohamed G. Hamed, Inés Ruedas-Torres, José

María Sánchez-Carvajal, Javier Domínguez, Francisco José Pallarés, Librado Carrasco, Irene M. Rodríguez-Gómez, Jaime Gómez-Laguna, 2024 (Internet). (cited 2025 May 2). Available from: https://journals.sagepub.com/doi/abs/10.1177/03009858241231606.

[28] Hunter L, Hingley-Wilson S, Stewart GR, Sharpe SA, Salguero FJ. Dynamics of macrophage, T and B cell infiltration within pulmonary granulomas induced by *Mycobacterium tuberculosis* in two non-human primate models of aerosol infection. Front Immunol January 6, 2022;12:776913.

[29] Immune mechanisms of granuloma formation in sarcoidosis and tuberculosis—Nuffield Department of Medicine (Internet). (cited 2025 May 2). Available from: https://www.ndm.ox.ac.uk/publications/1595617.

[30] Lyu J, Narum DE, Baldwin SL, Larsen SE, Bai X, Griffith DE, et al. Understanding the development of tuberculous granulomas: insights into host protection and pathogenesis, a review in humans and animals. Front Immunol December 9, 2024;15:1427559.

[31] Krueger G, Faisal S, Dorhoi A. Microenvironments of tuberculous granuloma: advances and opportunities for therapy. Front Immunol March 24, 2025;16:1575133.

[32] Kolloli A, Subbian S. Host-directed therapeutic strategies for tuberculosis. cited 2024 May 25 Front Med (Internet) October 18, 2017;4. Available from: https://www.frontiersin.org/articles/10.3389/fmed.2017.00171.

[33] Jaworska M, Szczudło J, Pietrzyk A, Shah J, Trojan SE, Ostrowska B, et al. The Warburg effect: a score for many instruments in the concert of cancer and cancer niche cells. Pharmacol Rep August 1, 2023;75(4):876–90.

[34] Ahn S, Jung J, Jang IA, Madsen EL, Park W. Role of glyoxylate shunt in oxidative stress response. J Biol Chem May 27, 2016;291(22):11928–38.

[35] Gleeson LE, Sheedy FJ, Palsson-McDermott EM, Triglia D, O'Leary SM, O'Sullivan MP, et al. Cutting edge: *Mycobacterium tuberculosis* induces aerobic glycolysis in human alveolar macrophages that is required for control of intracellular bacillary replication. J Immunol Baltim Md 1950 March 15, 2016;196(6):2444–9.

[36] Soto-Heredero G, Gómez de las Heras MM, Gabandé-Rodríguez E, Oller J, Mittelbrunn M. Glycolysis—a key player in the inflammatory response. FEBS J August 2020;287(16):3350–69.

[37] Galli G, Saleh M. Immunometabolism of macrophages in bacterial infections. Front Cell Infect Microbiol January 29, 2021;10:607650.

[38] DeMichele E, Buret AG, Taylor CT. Hypoxia-inducible factor-driven glycolytic adaptations in host-microbe interactions. Pflugers Arch 2024;476(9):1353–68.

[39] Li F, Feng L, Jin C, Wu X, Fan L, Xiong S, et al. LpqT improves mycobacteria survival in macrophages by inhibiting TLR2 mediated inflammatory cytokine expression and cell apoptosis. Tuberculosis July 1, 2018;111:57–66.

[40] Lipid droplet formation in Mycobacterium tuberculosis infected macrophages requires IFN-γ/HIF-1α signaling and supports host defense | PLOS Pathogens (Internet). (cited 2025 Apr 15). Available from: https://journals.plos.org/plospathogens/article?id=10.1371/journal.ppat.1006874.

[41] The Mycobacterium tuberculosis β-oxidation genes echA5 and fadB3 are dispensable for growth in vitro and in vivo—PMC (Internet). (cited 2025 Apr 15). Available from: https://pmc.ncbi.nlm.nih.gov/articles/PMC3220763/.

[42] Park JH, Shim D, Kim KES, Lee W, Shin SJ. Understanding metabolic regulation between host and pathogens: new opportunities for the development of improved therapeutic strategies against *Mycobacterium tuberculosis* infection. Front Cell Infect Microbiol March 16, 2021;11:635335.

[43] Stüve P, Minarrieta L, Erdmann H, Arnold-Schrauf C, Swallow M, Guderian M, et al. De Novo fatty acid synthesis during mycobacterial infection is a prerequisite for the function of highly proliferative T cells, but not for dendritic cells or macrophages. Front Immunol April 5, 2018;9:495.

[44] Gago G, Diacovich L, Gramajo H. Lipid metabolism and its implication in mycobacteria-host interaction. Curr Opin Microbiol February 2018;41:36–42.

[45] Dewan A, Jain C, Das M, Tripathi A, Sharma AK, Singh H, et al. Intracellular peroxynitrite perturbs redox balance, bioenergetics, and Fe–S cluster homeostasis in *Mycobacterium tuberculosis*. Redox Biol September 1, 2024;75:103285.

[46] Serafini A, Tan L, Horswell S, Howell S, Greenwood DJ, Hunt DM, et al. *Mycobacterium tuberculosis* requires glyoxylate shunt and reverse methylcitrate cycle for lactate and pyruvate metabolism. Mol Microbiol October 2019;112(4):1284–307.

[47] Yang P, Liu W, Chen Y, Gong AD. Engineering the glyoxylate cycle for chemical bioproduction. Front Bioeng Biotechnol December 2, 2022;10:1066651.

[48] Kim H, Shin SJ. Revolutionizing control strategies against *Mycobacterium tuberculosis* infection through selected targeting of lipid metabolism. Cell Mol Life Sci CMLS September 14, 2023;80(10):291.

[49] Torres N, Tobón-Cornejo S, Velazquez-Villegas LA, Noriega LG, Alemán-Escondrillas G, Tovar AR. Amino acid catabolism: an overlooked area of metabolism. Nutrients July 29, 2023;15(15):3378.

[50] The Mycobacterium tuberculosis H37Ra gene MRA_1916 causes growth defects upon down-regulation | Scientific Reports (Internet). (cited 2025 May 2). Available from: https://www.nature.com/articles/srep16131.

[51] Covarrubias AJ, Perrone R, Grozio A, Verdin E. NAD+ metabolism and its roles in cellular processes during ageing. Nat Rev Mol Cell Biol February 2021;22(2):119–41.

[52] Borah K, Beyß M, Theorell A, Wu H, Basu P, Mendum TA, et al. Intracellular *Mycobacterium tuberculosis* exploits multiple host nitrogen sources during growth in human macrophages. Cell Rep December 10, 2019; 29(11):3580–3591.e4.

[53] Chang DPS, Guan XL. Metabolic versatility of *Mycobacterium tuberculosis* during infection and dormancy. Metabolites February 2, 2021;11(2):88.

[54] Liu Y, Wu Z, Li Y, Chen Y, Zhao X, Wu M, et al. Metabolic reprogramming and interventions in angiogenesis. J Adv Res April 1, 2025;70:323–38.

[55] Bhat SA, Singh N, Trivedi A, Kansal P, Gupta P, Kumar A. The mechanism of redox sensing in *Mycobacterium tuberculosis*. Free Radic Biol Med October 15, 2012;53(8):1625–41.

[56] Dartois VA, Rubin EJ. Anti-tuberculosis treatment strategies and drug development: challenges and priorities. Nat Rev Microbiol 2022;20(11):685–701.

[57] Rankine Wilson LI, Shapira T, Sao Emani C, Av-Gay Y. From infection niche to therapeutic target: the intracellular lifestyle of *Mycobacterium tuberculosis*. Microbiology 2021;167(4):001041.

[58] Singha B, Murmu S, Nair T, Rawat RS, Sharma AK, Soni V. Metabolic rewiring of *Mycobacterium tuberculosis* upon drug treatment and antibiotics resistance. Metabolites January 18, 2024;14(1):63.

[59] Eoh H, Liu R, Lim J, Lee JJ, Sell P. Central carbon metabolism remodeling as a mechanism to develop drug tolerance and drug resistance in *Mycobacterium tuberculosis*. Front Cell Infect Microbiol August 22, 2022; 12:958240.

[60] Xu Y, Pooja BK. *Mycobacterium tuberculosis* carbon and nitrogen metabolic fluxes. Biosci Rep February 4, 2022;42(2). BSR20211215.

[61] Ofori-Anyinam B, Dolganov G, Van T, Davis JL, Walter ND, Garcia BJ, et al. Significant under expression of the DosR regulon in *M. tuberculosis* complex lineage 6 in sputum. Tuberc Edinb Scotl May 2017;104:58–64.

[62] Reichlen MJ, Leistikow RL, Scobey MS, Born SEM, Voskuil MI. Anaerobic *Mycobacterium tuberculosis* cell death stems from intracellular acidification mitigated by the DosR regulon. J Bacteriol October 31, 2017; 199(23). e00320-17.

[63] Goossens SN, Sampson SL, Van Rie A. Mechanisms of drug-induced tolerance in *Mycobacterium tuberculosis*. Clin Microbiol Rev October 14, 2020;34(1). e00141-20.

[64] Dookie N, Rambaran S, Padayatchi N, Mahomed S, Naidoo K. Evolution of drug resistance in *Mycobacterium tuberculosis*: a review on the molecular determinants of resistance and implications for personalized care. J Antimicrob Chemother January 19, 2018;73(5):1138.

[65] Quinonez CG, Lee JJ, Lim J, Odell M, Lawson CP, Anyogu A, et al. The role of fatty acid metabolism in drug tolerance of *Mycobacterium tuberculosis*. mBio February 22, 2022;13(1):e0355921.
[66] Peddireddy V, Doddam SN, Ahmed N. Mycobacterial dormancy systems and host responses in tuberculosis. Front Immunol February 15, 2017;8:84.
[67] Baker JJ, Johnson BK, Abramovitch RB. Slow growth of *Mycobacterium tuberculosis* at acidic pH is regulated by phoPR and host-associated carbon sources. Mol Microbiol October 2014;94(1):56–69.
[68] Zheng H, Williams JT, Aleiwi B, Ellsworth E, Abramovitch RB. Inhibiting *Mycobacterium tuberculosis* DosRST signaling by targeting response regulator DNA binding and sensor kinase heme. ACS Chem Biol January 17, 2020;15(1):52–62.
[69] Hasenoehrl EJ, Wiggins TJ, Berney M. Bioenergetic inhibitors: antibiotic efficacy and mechanisms of action in *Mycobacterium tuberculosis*. cited 2025 May 2 Front Cell Infect Microbiol (Internet) January 11, 2021;10. Available from: https://www.frontiersin.orghttps://www.frontiersin.org/journals/cellular-and-infection-microbiology/articles/10.3389/fcimb.2020.611683/full.
[70] Sharma A, Gupta VK, Pathania R. Efflux pump inhibitors for bacterial pathogens: from bench to bedside. Indian J Med Res February 2019;149(2):129–45.
[71] Antibiotic resistance and persistence—Implications for human health and treatment perspectives—PMC (Internet). (cited 2025 May 2). Available from: https://pmc.ncbi.nlm.nih.gov/articles/PMC7726816/.
[72] Remm S, Earp JC, Dick T, Dartois V, Seeger MA. Critical discussion on drug efflux in *Mycobacterium tuberculosis*. FEMS Microbiol Rev October 12, 2021;46(1):fuab050.
[73] Szumowski JD, Adams KN, Edelstein PH, Ramakrishnan L. Antimicrobial efflux pumps and *Mycobacterium tuberculosis* drug tolerance: evolutionary Considerations. Curr Top Microbiol Immunol 2013;374. https://doi.org/10.1007/82_2012_300.
[74] Mycobacterium tuberculosis Uses Host Triacylglycerol to Accumulate Lipid Droplets and Acquires a Dormancy-Like Phenotype in Lipid-Loaded Macrophages | PLOS Pathogens (Internet). (cited 2025 May 2). Available from: https://journals.plos.org/plospathogens/article?id=10.1371/journal.ppat.1002093.
[75] Jacobo-Delgado YM, Rodríguez-Carlos A, Serrano CJ, Rivas-Santiago B. *Mycobacterium tuberculosis* cell-wall and antimicrobial peptides: a mission impossible? Front Immunol May 17, 2023;14:1194923.
[76] Tyagi P, Dharmaraja AT, Bhaskar A, Chakrapani H, Singh A. *Mycobacterium tuberculosis* has diminished capacity to counteract redox stress induced by elevated levels of endogenous superoxide. Free Radic Biol Med July 2015;84:344–54.
[77] Pae HO, Chung HT. Heme oxygenase-1: its therapeutic roles in inflammatory diseases. Immune Netw Off J Korean Soc Immunol Biol Response Modif February 2009;9(1):12–9.
[78] Shukla S, Bhardwaj N, Singh A. Drug resistance in *Mycobacterium tuberculosis*: an evolutionary perspective and its adaptation to the lung microenvironment. Microbe December 1, 2024;5:100189.

CHAPTER

13 Revolutionizing TB therapy: Emerging opportunities and unresolved challenges

Nidhi Rani[1,2] **and Avadhesha Surolia**[1,3]

[1]*Molecular Biophysics Unit, Indian Institute of Science, Bangalore, Karnataka, India;* [2]*Jawaharlal Nehru Centre for Advanced Scientific Research, Bangalore, Karnataka, India;* [3]*Dr. Reddy's Institute of Life Science, Hyderabad, Telangana, India*

Highlights

1. Tuberculosis (TB) remains a major global health challenge, with 10.8 million new cases in 2023.
2. Development of anti-TB drugs requires political commitment, resources, and public-private partnerships for rapid TB elimination.
3. Bedaquiline, delamanid, and pretomanid show potential to combat drug-resistant TB when used in novel regimens.
4. Increasing incidence of resistance to recently identified anti-TB drugs is a significant concern.
5. *Mtb* employs complex metabolic processes resistant to anti-TB drugs, with inadequate treatment adherence causing resistance.
6. Exploring drugs with unique mechanisms of action, rather than conventional targets, is urgently needed for rapid treatment.
7. The review covers traditional and recently identified drug targets, their inhibitors, and screening approaches for drug discovery.

Abbreviations

AAPs Nα-aroyl-N-aryl-phenyl alaninamide
ADEP acyldepsipeptide
ADME absorption, distribution, metabolism, and excretion
AG arabinogalactan
AIDS acquired immune deficiency syndrome
ARVs antiretroviral agents
ATP adenosine triphosphate
BPaL bedaquiline, pretomanid, and linezolid
Clp caseinolytic proteases
CRISPR clustered regularly interspaced short palindromic repeats
DOTS directly observed short-course treatment

Navigating Tuberculosis. https://doi.org/10.1016/B978-0-443-27735-1.00008-3

DR drug-resistant
DS drug-sensitive
ETC electron transport chain
FAS I fatty acid synthase I
FAS II fatty acid synthase II
FDA food and drug administration
FDC fixed-dose combination
GFP green fluorescent protein
GSK glaxo smith kline
hERG ether-a-go-go-related gene
HIV human immunodeficiency virus
HTS high-throughput screening
IUATLD international union against tuberculosis and lung disease
LC-MS liquid chromatography-mass spectrometry
MA mycolic acid
MALDI matrix-assisted laser desorption ionization
MDR-TB multidrug-resistant tuberculosis
MIC minimum inhibitory concentration
Mtb mycobacterium tuberculosis
NADH nicotinamide adenine dinucleotide hydrogen
NAG N-acetylglucosamine
NAM N-acetylmuramic acid
NITD novartis institute for tropical diseases
NO nitric oxide
NTD N-terminal domain
PaMZ bedaquiline, pretomanid, and pyrazinamide
PBPs penicillin-binding proteins
PCR polymerase chain reaction
PG peptidoglycan
PMF proton motive force
RNAP DNA-dependent RNA polymerase
RR-TB rifampicin-resistant tuberculosis
SAR specific absorption rate
SDS-PAGE sodium dodecyl sulfate-polyacrylamide gel electrophoresis
SNP single-nucleotide polymorphism
TB tuberculosis
TCA tricarboxylic acid
USA United States of America
WGS whole-genome sequencing
WHO World Health Organization
XDR-TB extensively drug-resistant tuberculosis

1. Introduction

TB, caused by *Mtb*, is an ancient chronic inflammatory disease that infects approximately one-quarter of the world's population [1–3]. It is a global disease because of its prevalence in almost every country. *Mtb* often infects the lungs, leading to pulmonary TB; however, it can also affect other body parts (the kidney, lymph nodes, spine, brain, and other organs), causing extrapulmonary TB disease [4]. When

TB-infected individuals sneeze, cough, sing, talk, or spit, individuals with compromised immunity, including HIV-infected individuals, are more prone to contracting TB than healthy individuals [5].

1.1 An overview of TB cases based on the global tuberculosis report of 2024

According to the "Global Tuberculosis Report of 2024" released by the World Health Organization (WHO), approximately 10.8 million people were infected, and 1.25 million deaths were estimated from TB in 2023 [1]. The majority of individuals identified for TB infection each year belong to 30 high-TB-burden countries, assessed for 87% of the total global TB infections reported in 2023, while 56% of TB infections were reported in five countries, including India (26%), Indonesia (10%), China (6.8%), the Philippines (6.8%), and Pakistan (6.3%). Additionally, the infection percentage among men and women is 55% and 33%, respectively, while in children and young adolescents, it is 12% in 2023. During the period of the COVID-19 pandemic, i.e., during 2020 (1.40 million deaths) and 2021 (1.42 million deaths), TB death cases increased from 2019 (1.34 million deaths). A significant challenge in fighting TB is coinfection with human immunodeficiency virus (HIV). Approximately half of the people infected with HIV/acquired immune deficiency syndrome (AIDS) are diagnosed with active TB, and approximately 15% of AIDS patients worldwide die of TB every year. Globally, in 2023, TB deaths have declined to 1.25 million deaths, with 161,000 deaths among HIV-infected individuals and 1.09 million deaths among HIV-negative individuals. Overall, the global reduction rate in TB cases from 2015 to 2023 was 8.3%, below the WHO milestone for 2025, a 50% reduction. However, progress in TB reduction rates of 24% and 27% was observed in the African and European regions, respectively, while the other 79 countries achieved a 20% reduction in TB cases. Increasing drug resistance of *Mtb* is another major challenge in TB control globally. In 2023, 175923 people were diagnosed with multidrug-resistant or rifampicin-resistant TB (MDR-TB/RR-TB) (TB resistant to rifampicin and isoniazid). The lengthy and complex nature of first-line TB drug treatment, including misuse, incomplete, and insufficient use of frontline drugs, is also a contributing factor to the emergence of drug-resistant strains. Patient non-compliance with the treatment course, with limited access to diagnosis and treatment, increases the risk of the emergence of MDR-TB strains [6]. Additionally, poverty, overcrowding, malnutrition, complex diagnostics, and poor healthcare infrastructure remain hurdles in controlling TB. Overall, the TB report of 2024 indicated a positive sign against TB but remained distant from the goal of the WHO End TB Strategy milestone.

1.2 Novel anti-TB therapy

The antibiotics that constitute the first-line agents for TB treatment and are advocated by the WHO are isoniazid, rifampicin, pyrazinamide, ethambutol, and streptomycin [7,8]. Rifampicin and isoniazid are the first choices for the treatment because of their efficacy and low toxicity. In the United States of America (USA), due to its high resistance rate, streptomycin is not considered a first-line TB drug [9]. The second line of TB drugs is used when first-line drugs are ineffective or used to treat only resistant TB diseases, and are also considered first-line therapy for resistant TB cases (extensively drug-resistant TB (XDR-TB) or MDR-TB) [10]. Second-line TB drugs include aminoglycosides (amikacin and kanamycin), cycloserine, fluoroquinolones (ciprofloxacin, levofloxacin, and moxifloxacin), polypeptides (capreomycin, viomycin, and enviomycin), terizidone, and thioamides (ethionamide and prothionamide). Additionally, drugs with low efficacy and/or unproven efficacy are categorized as third-line TB drugs or special drugs, including rifabutin, macrolides, linezolid, thioacetazone, thioridazine, arginine, vitamin D, and bedaquiline. The third line of drugs is newly discovered drugs, which are also reserved for complex TB cases, such as MDR and/or XDR-TB. Among these, clarithromycin and linezolid are ineffective in treating TB. Although Rifabutin is effective because of its cost to developing countries, it is not included on the WHO list. However, the requirement for several drugs for treating TB also leads to non-adherence to prescriptions by the patient. Therefore, single-dose formulation or fixed-dose combination (FDC) tablets were introduced by the WHO and the International Union against Tuberculosis and Lung Disease (IUATLD) in 1994. FDC tablets simplified TB therapy and facilitated patient compliance with treatment without any errors, as recommended by the physician. The FDC improved drug supply management errors, costs, procurement, storage, and distribution. In 2011, the WHO included two FDC drug formulations (isoniazid + rifampicin and isoniazid + ethambutol), three drug formulations (isoniazid, rifampicin, ethambutol, and isoniazid + rifampicin + pyrazinamide), and four drug formulations (isoniazid + rifampicin + ethambutol + pyrazinamide). Despite the simplified TB therapy by anti-TB FDC formulation, queries related to their bioavailability and effectiveness have not been explained; therefore, the WHO, International Standards for TB Care, and the American Thoracic Society recommend that FDC formulations to be used only for active TB cases.

1.3 Emergence of drug resistance in the *Mtb*

The emergence of drug-resistant TB is a global hurdle in TB control. Based on the resistivity of various drugs, the WHO categorized drug-resistant TB into five groups [1]: isoniazid-resistant TB (resistant only to isoniazid) [2], rifampicin-resistant TB (resistant only to rifampicin, known as DR-TB) [3], multidrug-resistant TB (MDR-TB) [4], pre-extensively drug-resistant TB (pre-XDR-TB), and [5] extensively drug-resistant TB (XDR-TB) [11]. Additionally, resistance to both isoniazid and rifampicin was classified as MDR-TB [12]. The emergence of antibiotic resistance in *Mtb* generally follows a pattern in which resistance to isoniazid arises first, followed by rifampicin, pyrazinamide, and ethambutol resistance, and resistance to second-line TB drugs.

Over the years, *Mtb* has developed numerous resistance mechanisms to survive under extreme environmental conditions, including the host immune response and effects of anti-TB drugs. These mechanisms are classified as intrinsic and acquired resistance, which include impermeability of the cell envelope, drug degradation and modification, efflux pumps, phenotypic tolerance, epistasis, target mimicry, redox-based interventions, and drug target modification [13–15]. In *Mtb*, intrinsic resistance is mainly caused by its unique and complex waxy cell envelope, which reduces the cell wall permeability of drugs [16]. The effectiveness of certain anti-TB drugs, such as β-lactam antibiotics (hydrolyzed by β-lactamases) and various aminoglycoside/cyclic peptide antibiotics (altered by methylation or acetylation), is neutralized enzymatically by changing their structure. The mycobacterial efflux system is another significant challenge for drug resistance in *Mtb,* as its overexpression has been observed under various stress conditions [17]. Also, mycobacterial efflux pumps ability to extrude streptomycin, rifampicin, isoniazid, clofazimine, bedaquiline, fluoroquinolones, and ethambutol is reported. However, their role in generating resistance in *Mtb* is controversial.

Furthermore, due to the prolonged treatment of TB, the mycobacterial population known as persisters adapts to their stressful environments through physiological modifications, leading to phenotypic tolerance and hereditary resistance. However, acquired resistance is caused by a heritable change in the mycobacterial genome, contributing to the expansion of resistant *Mtb* strains. Resistance to front-line anti-TB drugs, such as isoniazid, which inhibits mycolic acid biosynthesis, is generated by mutations in the *InhA* and *KatG* genes encoding InhA and catalase-peroxidase, respectively. Similarly, a mutation in the *RpoB* gene, which encodes the β-subunit of bacterial RNA polymerase responsible for transcription, caused resistance to rifampicin. Additionally, mutations in *PncA*, *PanD*, and *RpsA* create resistance against pyrazinamide, and mutations in *EmbB* and *UbiA* cause resistance to ethambutol.

Resistivity against the recently identified drugs, bedaquiline, delamanid, and pretomanid, has emerged because of mutations in their molecular targets. Resistance against bedaquiline in *Mtb* was raised mainly because of mutations in the genes *AtpE*, *Rv0678*, and *PepQ*, with 17.3%, 5.26%, and 3.1%, respectively [18–21]. Target-based point mutation studies discovered A63P and I66M mutations in the membrane-spanning region of adenosine triphosphate (ATP) synthase that prevented bedaquiline binding to its c subunit, consequently increasing the minimum inhibitory concentration (MIC). A mutation in *Rv0678* causes non-target-based bedaquiline (2–8 times increased in MIC) and clofazimine resistance (2–8 times increased in MIC). The *Rv0678* gene of *Mtb* is a transcriptional regulator of the Mmps2-MmpL2, MmpS4-MmpL4, and MmpS5-MmpL5 efflux systems, and the *Rv0678* mutation leads to the overexpression of these efflux pumps, leading to bedaquiline resistance [18]. Similarly, *PepQ* increased non-target-based resistance to bedaquiline and MIC by 4-fold. A combination study of bedaquiline and clofazimine conducted in mouse models showed mutations in *PepQ* (frameshift mutations −C in Arg271 and +C in Ala14) [18,22]. Several new genes were identified for bedaquiline resistance, such as *Glyph, CobQ2, Pita, Gid, Cya, LppB, TrcR, Rv0071, Rv0405, Rv0953c, Rv1723, Rv2426c, Rv2477c, Rv2722, Rv2820c, Rv2823c, Rv3510c, and Rv3785* [23]. Resistance to delamanid develops due to mutations in the genes involved in the F420-dependent bioactivation pathway, namely, *FbiA, FbiB, FbiC, FbiD, Fgd1*, and *Ddn* [24,25]. The *Ddn* gene carries out the metabolism and activation of delamanid into its active form, which undergoes various mutations, including amino acid deletions, single missense mutations, and mutations in the stop codon. Impaired activity of these genes leads to an increased MIC in MDR-TB isolates for delamanid [26,27]. Resistance to pretomanid (PA-824) is associated with mutations in *FbiA, FbiB, FbiC, FbiD, Fgd1*, and *Ddn*, similar to delamanid [28,29].

2. Tuberculosis drug target and recent developments

2.1 The mycobacterium cell wall: Anti-TB targets, drugs, and mechanism of action

The mycobacterial cell wall is unique and one of the most complex bacterial envelopes. It plays a key role in survival under environmental stress conditions [30]. It consists of three major constituents: peptidoglycan (PG), arabinogalactan (AG), and mycolic acid (MA) [31]. The PG layer gives cells rigidity, integrity, and shape [32]. PG comprises N-acetylglucosamine (NAG) and N-acetylmuramic acid (NAM) residues linked by β [1–4] bondsin mycobacteria. Mur ligases (MurC/D/E/F) produce the pentapeptide L-alanyl-D-isoglutaminyl-meso-diaminopimel-yl-D-alanyl-D-alanine, which is acylated on the muramic acid component by polysaccharide strands. Cross-linkages in PG are formed by the penicillin-binding proteins (PBPs) PonA1 (*Rv0050*), PonA2 (*Rv3682*), and transpeptidases (L, D, and LdtMt1-5) [33,34]. The mycobacterial cell wall is a strong protective barrier against pathogens and is essential for bacterial viability. It provides structural strength to mycobacteria, is necessary for their growth, and responds to the host's immune response. Therefore, targeting the proteins involved in mycobacterial cell wall synthesis is an effective strategy for developing new and effective anti-TB drugs [35]. Targeting the mycobacterial cell wall is promising as a front-line anti-TB drug, including isoniazid and ethambutol, and several second-line anti-TB drugs, including ethionamide, thioacetazone, cycloserine, and terizidone, target it.

2.2 Novel anti-TB drugs for mycobacterial cell wall synthesis inhibition

The frontline anti-TB drugs isoniazid, ethambutol, and ethionamide are well-studied mycobacterial cell wall synthesis inhibitors [36,37]. Isonazid inhibits cell wall synthesis by blocking fatty acid synthases. Isoniazid is a prodrug activated by the mycobacterial catalase-peroxidase enzyme KatG. In the presence of isoniazid and nicotinamide adenine dinucleotide hydrogen (NADH), KatG synthesizes an isonicotinic acyl-NADH complex, which binds strongly to enoyl-acyl carrier protein reductase (InhA) and prevents its actual enoyl-AcpM substrate binding, ultimately inhibiting mycolic acid synthesis, a significant component of the mycobacterial cell wall. Resistance to isoniazid mainly occurs due to the activation failure caused by the point mutation (Arg463Leu and Ser315Thr with 62.1% and 55.2% frequency, respectively) in the KatG gene [38].

Additionally, mutations in the *InhA* gene and several other genes, such as *AhpC, FadE, FurA, IniABC, KasA, Ndh, Rv1592c,* and *Rv1772*, also contribute to isoniazid resistance. However, KatG and InhA mutation frequencies are comparatively higher and contribute primarily to lowering the efficacy of isoniazid. Ethambutol also belongs to the category of first-line drugs against TB. This bacteriostatic drug prevents the formation of key structural components of the mycobacterial cell wall, arabinogalactan, and lipoarabinomannan, by inhibiting arabinosyl transferases (*EmbA, EmbB*, and *EmbC*) and impeding mycobacterial cell division. Resistance to this drug arises primarily due to mutations in the *EmbB* gene, and the codons *EmbB306, EmbB406*, and *EmbB497* are considered hot spots for mutation [39,40]. These *EmbB* mutations alter the enzyme's catalytic activity, preventing the association of arabinogalactan with the mycobacterial cell wall. Ethambutol resistance also arises from the *Mtb* efflux system, which reduces drug concentration inside the mycobacterial cells. Another cell wall inhibitor drug is ethionamide, a second-line anti-TB drug commonly used in combination with other anti-TB drugs to treat active multidrug-resistant TB. Ethionamide is also a prodrug and structural analog of isoniazid, with a mechanism of action similar to isoniazid. EthA, a flavoprotein monooxygenase, activates ethionamide, which binds to $NAD+$ and inhibits InhA, similar to isoniazid. Mutations in the *EthA* gene lead to ethionamide resistance by reducing the enzyme activity responsible for activating the drug [40].

Additionally, some ethionamide-resistant strains developed because of the increased expression of efflux pumps, leading to a decreased drug concentration inside the mycobacterial cells. Another contributor to making ethionamide less effective is altering the target enzyme, InhA, due to mutations [40]. Several studies have reported that resistance may occur by activating alternative mycobacterial metabolic pathways that detour EthA, rendering ethionamide ineffective.

2.3 Recently identified mycobacterial cell wall synthesis inhibitors

2.3.1 Peptidoglycan biosynthesis and inhibitors

Inhibition of PG synthesis is a promising metabolic step for disrupting mycobacterial cell wall synthesis, which is completed in three stages. The first stage begins with the synthesis of precursor molecules, which starts in the cytoplasmic space, and enzymes such as GlmU, MurA, MurB, NamH, MurC, MurD, MurE, MurF, MurX, and MurG are involved. GlmU encodes a bifunctional acetyltransferase/uridyltransferase (acetyltransferase function absent in humans) and catalyzes the formation of UDP-GlcNAc for lipopolysaccharide and peptidoglycan synthesis. The substrate (GlcN-1-P) analog of GlmU, 2-amino-2,3-dideoxy-3-fluoro-α-D-glucopyranosyl phosphate, showed promising inhibitory activity. Similarly, Mur ligases (MurA, B, C, D, E, F, X, and G) are indispensable for mycobacterial cell integrity, and their lack of counterparts in eukaryotes makes them promising *Mtb* drug targets [41]. Multiple inhibitors have been identified for each class of Mur enzyme to date. Although fosfomycin is a clinically approved MurA inhibitor, it is ineffective against mycobacteria because of the structural differences in its active site. For MurX/MurY, nucleoside-peptide structural scaffolds bind to different enzyme sites, such as puromycin, liposidomycin, apramycin, and rapamycin. Using a synthetic library of over 7000 rapamycin analogs, SQ641 was identified as the best and is currently in preclinical development. Cycloserine, a structural analog of D-alanine, is a broad-spectrum antibiotic that inhibits PG synthesis by competitively binding to D-alanine-D-alanine ligase (Ddl). The second stage of PG synthesis involves the translocation of membrane-bound precursors to the periplasm. In the third stage, polymerization of the precursor and peptide cross-linking complete this process. Antibiotics such as ramoplanin (lipoglycodepsipeptide), enduracidin, and teixobactin prevent subsequent transglycosylation; teixobactin is known for its anti-*Mtb* activity, whereas others remain to be explored [42,43]. Generally, *Mtb* is resistant to β-lactam antibiotics due to the hydrolysis of the lactam ring by BlaC and redox potential in *Mtb* coordinated by *WhiB4*. However, a β-lactam antibiotic, carbapenem in combination with meropenem, and the β-lactamase inhibitor clavulanate demonstrated potent activity against *Mtb*. Additionally, carbapenems combined with rifampicin are effective against drug-sensitive (DS) and rifampicin-resistant *Mtb* strains [44].

2.3.2 Arabinogalactan biosynthesis inhibitors

The arabinogalactan layer of the mycobacterial cell wall consists of galactose and arabinose sugar moieties located in the central layer of the cell wall, covalently linked to both the PG (via linkage to MurNGlyc) and mycolic acid layers. Biosynthesis of the AG layer involves a multistep enzymatic process carried out by enzymes such as WecA, WbbL, GlfT1, GlfT2, Araf, DPA, PrsA, UbiA, DprE1, DprE2, DPX, AftA, EmbA, EmbB, AftB, AftC, and AftD. EmbB is a well-proven target, hallmarked by ethambutol, as a first-line TB drug to prevent AG assembly in the mycobacterial cell wall. Recently, a nucleoside antibiotic, CPZEN-45, has been in preclinical trials for its use against TB. CPZEN-45 targets WecA, indispensable for *Mtb* survival, and prevents the growth of both replicating and non-replicating *Mtb* by disrupting AG synthesis. Studies have also shown its efficacy against DS and DR, along with a synergistic effect when combined with other TB drugs [45]. A biochemical study by Peltier et al. identified derivatives of UDP-Galf modified at C-5 and C-6 that inhibited GlfT, an essential enzyme for *Mtb* AG synthesis. Several studies have also reported inhibitors of DprE1, such as nitrobenzothiazinone (BTZ) (BTZ043), benzothiazole, triazoles, nitrobenzamide benzothiazinone (PyrBTZ01, PBTZ169), thiophene (TCA1), quinoxaline, dinitrobenzamide, thiadiazole, azaindole, pyrazolopyridine, 3,4-dihydrocarbostyril derivative (OPC-167832), and piperidine amides [46–51]. Among these compounds, BTZ043, PyrBTZ01, PBTZ169, TBA7371, and OPC-167832 are currently undergoing clinical trials.

2.3.3 Mycolic acid biosynthesis and inhibitors

One of the essential and unique features of the *Mtb* is "mycolic acid," which provides a hydrophobic lipid layer that is impenetrable for small hydrophilic molecules, including antibiotics. It is the outermost layer of the organism and consists of long-chain fatty acids comprising α-alkyl (C24-C26) and β-hydroxy (C42-C62) chains attached covalently to the underlying AGs. Mycolic acid biosynthesis occurs in the cytoplasm via two interconnected enzymatic pathways: fatty acid synthase I (FAS I) and fatty acid synthase II (FAS II). FAS I synthesizes short chains of up to C24 fatty acyl CoA, which can either be used to synthesize the α-branch or C16/C18-CoA, which are directly used in FAS II for the production of meromycolic acid. Other enzymes involved in synthesis are FabH, FabD, MabA, β-ketoacyl-AcpM reductase, HadAB/BC, InhA, KasA, KasB, MmpL3, FadD32, and Pks13. InhA, enoyl-ACP reductase, an enzyme involved in the FAS II system of mycolic acid biosynthesis, is an established target for isoniazid and ethionamide [52]. A thiadiazole-based compound, GSK693, identified from a target-based screening approach using the Glaxo Smith Kline (GSK) compound collection, is at the preclinical development stage [52]. MmpL3, an essential enzyme that functions as a mycolic acid transporter and is mandatory for incorporating mycolic acid into the *Mtb* cell wall, has emerged as one of the most popular targets for TB drug discovery [53]. SQ109 (Sequella), identified as an MmpL3 inhibitor while screening a library of 63,238 ligands, was designed using the scaffold of the 1,2-ethylenediamine pharmacophore of ethambutol. SQ109 is considered a distinct mechanism of action drug, exhibiting activity against *Mtb*, MDR-TB, and XDR-TB with the potential to shorten the treatment duration in combination with other TB drugs [54]. SQ109 is undergoing Phase I clinical trials in the USA (NCT01358162, NCT01585636) and Phase II trials in South Africa (NCT01785186). Specific absorption rate (SAR) optimization of the chemical scaffold of 1,3,5-trisubstituted pyrazole led to the identification of BM212 and BM635 for MmpL3 inhibition with significant anti-tubercular activity. Similarly, compounds NITD-304 and NITD-349 (indolcarboxamides) developed by the Novartis Institute for Tropical Diseases (NITD) bind MmpL3 and inhibit DS and drug-resistant (DR) strains [55]. In addition, these compounds have shown anti-TB efficacy against acute and chronic infections in murine models [55,56]. Using a phenotypic whole-cell high-throughput screening campaign by industrial and academic collaboration resulted in the discovery of an indazole sulphonamide derivative, GSK 724 (DG167 or GSK3011724A) (Preclinical Development stage) against the KasA (β-ketoacyl-ACP synthase) enzyme with anti-tubercular activity [57]. Additionally, the X-ray crystal structure of the KasA-DG167 complex at 2.0 Å resolution revealed the structural details of their interaction at the molecular level, and whole-genome sequencing (WGS) of resistant mutants generated a single-nucleotide polymorphism (SNP) in the kasA gene, which confirmed this essential gene as a target [58]. In an in vitro study and an acute mouse model of TB, DG167 enhanced isoniazid activity and showed a synergistic effect [58]. Enzymes involved in mycolic acid synthesis, including FadD32 and Pks13, are also being explored as potential drug targets. FadD32 catalyzes the activation of the meromycolic acid branch from the FAS II system, followed by Pks13, which condenses it with the α-alkyl fatty acid branch from the FAS I system [59]. Recently, using a green fluorescent protein (GFP)-based high-throughput whole-cell assay, a coumarin derivative (4,6-diaryl-5,7-dimethyl coumarin) was identified from a library of 20,502 compounds and exhibited antimycobacterial potency in vitro and in vivo. Subsequently, high-throughput virtual screening of 19,271 isoxazole derivatives against FadD32 identified 2,4-dibromo-6-[3-(trifluoromethyl)-1,2-oxazol-5-yl]- phenol, demonstrating their inhibitory potential against *Mtb* in vitro, ex vivo, and in vivo [60].

In 2014, delamanid (OPC67683), a first-in-class nitro-dihydroimidazooxazole drug developed by Otsuka Pharmaceutical and approved by the European Medicines Agency (Europe, Japan, and South Korea), was undergoing phase II clinical trials (NCT01424670) in combination with standard TB treatment [61]. It is included in the list of "Essential Medicines" by the WHO. In vitro studies, the drug exhibited vigorous dose-dependent antimycobacterial activity against sensitive and resistant strains, with MICs ranging from 6 to 24 ng/mL. Delamanid is a prodrug activated by mycobacterial deazaflavin-dependent nitroreductase (Ddn), an enzyme involved in the F420-dependent antioxidant mechanism that shields *Mtb* from oxidative stress and bactericidal agents [62]. This enzyme uses dihydro-F420 (reduced form) and generates nitric oxide (NO) and a highly reactive metabolite that attacks DprE2, an enzyme essential for synthesizing the cell wall arabinogalactan [62]. Similarly, PA-824, also known as pretomanid, is a prodrug developed by the Global TB Alliance and is under the bedaquiline, pretomanid, and linezolid (BPaL) drug regimen. The activation and mechanism of action of pretomanid are similar to those of delamanid through Ddn and DprE2, respectively [62]. Approval for its use in the USA and the European Union was granted in August 2019 and July 2020, respectively. The US Food and Drug Administration (FDA) added a first-in-class medication to the list of essential medicines by the WHO.

3. Mycobacterial DNA replication, transcription, and translation: Anti-TB targets, drugs, and mechanism of action

Several drugs from the first- and second-line TB drug regimens, including rifampin (inhibits transcription), fluoroquinolones (DNA unwinding and replication), levofloxacin, moxifloxacin (DNA unwinding and replication), streptomycin, amikacin, kanamycin, and capreomycin (translation), target mycobacterial DNA replication and protein synthesis. These drugs target crucial enzymes involved in DNA replication and protein synthesis, which are required for *Mtb* growth. One of the promising targets of the mycobacterial DNA replication mechanism is DNA gyrase, a topoisomerase II [63]. In *Mtb* DNA, gyrase is encoded by two genes, *GyrA* and *GyrB*. DNA gyrase is an ATP-dependent enzyme that catalyzes transient double-stranded DNA breaks and facilitates DNA replication. The second line of TB drugs, fluoroquinolones, bind to the active site of DNA gyrase subunit A (GyrA), disrupting DNA replication and leading to permanent DNA breaks. Next-generation fluoroquinolones, moxifloxacin, levofloxacin, gatifloxacin, ciprofloxacin, and ofloxacin, follow a similar mechanism of action as fluoroquinolones by inhibiting GyrA [64]. Moxifloxacin exhibits activity against replicating and non-replicating *Mtb* by inactivating GyrA and inducing oxidative stress [65]. A natural cyclic depsidecapeptide compound, griselimycin, produced by *Streptomyces griseus*, is also effective against sensitive and drug-resistant *Mtb* by inhibiting the DNA polymerase sliding clamp, DnaN [66]. However, rifampin, a first-line TB drug, inhibits RNA synthesis by inhibiting DNA-dependent RNA polymerase (RNAP). RNAP is involved in both transcription initiation and elongation. Structurally, *Mtb* RNAP consists of a central core that is divided into five subunits ($\alpha 2\beta\beta'\omega$) and at least one additional σ-subunit. Rifampin prevents RNA elongation by binding to the RNAP β subunit, which is encoded by *RpoB* [67]. In rifampin resistance, the mutation in *the RpoB* gene reduces the binding affinity for rifampin, thereby generating resistance in *Mtb* to rifampin. This specific mutation in *RpoB* was used to diagnose MDR *Mtb*, and the broad-spectrum antibiotic linezolid targets mycobacterial RNA translation [68]. Unlike most β-lactams, which are semisynthetic, linezolid is an entirely synthetic drug, and the first oxazolidinone is endorsed for treating *Mtb*. It binds to the 23S subunit of the 50S subunit of the mycobacterial ribosome, impairing the formation of the initiation complex for protein synthesis [69].

3.1 Recently identified inhibitors of mycobacterial replication, transcription, and translation

Using biochemical and genetic approaches, a thiophene-based compound was identified that targets the allosteric pocket of DNA gyrase and transforms it into toxic DNA-cleavage complexes, similar to fluoroquinolones. This compound targets the allosteric pocket of DNA gyrase, thereby avoiding cross-resistance to fluoroquinolones. However, this allosteric pocket remains to be explored in *Mtb* DNA gyrase [70]. The aminobenzimidazole derivative SPR719 (developed by Bill & Melinda Gates Medical Research Institute), the active form of SPR720, has been identified as a DNA gyrB inhibitor [71]. SPR719 targets the ATPase subunit of mycobacterial DNA gyrase B. In vitro and in vivo murine infection models showed potency against *Mtb* [72]. An in vivo combination study of SPR720 with rifampicin and pyrazinamide showed comparable efficacy to that of the bedaquiline/rifampicin/pyrazinamide regimen (Stop TB Partnership) (https://www.newtbdrugs.org/pipeline/compound/spr720Accessed September 12, 2019). A recent study on the intrapulmonary pharmacokinetics of SPR719 following oral administration of SPR720 to healthy volunteers showed no unexpected safety findings (NCT05955586) [73]. Several chemical classes that target mycobacterial transcription mechanisms have also been identified. A high-throughput screening approach to screen a library of 114,000 synthetic compounds using a 384-well microplate-based fluorescence-detection assay identified an Nα-aroyl-N-aryl-phenyl alaninamide (AAPs) compound. This compound exhibited specific activity against mycobacterial RNAP [74]. AAPs bind to the RNAP bridge helix located at the N-terminus of mycobacterial RNAP.

Furthermore, a second-generation oxazolidinone, sutezolid (PNU-100480), a thiomorpholinyl moiety-containing drug developed by Sequella Inc. (TB Alliance), was used to inhibit protein synthesis. Sutezolid impedes the assembly of the 70S initiation complex by binding to the 23S rRNA of the 50S subunit. The antimycobacterial activity of sutezolid is better than that of linezolid, and currently, it is in a phase II clinical trial stage for use against TB. A combined study of sutezolid with bedaquiline, delamanid, and moxifloxacin against DS *Mtb* is at Phase II (NCT03959566). Delpazolid (LCB01-0371), another drug candidate, belongs to the oxazolidinone series developed by LegoChem Biosciences, Inc. Delpazolid, an amidrazone ring-containing drug, has broad-spectrum activity against various gram-positive pathogens, including potency against *Mtb*, and is in the clinical trial stage (NCT02836483) [75]. A previous study on *Mycobacterium abscessus* indicated that delpazolid shares a molecular target with other oxazolidinone drugs, and resistance to delpazolid arises because of mutations in the *RplC* gene, which encodes the 50S ribosomal protein [76]. However, mutations in the 23S ribosome conferred high resistance to both delpazolid and linezolid, whereas mutations in *the RplD* gene induced resistance only to linezolid, indicating that these two oxazolidinones have different mechanisms of action, possibly because of other binding sites in the ribosomal peptidyl transferase center [77].

3.2 Mycobacterial energy metabolism

To support the higher energy demand for survival under environmental stress, mycobacterial energy metabolism depends on two interlinked pathways, substrate-level phosphorylation and oxidative phosphorylation, to generate ATP. Substrate-level phosphorylation occurs under aerobic conditions, where glycolysis of carbohydrates and fatty acids provides acetyl-CoA to begin the tricarboxylic acid (TCA) cycle. This is followed by sequential enzymatic steps that transform acetyl-CoA to citrate and then to isocitrate, which finally breaks down into carbon dioxide. However, *Mtb* switches to a glyoxylate shunt under anaerobic conditions to preserve its carbon content. During this process, isocitrate is cleaved into glyoxylate and succinate by isocitrate lyase 1 (IcL1) [78]. Glyoxylate is converted to malate and fed back into the TCA cycle. Carbon is generated through substrate-level phosphorylation in the electron transport chain (ETC), which initiates the next metabolic pathway, oxidative phosphorylation. Oxidative phosphorylation is carried out by the ETC and ATP synthase, where electrons are harvested from central metabolic pathways, including substrate-level phosphorylation, used by NADP dehydrogenases (Type 1 and Type II), followed by transfer to the menaquinone pool. Quinones are re-oxidized by the cytochrome bc1-aa3 complex or the cytochrome bd-type terminal oxidase. The proton motive force (PMF) produced during various steps of the ETC is consumed by mycobacterial ATP synthase to generate ATP.

3.3 Novel anti-TB drug against mycobacterial energy metabolism, anti-TB targets, drugs, and mechanism of action

ATP synthase is a key enzyme in the mycobacterial oxidative phosphorylation process and is of special relevance because it has been discovered as a potential anti-TB drug target, bedaquiline [79]. Bedaquiline was sold under the Sirturo trade name and was approved for medical use in the USA in 2012. Bedaquiline was developed and identified as TMC207 and R207910 in 2005 during phenotypic screening of chemical compound libraries against *M. smegmatis*. Mutants resistant to bedaquiline generated mutations in the F0 domain of the ATP synthase enzyme and revealed its mechanism of action. Bedaquiline disrupts the ATP synthesis by binding to the c and ε subunit of F-ATP synthase, impairing the mycobacterial proton pumping function. Bedaquiline exhibits activity against both DS and DR *Mtb*. The FDA has approved this drug for the treatment of pulmonary TB in adults and pediatric patients [80]. The use of bedaquiline in combination therapy along with pretomanid and linezolid (BPaL) was evaluated in a clinical trial (NCT02333799) (developed by the Global TB Alliance and Mylan) and approved for use in the treatment of XDR-*Mtb* by the FDA. Moreover, a second multidrug regimen, in which bedaquiline was combined with pretomanid, bedaquiline, and pyrazinamide (BPaMZ), has been studied against DS and DR TB under Phase II/Phase III clinical trials (NCT03338621) (WHO Rapid Communication: Key changes to treatment of multidrug- and rifampicin-resistant tuberculosis (MDR/RR-TB) (WHO archived from the original on August 20, 2018. Retrieved April 24, 2019). Owing to the associated adverse effects of bedaquiline, the chemical scaffold of diarylquinolines was explored and identified as second-generation diarylquinolines TBAJ-587 and TBAJ-876, with similar or better potency and fewer side effects [81,82]. These two compounds are at the preclinical stages of development. Another imidazopyridine-based compound, Q203 (trade name Telacebec, developed by Qurient Co., Ltd.), was obtained by screening a library of 121,156 compounds for antimicrobial activity against *Mtb* within macrophages [83]. Genetic sequencing identified a spontaneous resistant mutant arising from Q203 in *the QcrB*, which encodes the cytochrome b subunit of the bc1 complex with a single amino acid substitution (T313A). The safety profiles of Q203 were comparatively more promising than those of bedaquiline, as multiple CYP450 targets and the human ether-a-go-go-related gene (hERG), potassium channel remained unaffected with a shorter terminal half-life. Both Phase 1a (NCT02530710) and Phase 1b (NCT02858973) clinical trials have demonstrated their tolerability in humans. A combination study of Q203, bedaquiline, and clofazimine assessed via *an* in vitro assay displayed a synergistic effect against intracellular *Mtb* [84]. Another compound, lansoprazole (prevacid), was identified as an inhibitor of the cytochrome bc1 complex in a drug repurposing study. Lansoprazole, an FDA-approved drug used as a gastric proton pump inhibitor, was discovered from high-throughput screening of an FDA-approved drug library against *Mtb*-infected MRC-5 lung fibroblast cells [85]. The ex vivo efficacy of this compound against *Mtb* is 22-fold higher than its in vitro efficacy. Pharmacokinetic and target identification studies have revealed the mechanism of action of lansoprazole in killing *Mtb* by targeting its cytochrome bc1 complex through intracellular sulfoxide reduction to lansoprazole sulfide [85]. Additionally, compounds such as phenoxyalkylbenzimidazole, 2-(quinolin-4-yloxy)acetamides (QOAs), imidazo[4,5-c] pyridine-type compounds, imidazo[1,2-a] pyridine-3-carboxamides, and arylvinylpiperazine amides have been explored as cytochrome bc1/aa3 super complex (QcrB) inhibitors [86–89].

Clofazimine (lamprene), a riminophenazine class of antibiotic used together with rifampicin to treat leprosy (*Mycobacterium leprae*), was repurposed against *Mtb* and tested in a Phase III trial in combination with other drugs (NCT03828201). Owing to its poor efficacy against *Mtb*, the WHO placed it on the second-tier list of recommended compounds for regimens to treat DR *Mtb*. The mechanism of action revealed that clofazimine binds to type II NADH dehydrogenase via a reductive reaction, followed by oxidation and release of ROS, leading to cell death. Another type II NADH dehydrogenase inhibitor, thioquinazoline (CBR-1825) and tetrahydroindazole, was identified by high-throughput screening [90]. Compound 2-mercapto-quinazolinones, with low potency (MIC = 0.3 uM) as antimycobacterial, were also identified as inhibitors of the type II NADH dehydrogenase pathway [91]. Enzymes involved in menaquinone biosynthesis (MenA-MenJ) are also promising targets against *Mtb*, as they lack their counterpart in humans [92]. However, clinical evidence of this pathway has not yet been provided. Moreover, compounds such as alkylamino-methanone-based scaffolds and 7-methoxy-2-naphthol have been identified as MenA inhibitors, and GSK1733953A (DG70) as MenG inhibitors [93]. The energy released due to the generation of protons across the periplasm produces the PMF required to initiate ATP synthesis. Studies have shown the role of PMF-dependent efflux pumps in creating innate resistance in *Mtb* against many first- and second-line anti-TB drugs. Therefore, inhibitors targeting ATP synthesis or PMF may help improve TB therapy [94]. SQ109 is speculated to have three mechanisms of action: disruption of PMF, interference in cell wall assembly by targeting MmpL, and inhibition of menaquinone biosynthesis. Owing to the proven genetic essentiality of mycobacterial ICL, which is based on the structural scaffold of its substrate inhibitors, such as itaconate, itaconic anhydride, 3-bromopyruvate, oxalate, malate, and 3-nitropropionate (3-NP) [95]. These inhibitors displayed activity against *Mtb* but failed to reach the clinical lead due to their associated cytotoxicity. Another set of compounds was also identified from high-throughput screening with improved cytotoxicity profile and exhibits inhibitory activity against ICL, such as salicylanilide derivatives, benzanilide derivatives, pthalazinyl derivatives, and copper complexes of pyruvate-isoniazid.

3.4 Mycobacterial proteolysis: Anti-TB targets, drugs, and mechanism of action

Proteostasis and proteolysis are *Mtb* shielding strategies to combat host cell-mediated responses. Proteostasis is the mechanism by which *Mtb* maintains an intricate network of chaperones to support its cellular proteins in their native conformations under adverse conditions. Proteolysis is a process by which damaged, overexpressed, or dysfunctional proteins are degraded by cellular proteases. There are two types of mycobacterial ATP-dependent proteases: caseinolytic proteases (Clp) and proteasomes. Targeting proteolysis is an emerging area of TB drug development [96]. Chemical scaffolds that target serine catalytic sites, where ClpP1P2 binds to the chaperone protein (ClpX/ClpC), include peptidyl boronate and β-lactones. At the same time, cyclic acyldepsipeptide (ADEP) prevents the binding of the chaperone protein (ClpX/ClpC) to ClpP1P2.

In contrast, actinomycetes derived cyclic peptides, including cyclosporin A, cumin, vancomycin, and rapamycin, inhibit ClpC and exhibit antimycobacterial activities in the nanomolar or micromolar range [97,98]. Ecumicin (MJM5123) was identified through high-throughput screening of natural compounds isolated from 7000 actinomycete strains, which led to the isolation of ecumicin (isolated from Nonomuraea sp.). Ecumicin enhanced the ATPase activity of ClpC by several fold, leading to reduced proteolysis by ClpP1P2 [97]. To date, ecumicin is the only known cyclic peptide with good in vitro activity against DS/DR *Mtb* and in vivo efficacy in mouse models of acute TB infection. Lassomycin (isolated from *Lactobacillus kentuckyensis* sp. IO0009804) is another natural cyclic peptide identified from high-throughput screening of soil bacterial extracts, with a mechanism of action similar to ecumicin. Another ClpC inhibitor, cyclomarin A (CNB-982), isolated from Streptomyces spp., binds to the ClpC protein at the N-terminal domain (NTD) and activates the ATPase activity of ClpC, resulting in uncontrolled proteolysis by ClpP1P2, ultimately causing cell death [99]. Despite their proven role in mycobacterial survival, pathogenicity, and currently identified potent inhibitors, none of the *Mtb* ClpP1P2 inhibitors have progressed into the clinical pipeline.

4. Conclusions and perspectives

One of the significant challenges in improving TB therapy is shortening the length and complexity of the TB treatment protocol, which adversely affects patient adherence to TB drugs and is a substantial cause of DR-TB emergence [100]. Therefore, the WHO introduced a cost-effective global health intervention, directly observed short-course treatment (DOTS or TB-DOTS), which includes direct observation by trained personnel for the consumption of TB medications [100]. This approach is known to be the best curative method for TB. However, the level of implementation varies, as it is demanding for patients and TB healthcare staff. Inadequate adherence to the TB drug regimen severely affects both infected individuals and public health, with increased TB resistance. When strictly treated according to the regulated program, TB cure rates with the standard drug regimen are pretty high, exceeding 90%. Therefore, the identification of drugs with a shortened treatment duration, along with a reduced number of drugs/dosage, thereby improving adherence, is key to improving the treatment of active TB. In addition, developing drugs with novel mechanisms of action will be effective against resistant TB cases (MDR-TB and XDR-TB). Another challenge in the treatment of TB is the high predominance of *Mtb* coinfection with HIV, where progression from latent to active TB is anticipated to be approximately fifty-fold higher in HIV-positive individuals than in HIV-negative individuals. TB treatment in HIV-positive patients is complicated by drug-drug interactions between antiretroviral agents (ARVs) and key TB drugs, particularly rifampin [101]. Hence, the development of drugs against TB coinfection with HIV that can be readily co-administered with antiretroviral agents is required.

The mandate for major drug discovery initiatives is to develop drugs that, in combination, create shorter, safer, and simpler regimens to cure all TB patients. Several organizations worldwide, including governmental research institutions, philanthropic organizations, academia, and the public or private sector, are working closely toward improving TB therapy. Additionally, in recent years, increased awareness of human health risks has attracted pharmaceutical biotechnology companies and government officials to include TB as their highest priority despite the lack of strong societal incentives. However, hurdles such as limited profitability and the need for significant investment from the government or private sector obstruct pharmaceutical companies' efforts to develop novel drug therapies for TB. Overcoming obstacles while conducting clinical trials, including discovering novel molecules effective against *Mtb* and/or MDR-TB and XDR-TB, and obtaining accurate laboratory specimens, is imperative. Furthermore, improving treatment efficacy by discovering new drugs and modifying existing drugs is essential in the current scenario.

Despite the availability of an extensive chemical library with several tools to predict their target and absorption, distribution, metabolism, and excretion (ADME) properties, compounds screened in a high-throughput manner for activity against the essential pathway/enzyme have failed to reach the clinical investigation stage, and drugs with novel mechanisms of action are particularly underrepresented. Only two new drugs, bedaquiline and pretomanid, have been approved by the US FDA for TB treatment. Understanding the mechanism of action of any identified drug plays an essential role in developing and discovering antibiotics. However, its mechanism of action is not necessary for its regulatory approval. Moreover, it provides the identity of the molecular target(s), which assists in lead optimization and understanding the resistance process. In recent years, advancements in WGS technology have been used to explore the molecular mechanisms of TB drugs during the drug screening process [102]. Spontaneous resistant mutants grown in vitro on solid media have become the most common and straightforward approach for finding the target gene using a putative inhibitor. This approach is widely used in high-throughput platforms to study the mechanism of action, where a mutation is introduced into mycobacteria and/or targets for further study at the metabolic and/or molecular levels, along with the conditional expression of a target gene. The discovery of bedaquiline and its mechanism of action via ATP synthase encoded by *AtpE* was identified using the WGS approach from in vitro-generated resistant strains of *M. smegmatis*.

Transcriptional profiling was introduced in TB research in the early 2000s and has since been widely used to explore *Mtb* physiology and for TB drug discovery. Sometimes, when generating resistant mutants while screening the drug candidate is unachievable, finding a target using WGS "Transcriptional profiling" is another approach to uncover mechanistic information on a broader scale (Fig. 13.1). Generally, bacteria respond to chemical or physical stress by altering their transcripts to rewire their cellular regulatory networks for survival. Transcriptional responses of *Mtb* at the genomic level in the presence of a drug can be used as supplementary information to reveal the expression levels of key genes to adapt to drug stress. This information helps explore the primary target and downstream perturbations caused by the drug. One such example is bedaquiline, where the upregulation of genes in the ATP synthase operon, which is the primary target of bedaquiline, was observed.

Additionally, two other non-traditional drug targets, *Rv0324* and *Rv0880,* were upregulated, contributing to the tolerance of *Mtb* to bedaquiline [103]. However, transcriptional profiling is an indirect approach requiring further experimental validation. Additionally, the transcriptional profile

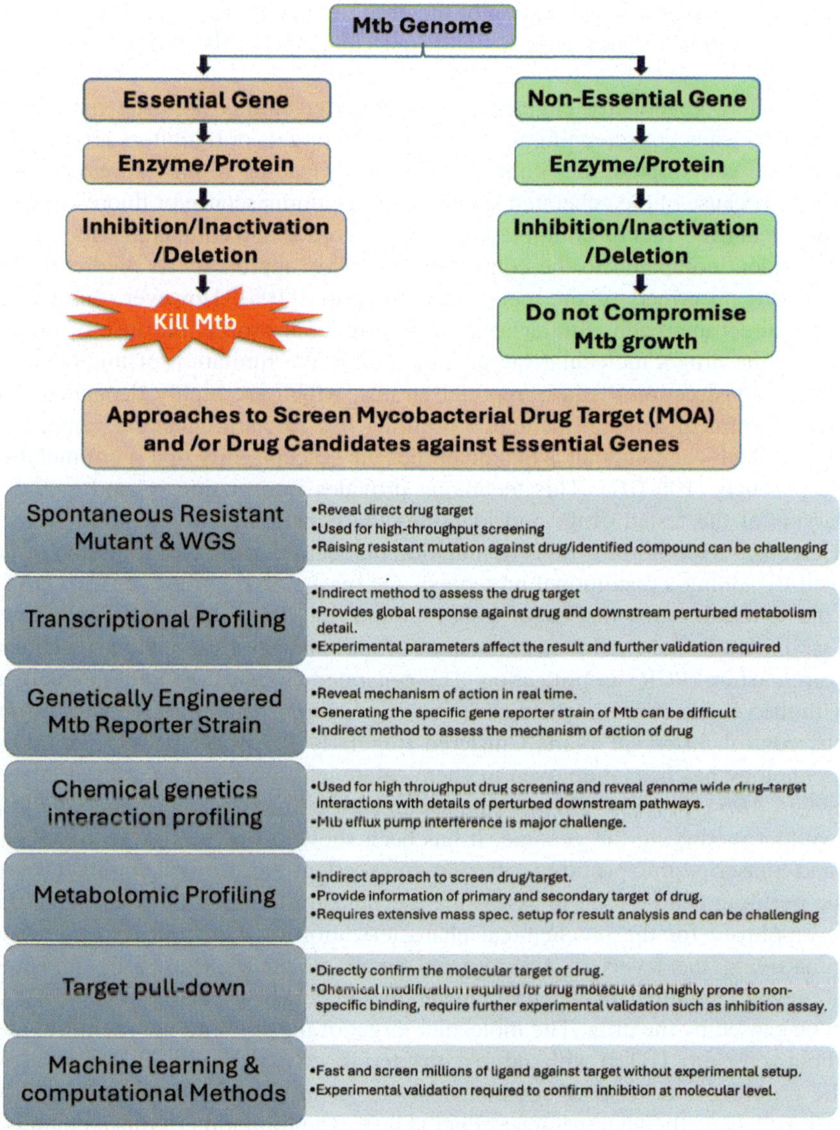

FIGURE 13.1 Schematic of the Approaches Used for Anti-Tubercular Drug Discovery.

Schematic representation of strategies for screening mycobacterial drug targets and candidates against *Mycobacterium tuberculosis* (Mtb) essential genes. Essential genes, when inhibited or deleted, lead to bacterial death, while non-essential genes do not compromise Mtb growth. The figure outlines six key approaches: (1) Spontaneous resistant mutants & whole-genome sequencing (WGS), which identify direct drug targets but face challenges in resistance induction; (2) Transcriptional profiling using engineered reporter strains to reveal real-time mechanisms of action; (3) Chemical-genetic interaction profiling for high-throughput screening of genome-wide drug-target interactions, limited by efflux pump interference; (4) Metabolomic profiling, an indirect method to elucidate primary and secondary drug targets, requiring advanced mass spectrometry; (5) Target pull-down assays, which directly confirm molecular targets but are prone to nonspecific binding; and (6) Machine learning/computational methods, enabling rapid virtual screening but requiring experimental validation. Each method is critically evaluated for its utility, limitations, and need for downstream validation in anti-TB drug discovery.

depends on various factors, such as duration of the drug treatment, different drug concentrations, and different growth conditions, which pose a significant challenge to the reproducibility of the results (Fig. 13.1).

Furthermore, using genetically engineered *Mtb* reporter strains for drug mechanism of action prediction is another aspect of screening for mycobacterial targets or inhibitors [104]. When a specific cellular metabolism is disrupted, the promoter is induced, leading to the expression of the reporter, which is readable because of the generated signals, such as luminescence or fluorescence. During drug treatment at different time intervals, luminescence was recorded and compared with the fold-change in luminescence readout compared to the control groups. This approach can predict the primary, secondary, or downstream effects of the mechanism of action [105]. However, this technique is also considered an indirect approach, similar to the transcriptional profile, because of the lack of information regarding the drug's molecular target. In addition, the limitation of this method is that it is dependent on the construction of these reporter strains, which are highly dependent on the chosen promoter genes. Chemical genetic interaction profiling is a forward chemical genetics approach for screening a set of chemical compounds that quantifies the susceptibility of a set of mutants depleted in specific gene products [106,107]. This technique provides information related to the primary and secondary targets of the tested drugs and identifies the genetic determinants of intrinsic drug susceptibility. Here, *Mtb* hypomorph strains (a mutation that causes reduced expression of protein/RNA) were generated by adding a conditional proteolysis tag (caseinolytic protease (Clp) DAS tag), with a barcode at the C-terminus of the target genes of interest. Drug libraries were screened against hypomorph strains, followed by the quantification of chromosomal barcodes and sequencing by polymerase chain reaction (PCR) amplification. The generated chemical-genetic interaction profile was compared with the vehicle control, where each compound represented the change in strain abundance relative to the control. In recent years, Clustered Regularly Interspaced Short Palindromic Repeats (CRISPR) technology has been employed in chemical genetics to advance drug discovery and target validation against *Mtb* [108,109]. This technique can be used to knock out genes and to control their expression without modifying the genome. It has been shown to be more effective than conditional proteolysis and transcriptomic profiling, offering an excellent tool for developing TB drugs.

Metabolic profiling can be used to assess the mechanism of action of drugs. Traditionally, it has been widely employed for the biochemical characterization of metabolic enzymes and pathways. However, quantifying the levels of accumulated small molecules caused by impaired metabolic pathways or enzymes signifies that bacterial metabolic processes are disrupted, directly or indirectly, which might be caused by the drug. The molecular target of the amidino-urea scaffold identified from high-throughput screening (HTS) with antimycobacterial activity (in vitro and in vivo) was explored via metabolic profiling and WGS of spontaneously resistant mutants, which revealed its molecular target to be phosphopantetheine transferase (PptT) [110]. The limitation of this method is that it traces metabolites upstream of the primary drug target, and analyzing the extensive data generated can be challenging. The target pull-down method in drug discovery is considered a direct method for capturing molecular targets. This approach has been used extensively in chemical biology. The identified inhibitor was immobilized onto a solid matrix through a linker, and the cell lysates were passed through this method. After washing the matrix, the immobilized protein was eluted, isolated, and characterized owing to its binding with the inhibitor. This requires liquid chromatography-mass spectrometry (LC-MS), sodium dodecyl sulphate-polyacrylamide gel electrophoresis (SDS-PAGE), native-PAGE, trypsin digestion, and identification of peptide fragments by Matrix-Assisted Laser

Desorption Ionization (MALDI), followed by computational software for analysis. However, chemical modification of the drug and non-specific binding interference are the limitations of this method. In recent years, machine learning and advancements in artificial intelligence have emerged in drug discovery, which is fast and can screen many chemical libraries in less time. Screening chemical libraries has also led to the identification of several drug candidates. Pranlukast was identified as a novel allosteric inhibitor for screening FDA-approved drugs against the ornithine acetyltransferase (MtArgJ) enzyme involved in the arginine biosynthesis pathway, with activity against DS and DR *Mtb* in vitro, with proven efficacy ex vivo and in vivo [111].

A comparative metabolic study of pranlukast-treated and untreated *Mtb* revealed pranlukast-mediated metabolic changes [112]. Subsequently, a preclinical mouse model of TB discovered pranlukast efficacy in reducing the *Mtb* burden in the lungs and aiding in the healing of the lungs by upregulating Ym^1/Ym^2 positive $Arg1^+$ macrophages. Thus, Prankulast is a double-edged weapon and an anti-tubercular agent [113]. Similarly, high-throughput virtual screening of 19,271 isoxazole derivatives against FadD32 led to the identification of 2,4-dibromo-6-[3-(trifluoromethyl)-1,2-oxazol-5-yl]- phenol with antimycobacterial activity in vitro, ex vivo, and in vivo [60]. Moreover, the computationally developed algorithm compares the structural similarity of input compounds or targets to compounds with experimentally validated targets (natural substrate or ligand) from the databases. It predicts the target characteristics based on existing biochemical assays. Machine learning will be at the forefront of TB drug discovery due to the availability of substantial experimental datasets, such as the *Mtb* metabolome, transcriptome, genome, and corresponding phenotypes.

Acknowledgments

AS is a Distinguished Visiting Professor at Dr. Reddy's Institute of Life Science, Hyderabad, India. NR is a Research Associate of Department of Biotechnology grant to AS (BT/ PR39006/BRB/10/1883/2020).

Declaration

We hereby confirm that the figure used in the manuscript is original (drawn for this book chapter).

References

[1] WH. Organization, Global tuberculosis report. World Health Organization (2024).

[2] Goletti D, Al-Abri S, Migliori GB, Coler R, Ong CWM, Esposito SMR, Tadolini M, Matteelli A, Cirillo D, Nemes E, Zumla A, Petersen E. World Tuberculosis Day 2023 theme "Yes! We Can End TB!". Int J Infect Dis 2023;130:S1. https://doi.org/10.1016/j.ijid.2023.04.006.

[3] Chakaya J, Khan M, Ntoumi F, Aklillu E, Fatima R, Mwaba P, Kapata N, Mfinanga S, Ehtesham Hasnain S, Katoto PDMC, Bulabula ANH, Sam-Agudu NA, Nachega JB, Tiberi S, McHugh TD, Abubakar I, Zumla A. Global tuberculosis report 2020—reflections on the global TB burden, treatment and prevention efforts. Int J Infect Dis 2021;113:S7. https://doi.org/10.1016/j.ijid.2021.02.107.

[4] Ravimohan S, Kornfeld H, Weissman D, Bisson GP. Tuberculosis and lung damage: from epidemiology to pathophysiology. Eur Respir Rev 2018;27(147). https://doi.org/10.1183/16000617.0077-2017.

[5] Isticato R, Ricca E. Spore surface display. The bacterial spore: from molecules to systems. Wiley; 2016.

[6] Monde N, Munyeme M, Chongwe G, Johansson Wensman J, Zulu M, Siziya S, Tembo R, Siame KK, Shambaba O, Malama S. First and second-line anti-tuberculosis drug-resistance patterns in pulmonary tuberculosis patients in Zambia. Antibiotics 2023;12(1):166. https://doi.org/10.3390/antibiotics12010166.

[7] WH. Organization, WHO treatment guidelines for isoniazid-resistant tuberculosis: supplement to the WHO treatment guidelines for drug-resistant tuberculosis (2018).

[8] Sileshi T, Tadesse E, Makonnen E, Aklillu E. The impact of first-line anti-tubercular drugs' pharmacokinetics on treatment outcome: a systematic review. Clin Pharmacol Adv Appl 2021;13:1–12. https://doi.org/10.2147/cpaa.s289714.

[9] Nahid P, Dorman SE, Alipanah N, Barry PM, Brozek JL, Cattamanchi A, Chaisson LH, Chaisson RE, Daley CL, Grzemska M, Higashi JM, Ho CS, Hopewell PC, Keshavjee SA, Lienhardt C, Menzies R, Merrifield C, Narita M, O'Brien R, Peloquin CA, Raftery A, Saukkonen J, Schaaf HS, Sotgiu G, Starke JR, Migliori GB, Vernon A. Official American Thoracic Society/Centers for Disease Control and Prevention/Infectious Diseases Society of America Clinical Practice Guidelines: Treatment of Drug-Susceptible Tuberculosis. Clin Infect Dis 2016;63(7):e147. https://doi.org/10.1093/cid/ciw376.

[10] Sotgiu G, Tiberi S, D'Ambrosio L, Centis R, Zumla A, Migliori GB. WHO recommendations on shorter treatment of multidrug-resistant tuberculosis. Lancet 2016;387(10037):2486–7. https://doi.org/10.1016/S0140-6736(16)30729-2.

[11] G. WHO, Global tuberculosis report (2020).

[12] WH. Organization, Diagnosis and notification of multidrug-resistant TB. PDF. WHO MDR TB Factsheet. Update (2016).

[13] Jones RM, Adams KN, Eldesouky HE, Sherman DR. The evolving biology of *Mycobacterium tuberculosis* drug resistance. Front Cell Infect Microbiol 2022;12. https://doi.org/10.3389/fcimb.2022.1027394.

[14] Poulton NC, Rock JM. Unraveling the mechanisms of intrinsic drug resistance in *Mycobacterium tuberculosis*. Front Cell Infect Microbiol 2022;12. https://doi.org/10.3389/fcimb.2022.997283.

[15] Mishra S, Shukla P, Bhaskar A, Anand K, Baloni P, Jha RK, Mohan A, Rajmani RS, Nagaraja V, Chandra N, Singh A. Efficacy of β-lactam/β-lactamase inhibitor combination is linked to WhiB4-mediated changes in redox physiology of *Mycobacterium tuberculosis*. eLife 2017;6. https://doi.org/10.7554/eLife.25624.001.

[16] Nguyen L. Antibiotic resistance mechanisms in *M. tuberculosis*: an update. Arch Toxicol 2016;90(7):1585–604. https://doi.org/10.1007/s00204-016-1727-6.

[17] Liu J, Shi W, Zhang S, Hao X, Maslov DA, Shur KV, Bekker OB, Danilenko VN, Zhang Y. Mutations in efflux pump Rv1258c (Tap) cause resistance to pyrazinamide, isoniazid, and streptomycin in *M. tuberculosis*. Front Microbiol 2019;10. https://doi.org/10.3389/fmicb.2019.00216.

[18] Nguyen TVA, Anthony RM, Bañuls AL, Vu DH, Alffenaar JWC. Bedaquiline resistance: its emergence, mechanism, and prevention. Clin Infect Dis 2018;66(10):1625–30. https://doi.org/10.1093/cid/cix992.

[19] An Q, Lin R, Yang Q, Wang C, Wang D. Evaluation of genetic mutations associated with phenotypic resistance to fluoroquinolones, bedaquiline, and linezolid in clinical *Mycobacterium tuberculosis*: a systematic review and meta-analysis. J Glob Antimicrob Resist 2023;34:214–26. https://doi.org/10.1016/j.jgar.2023.05.001.

[20] Ismail N, Omar S, Ismail N, Peters R. Collated data of mutation frequencies and associated genetic variants of bedaquiline, clofazimine and linezolid resistance in *Mycobacterium tuberculosis*. Data Brief 1975;20.

[21] Wu SH, Chan HH, Hsiao HC, Jou R. Primary bedaquiline resistance among cases of drug-resistant tuberculosis in Taiwan. Front Microbiol 2021;12. https://doi.org/10.3389/fmicb.2021.754249.

[22] Almeida D, Ioerger T, Tyagi S, Li SY, Mdluli K, Andries K, Grosset J, Sacchettini J, Nuermberger E. Mutations in pepQ confer low-level resistance to bedaquiline and clofazimine in *Mycobacterium tuberculosis*. Antimicrob Agents Chemother 2016;60(8):4590–9. https://doi.org/10.1128/AAC.00753-16.

[23] Guo Q, Bi J, Lin Q, Ye T, Wang Z, Wang Z, Liu L, Zhang G. Whole genome sequencing identifies novel mutations associated with bedaquiline resistance in *Mycobacterium tuberculosis*. Front Cell Infect Microbiol 2022;12. https://doi.org/10.3389/fcimb.2022.807095.

[24] Khoshnood S, Taki E, Sadeghifard N, Kaviar VH, Haddadi MH, Farshadzadeh Z, Kouhsari E, Goudarzi M, Heidary M. Mechanism of action, resistance, synergism, and clinical implications of delamanid against multidrug-resistant *Mycobacterium tuberculosis*. Front Microbiol 2021;12. https://doi.org/10.3389/fmicb.2021.717045.

[25] Fujiwara M, Kawasaki M, Hariguchi N, Liu Y, Matsumoto M. Mechanisms of resistance to delamanid, a drug for *Mycobacterium tuberculosis*. Tuberculosis 2018;108:186–94. https://doi.org/10.1016/j.tube.2017.12.006.

[26] Rancoita PMV, Cugnata F, Gibertoni Cruz AL, Borroni E, Hoosdally SJ, Walker TM, Grazian C, Davies TJ, Peto TEA, Crook DW, Fowler PW, Cirillo DM. Validating a 14-drug microtiter plate containing bedaquiline and delamanid for large-scale research susceptibility testing of mycobacterium tuberculosis. Antimicrob Agents Chemother 2018;62(9). https://doi.org/10.1128/AAC.00344-18.

[27] Schena E, Nedialkova L, Borroni E, Battaglia S, Cabibbe AM, Niemann S, Utpatel C, Merker M, Trovato A, Hofmann-Thiel S, Hoffmann H, Cirillo DM. Delamanid susceptibility testing of *Mycobacterium tuberculosis* using the resazurin microtitre assay and the BACTEC™ MGIT™ 960 system. J Antimicrob Chemother 2016;71(6):1532–9. https://doi.org/10.1093/jac/dkw044.

[28] Nguyen TVA, Nguyen QH, Nguyen TNT, Anthony RM, Vu DH, Alffenaar JWC. Pretomanid resistance: an update on emergence, mechanisms and relevance for clinical practice. Int J Antimicrob Agents 2023;62(4). https://doi.org/10.1016/j.ijantimicag.2023.106953.

[29] Rifat D, Li S-Y, Ioerger T, Shah K, Lanoix J-P, Lee J, Bashiri G, Sacchettini J, Nuermberger E. Mutations in fbiD (Rv2983) as a novel determinant of resistance to pretomanid and delamanid in *Mycobacterium tuberculosis*. Antimicrob Agents Chemother 2020;65(1). https://doi.org/10.1128/aac.01948-20.

[30] Bhat ZS, Rather MA, Maqbool M, Lah HU, Yousuf SK, Ahmad Z. Cell wall: a versatile fountain of drug targets in *Mycobacterium tuberculosis*. Biomed Pharmacother 2017;95:1520–34. https://doi.org/10.1016/j.biopha.2017.09.036.

[31] Batt SM, Burke CE, Moorey AR, Besra GS. Antibiotics and resistance: the two-sided coin of the mycobacterial cell wall. Cell Surf 2020;6:100044. https://doi.org/10.1016/j.tcsw.2020.100044.

[32] Alderwick LJ, Harrison J, Lloyd GS, Birch HL. The mycobacterial cell wall—peptidoglycan and arabinogalactan. Cold Spring Harb Perspect Med 2015;5.

[33] Filippova EV, Kiezer KJ, Luan CH, Wawrzak Z, Kiryukhina O, Rubin EJ, Anderson WF. Crystal structures of the transpeptidase domain of the *Mycobacterium tuberculosis* penicillin-binding protein PonA1 reveal potential mechanisms of antibiotic resistance. FEBS J 2016:2206–18. https://doi.org/10.1111/febs.13738.

[34] Basta LAB, Ghosh A, Pan Y, Jakoncic J, Lloyd EP, Townsend CA, Lamichhane G, Bianchet MA. Loss of a functionally and structurally distinct LD-transpeptidase, LdtMt5, compromises cell wall integrity in mycobacterium tuberculosis. J Biol Chem 2015;290(42):25670–85. https://doi.org/10.1074/jbc.M115.660753.

[35] Abrahams KA, Besra GS. Mycobacterial cell wall biosynthesis: a multifaceted antibiotic target. Parasitology 2018;145(2):116–33. https://doi.org/10.1017/S0031182016002377.

[36] Combrink M, Loots Du T, du Preez I. Metabolomics describes previously unknown toxicity mechanisms of isoniazid and rifampicin. Toxicol Lett 2020;322:104–10. https://doi.org/10.1016/j.toxlet.2020.01.018.

[37] Shetye GS, Franzblau SG, Cho S. New tuberculosis drug targets, their inhibitors, and potential therapeutic impact. Transl Res 2020;220:68–97. https://doi.org/10.1016/j.trsl.2020.03.007.

[38] Unissa AN, Subbian S, Hanna LE, Selvakumar N. Overview on mechanisms of isoniazid action and resistance in *Mycobacterium tuberculosis*. Infect Genet Evol 2016;45:474–92. https://doi.org/10.1016/j.meegid.2016.09.004.

[39] Zhao LL, Sun Q, Liu HC, Wu XC, Xiao TY, Zhao XQ, Li GL, Jiang Y, Zeng CY, Wan KL. Analysis of embCAB mutations associated with ethambutol resistance in multidrug-resistant mycobacterium tuberculosis isolates from China. Antimicrob Agents Chemother 2015;59(4):2045–50. https://doi.org/10.1128/AAC.04933-14.

[40] Brossier F, Sougakoff W, Bernard C, Petrou M, Adeyema K, Pham A, Breteque DAde la, Marine V, Vincent J, Sola C, Veziris N. Molecular analysis of the embCAB locus and embR gene involved in

ethambutol resistance in clinical isolates of *Mycobacterium tuberculosis* in France. Antimicrob Agents Chemother 2015;59(8):4800–8. https://doi.org/10.1128/aac.00150-15.

[41] Shinde Y, Ahmad I, Surana S, Patel H. The Mur enzymes chink in the armour of *Mycobacterium tuberculosis* cell wall. Eur J Med Chem 2021;222:113568. https://doi.org/10.1016/j.ejmech.2021.113568.

[42] Wu M-C, Styles MQ, Law BJC, Winona Struck A-, Nunns L, Micklefield J. Engineered biosynthesis of enduracidin lipoglycopeptide antibiotics using the ramoplanin mannosyltransferase Ram29. Microbiology 2015;161(7):1338–47. https://doi.org/10.1099/mic.0.000095.

[43] Piddock LJ. Teixobactin, the first of a new class of antibiotics discovered by iChip technology? J Antimicrob Chemother 2015;70(10).

[44] Kaushik A, Makkar N, Pandey P, Parrish N, Singh U, Lamichhane G. Carbapenems and rifampin exhibit synergy against *Mycobacterium tuberculosis* and Mycobacterium abscessus. Antimicrob Agents Chemother 2015;59(10):6561–7. https://doi.org/10.1128/aac.01158-15.

[45] Pitner RA, Durham PG, Stewart IE, Reed SG, Cassell GH, Hickey AJ, Carter D. A spray-dried combination of capreomycin and CPZEN-45 for inhaled tuberculosis therapy. J Pharmaceut Sci 2019;108(10):3302–11. https://doi.org/10.1016/j.xphs.2019.05.024.

[46] Chikhale RV, Barmade MA, Murumkar PR, Yadav MR. Overview of the development of DprE1 inhibitors for combating the menace of tuberculosis. J Med Chem 2018;61(19):8563–93. https://doi.org/10.1021/acs.jmedchem.8b00281.

[47] De Groote MA, Jarvis TC, Wong C, Graham J, Hoang T, Young CL, Ribble W, Day J, Li W, Jackson M, Gonzalez-Juarrero M, Sun X, Urs A. Ochsner, optimization and lead selection of benzothiazole amide analogs toward a novel antimycobacterial agent. Front Microbiol 2018;9. https://doi.org/10.3389/fmicb.2018.02231.

[48] Chirke SS, Krishna JS, Rathod BB, Bonam SR, Khedkar VM, Rao BV, Sampath Kumar HM, Shetty PR. Synthesis of triazole derivatives of 9-ethyl-9H-carbazole and dibenzo[b,d]furan and evaluation of their antimycobacterial and immunomodulatory activity. ChemistrySelect 2017;2(24):7309–18. https://doi.org/10.1002/slct.201701377.

[49] Makarov V, Neres J, Hartkoorn RC, Ryabova OB, Kazakova E, Šarkan M, Huszár S, Piton J, Kolly GS, Vocat A, Conroy TM, Mikušová K, Cole ST. The 8-pyrrole-benzothiazinones are noncovalent inhibitors of DprE1 from *Mycobacterium tuberculosis*. Antimicrob Agents Chemother 2015;59(8):4446–52. https://doi.org/10.1128/AAC.00778-15.

[50] Neres J, Hartkoorn RC, Chiarelli LR, Gadupudi R, Pasca MR, Mori G, Venturelli A, Savina S, Makarov V, Kolly GS, Molteni E, Binda C, Dhar N, Ferrari S, Brodin P, Delorme V, Landry V, De Jesus Lopes Ribeiro AL, Farina D, Saxena P, Pojer F, Carta A, Luciani R, Porta A, Zanoni G, De Rossi E, Costi MP, Riccardi G, Cole ST. 2-carboxyquinoxalines kill mycobacterium tuberculosis through noncovalent inhibition of DprE1. ACS Chem Biol 2015;10(3):705–14. https://doi.org/10.1021/cb5007163.

[51] Karabanovich G, Zemanová J, Smutny T, Székely R, et al. Development of 3, 5-dinitrobenzylsulfanyl-1, 3, 4-oxadiazoles and thiadiazoles as selective antitubercular agents active against replicating and nonreplicating *Mycobacterium tuberculosis*. J Med Chem 2016;59(6).

[52] Martínez-Hoyos M, Perez-Herran E, Gulten G, Encinas L, Álvarez-Gómez D, Alvarez E, Ferrer-Bazaga S, García-Pérez A, Ortega F, Angulo-Barturen I, Rullas-Trincado J, Ruano DB, Torres P, Castañeda P, Huss S, Menéndez RF, Valle SGdel, Ballell L, Barros D, Modha S, Dhar N, Signorino-Gelo F, McKinney JD, García-Bustos JF, Lavandera JL, Sacchettini JC, Soledad Jimenez M, Martín-Casabona N, Castro-Pichel J, Mendoza-Losana A. Antitubercular drugs for an old target: GSK693 as a promising InhA direct inhibitor. EBioMedicine 2016;8:291–301. https://doi.org/10.1016/j.ebiom.2016.05.006.

[53] Zhang B, Li J, Yang X, Wu L, Zhang J, Yang Y, Zhao Y, Zhang L, Yang X, Yang X, Cheng X, Liu Z, Jiang B, Jiang H, Guddat LW, Yang H, Rao Z. Crystal structures of membrane transporter MmpL3, an anti-TB drug target. Cell 2019;176(3):636. https://doi.org/10.1016/j.cell.2019.01.003.

[54] Laughon BE, Nacy CA. Tuberculosis-drugs in the 2016 development pipeline. Nat Rev Dis Primers 2017;3. https://doi.org/10.1038/nrdp.2017.15.

[55] Stec J, Onajole OK, Lun S, Guo H, Merenbloom B, Vistoli G, Bishai WR, Kozikowski AP. Indole-2-carboxamide-based MmpL3 inhibitors show exceptional antitubercular activity in an animal model of tuberculosis infection. J Med Chem 2016;59(13):6232–47. https://doi.org/10.1021/acs.jmedchem.6b00415.

[56] Li W, Sanchez-Hidalgo A, Jones V, De Moura VCN, North EJ, Jackson M. Synergistic interactions of MmpL3 inhibitors with antitubercular compounds in vitro. Antimicrob Agents Chemother 2017;61(4). https://doi.org/10.1128/AAC.02399-16.

[57] Abrahams KA, Chung CW, Ghidelli-Disse S, Rullas J, Rebollo-López MJ, Gurcha SS, Cox JAG, Mendoza A, Jiménez-Navarro E, Martínez-Martínez MS, Neu M, Shillings A, Homes P, Argyrou A, Casanueva R, Loman NJ, Moynihan PJ, Lelièvre J, Selenski C, Axtman M, Kremer L, Bantscheff M, Angulo-Barturen I, Izquierdo MC, Cammack NC, Drewes G, Ballell L, Barros D, Besra GS, Bates RH. Identification of KasA as the cellular target of an anti-tubercular scaffold. Nat Commun 2016;7. https://doi.org/10.1038/ncomms12581.

[58] Kumar P, Capodagli GC, Awasthi D, Shrestha R, Maharaja K, Sukheja P, Li SG, Inoyama D, Zimmerman M, Liang HPH, Sarathy J, Mina M, Rasic G, Russo R, Perryman AL, Richmann T, Gupta A, Singleton E, Verma S, Husain S, Soteropoulos P, Wang Z, Morris R, Porter G, Agnihotri G, Salgame P, Ekins S, Rhee KY, Connell N, Dartois V, Neiditch MB, Freundlich JS, Alland D. Synergistic lethality of a binary inhibitor of mycobacterium tuberculosis kasA. mBio 2018;9(6). https://doi.org/10.1128/mBio.02101-17.

[59] Kuhn ML, Alexander E, Minasov G, Page HJ, Warwrzak Z, Shuvalova L, Flores KJ, Wilson DJ, Shi C, Aldrich CC, Anderson WF. Structure of the essential Mtb FadD32 enzyme: a promising drug target for treating tuberculosis. ACS Infect Dis 2016;2(8):579–91. https://doi.org/10.1021/acsinfecdis.6b00082.

[60] Rani N, Rajmani RS, Surolia A. Identification of an isoxazole derivative as an antitubercular compound for targeting the FadD enzymes of *Mycobacterium tuberculosis*. J Med Chem 2025;68(1):270–86. https://doi.org/10.1021/acs.jmedchem.4c01844.

[61] Blair HA, Scott LJ. Delamanid: a review of its use in patients with multidrug-resistant tuberculosis. Drugs 2015;75(1):91–100. https://doi.org/10.1007/s40265-014-0331-4.

[62] Abrahams KA, Batt SM, Gurcha SS, Veerapen N, Bashiri G, Besra GS. DprE2 is a molecular target of the anti-tubercular nitroimidazole compounds pretomanid and delamanid. Nat Commun 2023;14(1). https://doi.org/10.1038/s41467-023-39300-z.

[63] Nagaraja V, Godbole AA, Henderson SR, Maxwell A. DNA topoisomerase I and DNA gyrase as targets for TB therapy. Drug Discov Today 2017;22(3):510–8. https://doi.org/10.1016/j.drudis.2016.11.006.

[64] Blower TR, Williamson BH, Kerns RJ, Berger JM. Crystal structure and stability of gyrase–fluoroquinolone cleaved complexes from *Mycobacterium tuberculosis*. Proc Natl Acad Sci 2016;113(7):1706–13. https://doi.org/10.1073/pnas.1525047113.

[65] Shee S, Singh S, Tripathi A, Thakur C, Kumar T A, Das M, Yadav V, Kohli S, Rajmani RS, Chandra N, Chakrapani H, Drlica K, Singh A. Moxifloxacin-mediated killing of *Mycobacterium tuberculosis* involves respiratory downshift, reductive stress, and accumulation of reactive oxygen species. Antimicrob Agents Chemother 2022;66(9). https://doi.org/10.1128/aac.00592-22.

[66] Kling A, Lukat P, Almeida DV, Bauer A, Fontaine E, Sordello S, Zaburannyi N, Herrmann J, Wenzel SC, König C, Ammerman NC, Barrio MB, Borchers K, Bordon-Pallier F, Brönstrup M, Courtemanche G, Gerlitz M, Geslin M, Hammann P, Heinz DW, Hoffmann H, Klieber S, Kohlmann M, Kurz M, Lair C, Matter H, Nuermberger E, Tyagi S, Fraisse L, Grosset JH, Lagrange S, Müller R. Targeting DnaN for tuberculosis therapy using novel griselimycins. Science 2015;348(6239):1106–12. https://doi.org/10.1126/science.aaa4690.

[67] Campbell EA, Korzheva N, Mustaev A, Murakami K, Nair S, Goldfarb A, Darst SA. Structural mechanism for rifampicin inhibition of bacterial RNA polymerase. Cell 2001;104(6):901–12. https://doi.org/10.1016/S0092-8674(01)00286-0.

[68] Singh B, Cocker D, Ryan H, Sloan DJ. Linezolid for drug-resistant tuberculosis. Cochrane Database Syst Rev 2017;2017(11). https://doi.org/10.1002/14651858.CD012836.

[69] Zhanel GG, Love R, Adam H, Golden A, Zelenitsky S, Schweizer F, Gorityala B, Lagacé-Wiens PRS, Rubinstein E, Walkty A, Gin AS, Gilmour M, Hoban DJ, Lynch JP, Karlowsky JA. Tedizolid: a novel oxazolidinone with potent activity against multidrug-resistant gram-positive pathogens. Drugs 2015;75(3):253–70. https://doi.org/10.1007/s40265-015-0352-7.

[70] Chan PF, Germe T, Bax BD, Huang J, Thalji RK, Bacqué E, Checchia A, Chen D, Cui H, Ding X, Ingraham K, McCloskey L, Raha K, Srikannathasan V, Maxwell A, Stavenger RA. Thiophene antibacterials that allosterically stabilize DNA-cleavage complexes with DNA gyrase. Proc Natl Acad Sci U S A 2017;114(22):E4492. https://doi.org/10.1073/pnas.1700721114.

[71] Brown-Elliott BA, Rubio A, Wallace RJ. In vitro susceptibility testing of a novel benzimidazole, SPR719, against nontuberculous mycobacteria. Antimicrob Agents Chemother 2018;62(11). https://doi.org/10.1128/AAC.01503-18.

[72] Locher CP, Jones SM, Hanzelka BL, Perola E, Shoen CM, Cynamon MH, Ngwane AH, Wiid IJ, Van Helden PD, Betoudji F, Nuermberger EL, Thomson JA. A novel inhibitor of gyrase B is a potent drug candidate for treatment of tuberculosis and nontuberculosis mycobacterial infections. Antimicrob Agents Chemother 2015;59(3):1455–65. https://doi.org/10.1128/AAC.04347-14.

[73] Rodvold KA, Gotfried MH, Ussery XT, Wong SL, Hamed KA. Intrapulmonary pharmacokinetics of SPR719 following oral administration of SPR720 to healthy volunteers. Antimicrob Agents Chemother 2024;68(11). https://doi.org/10.1128/aac.01103-24.

[74] Lin W, Mandal S, Degen D, Liu Y, Ebright YW, Li S, Feng Y, Zhang Y, Mandal S, Jiang Y, Liu S, Gigliotti M, Talaue M, Connell N, Das K, Arnold E, Ebright RH. Structural basis of *Mycobacterium tuberculosis* transcription and transcription inhibition. Mol Cell 2017;66(2):169. https://doi.org/10.1016/j.molcel.2017.03.001.

[75] Kim JS, Kim YH, Lee SH, Kim YH, Kim JW, Kang JY, Kim SK, Kim SJ, Kang YS, Kim TH, Mok J, Byun MK, Park HJ, Joh JS, Park YB, Lim HS, Choi H, Lee SH, Kim H, Yang J, Kim H, Shen X, Alsultan A, Cho IS, Geiter L, Shim TS. Early bactericidal activity of delpazolid (LCB01-0371) in patients with pulmonary tuberculosis. Antimicrob Agents Chemother 2022;66(2). https://doi.org/10.1128/aac.01684-21.

[76] Kim TS, Choe JH, Kim YJ, Yang CS, Kwon HJ, Jeong J, Kim G, Park DE, Jo EK, Cho YL, Jang J. Activity of LCB01-0371, a novel oxazolidinone, against mycobacterium abscessus. Antimicrob Agents Chemother 2017;61(9). https://doi.org/10.1128/AAC.02752-16.

[77] Zong Z, Jing W, Shi J, Wen S, Zhang T, Huo F, Shang Y, Liang Q, Huang H, Pang Y. Comparison of in vitro activity and MIC distributions between the novel oxazolidinone delpazolid and linezolid against multidrug-resistant and extensively drug-resistant mycobacterium tuberculosis in China. Antimicrob Agents Chemother 2018;62(8). https://doi.org/10.1128/AAC.00165-18.

[78] Ahn S, Jung J, Jang IA, Madsen EL, Park W. Role of glyoxylate shunt in oxidative stress response. J Biol Chem 2016;291(22):11928–38. https://doi.org/10.1074/jbc.M115.708149.

[79] Preiss L, Langer JD, Yildiz Ö, Eckhardt-Strelau L, Guillemont JEG, Koul A, Meier T. Structure of the mycobacterial ATP synthase Fo rotor ring in complex with the anti-TB drug bedaquiline. Sci Adv 2015;1(4). https://doi.org/10.1126/sciadv.1500106.

[80] Ahmad N, Ahuja SD, Akkerman OW, Alffenaar JWC, Anderson LF, Baghaei P, Bang D, Barry PM, Bastos ML, Behera D, Benedetti A, Bisson GP, Boeree MJ, Bonnet M, Brode SK, Brust JCM, Cai Y, Caumes E, Cegielski JP, Centis R, Chan PC, Chan ED, Chang KC, Charles M, Cirule A, Dalcolmo MP, D'Ambrosio L, de Vries G, Dheda K, Esmail A, Flood J, Fox GJ, Fréchet-Jachym M, Fregona G, Gayoso R, Gegia M, Gler MT, Gu S, Guglielmetti L, Holtz TH, Hughes J, Isaakidis P, Jarlsberg L, Kempker RR,

Keshavjee S, Khan FA, Kipiani M, Koenig SP, Koh WJ, Kritski A, Kuksa L, Kvasnovsky CL, Kwak N, Lan Z, Lange C, Laniado-Laborín R, Lee M, Leimane V, Leung CC, Leung ECC, Li PZ, Lowenthal P, Maciel EL, Marks SM, Mase S, Mbuagbaw L, Migliori GB, Milanov V, Miller AC, Mitnick CD, Modongo C, Mohr E, Monedero I, Nahid P, Ndjeka N, O'Donnell MR, Padayatchi N, Palmero D, Pape JW, Podewils LJ, Reynolds I, Riekstina V, Robert J, Rodriguez M, Seaworth B, Seung KJ, Schnippel K, Shim TS, Singla R, Smith SE, Sotgiu G, Sukhbaatar G, Tabarsi P, Tiberi S, Trajman A, Trieu L, Udwadia ZF, van der Werf TS, Veziris N, Viiklepp P, Vilbrun SC, Walsh K, Westenhouse J, Yew WW, Yim JJ, Zetola NM, Zignol M, Menzies D. Treatment correlates of successful outcomes in pulmonary multidrug-resistant tuberculosis: an individual patient data meta-analysis. Lancet 2018;392(10150): 821–34. https://doi.org/10.1016/S0140-6736(18)31644-1.

[81] Sarathy JP, Ragunathan P, Shin J, Cooper CB, Upton AM, Grüber G, Dick T. TBAJ-876 retains Bedaquiline's activity against subunits c and ε of *Mycobacterium tuberculosis* F-ATP synthase. Antimicrob Agents Chemother 2019;63(10). https://doi.org/10.1128/AAC.01191-19.

[82] Sutherland HS, Tong AS, Choi PJ, Blaser A, Conole D, Franzblau SG, et al. 3, 5-Dialkoxypyridine analogues of bedaquiline are potent antituberculosis agents with minimal inhibition of the hERG channel. Bioorg Med Chem 2019;27(7).

[83] Kang S, Kim YM, Kim RY, Seo MJ, No Z, Nam K, Kim S, Kim J. Synthesis and structure-activity studies of side chain analogues of the anti-tubercular agent, Q203. Eur J Med Chem 2017;125:807–15. https://doi.org/10.1016/j.ejmech.2016.09.082.

[84] Berube BJ, Parish T. Combinations of respiratory chain inhibitors have enhanced bactericidal activity against mycobacterium tuberculosis. Antimicrob Agents Chemother 2018;62(1). https://doi.org/10.1128/AAC.01677-17.

[85] Rybniker J, Vocat A, Sala C, Busso P, Pojer F, Benjak A, Cole ST. Lansoprazole is an antituberculous prodrug targeting cytochrome bc1. Nat Commun 2015;6(1). https://doi.org/10.1038/ncomms8659.

[86] Chandrasekera NS, Berube BJ, Shetye G, Chettiar S, O'Malley T, Manning A, Flint L, Awasthi D, Ioerger TR, Sacchettini J, Masquelin T, Hipskind PA, Odingo J, Parish T. Improved phenoxyalkylbenzimidazoles with activity against *Mycobacterium tuberculosis* appear to target QcrB. ACS Infect Dis 2017;3(12):898–916. https://doi.org/10.1021/acsinfecdis.7b00112.

[87] Subtil V, Abbadi BL, Rodrigues-Junior VS, Bizarro CV, Timmers LFSM, et al. Activity of 2-(quinolin-4-yloxy) acetamides in *Mycobacterium tuberculosis* clinical isolates and identification of their molecular target by whole-genome sequencing. Int J Antimicrob Agents 2018;51(3).

[88] Phummarin N, Boshoff HI, Tsang PS, Dalton J, Wiles S, Barry 3rd CE, et al. SAR and identification of 2-(quinolin-4-yloxy) acetamides as *Mycobacterium tuberculosis* cytochrome bc 1 inhibitors. Medchemcomm 2016;7(11).

[89] Foo CS, Lupien A, Kienle M, Vocat A, Benjak A, Sommer R, Lamprecht DA, Steyn AJC, Pethe K, Piton J, Altmann KH, Cole ST. Arylvinylpiperazine amides, a new class of potent inhibitors targeting QcrB of mycobacterium tuberculosis. mBio 2018;9(5). https://doi.org/10.1128/mBio.01276-18.

[90] Harbut MB, Yang B, Liu R, Yano T, Vilchèze C, Cheng B, Lockner J, Guo H, Yu C, Franzblau SG, Mike Petrassi H, Jacobs WR, Rubin H, Chatterjee AK, Wang F. Small molecules targeting *Mycobacterium tuberculosis* type II NADH dehydrogenase exhibit antimycobacterial activity. Angew Chem 2018;130(13): 3536–40. https://doi.org/10.1002/ange.201800260.

[91] Murugesan D, Ray PC, Bayliss T, Prosser GA, Harrison JR, Green K, Soares De Melo C, Feng TS, Street LJ, Chibale K, Warner DF, Mizrahi V, Epemolu O, Scullion P, Ellis L, Riley J, Shishikura Y, Ferguson L, Osuna-Cabello M, Read KD, Green SR, Lamprecht DA, Steyn AJC, Ioerger TR, Sacchettini J, Rhee KY, Arora K, Barry CE, Wyatt PG, Boshoff HIM. 2-Mercapto-Quinazolinones as inhibitors of type II NADH dehydrogenase and *Mycobacterium tuberculosis*: structure-activity relationships, mechanism of action and absorption, distribution, metabolism, and excretion characterization. ACS Infect Dis 2018;4(6): 954–69. https://doi.org/10.1021/acsinfecdis.7b00275.

[92] Boersch M, Rudrawar S, Grant G, Zunk M. Menaquinone biosynthesis inhibition: a review of advancements toward a new antibiotic mechanism. RSC Adv 2018;8(10):5099–105. https://doi.org/10.1039/c7ra12950e.
[93] Sukheja P, Kumar P, Mittal N, Li SG, Singleton E, Russo R, Perryman AL, Shrestha R, Awasthi D, Husain S, Soteropoulos P, Brukh R, Connell N, Freundlich JS, Alland D. A novel small-molecule inhibitor of the mycobacterium tuberculosis demethylmenaquinone methyltransferase meng is bactericidal to both growing and nutritionally deprived persister cells. mBio 2017;8(1). https://doi.org/10.1128/mBio.02022-16.
[94] Bald D, Villellas C, Lu P, Koul A, Rubin EJ, John Collier R. Targeting energy metabolism in *Mycobacterium tuberculosis* , a new paradigm in antimycobacterial drug discovery. mBio 2017;8(2). https://doi.org/10.1128/mbio.00272-17.
[95] Lee YV, Wahab HA, Choong YS. Potential inhibitors for isocitrate lyase of mycobacterium tuberculosis and Non- *M. tuberculosis*: a summary. BioMed Res Int 2015;2015. https://doi.org/10.1155/2015/895453.
[96] Lupoli TJ, Vaubourgeix J, Burns-Huang K, Gold B. Targeting the proteostasis network for mycobacterial drug discovery. ACS Infect Dis 2018;4(4):478–98. https://doi.org/10.1021/acsinfecdis.7b00231.
[97] Gao W, Kim JY, Anderson JR, Akopian T, Hong S, Jin YY, Kandror O, Kim JW, Lee IA, Lee SY, McAlpine JB, Mulugeta S, Sunoqrot S, Wang Y, Yang SH, Yoon TM, Goldberg AL, Pauli GF, Suh JW, Franzblau SG, Cho S. The cyclic peptide ecumicin targeting CLpC1 is active against *Mycobacterium tuberculosis* in vivo. Antimicrob Agents Chemother 2015;59(2):880–9. https://doi.org/10.1128/AAC.04054-14.
[98] Choules MP, Wolf NM, Lee H, Anderson JR, Grzelak EM, Wang Y, Ma R, Gao W, McAlpine JB, Jin YY, Cheng J, Lee H, Suh JW, Duc NM, Paik S, Choe JH, Jo EK, Chang CL, Lee JS, Jaki BU, Pauli GF, Franzblau SG, Cho S. Rufomycin targets clpc1 proteolysis in mycobacterium tuberculosis and M. Abscessus. Antimicrob Agents Chemother 2019;63(3). https://doi.org/10.1128/AAC.02204-18.
[99] Maurer M, Linder D, Franke KB, Jäger J, Taylor G, Gloge F, Gremer S, Le Breton L, Mayer MP, Weber-Ban E, Carroni M, Bukau B, Mogk A. Toxic activation of an AAA+ protease by the antibacterial drug cyclomarin A. Cell Chem Biol 2019;26(8):1169. https://doi.org/10.1016/j.chembiol.2019.05.008.
[100] Tousif S, Ahmad S. Challenges of tuberculosis treatment with DOTS: an immune impairment perspective. J Cell Sci Ther 2015;06(05). https://doi.org/10.4172/2157-7013.1000223.
[101] Cerrone M, Bracchi M, Wasserman S, Pozniak A, Meintjes G, Cohen K, Wilkinson RJ. Safety implications of combined antiretroviral and anti-tuberculosis drugs. Expert Opin Drug Saf 2020;19(1):23–41. https://doi.org/10.1080/14740338.2020.1694901.
[102] Witney AA, Cosgrove CA, Arnold A, Hinds J, Stoker NG, Butcher PD. Clinical use of whole genome sequencing for *Mycobacterium tuberculosis*. BMC Med 2016;14(1). https://doi.org/10.1186/s12916-016-0598-2.
[103] Peterson EJR, Ma S, Sherman DR, Baliga NS. Network analysis identifies Rv0324 and Rv0880 as regulators of bedaquiline tolerance in *Mycobacterium tuberculosis*. Nat Microbiol 2016;1(8). https://doi.org/10.1038/nmicrobiol.2016.78.
[104] Abramovitch RB. *Mycobacterium tuberculosis* reporter strains as tools for drug discovery and development. IUBMB Life 2018;70(9):818–25. https://doi.org/10.1002/iub.1862.
[105] Boot M, Commandeur S, Subudhi AK, Bahira M, Smith TC, Abdallah AM, van Gemert M, Lelièvre J, Ballell L, Aldridge BB, Pain A, Speer A, Bitter W. Accelerating early antituberculosis drug discovery by creating mycobacterial indicator strains that predict mode of action. Antimicrob Agents Chemother 2018;62(7). https://doi.org/10.1128/AAC.00083-18.
[106] Xu W, DeJesus MA, Rücker N, Engelhart CA, Wright MG, Healy C, Lin K, Wang R, Park SW, Ioerger TR, Schnappinger D, Ehrt S. Chemical genetic interaction profiling reveals determinants of intrinsic antibiotic resistance in *Mycobacterium tuberculosis*. Antimicrob Agents Chemother 2017;61(12). https://doi.org/10.1128/AAC.01334-17.

[107] Johnson LV, Office S, Meyer K, et al. Large-scale chemical–genetics yields new *M. tuberculosis* inhibitor classes. Nature 2019;571(7763):72–8.
[108] Rani N, Surolia A. Targeted suppression of MEP pathway genes DXS, IspD and IspF to explore the mycobacterial metabolism and survival. Int J Biol Macromol 2024;272(Pt 1):132727. https://doi.org/10.1016/j.ijbiomac.2024.132727.
[109] Rock JM, Hopkins FF, Chavez A, Diallo M, Chase MR, Gerrick ER, Pritchard JR, Church GM, Rubin EJ, Sassetti CM, Schnappinger D, Fortune SM. Programmable transcriptional repression in mycobacteria using an orthogonal CRISPR interference platform. Nat Microbiol 2017;2. https://doi.org/10.1038/nmicrobiol.2016.274.
[110] Ballinger E, Mosior J, Hartman T, Burns-Huang K, Gold B, Morris R, Goullieux L, Blanc I, Vaubourgeix J, Lagrange S, Fraisse L, Sans S, Couturier C, Bacqué E, Rhee K, Scarry SM, Aubé J, Yang G, Ouerfelli O, Schnappinger D, Ioerger TR, Engelhart CA, McConnell JA, McAulay K, Allison F, Roubert C, Sacchettini J, Nathan C. Opposing reactions in coenzyme A metabolism sensitize *Mycobacterium tuberculosis* to enzyme inhibition. Science 2019;363(6426). https://doi.org/10.1126/science.aau8959.
[111] Mishra A, Mamidi AS, Rajmani RS, Ray A, Roy R, Surolia A. An allosteric inhibitor of *Mycobacterium tuberculosis* ArgJ: implications to a novel combinatorial therapy. EMBO Mol Med 2018;10(4). https://doi.org/10.15252/emmm.201708038.
[112] Yelamanchi SD, Arun Kumar ST, Mishra A, Keshava Prasad TS, Surolia A. Metabolite dysregulation by pranlukast in *Mycobacterium tuberculosis*. Molecules 2022;27(5). https://doi.org/10.3390/molecules27051520.
[113] Rajmani RS, Surolia A. Antimycobacterial and healing effects of Pranlukast against MTB infection and pathogenesis in a preclinical mouse model of tuberculosis. Front Immunol 2024;15. https://doi.org/10.3389/fimmu.2024.1347045.

SECTION

IV

Nutrition, immunity, and broader mycobacterial contexts

CHAPTER

14

The enigma of food supplementation in TB: A role for unresolved inflammation

Kalpana Sriraman and Nerges Mistry

The Foundation for Medical Research, Mumbai, Maharashtra, India

Highlights

1. Malnutrition is a leading risk factor for tuberculosis (TB) infection and progression to active disease.
2. Systematic reviews show limited evidence for the impact of nutrition supplementation on TB treatment outcomes.
3. Inflammation driven by cytokines and metabolic perturbations impairs the ability to achieve optimal nutritional outcomes.
4. Persistent inflammation may explain the heterogeneous responses to nutritional supplementation among malnourished TB patients.
5. Anti-inflammatory diets, gut-directed therapies, and TB treatment may help resolve inflammation and improve nutritional outcomes.
6. A personalized nutrition approach considering people's socioeconomic realities is needed for optimal impact.
7. Mitigating malnutrition in young children is crucial to avoid future metabolic risks upon reaching adulthood.

Abbreviations

APP acute phase proteins
APR acute phase response
ATB active tuberculosis disease
BMI body mass index
CRP C-reactive protein
DHEA dehydroepiandrosterone
EFFORT early nutritional support on frailty, functional outcomes, and recovery of malnourished medical inpatients trial
HIV human immunodeficiency virus
HPA hypothalamic-pituitary-adrenal (axis)
HPT hypothalamic-pituitary-thyroid (axis)
IL-1 interleukin 1
IL-1a interleukin 1 alpha
IL-1B interleukin 1 beta

Navigating Tuberculosis. https://doi.org/10.1016/B978-0-443-27735-1.00018-6

IL-6 interleukin 6
LTBI latent tuberculosis infection
MDR multi-drug resistant
MoHFW Ministry of Health and Family Welfare (India)
Mtb mycobacterium tuberculosis
NSAIDs nonsteroidal anti-inflammatory drugs
RATIONS reducing activation of tuberculosis by improvement of nutritional status (trial)
TB tuberculosis
TNF tumor necrosis factor
UNICEF United Nations Children's Fund
WHO World Health Organization

1. Introduction

Tuberculosis(TB) has long been recognized as a social disease, with malnutrition as a leading attributable risk factor for *Mycobacterium tuberculosis* (*Mtb*) infection [1]. The risk of acquiring TB increases by 13.8% for each unit decrease in body mass index (BMI), although this is not maintained at the extremes of the BMI range [2]. Malnutrition is also a risk factor for the progression from latent TB infection (LTBI) (pre-existing but dormant) to active disease (ATB). Malnourished TB patients, particularly those presenting with drug-resistant TB, exhibit poor treatment responses and are twice as likely to die from TB.

The intergenerational nature of malnutrition globally makes recent numbers hard to ignore. Around the world, 148 million children under five experience stunted growth, 45 million suffer from wasting, and nearly half of the children who die each year die of malnutrition [3]. Those lucky to survive can never reach their full potential mentally and physically, or experience a good quality of life. Therefore, the assumption that TB will be eradicated from the face of the earth by 2050 is erroneous in light of unabated malnutrition in several vulnerable areas of the globe in Africa, Asia, and South America. India, with an annual prevalence of 2.5 million cases of TB [4] and an estimated 22.4 million undernourished individuals [5], highlights the futility of control measures even more strongly.

An estimated 108 million tribals, accounting for 8.6% of India's population, share a common feature of poor health indicators, a significant burden of morbidity and mortality, and limited access to health care. A dual morbidity of 10% malnutrition and TB (pooled estimate of 703/100,000 population) [6] has been recorded in the Indian tribals. Women with TB from such areas may be at higher risk for pre-eclampsia and other complications during pregnancy, such as the risk of premature birth, low birth weight, and perinatal death [7]. Though World Health Organization (WHO) guidelines recommend the provision of supplementary foods to achieve an average weekly weight gain of 300 g in the second and third trimesters [8], the evidence of the success of this intervention remains undisclosed to a great extent.

Systematic reviews of various macro- and micronutrition supplementation studies for TB patients show that free food/supplements may benefit drug therapy [9,10]. However, the evidence remains limited and heterogeneous, especially about treatment outcomes. Although recent evidence from large-scale food support trials in India has shown a positive impact on treatment outcomes and disease susceptibility [11,12,13], the need to understand biological factors that may explain the mixed effects of nutritional supplementation studies on disease and dietary outcomes in malnourished TB patients remains largely unaddressed.

While contracting TB infection by an individual is due to exogenous factors such as proximity to an infectious load of TB organisms, inhaled pollution levels, etc., the progression from infection to subclinical or preclinical infection is in the purview of innate immunity. Progressing to subsequently active symptomatic disease is due to endogenous factors, predominantly acquired host immunity [14]. Both natural and acquired cell-mediated immunity remain optimally functional with adequate and well-diversified nutrition, though both innate and adaptive immunity are damaged by malnutrition [15]. Malnutrition affects key immune functions such as phagocytosis, complement cascade, antigen presentation in blood, multiple tissue damage to the gastrointestinal mucosa, and deviated cytokine production. Malnutrition also manifests through epigenetic changes that regulate gene expression. Additionally, malnutrition influences the balance of the gut microbiota, leading to dysbiosis and hampering the immune regulatory function of the gut. Whether the multiple lesions in the immune system are caused by malnutrition resulting in miliary TB in under-five children is an indecisive finding due to miliary TB being grossly undetected, and data on its occurrence are not routinely reported or detected [16].

Water and sanitation also play a significant role in creating malnutrition and inflammation in vulnerable areas [17]. The high prevalence of helminthic diseases in several parts of India [18] lends itself to immune compromise and nutritional deviation. Frequent periods of infectious diarrhea through species such as Enterobacteriaceae also result in damage to the intestinal mucosa and malabsorption of food [19]. Therefore, any form of nutrition supplementation must go hand in hand with efforts to improve/mitigate gaps in access to clean water and sanitation.

In this chapter, we will review the impact of malnutrition on TB and the key findings of nutrition supplement trials in India and elsewhere. Further, we will discuss the heterogeneous responses among malnourished TB patients and factors that may contribute to the heterogeneity, especially inflammation, and highlight evidence around the need to address inflammation during nutrition supplementation to improve current nutritional practices. Lastly, we will share perspectives on potential ways to address inflammation and malnutrition to achieve better nutritional outcomes.

2. Attributable risk of malnutrition to TB susceptibility and severity

According to WHO global TB reports published annually, malnutrition is one of the leading risk factors [20], and for a long time, the estimates of TB attributable to malnutrition have been approximately 20%. This was based on the observed relative risk ratio of 3.2 for developing TB due to malnutrition [2]. A recent systematic review published in the Cochrane Database showed that malnutrition increases the risk of developing TB only by 2-fold (relative risk) in the short term (<10 years) [21], prompting the WHO to reduce its estimates of TB attributable to malnutrition in their 2024 report [22]. However, critical analysis and revised estimates published in 2024 showed that these recent relative risk estimates and subsequent population attributable fraction calculated by WHO could be significantly underestimated [23], given that the Cochrane study used pooled malnutrition estimates of different definitions and reference groups and excluded historical and most recent studies. The analysis showed that the revised estimates are consistent with the earlier reported risk of about three for the general population, and this risk varies only if other comorbidities like human immunodeficiency virus (HIV) or diabetes are also present [23].

Although the increased relative risk and attributable fraction globally are consistent across diverse populations [2,21,23], the actual risk and attributable fraction can substantially vary between populations or regions when the incidence of the disease, local prevalence of malnutrition, and the method by which malnutrition is measured are considered. A study that analyzed 30 high-burden TB countries showed that the attributable fraction could range from 24%–33%, with Asia having the highest [24]. For India, a country with the highest burden of TB in the world and a high prevalence of malnutrition, the relative risk has been observed to be as high as 6.2 [25], and the attributable fractions can vary from 30% to 50% in the general population [24]. This was higher in rural areas, tribal populations, and people in lower wealth index quartiles [26,27].

Malnutrition increases disease susceptibility, the severity of the disease, the response to treatment, and mortality. Radiological data shows that severe malnutrition is associated with a 4-5-fold increase in cavitation [28,29]. Malnutrition is also noted to cause delayed sputum conversion and unfavorable outcomes [30-33]. Many studies across several countries have pointed out that malnutrition is associated with approximately a 2-fold higher risk of mortality [31,32,34,35,36,37]. All these studies underscore the critical role of nutritional status as an independent determinant of TB, emphasizing the importance of social protection and other interventions to enhance nutrition as part of a comprehensive global strategy to combat TB.

3. Impact of macronutrient supplementation on nutritional status and treatment outcomes

Given the critical association between nutritional status and TB treatment outcomes, nutritional supplementation plays a supportive role in managing TB. Several studies have been undertaken to understand the impact of such supportive supplementation on patients' nutritional status and TB treatment outcomes. Table 14.1 summarizes numerous studies in the last few decades, including randomized controlled trials and prospective and retrospective programmatic studies. A 2016 systematic review and meta-analysis of food supplementation studies showed insufficient evidence to support improvement in treatment outcomes, preventing death, and increasing treatment adherence, while there is moderately good evidence that nutrition supplementation, along with treatment, improves weight and quality of life [9]. Trials and studies conducted subsequently in India [11,13,39,40] favor better treatment outcomes with food supplementation when compared to groups who did not receive any supplementation. As can be seen from the table, these observations and improvement in weight are especially notable in studies that included more holistic nutrition supplementation as food rations that provided the opportunity for a balanced diet rather than just improving particular nutrients like protein [41,42–44] or micronutrients [9].

While these studies report improved weight gain and nutritional status with food supplementation, it is critical to highlight that this represents only an average effect. A closer look at the studies conducted in settings with a high prevalence of malnutrition shows that a substantial portion of patients do not achieve the desired weight gain or continue to remain underweight even at the end of treatment. The largest trial conducted so far, the Reducing Activation of Tuberculosis by Improvement of Nutritional Status (RATIONS) trial [11], showed that 46% of the patients who were supported did not achieve even a 5% gain in weight at the end of two months of treatment, 54.6% of patients remained underweight even at the end of the treatment, and 8% of the patients did not achieve any weight gain

Table 14.1 Tabular representation of randomized control and clinical trials of macronutrient supplementation to TB patients.

Subjects and geography	BMI Kg/m2	N	Intervention	Nutritional status outcome	TB treatment outcome	Refs.
Adults, Singapore	17.9 ± 1.9 (CG) 16.7 ± 1.5 (IG)	36	Protein supplementation (Ensure plus powder) versus counseling alone	Improved weight gain at 6 weeks (2.57 Kg IG versus 0.84 kg CG) and grip strength	Not measured	[42]
Adult hospital TB patients, Mexico		21	2500 kcal diet with either 800 mg cholesterol (IG) or 250 mg cholesterol (CG) for 2 months		The high cholesterol group had 80% culture conversion at 2 weeks versus 9% in the low cholesterol group.	[45]
Adults, Timor-Leste	<18.5 in 80% of participants in both groups	270	Protein-rich meal once a day versus counseling alone	Improved weight gain at 8 weeks (3.5% CG versus 5.2 %IG) and end of treatment	No benefit on treatment outcome	[43]
Adults, India (Andhra Pradesh)	17.9 ± 2.1 (CG) 17.1 ± 2.8 (IG)	100	Diet plan, protein-rich balls and sprouts versus general nutrition advice	Improved weight gain at 12 weeks (2.8% in CG versus 8.6% in IG) and grip strength	Higher sputum conversion rate and treatment completion rate	[38]
Adults with HIV, Tanzania	52% of patients in IG versus 57% in CG were <18.5	377	1 (4.5 g protein; CG) versus 6 (27 g protein; IG) protein biscuits daily for 2 months	There is no improvement in weight but marginal improvement in hand grip strength	Not measured	[44]
Adults, India (Tamil Nadu)	17.2 (IG) 18.2 (CG)	103	930 Kcal Cereal-lentil mixture +Multivitamin for 6 months versus counseling alone		Marginal statistically non-significant improvement in good TB outcomes (cured or treatment completed)	[46]

Continued

Table 14.1 Tabular representation of randomized control and clinical trials of macronutrient supplementation to TB patients.—cont'd

Subjects and geography	BMI Kg/m2	N	Intervention	Nutritional status outcome	TB treatment outcome	Refs.
Adults, India (West Bengal) retrospective analysis	Not reported. But food insecure settings and BPL	573	Rations of Rice and lentils (protein) for 2–3 months versus standard care	Not studied	50% lower percentage of unsuccessful treatment outcome in IG	[47]
Adults with HIV-TB, Senegal, Africa		26	Food basket versus Ready to use therapeutic food for 6 months	Both groups showed improved weight gain and reduction in the proportion of malnourished patients at 6 months of treatment	Treatment outcome was not compared	[48]
Adults and >15 years, India (Madhya Pradesh)	17.7 (±3.8) in IG Not measured for CG	519	Food rations and counseling for 6 months versus standard care	90% in IG had weight gain >10% versus 17% in CG. 40.7% were severely underweight, reduced to 13% after 6 months, though overall, 35.8% remained underweight	The treatment success rate of 95% in IG versus 83.5% in CG	[40]
Adults, India (Jharkhand) RATIONS trial	82% < 18.5 (No CG)	2800	Rations of Rice, Lentils, oil, and RDA of multivitamin pills for 6 months	Median weight gain of 4.6 Kg at the end of treatment. 54.6% remained underweight at 6 months. 54% gained at least 5% at 2 months of treatment	93.7% successful outcome, which is better than similar settings without a nutrition support program 60% reduction in mortality if weight gain is >5% at 2 months of treatment	[11]

Table 14.1 Tabular representation of randomized control and clinical trials of macronutrient supplementation to TB patients.—cont'd

Subjects and geography	BMI Kg/m2	N	Intervention	Nutritional status outcome	TB treatment outcome	Refs.
Adults, India (Odisha)	16.4 (15.1–17.9) -CG 16.2 (14.9–17.8)-IG	761 (614 IG, 147 CG)	Rations of Rice, Lentils, oil, and Sathu powder versus standard care	~35% gained ≥5.0% weight and BMI in IG versus ~18% in CG Improved waist circumference, quality of life, respiratory status	87% favorable outcome in IG versus 77% in CG	[13]
Adults, India (Gujarat) Retrospective analysis	Not reported	1290	GOI-Nikshay Mitra Scheme-rations of Rice, Wheat flour, oil, lentils, chickpeas versus standard care	Weight gain over 6 months: 6.5 kg in IG versus 3.1 kg in CG	74.7% cure rate and 1.1% mortality in IG versus 50.9% cure rate and 5.7% mortality in CG	[39]
Adults and children, Madagascar Retrospective analysis	81.3% adults <18.5 55.4% of <5 years children malnourished 30% of >5 years children malnourished	1310	Enriched soy- and wheat-based flour and oil for <18.5 BMI	No significant weight gain from nutrition support. 61.7% remained malnourished adults after 6 months of support 70% of well-nourished adults not receiving support became malnourished at 6 months	Not studied	[41]

CG, *Control group;* IG, *intervention group;* RDA, *recommended daily allowance.*

from initiation to completion of treatment [11]. In another large food supplementation trial conducted in Odisha, India [13], only 35% showed more than 5% weight gain and Body Mass Index (BMI), even with supplementation. In comparison, 35% showed only 0%–5% change, and 15% showed decreased weight and BMI at the end of 2 months of treatment. Unchanged or decreased BMI after 2 months of treatment is associated with a 2.4–3.7-fold increased risk of unfavorable outcomes, depending on whether they received food supplementation [13], and a 5-fold higher risk of mortality in patients [31]. Studies in multidrug-resistant (MDR) TB patients have also shown that the most significant risk for all-cause mortality during as well as post-treatment is the lack of weight gain during treatment [49]. Moreover, it has also been noted that among persons underweight at diagnosis, weight gain of 5% or less during the first 2 months of treatment is associated with a 2.4-fold increased risk for relapse of TB. The significant number of patients who fail to achieve the desired weight gain of even 5% during treatment and remain underweight despite food supplementation in the aforementioned trials prompts us to question the underlying causes of non-responsiveness to nutritional support.

4. Endogenous factors that affect nutritional intake and nutrition status—The role of inflammation

A complex interplay of systems governing appetite, nutrient absorption, and metabolism links dietary intake to overall nutritional status and health. These endogenous systems and factors can be genetic, hormonal, and metabolic changes, physiological states, and immune status, particularly during an infection or disease. The impact of inflammation complicates the ability to delineate the role of nutrition in any disease [50]. A large body of literature describes the interactions between nutrition, the immune system, and inflammation. Various triggers like microbial invasion (external), tissue injury, and other inflammatory processes induce inflammation, and a central mechanism by which this inflammation interacts with nutrition, determining the neuroendocrine system, is the acute phase response (APR) [51]. The APR is initiated by activated macrophages resulting from external or internal triggers, leading to the secretion of various proinflammatory cytokines like interleukin 6 (IL-6), tumor necrosis factor (TNF), and interleukin 1 (IL-1) [51]. These cytokines have pleiotropic effects and elicit a coordinated response of immune, metabolic, and neuroendocrine systems supporting tissue repair and immune competence. They stimulate hypothalamic-pituitary-adrenal axes (HPA), suppress hypothalamic-pituitary-thyroid (HPT) and gonadal axes, and growth hormone secretion [52,53]. They also trigger hepatocytes to produce acute phase proteins (APP), which further modulate more than 200 proteins and effector molecules in the blood, impacting the immune system as well as the neuroendocrine system, which has significant nutritional implications [50].

In the context of nutrition and disease, these proinflammatory cytokines affect nutritional status through several mechanisms (Table 14.2). Firstly, in both acute and chronic disease, there are disturbances to energy

Table 14.2 Linkage of nutrition to acute-phase reaction.

Reaction	Nutrition effect
Release of cytokines Stimulation of HPA axis Suppression of HPT axis Production of acute phase proteins	Disturbed energy homeostasis Fat and muscle protein breakdown Appetite suppression Sequestration of micronutrients from circulation Change in intestinal morphology and barrier function

HPA, *Hypothalamic-pituitary-adrenal axes;* HPT, *Hypothalamic-pituitary-thyroid.*

homeostasis. Primarily there is an increased energy expenditure resulting from APP synthesis which is met by fat and muscle protein breakdown. This leads to fat and lean mass loss and eventually sarcopenia (progressive loss of muscle mass, strength, and function) and anorexia (appetite suppression and weight loss) as seen in chronic conditions [54]. Second, as a part of HPA axis responses, interleukin 1 beta (IL-1B), through direct action on the vagus nerve and IL-1 receptors on the hypothalamus, suppresses appetite and induces satiety, leading to reduced nutritional intake [54]. Studies also show that IL-1B and interleukin 1 alpha (IL-1A) induce leptin production in adipose tissue [55,56] and suppress ghrelin production in the stomach [57,58], both known to regulate food intake and body weight [59]. Leptin is a mediator of long-term energy balance regulation, suppressing food intake and causing weight loss, while ghrelin is a fast-acting hormone that triggers food intake and weight gain [59].

The third important effect of inflammation during infection and subsequent disease is that the micronutrients are sequestered or redistributed from the systemic circulation to starve the infecting pathogen and block its absorption [50,60,61]. This is most evident for iron [62] but also noted for zinc and vitamins A, B6, C, and D [50,61]. Consequently, the level of these micronutrients in the circulation decreases along with their binding partners, typically negative acute phase proteins such as transferrin (binding iron), albumin (binding 70% of all circulating zinc), and transthyretin (binding indirectly vitamin A) [61]. An important collateral effect of this phenomenon is that estimating true deficiency during disease or inflammation [60] becomes challenging. This may explain why studies involving supplementation of these micronutrients showed mixed results, with some studies showing benefits, while some studies showed adverse outcomes in select populations, mainly when provided to non-deficient individuals in population-based studies [50]. Therefore, determining micronutrient supplementation for patients with persistent inflammation solely based on serum/plasma concentrations is unlikely to improve disease outcomes.

Lastly, inflammation via proinflammatory cytokines could impact nutritional status by altering intestinal morphology and function, resulting in altered digestion, absorption, motility, and barrier function [63]. Animal model studies have shown that inflammation is associated with decreased villi height and crypt depth in ileal samples [63] altering the enterocytes through which macro and micronutrient absorption occurs [64]. Structural abnormalities and hence malabsorption have also been noted in the human population with protein-energy malnutrition conditions like chronic malnutrition [65]. Furthermore, it has been noted that intestinal repair after an infection is delayed in patients with chronic protein-energy malnutrition [66]. Inflammation also opens up tight junctions in an intestine, which blocks the passive absorption of hydrophilic molecules from the intestinal lumen and increases permeability [63,67]. Changes in barrier function may have implications for nutrient absorption and endotoxin-induced multi-organ failure.

5. Inflammation and endocrine-metabolic alterations in TB

A key hallmark of TB is chronic inflammation caused by sophisticated intracellular survival strategies adopted by the tubercle bacteria [68]. Since malnutrition also causes chronic inflammation [50,69,70,71], the levels of inflammation are expected to be more pronounced in TB patients with moderate to severe malnutrition. Studies from multiple countries that used BMI as the indicator for nutritional status showed that malnourished individuals infected with active TB disease present a different immunological profile tilted toward elevated proinflammatory response in comparison to well-nourished individuals [72–75]. In all these studies one common pro-inflammatory marker that was present in high concentrations was IL-6. Whereas in malnourished individuals with latent TB-infection, conflicting findings were noted [76,77]. A study that compared pro- and anti-inflammatory cytokines levels in a cross-section of latent TB-infection-positive (LTBI+) community populations with low and normal BMI found lower levels of proinflammatory cytokines in the low BMI group [76]. In contrast, another study that studied transcriptomic data in LTBI+ recently exposed household contacts noted elevated inflammatory response with upregulation of IL-1 and IL-6 signaling pathways [77]. While these differences could be attributed to variations in methodologies and study populations, both studies consistently indicate that malnourished LTBI+ individuals exhibit an immunological profile that increases susceptibility to TB.

Elevated inflammatory response through proinflammatory cytokines has been reported to correlate strongly with pathogenesis and adverse disease outcomes in TB. Protein-deficient animal model studies with guinea pigs and mice that reflected elevated inflammatory states showed compromised T-cell responses, altered cytokine production, diffuse necrotic granulomas, and lower survival rates [78,79,80]. Clinical studies have shown that severity and adverse outcomes in TB patients are associated with elevated proinflammatory cytokines before treatment initiation [81,72,73,82–84]. A multi-country cohort study showed high pre-treatment IL-6 concentrations were associated with higher odds of 3.32 for any unfavorable outcome, 2.16 higher odds for treatment failure, 5.36 higher odds for recurrence, and 4.62 higher odds for death [81].

Studies with TB patients have also shown that elevated proinflammatory cytokines are associated with dysregulated metabolic/endocrine hormone systems, which may be more pronounced in patients with low BMI [74,84,85,86,87]. This has implications for understanding the nutritional status of the patients during TB and its treatment. Santucci et al. [84] showed that treatment-naive TB patients had high concentrations of cortisol and ghrelin, while low levels of dehydroepiandrosterone (DHEA) and leptin. They suggested that because cortisol is known to cause lipolysis in fat cells, inhibition of protein synthesis, and proteolysis in muscle cells, high cortisol in TB patients probably contributes to wasting and weight loss during the disease. Further, altered DHEA hurts T-cell function and anti-inflammatory effect, potentially contributing to disease severity [88]. Other studies have shown altered levels of other nutrition-regulating metabolic hormones, i.e., adiponectin, Peptide YY, Insulin-like growth factor 1, and resistin in untreated pulmonary TB patients [89,90]. Peptide YY was noted to be an independent predictor of appetite in TB patients [89]. Adiponectin and Ghrelin were reported to be high in TB patients [84,90], while in contrast, a study that compared low and normal BMI TB patients showed ghrelin to be low in low BMI patients [74]. However, the exact mechanism of how ghrelin may affect body weight and energy intake in TB patients is unclear. Lastly, leptin is TB's most studied endocrine marker as it is essential in appetite and body weight management, and high leptin levels suppress appetite [59]. However, leptin studies in TB patients are conflicting. Some studies show high leptin in TB patients, consistent with appetite suppression [87,91], while a South Korean study found no significant difference between TB patients and controls [74]. Most studies indicate lower leptin levels in TB patients despite elevated inflammatory markers [84,89,92,93]. This inconsistency may be explained by leptin's strong correlation with body fat mass and BMI [93,94]. Schwenk et al. demonstrate that leptin levels are independent of cytokine activity and energy intake. It showed that leptin is unlikely to be directly responsible for appetite suppression and weight loss in TB.

Lastly, a study measuring whole-body protein turnover in TB patients, malnourished individuals without TB, and healthy people showed that inflammation, mainly IL-6-mediated response, is associated with protein anabolic block [95]. Essentially, the study showed that TB patients in a fed state with elevated inflammatory conditions diverted ingested amino acid away from utilization for protein synthesis toward oxidation, with consequent loss of the body protein pool. Longitudinal body composition measurements in TB patients during standard TB treatment showed that microbiological cure does not restore protein mass within 6 months and continues to have an anabolic block [96].

Overall, all the above studies show that the chronic inflammatory response in TB drives alterations in immune-endocrine-metabolic mechanisms that impact both treatment outcomes and nutritional status. As a result of inflammation-induced changes in appetite regulatory and nutrient metabolism mechanisms, it may be reasonable to conclude that providing energy alone through macro and micronutrient supplementation to malnourished TB patients is unlikely to achieve optimal outcomes. Therefore, to achieve optimal disease and nutritional outcomes in patients affected with TB, remediation of inflammation will need to be considered along with treatment.

6. Inflammation resolution and nutritional supplementation

One key process that reduces inflammation in TB is the anti-TB treatment itself. As the primary cause of inflammation is the persistent presence of bacteria, one would expect the elimination of bacteria to reduce inflammation. This is true to an extent. Studies that measured immune markers longitudinally during treatment have shown that the levels of inflammatory markers start reducing after treatment initiation and significantly reduce by the end of the treatment [72,74,84,87,97,98,99,100,101,102]. Interestingly, multiple studies have shown that some inflammatory markers, especially IL-6, remain substantially higher than those of the healthy population controls even at the end of treatment [72,82,84,87,100], suggesting that other endogenous factors continue to influence the inflammation levels. A critical matter to be noted in all these studies is that only the average trend in a sample population/cohort of TB patients is explained. However, at any given time, there appears to be a considerable variation in the markers at the individual levels, as evident from the standard deviation or spread of the data shown. Explaining a phenomenon based on an average effect is appropriate to understand the overall or predominant impact in a test population; considerable variations in individual levels of the markers suggest that the average response observed may not apply to all individuals within the same population. As can be seen from a study that reported individual-level changes in markers during treatment along with the average trend (Fig. 4 in the cited publication), in some patients, the levels decreased, reflecting the average trend. In contrast, many patients showed increased levels or no change [82]. This suggests that individual variations in inflammation may explain the differential response to nutritional supplementation observed in food supplementation trials mentioned in the earlier section, and why some patients may continue to have poor dietary status even at the end of treatment.

The direct evidence for the criticality of inflammation resolution for improving nutritional outcomes in TB patients comes from a study that measured inflammation and iron kinetics using tracer iron [103]. The study showed that iron absorption was negligible before treatment, and inflammation (as measured by IL-6 and C-reactive protein (CRP)) and hepcidin (an iron absorption inhibitor) were high. Treatment rapidly reduced the inflammatory markers and hepcidin levels with a concomitant increase in hemoglobin, iron absorption, and erythropoiesis. This supports an earlier observation during an iron supplementation trial, wherein the study showed that the TB patients' hematological status and recovery from anemia were only dependent on disease improvement and not iron supplementation [104]. Although these studies focused on particular micronutrients, indirect evidence for the resolution of inflammation leading to improved general nutritional status in TB comes from some preclinical and clinical studies utilizing anti-inflammatory or cytokine-suppressing agents as adjunct therapy to anti-TB treatment. Vitamin D is a well-known anti-inflammatory supplement strongly associated with inflammation reduction in TB [97]. Some Vitamin D adjunct therapy trials, specifically studies conducted in low and middle-income countries, showed that weight gain in TB patients who received Vitamin D along with TB treatment was significantly greater than in patients receiving standard TB therapy alone [105–107]. Thalidomide, a drug known to suppress proinflammatory cytokine TNF, when given to TB patients with anti-TB medications, enhanced weight gain compared to the control group [108]. Mouse studies in South Africa have demonstrated that omega-3 fatty acids (having known anti-inflammatory properties) supplementation and anti-TB drugs can reduce inflammation and anemia and induce better weight gain in TB-infected mice [109].

Evidence from other disease conditions also strongly supports the criticality of addressing inflammation to drive better patient outcomes during nutrition supplementation. A large clinical trial, Early Nutritional Support on Frailty, Functional Outcomes, and Recovery of Malnourished Medical Inpatients Trial (EFFORT), that tested individualized nutritional intervention in hospitalized elderly patients at risk for malnutrition in Switzerland, noted that patients' admission inflammatory status was associated with their response to nutritional support [110]. They observed that the nutritional intervention did not improve outcomes for patients with high inflammation, and only patients with low or moderate-level inflammation responded to the nutritional intervention. Resistance to nutritional support has also been observed in the clinical nutrition and cancer fields [111,112]. In clinical nutrition, the traditional approach of managing disease-related malnutrition by merely providing nutrients has been challenged due to increasing anomalous observations, prompting a shift toward developing new, tailored solutions based on deeper mechanistic understanding [113]. In cancer, a multimodal approach of chemotherapy along with the provision of nutrition, anti-inflammatory omega-3 fatty acids, drugs, and exercise has been shown to improve the nutritional status of cancer patients compared to standard care [114,115]. A systematic review of omega-3 fatty acid supplementation to cancer patients undergoing chemo/radiotherapy showed that preservation of the muscle mass was the most evident beneficial effect [116]. These observations further support the hypothesis that the effectiveness of any nutritional supplementation cannot be achieved unless inflammation remediation is considered along with disease therapy.

7. Conclusions and perspectives

With a host of efforts prioritizing solutions to mitigate malnutrition and nutritional requirements in TB, it is essential, on the one hand, to understand factors that are important for suboptimal outcomes of population-based feeding strategies and to identify the best approaches to make them better, on the other. Targeted nutrition supplementation effect can be marginal without knowing the causative factor behind conditions, e.g., anemia. A study in north Karnataka demonstrated a rise in anemia, not due to iron deficiency, when the midday meal was stopped. Wholesome and diverse nutrition serves better than the sum of its parts [117]. Secondly, a cautionary note on overfeeding in malnourished/stunted older children (>3 years of age) is worth mentioning since this appears to increase metabolic risk [118]. This may also be true of adults with an unintended increase in systemic inflammation.

Superimposed on malnutrition is the layer of chronic diseases such as TB, which, while conventionally linked to malnutrition as a whole, is vague on any specific deficiency of micronutrients, as well as the understanding of the underlying mechanisms of how specific nutrients affect the immune systems in large populations. It is often missed that malnourished adults are the successors of undernourished children and that inflammatory response/markers may be different in such adults from those that become malnourished later in life through a disease process itself. It becomes essential to determine the genesis of malnutrition in a patient when deciding on nutrition supplementation. This can be caused right from early life by poor socioeconomic conditions, poor maternal health, inadequate nutrition, frequent illness, and inappropriate infant and child feeding practices. In the future, well-planned prospective cohort studies should investigate the relationship between immunological markers and morbidity and mortality patterns, including specific details of nutrition status, viz., body composition, infections, enteropathy, and low grade inflammation. Additionally, innovative experimental methods for exploring gut microbiota, viriome, immune metabolomics, and immunological and nutritional sensing factors would provide novel chances for applying relevant interventions [119,120].

Nutritional deficiencies can trigger inflammation as an evolutionarily conserved innate defense mechanism. This holds for TB, where nutrient intake is compromised through the block in gut microbiota. A high-fat, low-carbohydrate (Ketogenic diet) diet has shown promise in reducing inflammation in type-2 diabetes [121] and obese patients [122]. Combined with digestion-resistant carrier proteins for micronutrients and antioxidants, this becomes important for promoting bioavailability through the lymphatic system [61] and avoiding high sugar or triglyceride-containing foods. The crucial lesson is that the make-up of any nutrient supplementation should be those foods that are in alignment with lymphatic absorption. Recent good-quality evidence also supports a Mediterranean diet with a low dietary inflammatory index as a potential anti-inflammatory diet. Its impact has been experienced in conditions ranging from atherosclerosis to coronavirus respiratory diseases, as well as modulation of gut microbiota [123]. Both diets are widely reflected in therapeutic guidelines for disease management but face individual, clinical health system, and cultural barriers to implementation.

Rapid reduction of inflammation and restoration of immune responses are critical for achieving TB disease control in a malnourished patient. How effective AKT treatment modulates inflammation and its time dynamics remains vague and needs further experimentation. If antigenic stimulation continues to increase due to late presentation in the disease cascade [124], or continuous compromised immunity, or development of acquired drug resistance, T-cell exhaustion is likely to occur [125].

The findings of Bhattarai et al. [126] emphasize that both pathogen sterilization and microbiome composition influence TB-related inflammation. Higher beneficial commensals and reduced dysbiosis, including bacteria, facilitate faster inflammatory resolution. However, it was also simultaneously reported by Wipperman et al. [127] that extended exposure to anti-TB drugs results in a dramatic loss of bacterial diversity in the gut and a disruption of the natural communication between the human gut and lungs, leading to pathological processes. Notably, the production of short-chain fatty acids and butyrate negatively correlates with the severity of anti-inflammatory pathways [128]. At the same time, a decreased abundance of Clostridia and Prevotella species with an increase in Enterobacteriaceae may lead to a reduced ability to synthesize amino acids and fatty acids and altered patterns of carbohydrate metabolism [129]. Therefore, the treatment may tilt the inflammation and nutrient absorption balance in specific circumstances. It is essential, thus, to define these in terms of opposing inflammation outcomes of anti-TB treatment to arrive at the most optimal time at which nutritional supplementation can be initiated. The timing of the nutritional intervention in TB, viz., at a time of declining anti-TB treatment-driven inflammation and its extended duration, may also be essential to maximize nutrition benefits to a broader range of TB patients for a more extended period. Such studies are cost and data-intensive and need to be implemented with full knowledge of the malnutrition syndrome, complete with the pattern of micro and macronutrient deficiencies.

Besides the composition of nutritional supplementation and timing, several approaches towards reducing inflammation have been tested. These include the intake of omega-3 fatty acids [130], vitamin D3 [131], and non-steroidal anti-inflammatory medicines such as ibuprofen [132], in addition to the use of probiotics. However, gut-focused therapies are regularly deactivated and flushed out. A recent study published in Nature Biotech shows how gut bacteria can be directed to produce and release anti-inflammatory proteins within the lower gastrointestinal tract, eliminating a significant roadblock [133]. This is undertaken using engineered phages that infect and reprogram the bacterial cell to produce and release a sustained flow of protein-based drugs. Eventually, the proteins could enter the systemic circulation.

Nutritional, metabolic, and immune factors could be highly diverse within and between different populations, indicating that one size of an anti-inflammatory nutrition supplementation may not fit all.

This would await the formatting of personalized nutrition [134], which would need to consider the complete picture of people's realities. To identify these, a tool has recently been proposed that brings together key realities that shape impact, including material, economic, and psychosocial realities and limitations faced by households, food environments, social interactions, and cultures [135]. Low-diversity gut microbiomes are common in persons with low socioeconomic status and persons receiving anti-TB medication. They may also result in colonization by MDR organisms in the gut [136].

Trials of nutrition supplementation in TB patients should be reinforced by testing diverse parameters to provide a fingerprint of comprehensive benefit. These should include serum biochemical parameters (Fe, Zn, Mg levels), immunological parameters reflecting innate and adaptive immunity, criteria highlighting metabolic risk, inflammatory markers, and gut integrity and microbiome. The sustenance of food supplementation benefits, if any, must be followed up for at least 1 year post-cessation of benefits with clinical and laboratory monitoring. If the benefits are short-lived, the risk of reinfection and relapse of TB and the occurrence of diverse comorbidities is very real. This could be avoided through persistent nutrition education and economic and food security assurance. Global nutrition targets to prevent adult malnutrition should also intensively target children under 2 years to avoid future metabolic risks in the child's life. This highlights countries' increasingly interdisciplinary and multipronged efforts to engender healthy food security for their citizens.

Key point approaches to remediating inflammation for improving nutritional outcomes

Addressing the infective trigger through chemotherapy (Anti-TB therapy)
Anti-inflammatory diet (maintaining dietary diversity, ketogenic diet, Mediterranean diet)
Adjunct therapies with anti-inflammatory medicines (NSAIDs, Thalidomide)
Adjunct therapies with anti-inflammatory nutritional supplements (Vitamin- D, Omega-3 fatty acids)
Gut-directed therapies (Probiotics, Engineered phages)

Footnote: Nonsteroidal anti-inflammatory drugs; (NSAIDs).

References

[1] WHO. Guideline: nutritional care and support for patients with tuberculosis. World Health Organization; 2013.

[2] Lönnroth K, Williams BG, Cegielski P, Dye C. A consistent log-linear relationship between tuberculosis incidence and body mass index. Int J Epidemiol 2010;39(1):149–55. https://doi.org/10.1093/ije/dyp308.

[3] Who. Factsheets. 2024.

[4] MoHFW (2024).

[5] UNICEF. The state of food security and nutrition in the world 2024: financing to end hunger, food insecurity and malnutrition in all its forms. 2024.

[6] Thomas BE, Adinarayanan S, Manogaran C, Swaminathan S. Pulmonary tuberculosis among tribals in India: a systematic review & meta-analysis. Indian J Med Res 2015;142(May):614–23. Available from: http://www.icmr.nic.in/ijmr/2015/may/0515.pdf.

[7] Madankar M, Kakade N, Basa L, Sabri B. Exploring maternal and child health among tribal communities in India: a life course perspective. Global J Health Sci 2024;16(2):31. https://doi.org/10.5539/gjhs.v16n2p31.

[8] Goldstein RF, Abell SK, Ranasinha S, Misso M, Boyle JA, Black MH, Li N, Hu G, Corrado F, Rode L, Kim YJ, Haugen M, Song WO, Kim MH, Bogaerts A, Devlieger R, Chung JH, Teede HJ. Association of gestational weight gain with maternal and infant outcomes: a systematic review and meta-analysis. JAMA 2017;317(21):2207–25. https://doi.org/10.1001/jama.2017.3635.

[9] Grobler L, Nagpal S, Sudarsanam TD, Sinclair D. Nutritional supplements for people being treated for active tuberculosis. Cochrane Database Syst Rev 2016;2016(6). https://doi.org/10.1002/14651858.CD006086.pub4.

[10] Si ZL, Kang LL, Shen XB, Zhou YZ. Adjuvant efficacy of nutrition support during pulmonary tuberculosis treating course: systematic review and meta-analysis. Chin Med J 2015;128(23):3219–30. https://doi.org/10.4103/0366-6999.170255.

[11] Bhargava A, Bhargava M, Meher A, Teja GS, Velayutham B, Watson B, Benedetti A, Barik G, Singh VP, Singh D, Madhukeshwar AK, Prasad R, Pathak RR, Chadha V, Joshi R. Nutritional support for adult patients with microbiologically confirmed pulmonary tuberculosis: outcomes in a programmatic cohort nested within the RATIONS trial in Jharkhand, India. Lancet Global Health 2023;11(9):e1402. https://doi.org/10.1016/S2214-109X(23)00324-8.

[12] Bhargava A, Bhargava M, Meher A, Benedetti A, Velayutham B, Teja GS, Watson B, Barik G, Pathak RR, Prasad R, Dayal R, Madhukeshwar AK, Chadha V, Pai M, Joshi R, Menzies D, Swaminathan S. Nutritional supplementation to prevent tuberculosis incidence in household contacts of patients with pulmonary tuberculosis in India (RATIONS): a field-based, open-label, cluster-randomised, controlled trial. Lancet 2023;402(10402):627–40. https://doi.org/10.1016/s0140-6736(23)01231-x.

[13] Mahapatra A, Thiruvengadam K, Nair D, Padmapriyadarsini C, Thomas B, Pati S, Bulliyya G, Das D, Chowdhury J, Bang A, Swaminathan S, Quinn F. Effectiveness of food supplement on treatment outcomes and quality of life in pulmonary tuberculosis: phased implementation approach. PLoS One 2024;19(7):e0305855. https://doi.org/10.1371/journal.pone.0305855.

[14] Koeken VACM, Verrall AJ, Netea MG, Hill PC, van Crevel R. Trained innate immunity and resistance to *Mycobacterium tuberculosis* infection. Clin Microbiol Infection 2019;25(12):1468–72. https://doi.org/10.1016/j.cmi.2019.02.015.

[15] Morales F, Montserrat-de la Paz S, Leon MJ, Rivero-Pino F. Effects of malnutrition on the immune system and infection and the role of nutritional strategies regarding improvements in children's health status: a literature review. Nutrients 2024;16(1). https://doi.org/10.3390/nu16010001.

[16] Marais BJ, Buck WC, Schaaf HS. The hidden burden of miliary TB in children and adolescents. Int J Tubercul Lung Dis 2024;28(9):407–9. https://doi.org/10.5588/ijtld.24.0232.
[17] Giacomin PR, Kraeuter AK, Albornoz EA, Jin S, Bengtsson M, Gordon R, Woodruff TM, Urich T, Sarnyai Z, Magalhães RJS. Chronic helminth infection perturbs the gut-brain axis, promotes neuropathology, and alters behavior. J Infect Dis 2018;218(9):1511–6. https://doi.org/10.1093/infdis/jiy092.
[18] Salam N, Azam S. Prevalence and distribution of soil-transmitted helminth infections in India. BMC Public Health 2017;17(1). https://doi.org/10.1186/s12889-017-4113-2.
[19] Moreira de Gouveia MI, Bernalier-Donadille A, Jubelin G. Enterobacteriaceae in the human gut: dynamics and ecological roles in health and disease. Biology 2024;13(3):142. https://doi.org/10.3390/biology13030142.
[20] WHO. World health organization: global tuberculosis report. 2016.
[21] Franco JV, Bongaerts B, Metzendorf M-I, Risso A, Guo Y, Silva LP, et al. Undernutrition as a risk factor for tuberculosis disease. Cochrane Database Syst Rev 2024;6.
[22] WHO. Global tuberculosis report 2024. World Health Organization; 2024.
[23] Saunders MJ, Cegielski JP, Clark, Houben RM, McQuaid F, Tuberculosis incidence and body mass index–an updated systematic literature review and global analysis.
[24] Bhargava A, Bhargava M, Beneditti A, Kurpad A. Attributable is preventable: corrected and revised estimates of population attributable fraction of TB related to undernutrition in 30 high TB burden countries. J Clin Tuberc Other Mycobact Dis 2022;27:100309. https://doi.org/10.1016/j.jctube.2022.100309.
[25] Paradkar M, Padmapriyadarsini C, Jain D, Yogendra Shivakumar SVB, Thiruvengadam K, Gupte AN, Thomas B, Kinikar A, Sekar K, Bharadwaj R, Dolla CK, Gaikwad S, Elilarasi S, Lokhande R, Reddy D, Murali L, Kulkarni V, Pradhan N, Hanna LE, Sathyamurthi P, Rewa K, Rani S, Nishi S, Shrinivasa BM, Cox SR, Selvaraju S, Gupte N, Mave V, Gupta A, Bollinger RC, Neyrolles O. Tuberculosis preventive treatment should be considered for all household contacts of pulmonary tuberculosis patients in India. PLoS One 2020;15(7):e0236743. https://doi.org/10.1371/journal.pone.0236743.
[26] Bhargava A, Benedetti A, Oxlade O, Pai M, Menzies D. Undernutrition and the incidence of tuberculosis in India: national and subnational estimates of the population-attributable fraction related to undernutrition. Natl Med J India 2014;27(3):128–33. Available from: http://www.nmji.in/archives/Volume-27/Issue-3/02-27-3-OA-I.pdf.
[27] Kannayan SC, Watson B, Thiruvengadam K, Shamail MHG, Velayutham B, Krishnan R, Natarajan S. Malnutrition in tribal people living with tuberculosis in India: a systematic review & meta-analysis. Indian J Med Res 2024;160(1):23–30. https://doi.org/10.25259/IJMR_2188_23.
[28] Hoyt KJ, Sarkar S, White L, Joseph NM, Salgame P, Lakshminarayanan S, Muthaiah M, Kumar SV, Ellner JJ, Roy G, Horsburgh CR, Hochberg NS, Rottenberg ME. Effect of malnutrition on radiographic findings and mycobacterial burden in pulmonary tuberculosis. PLoS One 2019;14(3):e0214011. https://doi.org/10.1371/journal.pone.0214011.
[29] Podewils LJ, Holtz T, Riekstina V, Skripconoka V, Zarovska E, Kirvelaite G, Kreigere E, Leimane V. Impact of malnutrition on clinical presentation, clinical course, and mortality in MDR-TB patients. Epidemiol Infect 2011;139(1):113–20. https://doi.org/10.1017/s0950268810000907.
[30] Agrawal A. Correlation between BMI and sputum conversion in sputum positive pulmonary tuberculosis. Int J Life Sci 2024;13(10). 229–3.
[31] Sinha P, Ponnuraja C, Gupte N, Babu SP, Cox SR, Sarkar S, Mave V, Paradkar M, Cintron C, Govindarajan S, Kinikar A, Priya N, Gaikwad S, Thangakunam B, Devarajan A, Dhanasekaran M, Tornheim JA, Gupta A, Salgame P, Christopher DJ, Kornfeld H, Viswanathan V, Ellner JJ, Horsburgh CR, Gupte AN, Padmapriyadarsini C, Hochberg NS. Impact of undernutrition on tuberculosis treatment outcomes in India: a multicenter, prospective, cohort analysis. Clin Infect Dis 2023;76(8):1483–91. https://doi.org/10.1093/cid/ciac915.

[32] Wagnew F, Alene KA, Kelly M, Gray D. The effect of undernutrition on sputum culture conversion and treatment outcomes among people with multidrug-resistant tuberculosis: a systematic review and meta-analysis. Int J Infect Dis 2023;127:93–105. https://doi.org/10.1016/j.ijid.2022.11.043.

[33] Wagnew F, Addis Alene K, Kelly M, Gray D. Undernutrition increases the risk of unsuccessful treatment outcomes of patients with tuberculosis in Ethiopia: a multicenter retrospective cohort study. J Infect 2024; 89(1):106175. https://doi.org/10.1016/j.jinf.2024.106175.

[34] Alebel A, Demant D, Petrucka P, Sibbritt D. Effects of undernutrition on mortality and morbidity among adults living with HIV in sub-Saharan Africa: a systematic review and meta-analysis. BMC Infect Dis 2021; 21(1):1471–2334. https://doi.org/10.1186/s12879-020-05706-z.

[35] Bhargava A, Chatterjee M, Jain Y, Chatterjee B, Kataria A, Bhargava M, Kataria R, D'Souza R, Jain R, Benedetti A, Pai M, Menzies D, Noymer A. Nutritional status of adult patients with pulmonary tuberculosis in rural Central India and its association with mortality. PLoS One 2013;8(10):e77979. https://doi.org/10.1371/journal.pone.0077979.

[36] Min J, Kim JS, Kim HW, Ko Y, Oh JY, Jeong YJ, Lee EH, Yang B, Lee KM, Ahn JH, Kim JW, Hwang YI, Lee SS, Park JS, Koo HK. Effects of underweight and overweight on mortality in patients with pulmonary tuberculosis. Front Public Health 2023;11. https://doi.org/10.3389/fpubh.2023.1236099.

[37] Zachariah R, Spielmann MP, Harries AD, Salaniponi FML. Moderate to severe malnutrition in patients with tuberculosis is a risk factor associated with early death. Trans R Soc Trop Med Hyg 2002;96(3):291–4. https://doi.org/10.1016/S0035-9203(02)90103-3.

[38] Jahnavi G, Sudha CH. Randomised controlled trial of food supplements in patients with newly diagnosed tuberculosis and wasting. Singapore Med J 2010;51(12):957.

[39] Shah V, Murugan Y, Patel SS, Trivedi NS, Pithadiya D, Makwana N, Parmar D. Nutritional supplementation in tuberculosis treatment: a mixed methods study of clinical outcomes and patient perceptions in Jamnagar, India. Cureus 2024;16(9). https://doi.org/10.7759/cureus.70300.

[40] Singh AK, Siddhanta A, Goswami L. Improving tuberculosis treatment success rate through nutrition supplements and counselling: findings from a pilot intervention in India. Clin Epidemiol Glob Health 2021; 11:100782. https://doi.org/10.1016/j.cegh.2021.100782.

[41] Franke MA, Emmrich JV, Ranjaharinony F, Ravololohanitra OG, Andriamasy HE, Knauss S, Muller N. A cross-sectional analysis of the effectiveness of a nutritional support programme for people with tuberculosis in Southern Madagascar using secondary data from a non-governmental organisation. Infect Dis Poverty 2024;13(1). https://doi.org/10.1186/s40249-024-01182-8.

[42] Paton NI, Chua YK, Earnest A, Chee CBE. Randomized controlled trial of nutritional supplementation in patients with newly diagnosed tuberculosis and wasting. Am J Clin Nutr 2004;80(2):460–5. https://doi.org/10.1093/ajcn/80.2.460.

[43] Martins N, Morris P, Kelly PM. Food incentives to improve completion of tuberculosis treatment: randomised controlled trial in Dili, Timor-Leste. BMJ 2009;339(Oct 26):b4248. https://doi.org/10.1136/bmj.b4248.

[44] Praygod G, Range N, Faurholt-Jepsen D, Jeremiah K, Faurholt-Jepsen M, Aabye MG, Jensen L, Jensen AV, Grewal HMS, Magnussen P, Changalucha J, Andersen AB, Friis H. The effect of energy-protein supplementation on weight, body composition and handgrip strength among pulmonary tuberculosis HIV-co-infected patients: randomised controlled trial in Mwanza, Tanzania. Br J Nutr 2012;107(2):263–71. https://doi.org/10.1017/S0007114511002832.

[45] Pérez-Guzmán C, Vargas MH, Quiñonez F, Bazavilvazo N, Aguilar A, Outpatient Service Team. A cholesterol-rich diet accelerates bacteriologic sterilization in pulmonary tuberculosis. Chest 2005;127(2): 643–51.

[46] Sudarsanam TD, John J, Kang G, Mahendri V, Gerrior J, Franciosa M, et al. Pilot randomized trial of nutritional supplementation in patients with tuberculosis and HIV–tuberculosis coinfection receiving directly observed short-course chemotherapy for tuberculosis. Trop Med Int Health 2011;16:699–706.

[47] Samuel B, Volkmann T, Conelius S, Mukhopadhay S, Mitra K, Kumar AMV, et al. Relationship between nutritional support and tuberculosis treatment outcomes in West Bengal, India. J Tuberc Res 2016;4(4): 213–9.
[48] Benzekri NA, Sambou JF, Tito Tamba I, Jean Philippe D, Sall I, Cisse O, et al. Nutrition support for HIV-TB co-infected adults in Senegal, West Africa: a randomized pilot implementation study. PLoS One 2019;14 (7):p.e0219118.
[49] Chakhaia T, Blumberg HM, Kempker RR, Luo R, Dzidzikashvili N, Chincharauli M, et al. Lack of weight gain and increased mortality during and after treatment among adults with drug-resistant tuberculosis: a retrospective cohort study in Georgia, 2009–2020. ERJ Open Res 2025;11(3):00839-2024. https://doi.org/10.1183/23120541.00839-2024.
[50] Raiten DJ, Ashour FAS, Ross AC, Meydani SN, Dawson HD, Stephensen CB, Brabin BJ, Suchdev PS, van Ommen B. Inflammation and nutritional science for programs/policies and interpretation of research evidence (INSPIRE). J Nutr 2015;145(5):1039S. https://doi.org/10.3945/jn.114.194571.
[51] Baumann H, Gauldie J. The acute phase response. Immunol Today 1994;15(2):74–80. https://doi.org/10.1016/0167-5699(94)90137-6.
[52] Haddad JJ, Saadé NE, Safieh-Garabedian B. Cytokines and neuro-immune-endocrine interactions: a role for the hypothalamic-pituitary-adrenal revolving axis. J Neuroimmunol 2002;133(1–2):1–19. https://doi.org/10.1016/S0165-5728(02)00357-0.
[53] Konsman JP, Dantzer R. How the immune and nervous systems interact during disease-associated anorexia. Nutrition 2001;17:664–8. https://doi.org/10.1016/S0899-9007(01)00602-5.
[54] Richardson RA, Davidson HIM. Nutritional demands in acute and chronic illness. Proc Nutr Soc 2003;62: 777–81. https://doi.org/10.1079/PNS2003302.
[55] Faggioni R, Fantuzzi G, Fuller J, Dinarello CA, Feingold KR, Grunfeld C. Il-1β mediates leptin induction during inflammation. Am J Physiol Regul Integr Comp Physiol 1998;274(1):R204. https://doi.org/10.1152/ajpregu.1998.274.1.r204.
[56] Janik JE, Curti BD, Considine RV, Rager HC, Powers GC, Alvord WG, Smith JW, Gause BL, Kopp WC. Interleukin 1α increases serum leptin concentrations in humans. J Clin Endocrinol Metab 1997;82(9): 3084–6. https://doi.org/10.1210/jc.82.9.3084.
[57] Bando M, Iwakura H, Ueda Y, Ariyasu H, Inaba H, Furukawa Y, Furuta H, Nishi M, Akamizu T. IL-1β directly suppress ghrelin mRNA expression in ghrelin-producing cells. Mol Cell Endocrinol 2017;447: 45–51. https://doi.org/10.1016/j.mce.2017.02.032.
[58] Iwakura H, Bando M, Ueda Y, Akamizu T. The effects of inflammatory cytokines on the expression of ghrelin. Endocr J 2017;64(Suppl. l):S25. https://doi.org/10.1507/endocrj.64.s25.
[59] Klok MD, Jakobsdottir S, Drent ML. The role of leptin and ghrelin in the regulation of food intake and body weight in humans: a review. Obes Rev 2007;8(1):21–34. https://doi.org/10.1111/j.1467-789x.2006.00270.x.
[60] McMillan DC, Maguire D, Talwar D. Relationship between nutritional status and the systemic inflammatory response: micronutrients. Proc Nutr Soc 2019;78(1):56–67. https://doi.org/10.1017/s0029665118002501.
[61] Roth-Walter F, Berni Canani R, O'Mahony L, Peroni D, Sokolowska M, Vassilopoulou E, Venter C. Nutrition in chronic inflammatory conditions: bypassing the mucosal block for micronutrients. Allergy Eur J Allergy Clin Immunol 2024;79(2):353–83. https://doi.org/10.1111/all.15972.
[62] Drakesmith H, Prentice AM. Hepcidin and the iron-infection axis. Science 2012;338(6108):768–72. https://doi.org/10.1126/science.1224577.
[63] Peuhkuri K, Vapaatalo H, Korpela R. Even low-grade inflammation impacts on small intestinal function. World J Gastroenterol 2010;16(9):1057–62. https://doi.org/10.3748/wjg.v16.i9.1057.
[64] McCauley HA. Enteroendocrine regulation of nutrient absorption. J Nutr 2020;150(1):10–21. https://doi.org/10.1093/jn/nxz191.

[65] Shiner M, Redmond AOB, Hansen JDL. The jejunal mucosa in protein-energy malnutrition. A clinical, histological, and ultrastructural study. Exp Mol Pathol 1973;19(1):61–78. https://doi.org/10.1016/0014-4800(73)90041-5.

[66] Butzner JD, Butler DG, Miniats OP, Hamilton JR. Impact of chronic protein-calorie malnutrition on small intestinal repair after acute viral enteritis: a study in gnotobiotic piglets. Pediatr Res 1985;19(5):476–81. https://doi.org/10.1203/00006450-198505000-00014.

[67] Welsh FKS, Farmery SM, MacLennan K, Sheridan MB, Barclay GR, Guillou PJ, Reynolds JV. Gut barrier function in malnourished patients. Gut 1998;42(3):396–401. https://doi.org/10.1136/gut.42.3.396.

[68] Kaufmann SHE, Dorhoi A. Inflammation in tuberculosis: interactions, imbalances and interventions. Curr Opin Immunol 2013;25(4):441–9. https://doi.org/10.1016/j.coi.2013.05.005.

[69] Bresnahan KA, Tanumihardjo SA. Undernutrition, the acute phase response to infection, and its effects on micronutrient status indicators. Adv Nutr 2014;5(6):702–11. https://doi.org/10.3945/an.114.006361.

[70] Sauerwein RW, Mulder JA, Mulder L, Lowe B, Peshu N, Demacker PN, van der Meer JW, Marsh K. Inflammatory mediators in children with protein-energy malnutrition. Am J Clin Nutr 1997;65(5):1534–9. https://doi.org/10.1093/ajcn/65.5.1534.

[71] Sturgeon JP, Njunge JM, Bourke CD, Gonzales GB, Robertson RC, Bwakura-Dangarembizi M, Berkley JA, Kelly P, Prendergast AJ. Inflammation: the driver of poor outcomes among children with severe acute malnutrition? Nutr Rev 2023;81(12):1636–52. https://doi.org/10.1093/nutrit/nuad030.

[72] Chowdhury IH, Ahmed AM, Choudhuri S, Sen A, Hazra A, Pal NK, Bhattacharya B, Bahar B. Alteration of serum inflammatory cytokines in active pulmonary tuberculosis following anti-tuberculosis drug therapy. Mol Immunol 2014;62(1):159–68. https://doi.org/10.1016/j.molimm.2014.06.002.

[73] Karyadi E, Dolmans WMV, West CE, Van Crevel R, Nelwan RHH, Amin Z, Gross R, Van Der Ven-Jongekrijg J, Van Der Meer JWM. Cytokines related to nutritional status in patients with untreated pulmonary tuberculosis in Indonesia. Asia Pac J Clin Nutr 2007;16(2):218–26. Available from: http://www.healthyeatingclub.org/APJCN/Volume16/vol16.2/Finished/ElvinaKaryadi(218-226).pdf.

[74] Kim JH, Lee CT, Yoon HI, Song J, Shin WG, Lee JH. Relation of ghrelin, leptin and inflammatory markers to nutritional status in active pulmonary tuberculosis. Clin Nutr 2010;29(4):512–8. https://doi.org/10.1016/j.clnu.2010.01.008.

[75] Richterman A, Sinha P, Ivers LC, Gross R, Rantleru T, Tamuhla N, et al. Food insecurity and undernutrition are associated with distinct immunologic profiles in people with tuberculosis and advanced HIV starting antiretroviral therapy. J Acquir Immune Defic Syndr 2024;95:494–504.

[76] Anuradha R, Munisankar S, Bhootra Y, Kumar NP, Dolla C, Kumaran P, Babu S. Coexistent malnutrition is associated with perturbations in systemic and antigen-specific cytokine responses in latent tuberculosis infection. Clin Vaccine Immunol 2016;23(4):339–45. https://doi.org/10.1128/CVI.00009-16

[77] VanValkenburg A, Kaipilyawar V, Sarkar S, Lakshminarayanan S, Cintron C, Prakash Babu S, Knudsen S, Joseph NM, Robert Horsburgh C, Sinha P, Ellner JJ, Babu Narasimhan P, Evan Johnson W, Hochberg NS, Salgame P. Malnutrition leads to increased inflammation and expression of tuberculosis risk signatures in recently exposed household contacts of pulmonary tuberculosis. Front Immunol 2022;13. https://doi.org/10.3389/fimmu.2022.1011166.

[78] Chan J, Tian Y, Tanaka KE, Tsang MS, Yu K, Salgame P, Carroll D, Kress Y, Teitelbaum R, Bloom BR. Effects of protein calorie malnutrition on tuberculosis in mice. Proc Natl Acad Sci 1996;93(25):14857–61. https://doi.org/10.1073/pnas.93.25.14857.

[79] Dai G, McMurray DN. Altered cytokine production and impaired antimycobacterial immunity in protein-malnourished Guinea pigs. Infect Immun 1998;66(8):3562–8. https://doi.org/10.1128/iai.66.8.3562-3568.1998.

[80] McMurray DN, Carlomagno MA, Mintzer CL, Tetzlaff CL. Mycobacterium bovis BCG vaccine fails to protect protein-deficient Guinea pigs against respiratory challenge with virulent *Mycobacterium tuberculosis*. Infect Immun 1985;50(2):555–9. https://doi.org/10.1128/iai.50.2.555-559.1985.

[81] Gupte AN, Kumar P, Araújo-Pereira M, Kulkarni V, Paradkar M, Pradhan N, Menon P, Padmapriyadarsini C, Hanna L-E, Shivakumar SVBY, Rockwood N, Du Bruyn E, Karyakarte R, Gaikwad S, Bollinger R, Golub J, Gupte N, Viswanathan V, Wilkinson RJ, Mave V, Babu S, Kornfeld H, Andrade BB, Gupta A. Baseline IL-6 is a biomarker for unfavourable tuberculosis treatment outcomes: a multisite discovery and validation study. Eur Respir J 2022;59(4):2100905. https://doi.org/10.1183/13993003.00905-2021.

[82] Kumar NP, Moideen K, Banurekha VV, Nair D, Babu S. Plasma proinflammatory cytokines are markers of disease severity and bacterial burden in pulmonary tuberculosis. Open Forum Infect Dis 2019;6(7). https://doi.org/10.1093/ofid/ofz257.

[83] Rambaran S, Naidoo K, Lara L, Hassan-Moosa R, Govender D, Samsunder N, Scriba TJ, Padayatchi N, Sivro A. Effect of inflammatory cytokines/chemokines on pulmonary tuberculosis culture conversion and disease severity in HIV-infected and -uninfected individuals from South Africa. Front Immunol 2021;12. https://doi.org/10.3389/fimmu.2021.641065.

[84] Santucci N, D'Attilio L, Kovalevski L, Bozza V, Besedovsky H, del Rey A, Bay ML, Bottasso O, Giambartolomei GH. A multifaceted analysis of immune-endocrine-metabolic alterations in patients with pulmonary tuberculosis. PLoS One 2011;6(10):e26363. https://doi.org/10.1371/journal.pone.0026363.

[85] Mahuad C, Bozza V, Pezzotto SM, Bay ML, Besedovsky H, Del Rey A, Bottasso O. Impaired immune responses in tuberculosis patients are related to weight loss that coexists with an immunoendocrine imbalance. Neuro Immuno Modulation 2007;14:193–9. https://doi.org/10.1159/000110646.

[86] Rey AD, Mahuad CV, Bozza VV, Bogue C, Farroni MA, Bay ML, Bottasso OA, Besedovsky HO. Endocrine and cytokine responses in humans with pulmonary tuberculosis. Brain Behav Immun 2007;21(2): 171–9. https://doi.org/10.1016/j.bbi.2006.06.005.

[87] Zheng Y, Ma A, Wang Q, Han X, Cai J, Schouten EG, Kok FJ, Li Y, Amar S. Relation of leptin, ghrelin and inflammatory cytokines with body mass index in pulmonary tuberculosis patients with and without type 2 diabetes mellitus. PLoS One 2013;8(11):e80122. https://doi.org/10.1371/journal.pone.0080122.

[88] Dillon JS. Dehydroepiandrosterone, dehydroepiandrosterone sulfate and related steroids: their role in inflammatory, allergic and immunological disorders. Curr Drug Targets Inflamm Allergy 2005;4(3):377–85. https://doi.org/10.2174/1568010054022079.

[89] Chang SW, Pan WS, Lozano Beltran D, Oleyda Baldelomar L, Solano MA, Tuero I, Friedland JS, Torrico F, Gilman RH. Gut hormones, appetite suppression and cachexia in patients with pulmonary TB. PLoS One 2013;8(1). https://doi.org/10.1371/journal.pone.0054564.

[90] Moideen K, Kumar NP, Nair D, Banurekha VV, Babu S. Altered systemic adipokine levels in pulmonary tuberculosis and changes following treatment. Am J Trop Med Hyg 2018;99(4):875–80. https://doi.org/10.4269/ajtmh.18-0206.

[91] Çakir B, Yönem A, Güler S, Odabasi E, Demirbas B, Gürsoy G, Aral Y. Relation of leptin and tumor necrosis factor α to body weight changes in patients with pulmonary tuberculosis. Horm Res 1999;52(6): 279–83. https://doi.org/10.1159/000023495.

[92] Buyukoglan H, Gulmez I, Kelestimur F, Kart L, Oymak FS, Demir R, Ozesmi M. Leptin levels in various manifestations of pulmonary tuberculosis. Mediat Inflamm 2007;2007. https://doi.org/10.1155/2007/64859.

[93] Van Crevel R, Karyadi E, Netea MG, Verhoef H, Nelwan RHH, West CE, Van Der Meer JWM. Decreased plasma leptin concentrations in tuberculosis patients are associated with wasting and inflammation. J Clin Endocrinol Metab 2002;87(2):758–63. https://doi.org/10.1210/jcem.87.2.8228.

[94] Schwenk A, Hodgson L, Rayner CFJ, Griffin GE, Macallan DC. Leptin and energy metabolism in pulmonary tuberculosis. Am J Clin Nutr 2003;77(2):392–8. https://doi.org/10.1093/ajcn/77.2.392.

[95] Macallan DC, McNurlan MA, Kurpad AV, De Souza G, Shetty PS, Calder AG, Griffin GE. Whole body protein metabolism in human pulmonary tuberculosis and undernutrition: evidence for anabolic block in tuberculosis. Clin Sci 1998;94(3):321–31. https://doi.org/10.1042/cs0940321.

[96] Schwenk A, Hodgson L, Wright A, Ward LC, Rayner CFJ, Grubnic S, Griffin GE, Macallan DC. Nutrient partitioning during treatment of tuberculosis: gain in body fat mass but not in protein mass. Am J Clin Nutr 2004;79(6):1006–12. https://doi.org/10.1093/ajcn/79.6.1006.

[97] Coussens AK, Wilkinson RJ, Hanifa Y, Nikolayevskyy V, Elkington PT, Islam K, Timms PM, Venton TR, Bothamley GH, Packe GE, Darmalingam M, Davidson RN, Milburn HJ, Baker LV, Barker RD, Mein CA, Bhaw-Rosun L, Nuamah R, Young DB, Drobniewski FA, Griffiths CJ, Martineau AR. Vitamin D accelerates resolution of inflammatory responses during tuberculosis treatment. Proc Natl Acad Sci 2012;109 (38):15449–54. https://doi.org/10.1073/pnas.1200072109.

[98] Djoba Siawaya JF, Beyers N, Van Helden P, Walzl G. Differential cytokine secretion and early treatment response in patients with pulmonary tuberculosis. Clin Exp Immunol 2009;156(1):69–77. https://doi.org/10.1111/j.1365-2249.2009.03875.x.

[99] Krivošová M, Dohál M, Mäsiarová S, Peršo K, Gondáš E, Murín R, Fraňová S, Porvazník I, Solovič I, Mokrý J, Pinzon RT. Exploring cytokine dynamics in tuberculosis: a comparative analysis of patients and controls with insights from three-week antituberculosis intervention. PLoS One 2024;19(8):e0305158. https://doi.org/10.1371/journal.pone.0305158.

[100] Kumar NP, Fukutani KF, Shruthi BS, Alves T, Silveira-Mattos PS, Rocha MS, West K, Natarajan M, Viswanathan V, Babu S, Andrade BB, Kornfeld H. Persistent inflammation during anti-tuberculosis treatment with diabetes comorbidity. eLife 2019;8. https://doi.org/10.7554/eLife.46477.

[101] Peresi E, Silva SMUR, Calvi SA, Marcondes-Machado J. Cytokines and acute phase serum proteins as markers of inflammatory regression during the treatment of pulmonary tuberculosis. J Bras Pneumol 2008; 34:942–9.

[102] Vivekanandan MM, Adankwah E, Aniagyei W, Acheampong I, Yeboah A, Arthur JF, Lamptey MNK, Abass MK, Gawusu A, Kumbel F, Osei-Yeboah F, Debrah LB, Owusu DO, Debrah A, Mayatepek E, Seyfarth J, Phillips RO, Jacobsen M. Plasma cytokine levels characterize disease pathogenesis and treatment response in tuberculosis patients. Infection 2023;51(1):169–79. https://doi.org/10.1007/s15010-022-01870-3.

[103] Cercamondi CI, Stoffel NU, Moretti D, Zoller T, Swinkels DW, Zeder C, Mhimibra F, Hella J, Fenner L, Zimmermann MB. Iron homeostasis during anemia of inflammation: a prospective study of patients with tuberculosis. Blood 2021;138(15):1293–303. https://doi.org/10.1182/blood.2020010562.

[104] Devi U, Rao CM, Srivastava VK, Rath PK, Das BS. Effect of iron supplementation on mild to moderate anaemia in pulmonary tuberculosis. Br J Nutr 2003;90(3):541–50. https://doi.org/10.1079/BJN2003936.

[105] Chandra H, Rahman A, Yadav P, Maurya G, Kumar Shukla S. Effect of adjunct vitamin D treatment in vitamin D deficient pulmonary tuberculosis patients: a randomized, double blind, active controlled clinical trial. Indian J Tubercul 2024;71(2):170–8. https://doi.org/10.1016/j.ijtb.2023.04.026.

[106] Farazi A, Didgar F, Sarafraz A. The effect of vitamin D on clinical outcomes in tuberculosis. Egypt J Chest Dis Tuberc 2017;66(3):419–23. https://doi.org/10.1016/j.ejcdt.2017.01.004.

[107] Salahuddin N, Ali F, Hasan Z, Rao N, Aqeel M, Mahmood F. Vitamin D accelerates clinical recovery from tuberculosis: results of the SUCCINCT Study [Supplementary Cholecalciferol in recovery from tuberculosis]. A randomized, placebo-controlled, clinical trial of vitamin D supplementation in patients with pulmonary tuberculosis'. BMC Infect Dis 2013;13(1). https://doi.org/10.1186/1471-2334-13-22.

[108] Tramontana JM, Utaipat U, Molloy A, Akarasewi P, Burroughs M, Makonkawkeyoon S, Johnson B, Klausner JD, Rom W, Kaplan G. Thalidomide treatment reduces tumor necrosis factor α production and enhances weight gain in patients with pulmonary tuberculosis. Mol Med 1995;1(4):384–97. https://doi.org/10.1007/BF03401576.

[109] Nienaber A, Baumgartner J, Dolman RC, Ozturk M, Zandberg L, Hayford FEA, Brombacher F, Blaauw R, Parihar SP, Smuts CM, Malan L. Omega-3 fatty acid and iron supplementation alone, but not in combination, lower inflammation and anemia of infection in *Mycobacterium tuberculosis*-infected mice. Nutrients 2020;12(9):2897. https://doi.org/10.3390/nu12092897.

[110] Merker M, Felder M, Gueissaz L, Bolliger R, Tribolet P, Kägi-Braun N, Gomes F, Hoess C, Pavlicek V, Bilz S, Sigrist S, Brändle M, Henzen C, Thomann R, Rutishauser J, Aujesky D, Rodondi N, Donzé J, Stanga Z, Mueller B, Schuetz P. Association of baseline inflammation with effectiveness of nutritional support among patients with disease-related malnutrition a secondary analysis of a randomized clinical trial. JAMA Netw Open 2020;3(3):E200663. https://doi.org/10.1001/jamanetworkopen.2020.0663.
[111] Moore FA, Phillips SM, McClain CJ, Patel JJ, Martindale RG. Nutrition support for persistent inflammation, immunosuppression, and catabolism syndrome. Nutr Clin Pract 2017;32:121–7. https://doi.org/10.1177/0884533616687502.
[112] Schiessel DL, Baracos VE. Barriers to cancer nutrition therapy: excess catabolism of muscle and adipose tissues induced by tumour products and chemotherapy. Proc Nutr Soc 2018;77:394–402. https://doi.org/10.1017/S0029665118000186.
[113] Cardenas D, Ochoa JB. A paradigm shift in clinical nutrition. Clin Nutr 2023;42(3):380–3. https://doi.org/10.1016/j.clnu.2023.01.014.
[114] Kaasa S, Solheim T, Laird BJA, Balstad T, Birgitte Stene G, Bye A, Fallon MT, Fayers P, Fearon K. A randomised, open-label trial of a multimodal intervention (exercise, nutrition and anti-inflammatory medication) plus standard care versus standard care alone to prevent/attenuate cachexia in advanced cancer patients undergoing chemotherapy. J Clin Oncol 2015;33(15_Suppl. l):9628. https://doi.org/10.1200/jco.2015.33.15_suppl.9628.
[115] Solheim TS, Laird BJA, Balstad TR, Bye A, Stene G, Baracos V, Strasser F, Griffiths G, Maddocks M, Fallon M, Kaasa S, Fearon K. Cancer cachexia: rationale for the MENAC (Multimodal-Exercise, nutrition and anti-inflammatory medication for cachexia) trial. BMJ Support Palliat Care 2018;8(3):258–65. https://doi.org/10.1136/bmjspcare-2017-001440.
[116] Silva JDAP, Fabre MEDS, Waitzberg DL. Omega-3 supplements for patients in chemotherapy and/or radiotherapy: a systematic review. Clin Nutr 2015;34(3).
[117] Anura Kurpad HSS. Why public health nutrition policies need novelty, precise targeting. The Hindu; 2025.
[118] Arndt MB, Abate YH, Abbasi-Kangevari M, Abd ElHafeez S, Abdelmasseh M, Abd-Elsalam S, Abdulah DM, Abdulkader RS, Abidi H, Abiodun O, Aboagye RG, Abolhassani H, Abtew YD, Abu-Gharbieh E, Abu-Rmeileh NM, Acuna JM, Adamu K, Adane DE, Addo IY, Adeyinka DA, Adnani QES, Afolabi AA, Afrashteh F, Afzal S, Agodi A, Ahinkorah BO, Ahmad A, Ahmad S, Ahmad T, Ahmadi A, Ahmed A, Ahmed LAA, Ajami M, Aji B, Akbarialiabad H, Akonde M, Al Hamad H, Al Thaher Y, Al-Aly Z, Alhabib KF, Alhassan RK, Ali BA, Ali SS, Alimohamadi Y, Aljunid SM, Al-Mekhlafi HM, Almustanyir S, Alomari MA, Al-Tammemi AB, Altirkawi KA, Alvis-Guzman N, Alvis Zakzul N, Ameyaw EK, Amin TT, Amiri S, Amu H, Amugsi DA, Anagaw TFF, Ancuceanu R, Angappan D, Ansari-Moghaddam A, Antriyandarti E, Anvari D, Anyasodor AE, Arabloo J, Aravkin AY, Ariffin H, Aripov T, Arkew M, Armocida B, Arumugam A, Aryastami NK, Asaad M, Asemi Z, Asemu MT, Asghari-Jafarabadi M, Astell-Burt T, Athari SS, Atomsa GH, Atorkey P, Atout MMdW, Aujayeb A, Awoke MA, Azadnajafabad S, Azevedo RMS, Badiye AD, Baghcheghi N, Bagheri N, Bagherieh S, Baig AA, Baker JL, Balasubramanian M, Baltatu OC, Banach M, Banik PC, Barchitta M, Bärnighausen TW, Barr RD, Barrow A, Barua L, Bashiri A, Baskaran P, Basu S, Bekele A, Belay SA, Belgaumi UI, Bell SL, Belo L, Bennett DA, Bensenor IM, Beressa G, Bermudez ANC, Beyene HB, Bhagavathula AS, Bhardwaj N, Bhardwaj P, Bhaskar S, Bhattacharjee NV, Bhutta ZA, Bitaraf S, Bodolica V, Bonakdar Hashemi M, Braithwaite D, Butt MH, Butt ZA, Calina D, Cámera LA, Campos LA, Cao C, Cárdenas R, Carvalho M, Castañeda-Orjuela CA, Catapano AL, Cattaruzza MS, Cembranel F, Cerin E, Chadwick J, Chalek J, Chandrasekar EK, Charan J, Chattu VK, Chauhan K, Chien JH, Chitheer A, Choudhari SG, Chowdhury EK, Chu DT, Chukwu IS, Chung SC, Claro RM, Columbus A, Cortese S, Cruz-Martins N, Dabo B, Dadras O, Dai X, D'Amico E, Dandona L, Dandona R, Darban I, Darmstadt GL, Darwesh AM, Darwish AH, Das JK, Das S, Davletov K, De la Hoz FP, Debele AT, Demeke D, Demissie S, Denova-Gutiérrez E, Desai HD, Desta AA, Dharmaratne SD, Dhimal M, Dias da Silva D, Diaz D, Diress M,

Djalalinia S, Doaei S, Dongarwar D, Dsouza HL, Edalati S, Edinur HA, Ekholuenetale M, Ekundayo TC, Elbarazi I, Elgendy IY, Elhadi M, Elmeligy OAA, Eshetu HB, Espinosa-Montero J, Esubalew H, Etaee F, Etafa W, Fagbamigbe AF, Fakhradiyev IR, Falzone L, Farinha CSeS, Farmer S, Fasanmi AO, Fatehizadeh A, Feigin VL, Feizkhah A, Feng X, Ferrara P, Fetensa G, Fischer F, Fitzgerald R, Flood D, Foigt NA, Folayan MO, Fowobaje KR, Franklin RC, Fukumoto T, Gadanya MA, Gaidhane AM, Gaihre S, Gakidou E, Galali Y, Galehdar N, Gardner WM, Garg P, Gebremeskel TG, Gerema U, Getacher L, Getachew ME, Getawa S, Ghaffari K, Ghamari SH, Ghasemi Nour M, Ghassemi F, Ghith N, Gholamalizadeh M, Gholami A, Gholamrezanezhad A, Ghozy S, Gill PS, Gill TK, Glasbey JC, Golechha M, Goleij P, Golinelli D, Goudarzi H, Grivna M, Guadie HA, Gubari MIM, Gudayu TW, Guha A, Gunawardane DA, Gupta AK, Gupta B, Gupta R, Gupta S, Gupta VB, Gupta VK, Hagins H, Haj-Mirzaian A, Handal AJ, Hanif A, Hankey GJ, Harapan H, Hargono A, Haro JM, Hasaballah AI, Hasan MM, Hasani H, Hashi A, Hassanipour S, Havmoeller RJ, Hay SI, Hayat K, He J, Heidari-Foroozan M, Herteliu C, Hessami K, Heyi DZ, Hezam K, Hiraike Y, Holla R, Hoogar P, Hossain SJ, Hosseinzadeh M, Hostiuc M, Hostiuc S, Hoveidamanesh S, Huang J, Humphrey KM, Hussain S, Hussien FM, Hwang BF, Iacoviello L, Iftikhar PM, Ilesanmi OS, Ilic IM, Ilic MD, Immurana M, Inbaraj LR, Iravanpour F, Islam SMS, Islami F, Ismail NE, Iso H, Isola G, Iwagami M, Iwu CD, Jacob L, Jahrami H, Jakovljevic M, Jamshidi E, Janodia MD, Jayanna K, Jayapal SK, Jayaram S, Jebai R, Jema AT, Jeswani BM, Jonas JB, Joseph A, Joseph N, Joshua CE, Jozwiak JJ, Jürisson M, Kaambwa B, Kabir A, Kabir Z, Kadashetti V, Kamal VK, Kamble BD, Kandel H, Kapoor N, Karaye IM, Katoto PD, Kauppila JH, Kaur H, Kayode GA, Kebede WM, Kebira JY, Keflie TS, Kerr JA, Keykhaei M, Khader YS, Khajuria H, Khalid N, Khammarnia M, Khan MN, Khan MA, Khan T, Khan YH, Khanali J, Khanmohammadi S, Khatab K, Khatatbeh MM, Khateri S, Khatib MN, Khayat Kashani HR, Khubchandani J, Kifle ZD, Kim GR, Kimokoti RW, Kisa A, Kisa S, Kompani F, Kondlahalli SKM, Koohestani HR, Korzh O, Koulmane Laxminarayana SL, Koyanagi A, Krishan K, Krishnamoorthy V, Kuate Defo B, Kucuk Bicer B, Kuddus M, Kumar GA, Kumar M, Kumar N, Kurmanova A, Kurmi OP, Kusuma D, La Vecchia C, Lacey B, Lal DK, Larsson AO, Latief K, Ledda C, Lee PH, Lee Sw, Lee WC, Lee YH, Lenzi J, Li MC, Li W, Ligade VS, Lim SS, Lindstedt PA, Lo CH, Lo J, Lodha R, Loreche AM, Lorenzovici L, Lorkowski S, Madadizadeh F, Madureira-Carvalho ÁM, Mahajan PB, Makris KC, Malakan Rad E, Malik AA, Mallhi TH, Malta DC, Manguerra H, Marjani A, Martini S, Martorell M, Masrie A, Mathews E, Maugeri A, Mazaheri M, Mediratta RP, Mehndiratta MM, Melaku YA, Mendoza W, Menezes RG, Mensah GA, Mentis AFA, Meretoja TJ, Mestrovic T, Miazgowski T, Miller TR, Mini GK, Mirghafourvand M, Mirica A, Mirrakhimov EM, Mirza M, Misra S, Mithra P, Mohammad KA, Mohammadian-Hafshejani A, Mohammed S, Mohseni M, Mokdad AH, Monasta L, Moni MA, Moradi M, Moradi Y, Morrison SD, Mougin V, Mubarik S, Mueller UO, Mulita F, Munblit D, Murillo-Zamora E, Murray CJL, Mustafa G, Nagarajan AJ, Nangia V, Narasimha Swamy S, Natto ZS, Naveed M, Nayak BP, Nejadghaderi SA, Nguefack Tsague G, Ngunjiri JW, Nguyen PT, Nguyen QAP, Niazi RK, Nnaji CA, Noor NM, Noubiap JJ, Nri-Ezedi CA, Nurrika D, Nwatah VE, Oancea B, Obamiro KO, Oghenetega OB, Ogunsakin RE, Okati-Aliabad H, Okekunle AP, Okello DM, Okonji OC, Olagunju AT, Olana DD, Oliveira GMM, Olusanya BO, Olusanya JO, Ong SK, Ortega-Altamirano DV, Ortiz A, Ostojic SM, Otoiu A, Oumer A, Padron-Monedero A, Padubidri JR, Pana A, Panda-Jonas S, Pandey A, Pandi-Perumal SR, Papadopoulou P, Pardhan S, Pasovic M, Patel J, Pathan AR, Paudel D, Pawar S, Pepito VCF, Pereira G, Pereira M, Perico N, Perna S, Petcu IR, Petermann-Rocha FE, Piracha ZZ, Plakkal N, Pourtaheri N, Radfar A, Radhakrishnan V, Raggi C, Raghav P, Rahim F, Rahimi-Movaghar V, Rahman A, Rahman MM, Rahman MO, Rahman M, Rahman MA, Rahmani AM, Rahmanian V, Rahmawaty S, Rai RK, Raimondo I, Rajaa S, Rajput P, Ram P, Ramasamy SK, Ramazanu S, Rao CR, Rao IR, Rao SJ, Rasali DP, Rashid AM, Rashidi MM, Ratan ZA, Rawaf S, Rawal L, Redwan EMMM, Remuzzi G, Rengasamy KR, Renzaho AMN, Rezaee M, Rezaei N, Rezaeian M, Riad A, Rickard J, Rodriguez A, Rodriguez JAB, Roever L, Rohloff P, Roy B, Rwegerera GM, Saad AMA, Saber-Ayad MM, Sabour S, Sachdeva Dhingra M, Saddik BA, Sadeghi E, Sadeghi M,

Sadeghian S, Saeed U, Saeedi Moghaddam S, Safi SZ, Saheb Sharif-Askari F, Sahebkar A, Sahoo H, Sahoo SS, Sajid MR, Salem MR, Samy AM, Sanabria J, Sanjeev RK, Sankararaman S, Santos IS, Santric-Milicevic MM, Saraswathy SYI, Sargazi S, Sarikhani Y, Satpathy M, Sawhney M, Saya GK, Sayeed A, Scarmeas N, Schlaich MP, Schneider RD, Schutte AE, Senthilkumaran S, Sepanlou SG, Serban D, Seylani A, Shafie M, Shah PA, Shahbandi A, Shaikh MA, Shama ATT, Shams-Beyranvand M, Shanawaz M, Sharew MM, Shetty PH, Shiri R, Shivarov V, Shorofi SA, Shuval K, Sibhat MM, Silva LMLR, Singh JA, Singh NP, Singh P, Singh S, Skryabina AA, Smith AE, Solomon Y, Song Y, Sorensen RJD, Stanaway JD, Sufiyan MB, Suleman M, Sun J, Sunuwar DR, Szeto MD, Tabarés-Seisdedos R, Tabatabaeizadeh SA, Tabatabai S, Taheri Soodejani M, Tamuzi JLJ, Tan KK, Tarigan IU, Tariku Z, Tariqujjaman M, Tarkang EE, Tat NY, Taye BT, Taylor HJ, Tefera YM, Tehrani-Banihashemi A, Temsah MH, Teramoto M, Thangaraju P, Thapar R, Thiyagarajan A, Thrift AG, Tichopad A, Ticoalu JHV, Tillawi T, Tiruye TY, Tonelli M, Topor-Madry R, Touvier M, Tovani-Palone MR, Tran MTN, Ullah S, Undurraga EA, Unnikrishnan B, Ushula TW, Vahabi SM, Vakilian A, Valadan Tahbaz S, Valizadeh R, Van den Eynde J, Varthya SB, Vasankari TJ, Venketasubramanian N, Verma M, Veroux M, Vervoort D, Vlassov V, Vollset SE, Vukovic R, Waheed Y, Wang C, Wang F, Wassie MM, Weerakoon KG, Wei MY, Werdecker A, Wickramasinghe ND, Wolde AA, Wubetie GA, Wulandari RD, Xu R, Xu S, Xu X, Yadav L, Yamagishi K, Yang L, Yano Y, Yaya S, Yazdanpanah F, Yehualashet SS, Yiğit A, Yiğit V, Yon DK, Yu C, Yuan CW, Zamagni G, Zaman SB, Zanghì A, Zangiabadian M, Zare I, Zastrozhin M, Zigler B, Zoladl M, Zou Z, Kassebaum NJ, Reiner RC. Global, regional, and national progress towards the 2030 global nutrition targets and forecasts to 2050: a systematic analysis for the Global Burden of disease study 2021. Lancet 2024;404(10471):2543–83. https://doi.org/10.1016/S0140-6736(24)01821-X.

[119] Bourke CD, Berkley JA, Prendergast AJ. Immune dysfunction as a cause and consequence of malnutrition. Trends Immunol 2016;37(6):386–98. https://doi.org/10.1016/j.it.2016.04.003.

[120] Rytter MJH, Kolte L, Briend A, Friis H, Christensen VB, Akiyama T. The immune system in children with malnutrition—a systematic review. PLoS One 2014;9(8):e105017. https://doi.org/10.1371/journal.pone.0105017.

[121] Gram-Kampmann EM, Olesen TB, Hansen CD, Hugger MB, Jensen JM, Handberg A, Beck-Nielsen H, Krag A, Olsen MH, Højlund K. A six-month low-carbohydrate diet high in fat does not adversely affect endothelial function or markers of low-grade inflammation in patients with type 2 diabetes: an open-label randomized controlled trial. Cardiovasc Diabetol 2023;22(1). https://doi.org/10.1186/s12933-023-01956-8.

[122] Esposito K, Marfella R, Ciotola M, Di Palo C, Giugliano F, Giugliano G, D'Armiento M, D'Andrea F, Giugliano D. Effect of a Mediterranean-style diet on endothelial dysfunction and markers of vascular inflammation in the metabolic syndrome: a randomized trial. JAMA 2004;292(12):1440–6. https://doi.org/10.1001/jama.292.12.1440.

[123] Itsiopoulos C, Mayr HL, Thomas CJ. The anti-inflammatory effects of a Mediterranean diet: a review. Curr Opin Clin Nutr Metab Care 2022;25(6):415–22. https://doi.org/10.1097/MCO.0000000000000872.

[124] Mistry N, Rangan S, Dholakia Y, Lobo E, Shah S, Patil A, Hozbor DF. Durations and delays in care seeking, diagnosis and treatment initiation in uncomplicated pulmonary tuberculosis patients in Mumbai, India. PLoS One 2016;11(3):e0152287. https://doi.org/10.1371/journal.pone.0152287.

[125] Vats D, Rani G, Arora A, Sharma V, Rathore I, Abdul Mubeen S, Singh A. Tuberculosis and T cells: impact of T cell diversity in tuberculosis infection. Tuberculosis 2024;149:102567. https://doi.org/10.1016/j.tube.2024.102567.

[126] Bhattarai SK, Du M, Zeamer AL, Morzfeld BM, Kellogg TD, Firat K, Benjamin A, Bean JM, Zimmerman M, Mardi G, Vilbrun SC, Walsh KF, Fitzgerald DW, Glickman MS, Bucci V. Commensal antimicrobial resistance mediates microbiome resilience to antibiotic disruption. Sci Transl Med 2024;16(730). https://doi.org/10.1126/scitranslmed.adi9711.

[127] Wipperman MF, Bhattarai SK, Vorkas CK, Maringati VS, Taur Y, Mathurin L, McAulay K, Vilbrun SC, Francois D, Bean J, Walsh KF, Nathan C, Fitzgerald DW, Glickman MS, Bucci V. Gastrointestinal

microbiota composition predicts peripheral inflammatory state during treatment of human tuberculosis. Nat Commun 2021;12(1). https://doi.org/10.1038/s41467-021-21475-y.

[128] van Deuren T, Blaak EE, Canfora EE. Butyrate to combat obesity and obesity-associated metabolic disorders: current status and future implications for therapeutic use. Obes Rev 2022;23(10). https://doi.org/10.1111/obr.13498.

[129] Eribo OA, du Plessis N, Ozturk M, Guler R, Walzl G, Chegou NN. The gut microbiome in tuberculosis susceptibility and treatment response: guilty or not guilty? Cell Mol Life Sci 2020;77(8):1497–509. https://doi.org/10.1007/s00018-019-03370-4.

[130] Mori TA, Beilin LJ. Omega-3 fatty acids and inflammation. Curr Atheroscler Rep 2004;6(6):461–7. https://doi.org/10.1007/s11883-004-0087-5.

[131] Coussens AK, Martineau AR, Wilkinson RJ. Anti-inflammatory and antimicrobial actions of vitamin D in combating TB/HIV. Scientifica 2014;2014:1–13. https://doi.org/10.1155/2014/903680.

[132] Zumla A, Rao M, Parida SK, Keshavjee S, Cassell G, Wallis R, Axelsson-Robertsson R, Doherty M, Andersson J, Maeurer MJ. Inflammation and tuberculosis: host-directed therapies. J Intern Med 2015;277(4):373–87. https://doi.org/10.1111/joim.12256.

[133] Baker ZR, Zhang Y, Zhang H, Franklin HC, Serpa PB, Southard T, et al. Sustained in situ protein production and release in the mammalian gut by an engineered bacteriophage. Nat Biotechnol 2025:1–10.

[134] Neu J. The evolution of personalized nutrition. Nat Med 2024;30(7):1826–7. https://doi.org/10.1038/s41591-024-03000-y.

[135] Hawkes C, Gallagher-Squires C, Spires M, Hawkins N, Neve K, Brock J, Isaacs A, Parrish S, Coleman P. The full picture of people's realities must be considered to deliver better diets for all. Nat Food 2024;5(11):894–900. https://doi.org/10.1038/s43016-024-01064-0.

[136] Zuniga-Chaves I, Eggers S, Kates AE, Safdar N, Suen G, Malecki KMC. Neighborhood socioeconomic status is associated with low diversity gut microbiomes and multi-drug resistant microorganism colonization. Npj Biofilms Microbiomes 2023;9(1). https://doi.org/10.1038/s41522-023-00430-3.

CHAPTER

15

Immunological insights: How macrophages combat non-tuberculous mycobacteria

Zin Mar Htun[1, a], Muhammad H. Gul[1, a] and Ruxana T. Sadikot[1, 2]

[1]*Division of Pulmonary, Critical Care & Sleep, Department of Internal Medicine, University of Nebraska Medical Center, Omaha, NE, United States;* [2]*VA Nebraska Western Iowa Health Care System, Omaha, NE, United States*

Highlights

1. Non-tuberculous mycobacterial (NTM) infections are increasing globally and can cause significant disease.
2. Macrophages are the first line of defense against NTM and orchestrate the immune response.
3. Pattern recognition receptors on macrophages bind to pathogen-associated molecular patterns on the NTM, triggering phagocytosis.
4. M1 macrophage polarization promotes bacterial death, whereas M2 polarization allows NTM proliferation in granulomas.
5. Apoptosis and autophagy help eliminate NTM without excessive inflammation, whereas necroptosis and pyroptosis cause damage to inflammation.
6. The IL-12/IFN-γ axis is crucial for the presence of NTM, and its disruption leads to disseminated infection.
7. Understanding NTM pathogenesis provides opportunities for the development of host-directed therapies to improve the treatment outcomes.

1. Introduction

Non-tuberculous mycobacteria (NTM) refer to all mycobacteria belonging to the genus *Mycobacterium*, excluding the *Mycobacterium tuberculosis* complex and *Mycobacterium leprae* complex. NTM are ubiquitous in soil, water, and other environmental sources and are usually not transmitted from person to person. Environmental exposure to NTM typically does not result in clinically significant diseases in the general population; however, NTM can be pathogenic in both immunocompetent and immunocompromised patients. The incidence and prevalence of NTM infections have been increasing both in North America and globally [1,2]. Hence, it is increasingly important to study the pathogenesis of NTM and the host immune response against NTM to serve as the cornerstone for drug discovery and the development of novel therapeutic options.

Variability in the virulence of NTM infections is dependent on host susceptibility. Once infected, both the innate and adaptive immune systems work together to combat organisms. Macrophages act as

[a]Authors contributed equally.

Navigating Tuberculosis. https://doi.org/10.1016/B978-0-443-27735-1.00017-4

the frontline defenders of the innate immune system by engulfing pathogens through phagocytosis, secreting cytokines upon activation, and serving as antigen-presenting cells that initiate both cellular and humoral arms of the adaptive immune response. In this book chapter, we discuss the unique role macrophages play in orchestrating a coordinated host immune response to combat NTM bacteria.

2. Pathogenesis of NTM

NTM are acid-fast, aerobic bacilli, and genomic sequencing has identified approximately 200 species, of which approximately 140 are pathogenic to humans and animals, and the remaining 60 species are potentially pathogenic [3]. NTM-induced infections in humans are classified into four distinct groups: 1. NTM-pulmonary disease (NTM-PD); 2. superficial lymphadenitis, 3. disseminated disease in severely immunocompromised patients and 4. Soft tissue infections secondary to direct inoculation [4]. 80%–90% of recorded NTM infections manifest as NTM-PD and are typically caused by *Mycobacterium avium complex* (MAC), *Mycobacterium abscessus subsp. abscessus*, and *Mycobacterium kansasii* [5,6].

Microbiologically, the Centers for Disease Control and Prevention (CDC) categorizes NTM into two groups: rapid-growing species, which form colonies in culture within 7–10 days, and slow-growing species, which take more than 14 days to grow in culture [7]. Among the rapidly growing mycobacteria, the *M. abscessus* complex, *M. chelonae,* and *M. fortuitum* are the most clinically relevant. The *M. avium complex* (MAC), *M. xenopi*, *M. kansasii*, *M. simiae*, *M. malmoense*, and *M. szulgai* play the most significant roles in human diseases [8].

Mycobacteria have a unique cell envelope that distinguishes them from other bacteria and contributes to their environmental and intracellular survival. It has an outermost capsule composed of glucan and proteins; a cell wall composed of polysaccharides, including arabinogalactan, peptidoglycan, and large quantities of mycolic acid; and a conventional plasma membrane [9]. Mycolic acid in its cell wall is responsible for its acid-fast staining properties and resistance to broad-spectrum antibiotics, making NTM infections challenging to eradicate. The pathogenesis of NTM is primarily due to its ability to colonize the intestinal or respiratory mucosa and resist intracellular killing in macrophages, as observed in species such as *M. avium*, where biofilm formation plays a role [6].

3. Overview of host responses to NTM infection

NTM is usually acquired by the inhalation of bacterial aerosols via the respiratory tract. The host immune response against NTM invasion involves a coordinated effort between multiple cell types in innate and adaptive immunity. Once invaded, alveolar epithelial cells and alveolar macrophages serve as the first line of defense in the innate immune system. NTM possesses different PAMPs, which are recognized by PRRs expressed by innate immune cells such as macrophages. This results in the phagocytosis of NTM by macrophages and dendritic cells, the formation of intracellular phagolysosomes that degrade microbes, and the release of toxic nitrogen and oxygen radicals. Macrophage activation also releases both pro-inflammatory and anti-inflammatory cytokines. It presents antigens to adaptive immune cells, which in turn triggers downstream cellular and humoral immune reactions, ultimately leading to granuloma formation [10]. Understanding the complex interaction between NTM and the host, orchestrated by macrophages, is crucial for identifying factors that contribute to NTM virulence and host susceptibility, ultimately leading to disseminated NTM disease. Defining the role of various pathways may also provide opportunities to develop host-directed therapies for resistant infections.

4. PRRs binding with PAMPs

PAMPs are molecular structures found on pathogens, such as bacteria, viruses, or fungi, which the immune system recognizes to trigger an inflammatory response [11]. Mycobacterial PAMPs are components of the cell wall and nucleic acids. Lipomannan (LM), lipoarabinomannan (LAM), phosphatidylinositol mannosides (PIMs), and mycolic acid are mycobacterial cell wall components recognized by PRRs. Among nucleic acid components, unmethylated CpG motifs in mycobacterial DNA are recognized as PAMPs [12]. PRRs on innate immune cells, such as macrophages, sense PAMPs. Alveolar macrophages are usually in the surveillance mode, maintaining a low reactive oxygen species environment until invasive bacteria appear [13]. PRR are a large family of cell surface and intracellular receptors in macrophages and dendritic cells, including Toll-like receptors (TLRs), nucleotide-binding oligomerization domain (NOD)-like receptors (NLRs), C-type lectins, and scavenger receptors [14].

Triacylated lipopeptides on bacterial cell walls are assumed to be recognized by TLR1/TLR2, whereas TLR2/TLR6 activates diacylated lipopeptides [15]. TLR1, 2, 4, 5, and 6 are present on the cell surface where they bind to mycobacterial lipoproteins and glycopeptides. TLR 3, 7, 8, and 9 are found intracellularly and trafficked through the endoplasmic reticulum to endolysosomes, where they recognize mycobacterial DNA [10,16]. All TLRs utilize MyD88 (myeloid differentiation primary response gene 88) adaptor protein for downstream signaling, except for TLR3. TLR3 and TLR4 also have TLR-MyD88-independent signaling pathways [12]. *M. avium* expresses serovar-specific glycopeptides that activate macrophages in a TLR2-and MyD88-dependent manner [17]. TLR2, TLR4, TLR9, and MyD88 play the most significant roles in mycobacterial infections [14]. TLR2-deficient mice have impaired *M. abscessus* induced signaling pathways, and MyD88-deficient mice are more susceptible to *M. avium* infection [18,19]. In humans, mRNA expression of TLR2, TNF-α, and IL-12 p40 (the beta subunit of the IL-12 family) was lower in NTM-infected patients than in healthy controls [20].

Dectin-1 (dendritic cell-associated C-type lectin) is a PRR that recognizes microbial β-glucans, facilitates phagocytosis, cooperates with TLR-2, and promotes the production of interleukin 6 (IL-6), RANTES (regulated on activation, normal T expressed and secreted), and granulocyte-colony stimulating factor (G-CSF) by mycobacterium-infected macrophages [21]. Dectin-1, originally known for its antifungal role, recognizes *M. tuberculosis, M. bovis, M. avium, and M. abscessus* for phagocytosis [18]. However, Dectin-1-independent phagocytosis by *M. abscessus* has also been reported [22]. Dectin-2 recognizes mannose-capped lipoarabinomannan (Man-LAM) found in *M. tuberculosis* and *M. intracellulare* but does not recognize *M. abscessus* or *M. smegmatis* because it does not produce Man-LAM [23]. Dectin-1 signaling has been extensively studied in the context of antifungal activity. It has been shown to activate phagocytosis, ROS production, cytokine release (mainly TNF-alpha, IL-12, IL-18) to initiate an inflammatory response, and collaborate with TLR-2 receptors to ignite the differentiation of naïve $CD4^+$ T cells to T helper Th 1 or Th 17 phenotype, and activation of $CD8^+$ T cells [24].

5. Phagocytosis and oxidative stress-mediated killing

Once PRRs are activated, NTMs are phagocytosed by macrophages to form intracellular phagosomes. Opsonization is a process in which microbes are coated with opsonins such as antibodies and complement components to enhance immune cell recognition via PRR and phagocytosis. Complement component C3b-mediated opsonization is another important mechanism for the adherence and phagocytosis of mycobacteria [25]. Phagocytosis of NTM occurs via opsonic and nonopsonic receptor-mediated pathways. Opsonic receptors include Fc, integrin, and complement receptors. Nonopsonic receptors include CD169 and CD33 as well as C-type lectins such as Dectin-1 and Dectin-2, and scavenger receptors [26]. Internalization of NTM occurs via actin polymerization, which causes the plasma membrane to wrap around the organism or through aggregation of lipid rafts [27]. The engagement of both cell surface and intracellular TLRs and cytosolic NLRs with their ligands activates the NF-κB and MAP kinase (mitogen-activated protein kinase) pathways intracellularly, which in turn produce inflammatory cytokines that promote phagosome maturation from early to late stages, decreasing luminal pH by recruiting proton pump v-ATPase to phagosomes. The fusion of lysosomes with late phagosomes further acidifies the environment, creating phagolysosomes.

Within minutes of phagocytosis, RNS and ROS are produced through enzymatic activities of iNOS and NADPH oxidase (NOX2), respectively, via an r-IFN-mediated mechanism [26,28]. Oxidative stress induced by ROS is responsible for the degradation of NTM within phagolysosomes and triggers apoptosis. The host cells also possess anti-inflammatory and counter-regulatory mechanisms, such as Keap1 (Kelch-like ECH-associated Protein 1), to limit damage. ROS and RNS activate and regulate apoptosis by activating and inhibiting kinases, caspases, transcription factors, and cytokines [10]. Nitric oxide (NO) is known to deaminate and directly damage bacterial DNA and interact with iron-sulfur groups, heme groups, thiols, and phenol residues, causing enzymatic protein malfunctions [28]. The role of NO in mycobacterial infections has been a subject of debate and has been explored as a potential therapeutic target [29,30].

6. Mycobacterial survival mechanisms

Both *M. tuberculosis* and NTM possess fascinating intrinsic capacities to inhibit the host's phagocytosis and intracellular killing mechanisms, which are essential for their survival within host macrophages and cell-to-cell transmission [31]. These properties vary among the different *Mycobacterium* species. *M. avium* disrupts macrophage actin filament networks, inhibits NF-κB activation and TLR signaling, blocks phagolysosome fusion, and prevents acidification of the phagosome environment by inhibiting the vesicular proton-ATPase pump (v-ATPase) [32–35]. *M abscessus* has a high aggregation property and can exist in clumps in phagocytic cups without being internalized. Mycobacterial membrane protein large (MmpL) proteins are responsible for the transport of glycopeptidolipids (GPL) to the outer mycobacterial surface. Modification of these proteins enables *M. abscessus* and other NTMs to exist in two forms: the smooth morphotype (S) and rough morphotype (R) [36]. The S variant allows for tight opposition between the phagosomal membrane and mycobacterial cell wall, thereby preventing phagosome maturation, autophagy, and cell apoptosis. The R variant tends to aggregate into social phagosomes, which contain more than one bacillus in a single phagosome, and encourages phagolysosome fusion and cell apoptosis. Once the macrophage dies, NTM is released into the extracellular space, forming serpentine cords that resist phagocytosis, promote active inflammation, and facilitate granuloma formation and transmission to the next host cell [37,38].

7. Cell death and autophagy

Once the NTM is inside a phagocyte, it can undergo apoptosis, pyroptosis, or necroptosis. Apoptosis facilitates antigen presentation and bacterial clearance without excessive inflammation. Pyroptosis and necroptosis are known to cause excessive inflammation, resulting in host tissue damage [39]. Apoptosis in *M. avium* is mediated by TNF-α and Fas and involves the activation of caspases [40]. Mycobacteria are packed inside apoptotic bodies, which are subsequently phagocytosed [41]. *M. avium* has developed an apoptosis inhibitor of macrophages (AIM) that increases resistance to macrophage apoptosis, thereby ensuring its survival [42]. Pyroptosis is a programmed cell death mediated by inflammasome macromolecular complexes. The inflammasome is a multiprotein complex that is assembled after pattern recognition receptor activation. *M. intracellulare* infection activates the inflammasome complex via absent in melanoma 2 (AIM2) and NLRP-3, which are regulated by caspases [43]. NLRP3 inflammasome contributes to lung inflammation. However, it did not affect the growth or clearance of *M. abscessus,* suggesting that inflammation contributes to more pathology than bacteria [44]. A study on *M. marinum* revealed that the host can stimulate macrophages infected with bacteria through a programmed form of cell death known as necroptosis [45]. Necroptosis is characterized by plasma membrane rupture and uncontrolled release of intracellular contents, which elicit a strong inflammatory response [39]. This results in unstructured necrosis with excessive inflammation, causing damage to the host.

Autophagy is a conserved process in eukaryotes that degrades cytoplasmic components via lysosomes to maintain cellular homeostasis, especially under stressful conditions, such as nutrient deprivation, DNA damage, and oxidative stress [46]. It also provides an independent mechanism for eliminating intracellular organisms that is distinct from apoptosis or necroptosis. It is a controlled process that dampens inflammation. However, it can occasionally become uncontrolled, destroying the essential components of the cells and ultimately resulting in cell death [39].

8. Polarization and biphasic metabolic response of macrophages

The diversity in the responses of macrophages to NTM infections has been attributed to the temporal biphasic metabolic response of macrophages, resulting in M1 (classically activated macrophages) and M2 (alternatively activated macrophages) subtypes [47]. Metabolic rewiring results in the polarization of macrophages into two distinct forms.

The early phase (4–8 hours postinfection), represented by M1 polarization, is marked by a shift from oxidative phosphorylation to the hypoxia-inducible factor alpha-1 (HIF-1α) glycolysis pathway (also referred to as the Warburg effect in cancer cells). The up-regulation of HIF-1α is triggered by TLR2 activation. During glycolysis, M1 macrophages produce the rapid energy and metabolic intermediates required for the upregulation of pro-inflammatory cytokines (IL-1β, IL-12, and TNF-α), ROS, and iNOS/NO [48].

The early pro-inflammatory metabolic response is followed by a late adaptive and resolution anti-inflammatory phase, which is marked by polarization into M2 macrophages. This is characterized by the switch to oxidative phosphorylation and dampening of pro-inflammatory and antimicrobial responses [47]. Mannose-capped lipoarabinomannan (ManLAM), which is shed from the cell wall during microbial replication, induces PPARγ expression. PPARγ is an important member of the lipid-activated nuclear receptor family. PPARγ expression downregulates pro-inflammatory cytokines and iNOS. PPARγ binds to DNA and acts as a transcriptional repressor of the ligand-mediated NF-κB target genes. PPARγ activation leads to increased cyclooxygenase-2 and PGE2 expression, which downregulate the Th-1 response toward intracellular organisms [47,49–51]. PPARγ also induces the production of lipid droplets through adipose differentiation-related protein (ADRP) [51]. ADRP is believed to play a crucial role in adipocyte production, differentiation, and modulation [52]. *M. avium* and *M. abscessus* position themselves close to lipid-laden foamy M2 macrophages, which siphon off intralipid inclusions, thereby improving intracellular mycobacterial survival [53,54]. IL-4 and IL-13 from Th1 cells activate STAT-6, which together with PPARγ coactivator-1 beta (PGC-1β), also plays a role in M2 transformation, contributing toward the switch from anaerobic glycolysis to oxidative phosphorylation [55,56]. STAT6/PPARγ induces ARG-1+ macrophages, which limit L-arginine, which serves as a substrate for iNOS. ARG-1+ macrophage density has been correlated with the progression of *M. avium* [57]. M2 macrophages secrete IL-10 and transforming growth factor beta (TGF-β), which suppresses IL-12 and IL-1β and attenuate mycobacterial apoptosis by dampening the Th1-IFNγ response [58].

Studies of *M. avium subsp. Paratuberculosis* (MAP) indicates that MAP can regulate the expression of host microRNA to mediate the shift toward M2 transformation, thus favoring intracellular survival [59]. Silencing high-mobility group N2 (HMGN2) decreases the survival of NTM by diverting the shift toward M1 polarization [60]. Patients with cystic fibrosis are predisposed to NTM and have been reported to have a defect in switching between M1 and M2 macrophages [61]. Thus, increasing evidence suggests that NTM exploits two-way traffic from the M1 pro-inflammatory state to the lipid-fueled anti-inflammatory M2 state to improve its chances of intracellular survival.

9. Cell-mediated immunity and the cytokine release

Once NTM is sensed by the innate immune system, resulting in phagocytosis, it forms a signaling hub for the release of cytokines and chemokines. PRRs, such as TLR2/TLR4, dectin-2, and NOD-2 cooperate to induce NF-κB transcriptional priming and phagosome production, leading to the release of cytokines [18,62,63]. NF-κB is a ubiquitous transcriptional protein that is normally found in the cytoplasm and tightly regulated by cytokines and chemokines [64]. This results in the early release of TNF-α, IL-1β, IL-12, CXCL10, CXCL9 [63,65] via TLR2/Dectin cooperation [18].

TNF-α plays a crucial role in early granuloma formation [66]. It also stabilizes the architecture of the granuloma, and its absence in knockout studies causes fatal granuloma necrosis [67]. NLRP3-inflammasome activation mediates IL-1β production, which, together with TNF-α, maintains the chemokine gradient [68]. MAP exploits IL-1β processing and phagosome acidification to facilitate transmigration across epithelial cells, thereby enabling their entry into macrophages [69]. NF-κB transcriptome-driven CXCL9 and CXCL10 contribute to granuloma formation by attracting Th1 cells.

CXCL10 concentrations may serve as biomarkers in fibrocavitary *M. avium* lung disease [70,71]. IL 12 serves as a bridge between the innate and cell-mediated immunity [12]. IL-12 promotes differentiation of $CD4^+$ cells into Th1 cells. Th1 cells and natural killer cells are activated by IL-12, resulting in the production of IFN-γ, which is essential for granuloma formation [12,72]. IL-12-IFN-γ axis deficits have significant clinical implications in the pathogenesis of disseminated NTM, as explained in the disseminated NTM section [73].

10. Granuloma formation

Granuloma formation is one of the primary methods for preventing the dissemination of NTMs by the host defense system. The early form includes infectious forms of macrophages, which evolve into epithelioid cells, multinucleated giant cells, and foamy macrophages, surrounded by T and B lymphocytes [74]. This progresses into a mature granuloma, which involves the recruitment of fibroblasts that may develop caseous necrosis and the recruitment of blood vessels through angiogenesis [75]. In a mature granuloma, the core is composed of M2-polarized macrophages that are transformed into foamy macrophages. The mycobacterium moves into foamy macrophages, survives within the Arg1+ lipid-rich necrotic, hypoxic core, and becomes dormant, resulting in replication arrest [53]. STAT6-dependent signaling is essential for epithelialization of macrophages and formation of necrotic granulomas. The outer layer is primarily composed of iNOS-rich M1 macrophages, along with lymphocytes [76,77].

11. Pathology spectrum in NTM

The pathological spectrum of NTM infection determines the clinical course of the infection, which can generally be divided into three pathways.

M1 transformation with autophagy is the ideal outcome, in which the NTM is contained early or cleared without significant damage to the host. Intact TNF-α, IL-12, and TLR signaling, followed by phagocytosis by phagolysosomes and high iNOS and ROS levels, result in the early eradication of mycobacteria with minimal damage to the host cell [10,66]. Clinically, this condition is asymptomatic or presents with mild symptoms. A low bacterial inoculum combined with an immunocompetent host favors this outcome.

Mixed transformation of M1 and M2, along with an intact IL-12-IFN-γ axis, results in granuloma formation containing mycobacteria in the lipid-rich core without further replication. The outer layer of the granuloma served as a containment layer for mycobacteria. Apoptosis balances low-grade replication in the granuloma core. This results in a subclinical or latent form of mycobacterial infection, where the mycobacterial infection survives without actively increasing the bacterial burden and awaits the right circumstances for further dissemination and proliferation [78].

Dominant M2 transformation or disruption of the IL-12-IFN-γ axis results in disease progression. IL-4/IL-13, along with STAT6/PPARγ and ARG1, stimulates the development of lipid-rich M2 transformation, where mycobacteria thrive [79]. In the setting of low or compromised IL-12-IFN γ, with high TGF-β and IL-10 levels in the milieu, mycobacteria can disseminate [80,81].

12. Clinical relevance of NTM infection pathophysiology and host-directed therapies for NTM infections

NTM infections are difficult to eradicate because of their intrinsic antimicrobial resistance, ability to persist in biofilms, and intracellular niches, especially in macrophages, as discussed above, and the need for prolonged, multi-drug therapy with frequent toxicity. Given these challenges, host-directed therapies aim to enhance the host immune response or modulate pathological inflammation to improve the outcomes. Here, we discuss the clinical relevance of understanding NTM pathophysiology that may aid in the development of host-directed therapies.

12.1 The IL-12/IFN-γ disruption results in disseminated NTM

Disseminated NTM is observed in immunocompromised patients with human immune deficiency virus (HIV) (CD4 count less than 50), cystic fibrosis, transplant patients, and patients with deficits in the IL-12–IFN-γ axis [82–84]. The IL-12-IFN-γ axis deficits caused by genetic defects are known as Mendelian susceptibility to mycobacterial diseases [73]. These include disease-causing mutations in 11 genes that underlie 21 different genetic disorders through allelic heterogeneity [73]. Autoantibodies against IFN-γ confer increased susceptibility to NTM in patients with IFN-γ deficits, even in the absence of a genetic disorder. In a retrospective review of HIV-negative disseminated NTM patients, 81% of patients had increased IFN-γ -autoantibody titers [84]. Thus, IL-12–IFN-γ axis deficits should be screened for in all cases of unexplained NTM dissemination.

12.2 Increased autophagy with metformin and vitamin D use

Metformin activates AMP-activated protein kinase (AMPK), thereby increasing autophagy by promoting phagolysosome fusion and enhancing mitochondrial ROS function [85]. In a population-based retrospective study of *M. tuberculosis*, metformin use was associated with a decreased risk of tuberculosis [86]. In another study, the combined use of metformin and the existing regimen resulted in improved treatment success and decreased risk of relapse [87]. However, a case-controlled matched analysis from South Korea did not show any significant effect of metformin on incident NTM infections [88].

Vitamin D3 promotes autophagy by upregulating human cathelicidin antimicrobial peptide (CAMP) in human monocytes upon infection with *M. marinum* [89]. In a retrospective cohort study and another prospective study, Vitamin D deficiency was associated with an increased risk of NTM infection [90,91].

12.3 Immunosuppression with steroids and TNF-α inhibitors may increase the NTM risk

The use of TNF-α inhibitors or high-dose steroids favors M2 macrophage polarization over M1 polarization, which provides a more favorable environment for NTM proliferation. In a retrospective cohort study of TNF-α inhibitors in rheumatoid arthritis patients, the use of TNF-α inhibitors was associated with a higher risk of NTM infection [92].

12.4 Amikacin liposomal inhalational suspension penetrates biofilms/cords and improves outcomes in refractory NTM

Amikacin liposomal inhalational suspension (ALIS) enhances penetration of the extracellular biofilm or cords formed by NTMs and increases the uptake of amikacin into macrophages [93]. In a prospective open-label randomized (CONVERT) trial, the addition of ALIS to standard guideline-based therapy achieved greater culture conversion by the sixthmonth than guideline-based therapy alone, with similar rates of serious adverse events [94].

Some of the challenges for developing host-directed therapies include the fact that NTM species are diverse and have variable host-pathogen interactions. Many species pose the risk of exacerbating immune-mediated lung injury. Furthermore, there is a lack of robust animal models that mimic human diseases to test these therapies in preclinical models. Most importantly, even in human diseases, it is often difficult to measure meaningful clinical endpoints in NTM trials, such as culture conversion and functional outcomes. Nonetheless, investigating these systematically provides opportunities to develop novel approaches, especially for cases of NTM infections that become resistant and difficult to treat.

13. Summary and conclusions

NTM can evade host immunity by residing within the macrophages. Macrophage host response is complex and involves the activation of several pathways. Excessive or dysregulated immune responses may contribute to lung damage. Defining the role of pathways that contribute to the host response provides opportunities to develop novel therapies to treat infections that are difficult to treat. Boosting protective immunity or controlling inflammation may improve clearance and limit lung injury.

References

[1] Pedersen AA, et al. Nationwide increasing incidence of nontuberculous mycobacterial diseases among adults in Denmark: eighteen years of follow-up. Chest 2024;166(2):271–80.

[2] Winthrop KL, et al. Incidence and prevalence of nontuberculous mycobacterial lung disease in a large U.S. Managed Care Health Plan, 2008-2015. Ann Am Thorac Soc 2020;17(2):178–85.

[3] Matsumoto Y, et al. Comprehensive subspecies identification of 175 nontuberculous mycobacteria species based on 7547 genomic profiles. Emerg Microbes Infect 2019;8(1):1043–53.

[4] Griffith DE, et al. An official ATS/IDSA statement: diagnosis, treatment, and prevention of nontuberculous mycobacterial diseases. Am J Respir Crit Care Med 2007;175(4):367–416.

[5] Tortoli E. Clinical manifestations of nontuberculous mycobacteria infections. Clin Microbiol Infection 2009; 15(10):906–10.

[6] To K, et al. General overview of nontuberculous mycobacteria opportunistic pathogens: *Mycobacterium avium* and Mycobacterium abscessus. J Clin Med 2020;9(8).

[7] CDC. Clinical overview of nontuberculous mycobacteria (NTM). April 15, 2024. Available from: https://www.cdc.gov/nontuberculous-mycobacteria/hcp/clinical-overview/index.html.

[8] Cowman S, et al. Non-tuberculous mycobacterial pulmonary disease. Eur Respir J 2019;54(1):1900250.

[9] Chiaradia L, et al. Dissecting the mycobacterial cell envelope and defining the composition of the native mycomembrane. Sci Rep 2017;7(1):12807.

[10] Prasla Z, Sutliff RL, Sadikot RT. Macrophage signaling pathways in pulmonary nontuberculous mycobacteria infections. Am J Respir Cell Mol Biol 2020;63(2):144–51.
[11] Tang D, et al. PAMPs and DAMPs: signal 0s that spur autophagy and immunity. Immunol Rev 2012;249(1): 158–75.
[12] Abe Y, et al. Host immune response and novel diagnostic approach to NTM infections. Int J Mol Sci 2020;21 (12):4351.
[13] Sweet MJ, et al. Inducible antibacterial responses in macrophages. Nat Rev Immunol 2024:1–16.
[14] Shamaei M, Mirsaeidi M. Nontuberculous mycobacteria, macrophages, and host innate immune response. Infect Immun 2021;89(8).
[15] Buwitt-Beckmann U, et al. TLR1- and TLR6-independent recognition of bacterial lipopeptides. J Biol Chem 2006;281(14):9049–57.
[16] Akira S, Takeda K. Toll-like receptor signalling. Nat Rev Immunol 2004;4(7):499–511.
[17] Sweet L, Schorey JS. Glycopeptidolipids from *Mycobacterium avium* promote macrophage activation in a TLR2- and MyD88-dependent manner. J Leukoc Biol 2006;80(2):415–23.
[18] Shin D-M, et al. Mycobacterium abscessus activates the macrophage innate immune response via a physical and functional interaction between TLR2 and dectin-1. Cell Microbiol 2008;10(8):1608–21.
[19] Feng CG, et al. Mice lacking myeloid differentiation factor 88 display profound defects in host resistance and immune responses to *Mycobacterium avium* infection not exhibited by Toll-like receptor 2 (TLR2)- and TLR4-deficient animals. J Immunol 2003;171(9):4758–64.
[20] Ryu YJ, et al. Impaired expression of Toll-like receptor 2 in nontuberculous mycobacterial lung disease. Eur Respir J 2007;30(4):736–42.
[21] Yadav M, Schorey JS. The β-glucan receptor dectin-1 functions together with TLR2 to mediate macrophage activation by mycobacteria. Blood 2006;108(9):3168–75.
[22] Ochoa AE, et al. Dectin-1-independent macrophage phagocytosis of Mycobacterium abscessus. Int J Mol Sci 2023;24(13).
[23] Yonekawa A, et al. Dectin-2 is a direct receptor for mannose-capped lipoarabinomannan of mycobacteria. Immunity 2014;41(3):402–13.
[24] Mata-Martínez P, Bergón-Gutiérrez M, Del Fresno C. Dectin-1 signaling update: new perspectives for trained immunity. Front Immunol 2022;13:812148.
[25] Mueller-Ortiz SL, Wanger AR, Norris SJ. Mycobacterial protein HbhA binds human complement component C3. Infect Immun 2001;69(12):7501–11.
[26] Awuh JA, Flo TH. Molecular basis of mycobacterial survival in macrophages. Cell Mol Life Sci 2017;74(9): 1625–48.
[27] Anes E, et al. Selected lipids activate phagosome actin assembly and maturation resulting in killing of pathogenic mycobacteria. Nat Cell Biol 2003;5(9):793–802.
[28] Yang C-S, Yuk J-M, Jo E-K. The role of nitric oxide in mycobacterial infections. Immune Netw 2009;9(2): 46.
[29] Flume PA, et al. Inhaled nitric oxide for adults with pulmonary non-tuberculous mycobacterial infection. Respir Med 2023;206:107069.
[30] Goldbart A, Gatt D, Golan Tripto I. Non-tuberculous mycobacteria infection treated with intermittently inhaled high-dose nitric oxide. BMJ Case Rep 2021;14(10).
[31] von Both U, et al. *Mycobacterium tuberculosis* exploits a molecular off switch of the immune system for intracellular survival. Sci Rep 2018;8(1):661.
[32] Frehel C, et al. Evidence for inhibition of fusion of lysosomal and prelysosomal compartments with phagosomes in macrophages infected with pathogenic *Mycobacterium avium*. Infect Immun 1986;52(1): 252–62.
[33] Sturgill-Koszycki S, et al. Lack of acidification in Mycobacterium phagosomes produced by exclusion of the vesicular proton-ATPase. Science 1994;263(5147):678–81.

[34] Guérin I, de Chastellier C. Pathogenic mycobacteria disrupt the macrophage actin filament network. Infect Immun 2000;68(5):2655–62.
[35] Ikegami H, et al. Reduced phagocytic activity of human alveolar macrophages infected with *Mycobacterium avium* complex. J Infect Chemother 2022;28(11):1506–12.
[36] Ferrell KC, et al. Virulence mechanisms of Mycobacterium abscessus: current knowledge and implications for vaccine design. Front Microbiol 2022;13:842017.
[37] López-Roa P, Esteban J, Muñoz-Egea MC. Updated review on the mechanisms of pathogenicity in Mycobacterium abscessus, a rapidly growing emerging pathogen. Microorganisms 2022;11(1).
[38] Kam JY, et al. Rough and smooth variants of Mycobacterium abscessus are differentially controlled by host immunity during chronic infection of adult zebrafish. Nat Commun 2022;13(1):952.
[39] Prata R, Pinheiro RO. Cell death mechanisms in Mycobacterium abscessus infection: a double-edged sword. Pathogens 2025;14(4).
[40] Bermudez LE, Parker A, Petrofsky M. Apoptosis of Mycobacterium avium-infected macrophages is mediated by both tumour necrosis factor (TNF) and Fas, and involves the activation of caspases. Clin Exp Immunol 1999;116(1):94–9.
[41] Shundo Y, et al. TNFR1 mediated apoptosis is protective against *Mycobacterium avium* in mice. Microorganisms 2023;11(3):778.
[42] Kajiwara C, et al. Apoptosis inhibitor of macrophages contributes to the chronicity of *Mycobacterium avium* infection by promoting foamy macrophage formation. J Immunol 2023;210(4):431–41.
[43] Ying S, et al. Mycobacterium intracellulare mediates macrophage pyroptosis by activating AIM2 and NLRP3 inflammasomes. Vet Res Commun 2024;48(5):3445–54.
[44] Ahn J-H, et al. Impact of IL-1β on lung pathology caused by Mycobacterium abscessus infection and its association with IL-17 production. Microb Infect 2024;26(5):105351.
[45] Roca FJ, Ramakrishnan L. TNF dually mediates resistance and susceptibility to mycobacteria via mitochondrial reactive oxygen species. Cell 2013;153(3):521–34.
[46] Wang J, Lee S. Targeting autophagy as a strategy for developing new host-directed therapeutics against nontuberculous mycobacteria. Pathogens 2025;14(5):472.
[47] Shi L, et al. Biphasic dynamics of macrophage immunometabolism during *Mycobacterium tuberculosis* infection. mBio 2019;10(2). https://doi.org/10.1128/mbio.02550-18.
[48] Howard NC, Khader SA. Immunometabolism during *Mycobacterium tuberculosis* infection. Trends Microbiol 2020;28(10):832–50.
[49] Rajaram MV, et al. *Mycobacterium tuberculosis* activates human macrophage peroxisome proliferator-activated receptor γ linking mannose receptor recognition to regulation of immune responses. J Immunol 2010;185(2):929–42.
[50] Mahajan S, et al. *Mycobacterium tuberculosis* modulates macrophage lipid-sensing nuclear receptors PPARγ and TR4 for survival. J Immunol 2012;188(11):5593–603.
[51] Almeida PE, et al. PPARγ expression and function in mycobacterial infection: roles in lipid metabolism, immunity, and bacterial killing. PPAR Res 2012;2012(1):383829.
[52] Brasaemle DL. Thematic review series: adipocyte biology. The perilipin family of structural lipid droplet proteins: stabilization of lipid droplets and control of lipolysis. J Lipid Res 2007;48(12):2547–59.
[53] Caire-Brändli I, et al. Reversible lipid accumulation and associated division arrest of *Mycobacterium avium* in lipoprotein-induced foamy macrophages may resemble key events during latency and reactivation of tuberculosis. Infect Immun 2014;82(2):476–90.
[54] Dargham T, et al. A proteomic and functional view of intrabacterial lipid inclusion biogenesis in mycobacteria. mBio 2025;16(4):e0147524.
[55] Vats D, et al. Oxidative metabolism and PGC-1beta attenuate macrophage-mediated inflammation. Cell Metab 2006;4(1):13–24.

[56] Azouaou N, et al. *Mycobacterium avium* infection in mice is associated with time-related expression of Th1 and Th2 CD4+ T-lymphocyte response. Immunology 1997;91(3):414–20.

[57] Rosenbloom R, et al. Progression and dissemination of pulmonary *Mycobacterium avium* infection in a susceptible immunocompetent mouse model. Int J Mol Sci 2022;23(11):5999.

[58] Balcewicz-Sablinska MK, Gan H, Remold HG. Interleukin 10 produced by macrophages inoculated with *Mycobacterium avium* attenuates mycobacteria-induced apoptosis by reduction of TNF-alpha activity. J Infect Dis 1999;180(4):1230–7.

[59] Wright K, et al. *Mycobacterium avium* subsp. paratuberculosis exploits miRNA expression to modulate lipid metabolism and macrophage polarisation pathways during infection. Sci Rep 2022;12(1):9681.

[60] Wang X, et al. HMGN2 regulates non-tuberculous mycobacteria survival via modulation of M1 macrophage polarization. J Cell Mol Med 2019;23(12):7985–98.

[61] Abdelaal HFM, et al. Mycobacterium abscessus: it's Complex. Microorganisms 2022;10(7).

[62] Shandilya Umesh K, et al. Role of toll-like receptor 4 in *Mycobacterium avium* subsp. paratuberculosis infection of bovine mammary epithelial (MAC-T) cells in vitro. Microbiol Spectr 2023;11(2). p. e04393-22.

[63] Lee J-Y, et al. Nucleotide-binding oligomerization domain 2 contributes to limiting growth of Mycobacterium abscessus in the lung of mice by regulating cytokines and nitric oxide production. Front Immunol 2017;8.

[64] Giri DK, et al. Mycobacterium avium-intracellulare complex activates nuclear transcription factor-κB in different cell types through reactive oxygen intermediates. J Immunol 1998;161(9):4834–41.

[65] Sampaio EP, et al. Mycobacterium abscessus and *M. avium* trigger Toll-like receptor 2 and distinct cytokine response in human cells. Am J Respir Cell Mol Biol 2008;39(4):431–9.

[66] Smith D, et al. T-cell-independent granuloma formation in response to *Mycobacterium avium*: role of tumour necrosis factor-α and interferon-γ. Immunology 1997;92(4):413–21.

[67] Ehlers S, et al. Fatal granuloma necrosis without exacerbated mycobacterial growth in tumor necrosis factor receptor p55 gene-deficient mice intravenously infected with *Mycobacterium avium*. Infect Immun 1999;67 (7):3571–9.

[68] Kim BR, et al. Mycobacterium abscessus infection leads to enhanced production of type 1 interferon and NLRP3 inflammasome activation in murine macrophages via mitochondrial oxidative stress. PLoS Pathog 2020;16(3):e1008294.

[69] Lamont EA, et al. Infection with *Mycobacterium avium* subsp. paratuberculosis results in rapid interleukin-1β release and macrophage transepithelial migration. Infect Immun 2012;80(9):3225–35.

[70] Bamba Y, et al. Multiplex cytokine analysis in *Mycobacterium avium* complex lung disease: relationship between CXCL10 and poor prognostic factors. BMC Infect Dis 2019;19(1):263.

[71] Awuh JA, et al. Keap1 regulates inflammatory signaling in Mycobacterium avium-infected human macrophages. Proc Natl Acad Sci 2015;112(31):E4272–80.

[72] Serbina NV, Flynn JL. CD8+ T cells participate in the memory immune response to *Mycobacterium tuberculosis*. Infect Immun 2001;69(7):4320–8.

[73] Rosain J, et al. Mendelian susceptibility to mycobacterial disease: 2014–2018 update. Immunol Cell Biol 2019;97(4):360–7.

[74] Silva Miranda M, et al. The tuberculous granuloma: an unsuccessful host defence mechanism providing a safety shelter for the bacteria? J Immunol Res 2012;2012(1):139127.

[75] Je S, et al. An in vitro model of granuloma-like cell aggregates substantiates early host immune responses against Mycobacterium massiliense infection. Biology Open 2016;5(8):1118–27.

[76] Cronan MR, et al. A non-canonical type 2 immune response coordinates tuberculous granuloma formation and epithelialization. Cell 2021;184(7):1757–1774.e14.

[77] Seto S, et al. Spatial multiomic profiling reveals the novel polarization of foamy macrophages within necrotic granulomatous lesions developed in lungs of C3HeB/FeJ mice infected with *Mycobacterium tuberculosis*. Front Cell Infect Microbiol 2022;12.

[78] Ceccato J, et al. Shaping rare granulomatous diseases in the lab: how new models are changing the game. Cells 2025;14(4).

[79] Duque-Correa MA, et al. Macrophage arginase-1 controls bacterial growth and pathology in hypoxic tuberculosis granulomas. Proc Natl Acad Sci U S A 2014;111(38):E4024–32.

[80] Liang H, et al. Clinical characteristics of acquired anti-IFN-γ autoantibodies in patients infected with non-tuberculous mycobacteria: a prospective cohort study. BMC Pulm Med 2025;25(1):95.

[81] Schreiber T, et al. Autocrine IL-10 induces hallmarks of alternative activation in macrophages and suppresses antituberculosis effector mechanisms without compromising T cell immunity. J Immunol 2009;183 (2):1301–12.

[82] Wetzstein N, et al. Disseminated disease due to non-tuberculous mycobacteria in HIV positive patients: a retrospective case control study. PLoS One 2021;16(7):e0254607.

[83] Martiniano SL, Nick JA. Nontuberculous mycobacterial infections in cystic fibrosis. Clin Chest Med 2015; 36(1):101–15.

[84] Aoki A, et al. Clinical significance of interferon-γ neutralizing autoantibodies against disseminated non-tuberculous mycobacterial disease. Clin Infect Dis 2018;66(8):1239–45.

[85] Lachmandas E, et al. Metformin alters human host responses to *Mycobacterium tuberculosis* in healthy subjects. J Infect Dis 2019;220(1):139–50.

[86] Pan S-W, et al. The risk of TB in patients with type 2 diabetes initiating metformin vs sulfonylurea treatment. Chest 2018;153(6):1347–57.

[87] Ma Y, et al. Metformin reduces the relapse rate of tuberculosis patients with diabetes mellitus: experiences from 3-year follow-up. Eur J Clin Microbiol Infect Dis 2018;37(7):1259–63.

[88] Kim HH, et al. Stratification of nontuberculous mycobacterial disease risk in type 2 diabetes based on metformin use: a population-based cohort study in South Korea. Antimicrob Agents Chemother 2023;67(6): e0167522.

[89] Sato E, et al. Vitamin D-dependent cathelicidin inhibits Mycobacterium marinum infection in human monocytic cells. J Dermatol Sci 2013;70(3):166–72.

[90] Richter WJ, et al. Vitamin D deficiency is associated with increased nontuberculous mycobacteria risk in cystic fibrosis. Ann Am Thorac Soc 2021;18(5):913–6.

[91] Kwon BS, et al. A prospective cohort study of bioavailable 25-hydroxyvitamin D levels as a marker of Vitamin D status in nontuberculous mycobacterial pulmonary disease. Nutrients 2021;13(8).

[92] Park DW, et al. TNF inhibitors increase the risk of nontuberculous mycobacteria in patients with seropositive rheumatoid arthritis in a mycobacterium tuberculosis endemic area. Sci Rep 2022;12(1):4003.

[93] Zhang J, et al. Amikacin liposome inhalation suspension (ALIS) penetrates non-tuberculous mycobacterial biofilms and enhances amikacin uptake into macrophages. Front Microbiol 2018;9.

[94] Griffith DE, et al. Amikacin liposome inhalation suspension for treatment-refractory lung disease caused by *Mycobacterium avium* complex (CONVERT). A prospective, open-label, randomized study. Am J Respir Crit Care Med 2018;198(12):1559–69.

CHAPTER

Extra-pulmonary tuberculosis: Insights into mitigation and control 16

Vishwanath Upadhyay[1,2], Alladi Mohan[3] and Surendra Kumar Sharma[1,4]

[1]*Department of Molecular Medicine, Jamia Hamdard (Deemed to Be University), New Delhi, India;* [2]*Department of Life Science, Sharda University, Noida, Uttar Pradesh, India;* [3]*Department of Medicine, Sri Venkateswara Institute of Medical Sciences, Tirupati, Andhra Pradesh, India;* [4]*Faculty of Medicine, Datta Meghe Institute of Medical Sciences (Deemed to Be University), Wardha, Maharashtra, India*

1. Introduction

Tuberculosis (TB), caused by the bacillus *Mycobacterium tuberculosis* (*Mtb*), remains a preeminent cause of morbidity and mortality globally, particularly in low- and middle-income countries (1). While the pulmonary manifestation of the disease is the most common and infectious form, facilitating transmission via aerosolized droplets, the pathogen possesses a formidable ability to disseminate beyond the lungs. Extra-pulmonary TB (EPTB) constitutes approximately 15%–20% of all notified cases in immunocompetent individuals and up to 50% in those with human immunodeficiency virus (HIV) co-infection [1–4]. In non-HIV-infected individuals, lymph node TB (LNTB) is the most common form (~35%), followed by pleural disease (~20%), skeletal involvement (~10%), genito-urinary TB (~9%), disseminated TB [DTB]/miliaryTB (~8%), and TB meningitis (TBM~5%) [1]. In immunosuppressed individuals and in people living with HIV [PLHIV], DTB/miliary TB accounts for up to one-half of cases, alongside frequent LN (~35%) and pleural (~20%) involvement, reflecting the impact of advanced immunosuppression on disease dissemination [1]. According to the Global TB report 2025, the overall global prevalence of extrapulmonary TB [EPTB] is 16% among all notified cases in 2024, with maximum cases in the Southeast-Asia region (25%) and minimum in the Western Pacific region (8%) [5]. EPTB presents a significant diagnostic and therapeutic challenge due to the paucibacillary nature of the lesions, the difficulty in obtaining representative tissue samples, and the vast heterogeneity of clinical presentations that mimic other inflammatory and neoplastic disorders [6]. Beyond HIV co-infection, several host and environmental factors increase the risk of extra-pulmonary TB, including extremes of age, female sex, malnutrition, diabetes mellitus, chronic kidney or liver disease, malignancy, and the use of immunosuppressive therapies such as corticosteroids or biologics. Socioeconomic determinants, overcrowding, poverty, and delayed access to care further predispose individuals to disseminated and atypical presentations, compounding the challenge of timely diagnosis and control [1,7].

Among the various forms of disseminated disease, miliary TB represents a particularly severe manifestation resulting from massive lymphohaematogenous dissemination of bacilli. Historically

Navigating Tuberculosis. https://doi.org/10.1016/B978-0-443-27735-1.00011-3

described by John Jacob Manget in 1700 due to the resemblance of the lesions to millet seeds, miliary TB is potentially lethal if not promptly diagnosed and treated [4]. The pathogenesis of EPTB and miliary TB involves a complex interplay between bacterial virulence factors and host immune responses. Understanding the molecular mechanisms of dissemination, ranging from angiogenesis and granuloma formation to host genetic susceptibility and proteomic alterations, is essential for developing novel mitigation strategies.

This chapter synthesizes recent advances in the understanding of EPTB and miliary TB. It explores the risk factors, utility of inflammatory biomarkers, antimicrobial peptides, and specific cell-surface indices for diagnosis and treatment monitoring. Furthermore, it delves into the genetic and epigenetic determinants of susceptibility, the role of adipose tissue as a reservoir, and the mechanisms of angiogenesis that facilitate bacterial spread. By integrating findings from proteomics, genomics, and immunology, we aim to provide comprehensive insights into the control of these complex forms of TB.

2. Risk factors for extra-pulmonary tuberculosis

2.1 Human immunodeficiency virus and immunosuppression

HIV co-infection significantly increases EPTB risk, accounting for up to 50% of all tuberculosis cases among affected individuals, with cluster of differentiation four counts below 200 cells per μL serving as a robust predictor of hematogenous dissemination [1–4]. Additional factors include immunosuppressive agents, including systemic corticosteroids, tumor necrosis factor-alpha (TNF-α) inhibitors such as infliximab, calcineurin inhibitors, and biologic immunomodulators, which produce analogous effects, with solid organ transplant recipients demonstrating 20–74-fold elevated EPTB incidence rates [1–4].

2.2 Host demographic factors

Extremes of age represent critical vulnerability periods for the occurrence of EPTB [7–9]. Children younger than 5 years exhibit two-to three-fold higher EPTB risk, manifesting predominantly as cervical lymphadenitis, abdominal TB, and TBM of pediatric TB cases [9], while elderly individuals [>65 years] frequently present with cryptic MTB, fever of unknown origin due to progressive immunosenescence [7,8]. Female predominance is evident in certain forms of EPTB like LNTB, implicating the role of hormonal and immunological factors in enhanced susceptibility. More research is needed for clarifying these issues [8].

3. Comorbidities and metabolic factors

Undernutrition, (body mass index, [kg/m^2] < 16 kg/m^2, elevates EPTB risk three-to five-fold through profound impairment of cell-mediated immunity, representing the predominant risk factor among both pediatric and geriatric populations [1,7,8]. Although diabetes mellitus is associated more strongly with pulmonary TB [PTB], conferring a threefold risk elevation, poorly controlled disease doubles TBM and MTB incidence through neutrophil dysfunction and altered cytokine profiles [7]. In end-stage renal disease [ESRD], patients on hemodialysis, silicosis, hematologic malignancies such as lymphoma and leukemia, demonstrate a several-fold increase in occurrence of EPTB [7].

3.1 Socio-economic and environmental risks

Household overcrowding, entrenched poverty, and rural-to-urban migration patterns elevate EPTB risk primarily through diagnostic delays and emergence of atypical presentations [10,11]. In high-burden settings such as India, socioeconomic deprivation strongly correlates with increased EPTB notification rates, underscoring structural barriers to early detection and intervention [11].

4. Extra-pulmonary tuberculosis in children

Children younger than 15 years constitute 10%–15% of total TB cases globally, yet account for 20%–40% of EPTB burden [12–15]. Predominant risk factors include malnutrition, recent household contact with TB patients, HIV co-infection, and age below 2 years [13,14]. Diagnostic challenges arise from paucibacillary disease, limiting acid-fast bacilli (AFB) smear and culture yields, necessitating reliance upon contact history, tuberculin skin test (TST), or interferon-gamma (IFN-γ) release assays (IGRA) [12–14]. The bacille Calmette-Guérin (BCG) vaccination confers 70%–80% protection against severe disseminated childhood forms [15].

5. Pathogenesis and mechanisms of dissemination

The transition from a localized pulmonary infection to disseminated EPTB is a multifactorial process. It involves the failure of the primary granuloma to contain the infection, the exploitation of host physiological processes by the bacteria, and the specific predilections of different *Mtb* lineages.

5.1 Angiogenesis and bacterial spread

Traditionally, the granuloma is viewed as a host-protective structure intended to wall off infection. However, recent evidence suggests that the granuloma is a dynamic environment that Mtb actively manipulates to favor its own survival and dissemination. A critical mechanism in this process is angiogenesis. The formation of new blood vessels, primarily mediated by vascular endothelial growth factor (VEGF), is exploited by the pathogen. Transcriptomic analyses of human monocyte-derived macrophages infected with *Mtb* have identified an "angiogenic signature," characterized by the upregulation of genes such as *VEGF-A*, interleukin (IL)-8 (CXCL8), and placental growth factor (PGF) [16].

Mtb induces macrophages to secrete VEGF in an RD1-dependent manner. The RD1 region encodes the ESX-1 secretion system, which exports virulence factors like ESAT-6 and CFP-10. Strains lacking RD1, such as *M. bovis* BCG, are poor inducers of VEGF, whereas virulent strains, particularly those of the Beijing lineage, induce high levels of VEGF secretion [16]. This neovascularization facilitates the recruitment of new, uninfected macrophages to the granuloma, which the bacteria then infect and use as vehicles for dissemination—a "Trojan horse" mechanism. In murine models, the inhibition of VEGF signaling via monoclonal antibodies or receptor blockade significantly reduces bacterial dissemination to the spleen and liver, confirming that *Mtb* exploits host angiogenesis to escape the primary site of infection [16,17].

5.2 The role of adipose tissue

While the lungs and lymph nodes are primary sites of infection, the question of where *Mtb* persists in latent and extrapulmonary cases remains critical. Recent research has identified adipose tissue as a potential reservoir. Adipose tissue is abundant, widely distributed, and contains a variety of cell types, including adipocytes and macrophages. *Mtb* can bind to adipocytes via scavenger receptors and enter the cells, where it accumulates intracytoplasmic lipid inclusions [18].

Inside the adipocyte, the bacterium enters a non-replicating state that is notably insensitive to isoniazid, a drug that targets actively dividing cells. However, these persisters remain susceptible to pyrazinamide [18]. The detection of *Mtb* DNA in adipose tissue surrounding the kidneys, heart, and lymph nodes in individuals without active TB suggests that this tissue may serve as a sanctuary site, protecting the bacillus from the host immune system and antibiotics, and potentially acting as a source for reactivation and extrapulmonary dissemination [18].

5.3 Host and bacterial genetics in dissemination

The outcome of *Mtb* infection, whether it remains latent, manifests as pulmonary TB, or disseminates to cause miliary TB or TBM, is heavily influenced by the interaction between host and bacterial genotypes. Host and bacterial genetic factors influencing dissemination are presented in Table 16.1.

5.4 Bacterial lineages

Mtb is not a monolith; different lineages exhibit distinct virulence profiles. The Euro-American lineage of Mtb has been associated with a significant protective effect against TBM compared to

Table 16.1 Host and bacterial genetic factors influencing dissemination.

Factor type	Gene/Lineage	Effect on disease phenotype
Bacterial Lineage	Euro-American lineage	Protective association against TBM (more likely to cause pulmonary TB).
Bacterial lineage	East-Asian (Beijing)	Associated with hypervirulence and increased dissemination (TBM) when combined with host susceptibility.
Bacterial gene	*Rv0218*	SNP associated with TB phenotype; encodes a secreted protein involved in host-pathogen interaction.
Bacterial gene	*nanK*, *Rv3433c*	Absence of these genes associated with TBM phenotype.
Host gene	*TLR2* (T597C polymorphism)	C Allele increases susceptibility to TB caused by Beijing lineage strains.
Host gene	*TIRAP* (C558T polymorphism)	Associated with susceptibility to meningeal TB.
Host gene	*NLRC4* (rs479333 variant)	Loss-of-function variant associated with extra-pulmonary TB; defective IL-18 axis.
Host gene	*VEGF* pathway	Upregulation promotes angiogenesis and dissemination; inhibition prevents spread in models.

the East-Asian (Beijing) and Indo-Oceanic lineages. In a Vietnamese cohort, the Euro-American lineage was strongly associated with pulmonary rather than meningeal TB, suggesting these strains are less capable of extrapulmonary dissemination in that specific population [19]. Conversely, the Beijing genotype is often associated with hypervirulence and increased dissemination, potentially due to the production of phenolic glycolipids that dampen the host innate immune response [19].

5.4.1 Host susceptibility

On the host side, polymorphisms in genes encoding the innate immune system play a crucial role. Polymorphisms in Toll-like receptor 2 (TLR2) and TIRAP (Toll-interleukin 1 receptor domain containing adaptor protein) have been linked to susceptibility to TBM. Specifically, the *TLR2* T597C polymorphism shows a significant interaction with bacterial genotype; individuals with the C allele are more likely to develop TB caused by the Beijing genotype [19].

Furthermore, proteomic analysis of spinal TB (Pott's disease) has identified specific host molecular mechanisms involved in the focal immune response. Upregulation of proteins such as cytochrome B-245 beta chain (CYBB), matrix Metallopeptidase 9 (MMP9), and CXCL10 in macrophages within the tubercular disc suggests a robust, albeit often destructive, inflammatory response. MMP9, in particular, contributes to tissue destruction and bone resorption, characteristic of spinal TB [20]. Additionally, in TBM, specific bacterial genetic variations, such as mutations in the *Rv0218* gene, have been associated with the meningeal phenotype, highlighting the role of bacterial secretome proteins in tissue tropism [20].

5.5 Gut dysbiosis and the gut-lung axis

Gut dysbiosis, defined by reduced microbial α-diversity and inversion of the Firmicutes to Bacteroidetes ratio [F:B ratio], critically contributes to EPTB pathogenesis through disruption of the gut-lung immunological axis [21]. *Mtb* infection induces profound microbiota alterations characterized by depletion of short-chain fatty acid-producing Prevotella and Bifidobacterium species, concurrent with enrichment of Enterobacteriaceae family members, resulting in diminished butyrate production essential for supporting IFN-γ and IL-22 mediated mucosal and systemic immunity [21]. This dysbiotic state precedes and facilitates hematogenous dissemination, particularly enabling ileocecal junction invasion characteristic of abdominal tuberculosis while promoting systemic miliary patterns through defective granuloma maintenance and impaired Th17 cell differentiation [21]. Fecal 16S ribosomal ribonucleic acid (16SrRNA) profiling consistently identifies reduced Bifidobacterium abundance and elevated Proteobacteria levels correlating directly with EPTB severity, with microbial perturbations persisting beyond 6 months following completion of anti tuberculosis chemotherapy [21]. Probiotic interventions utilizing Lactobacillus and Bifidobacterium species restore short-chain fatty acid homeostasis, enhance Th17 cell populations, and significantly attenuate dissemination in preclinical models, positioning microbiome modulation as a promising adjunctive host-directed therapeutic strategy [22].

5.6 Proline-glutamic acid and proline-proline-glutamic acid protein interactions

The proline-glutamic acid (PE) and proline-proline-glutamic acid (PPE) protein superfamily, constituting approximately 10% of the *Mtb* genome, represents a mycobacteria-specific expansion mediating sophisticated host-pathogen protein interactions essential for extra-pulmonary TB [22,23]. PE_Polymorphic GC-Rich Sequence (PGRS) domain-containing proteins such as PE_PGRS11 (Rv3514) directly engage TLR-2 and TLR-4 receptors, subverting innate immunity through oxidative stress evasion, inhibition of macrophage apoptosis, and manipulation of phagosome maturation arrest. PPE38 modulates host cytokine responses while PE_PGRS33 promotes lysosomal escape, enhancing intracellular persistence critical for hematogenous transit to extrapulmonary sites [23]. Tissue tropism specificity emerges through differential expression; PE_PGRS33(Rv1917c) upregulation characterizes TBM lesions, suggesting secretome-mediated meningeal targeting [23]. Host TLR-2 T597C polymorphisms exhibit epistatic interactions with Beijing lineage PE proteins, explaining lineage-specific extra-pulmonary predominance [24]. High guanine-cytosine content confers environmental stress-responsive regulation, including hypoxia and nutrient deprivation encountered during dissemination. PE/PPE proteins associate with ESX secretion systems, facilitating translocation via EccC ATPase interactions with YxxxD/E motifs of PPE protein [22]. Therapeutic targeting of PE/PPE-toll-like receptor interfaces offers novel host-directed opportunities aligning directly with host-pathogen protein interaction paradigms.

5.7 Miliary tuberculosis: Clinical and pathological features

Miliary TB is defined by the presence of diffuse miliary infiltrates on chest radiography or high-resolution computed tomography (HRCT), or evidence of miliary tubercles in multiple organs [4]. It accounts for 1%–2% of all TB cases and roughly 8% of EPTB cases in immunocompetent individuals, with higher rates in HIV-infected populations.

5.8 Clinical manifestations

The clinical presentation of miliary TB is protean and often non-specific, leading to diagnostic delays. Common symptoms include fever, anorexia, weight loss, and cough. However, "cryptic" miliary TB, common in the elderly, and may present as fever of unknown origin (FUO) or unexplained wasting without localizing signs, sometimes mimicking metastatic malignancy [4].

A pathognomonic feature of miliary TB is the presence of choroidal tubercles—granulomas in the choroid of the eye. These appear as yellowish nodules and are a valuable diagnostic clue, necessitating systematic ophthalmoscopic examination in suspected cases [26. Miliary TB can also involve the central nervous system (CNS), with TBM occurring in 10%–30% of adult miliary TB patients [4,25].

5.8.1 Histopathology

The hallmark of miliary TB is the granuloma. In immunocompetent hosts, these are typically caseating granulomas composed of epithelioid histiocytes and Langhans giant cells. However, in immunosuppressed patients, such as those with advanced HIV, the pathological picture changes. Granulomas may be poorly formed, necrotic, and contain abundant acid-fast bacilli (AFB), reflecting an inability to contain the pathogen [4,25].

The "soft" or exudative lesions associated with acute massive dissemination often contain demonstrable bacilli, whereas "hard" tubercles found in more protracted disease courses rarely reveal organisms on staining [4]. This variability underscores the importance of combining histological examination with microbiological culture and molecular testing.

6. Biomarkers for diagnosis and therapeutic monitoring

Given the diagnostic challenges of EPTB and miliary TB, stemming from paucibacillary samples and difficult-to-access infection sites, there is an urgent need for non-invasive biomarkers. These markers not only aid in diagnosis but also in monitoring therapeutic response, which is critical for preventing relapse and managing drug toxicity.

6.1 Inflammatory cytokines and chemokines

The systemic immune response to *Mtb* infection alters the profile of circulating cytokines and chemokines. In patients with TB lymphadenitis and pleuritis, specific inflammatory biomarkers have been shown to correlate with treatment success. A biosignature comprising five biomarkers, MIG (CXCL9), IP-10 (CXCL10), MIF (Macrophage Migration Inhibitory Factor), CCL22, and CCL23, has demonstrated high utility. Levels of MIG, IP-10, and CCL23 significantly decrease following successful treatment [6].

Conversely, in patients with partial or poor response to treatment, these markers fail to decline significantly. For instance, a biosignature defined by a significant decrease in any one of these five biomarkers corresponded with a satisfactory clinical response in 97% of patients at 2 months and 99% at 6 months [6]. This suggests that monitoring plasma levels of these chemokines could serve as an objective criterion for shortening or extending treatment duration, moving toward personalized medicine in EPTB.

6.1.1 Antimicrobial peptides

Antimicrobial peptides (AMPs) are critical components of the innate immune defense. In tuberculous lymphadenitis (TBL), systemic levels of specific AMPs, including Granulysin and Human Neutrophil Peptides 1 3 (HNP1-3), are significantly diminished compared to latent TB and healthy controls [10]. This suppression of innate immunity may facilitate bacterial survival. Importantly, successful anti TB treatment (ATT) reverses this suppression, restoring levels of HNP1-3 and Human Beta Defensin-2 (HBD-2) toward normal [27]. Therefore, measuring circulating AMPs offers another avenue for assessing disease activity and recovery.

6.2 The nCD64 index

CD64 (FcγRI) is a high-affinity immunoglobulin receptor typically expressed at low levels on resting neutrophils but upregulated during infection. The neutrophil CD64 (nCD64) index has emerged as a rapid, non-invasive biomarker for active TB. It is significantly elevated in patients with active TB compared to those with latent infection or healthy controls, with particularly high levels observed in disseminated and extrapulmonary forms [28].

The nCD64 index demonstrates high specificity (100%) for diagnosing TBM, outperforming the T-SPOT.TB assay in distinguishing TBM from non-tuberculous CNS infections [26]. Furthermore, the index declines rapidly within the first weeks of effective therapy, making it a promising tool for early therapeutic monitoring and prognosis.

Table 16.2 Potential biomarkers for diagnosis and monitoring of EPTB and miliary TB.

Biomarker category	Specific markers	Clinical utility
Chemokines/cytokines	MIG (CXCL9), IP-10 (CXCL10), MIF, CCL22, CCL23	Monitoring therapeutic response; a decrease indicates success. Predictive of cure at 2 and 6 months.
Cell surface receptors	Neutrophil CD64 (nCD64) index	High sensitivity/specificity for active TB and TBM. Distinguishes TBM from other CNS infections.
Antimicrobial peptides	HNP1-3, Granulysin	Reduced in active TBL; levels normalize with effective treatment. Diagnostic potential.
Cytokines (cell-associated)	Membrane-bound IL-22	Present on $CD4^+$ T cells in active TB; decreases with treatment.
Proteins (tissue-specific)	CYBB, MMP9, CXCL10	Highly expressed in spinal TB tissues; potential targets for inhibition to prevent bone destruction.
MicroRNAs	miR-29a-3p	Downregulated in spinal TB; associated with inflammation regulation via METTL3.

6.3 Epigenetic markers: MicroRNAs

Epigenetic regulation, specifically through microRNAs (miRNAs), plays a pivotal role in the host response to TB. In spinal TB, miR-29a-3p is significantly downregulated in vertebral lesions [27]. This miRNA normally regulates the expression of pro-inflammatory cytokines; its downregulation leads to excessive inflammation and tissue damage. Additionally, miRNAs regulate angiogenesis. Mtb infection alters the expression of host miRNAs to modulate VEGF signaling, further linking epigenetic changes to the physical dissemination of the bacteria [28]. Circulating miRNAs, therefore, hold potential as biomarkers for disease progression and tissue-specific involvement.

6.4 Interleukin-22

Interleukin-22 (IL-22), a cytokine produced by T-helper 17 (Th17) and Th22 cells, plays a dual role in TB. While soluble IL-22 is found at high levels in pleural fluid and plasma of active TB patients, a membrane-bound form of IL-22 on $CD4^+$ T cells has been identified [29]. This membrane-bound IL-22 appears to inhibit intracellular Mtb replication. The levels of membrane-bound IL-22+ T cells decrease significantly after drug therapy, tracking with bacterial clearance [25]. Table 16.2 summarizes potential biomarkers for EPTB diagnosis and management.

7. Some forms of extra-pulmonary tuberculosis

7.1 Tuberculosis lymphadenitis

Peripheral TB lymphadenitis is the most common form of EPTB, particularly in children and HIV-infected individuals. It typically presents as painless cervical lymphadenopathy (scrofula). The immune response in the lymph node is granulomatous, but as noted, systemic immunity is often depressed, characterized by low AMP levels [19]. Diagnosis relies on fine-needle aspiration cytology (FNAC), biopsy, and molecular testing (GeneXpert).

7.2 Spinal tuberculosis (Pott's disease)

Spinal TB accounts for half of all skeletal TB cases and can lead to catastrophic neurological deficits. Proteomic analysis of intervertebral discs from spinal TB patients reveals a distinct immune microenvironment dominated by macrophages and characterized by the upregulation of CYBB, MMP9, and CXCL10 [28]. The enzyme MMP9 is particularly relevant as it degrades the extracellular matrix, facilitating the destruction of the vertebral disc and bone, leading to the characteristic kyphotic deformity. Epigenetic mechanisms, such as the METTL3-mediated downregulation of miR-29a-3p, further drive the inflammatory destruction in spinal lesions [27].

7.3 Tuberculosis meningitis

TBM is the most lethal form of TB. It arises from the rupture of a subependymal tubercle (Rich focus) into the subarachnoid space. The inflammatory exudate, concentrated at the base of the brain, can cause cranial nerve palsies and hydrocephalus. Recent genomic studies have identified bacterial genetic loci associated with TBM, including *Rv0218*, a secreted protein potentially involved in host-pathogen interaction [20]. Host genetics (TLR2 polymorphisms) and the infecting lineage (Beijing strain) significantly influence the risk of developing TBM [19].

7.4 Inflammasome activation

The activation of the inflammasome, a multiprotein oligomer responsible for the activation of inflammatory responses, is crucial in the defense against Mtb. In EPTB, particularly in endemic areas, susceptibility has been linked to defects in the NLRC4 inflammasome pathway. Loss-of-function variants in *NLRC4* are associated with extra-pulmonary dissemination, suggesting that a failure to mount an appropriate IL-18 mediated response contributes to the inability to contain the infection within the lungs [30]. Furthermore, the balance between IL-1β and Type I interferons is critical; TB patients often exhibit a suppressed NLRP3 inflammasome response, leading to lower IL-1β levels and higher Type I interferons, a profile associated with disease severity [30,31].

8. Mitigation and control strategies

Controlling EPTB and miliary TB requires a multi-pronged approach that goes beyond standard pulmonary TB control measures. Effective management integrates heightened clinical vigilance, advanced diagnostic algorithms, optimized anti TB treatment, adjunctive host-directed therapies, surgical interventions, rigorous therapeutic monitoring, and innovative vaccination strategies.

9. Early diagnosis and biomarker integration

The non-specific presentation of EPTB necessitates a high index of suspicion. Integrating novel biomarkers into clinical practice is essential. Implementation of flow cytometric assessment of neutrophil CD64 (nCD64 Index) can provide rapid confirmation of active TB in smear-negative cases and TBM suspects [26]. Utilizing multiplex assays to measure chemokine biosignatures MIG, IP-10, and CCL22 can help monitor treatment response, distinguishing slow responders who may require extended therapy from those who are cured [6]. Lower thresholds for using multi-detector computed tomography (MDCT), magnetic Resonance Imaging (MRI), positron emission tomography-computed tomography (PET-CT), and positron emission tomography-magnetic resonance imaging (PET-MRI) in patients with FUO or unexplained lymphadenopathy can detect cryptic miliary TB and early spinal involvement. Fig. 16.1 illustrates the diagnostic algorithm commencing with clinical risk stratification [3]. Usefulness of first-line biomarkers (Table 16.2), other tests like fecal 16S rRNA that tracks dysbiosis resolution, merit further study [21]. Extrapulmonary nontuberculous mycobacterial (NTM) disease presents clinically similar manifestations to EPTB, including lymphadenitis, pleural effusions,

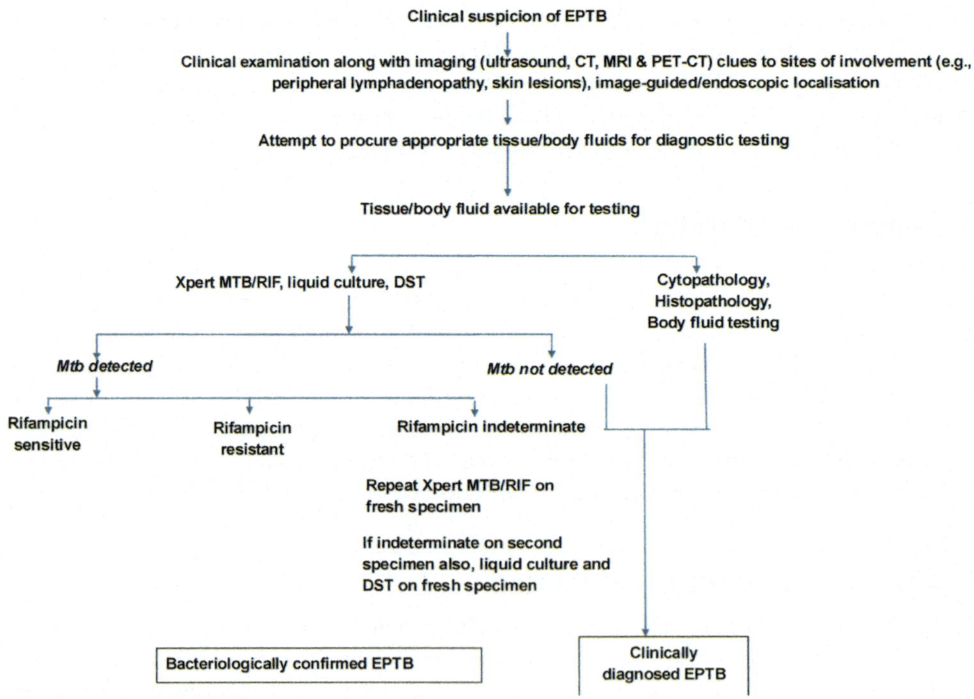

FIGURE 16.1

Diagnostic algorithm for EPTB.

CNS, *central nervous system;* CT, *computed tomography;* DST, *Drug Susceptibility Testing;* EPTB, *Extra-Pulmonary Tuberculosis;* MRI, *magnetic resonance imaging,* MTB, Mycobacterium tuberculosis, PET-CT, *positron emission tomography-computed tomography,* Xpert MTB/RIF, *Cartridge-based nucleic acid amplification test (GeneXpert),* USG, *Ultrasonography.*

skeletal involvement, and peritoneal disease, often indistinguishable on initial presentation. Laboratory differentiation requires mycobacterial culture speciation/subspeciation (LJ media, 6–8 weeks) and molecular identification and characterization [32].

9.1 Therapeutic interventions

9.1.1 General measures

Ensuring adequate nutritional support, improving the general condition, and optimizing treatment of co-morbid conditions are mandatory in the treatment of EPTB.

9.1.2 Anti tuberculosis treatment

Standard anti TB regimens are used to treat EPTB, with site-specific duration extensions required for central nervous system and skeletal disease, anti TB drug penetration challenges (Table 16.3) [33–36].

9.1.3 Surgery

Surgical interventions are required for procuring tissue to confirm diagnosis; or in specific situations like decompression/instrumentation in spinal TB, pericardiectomy relieving constriction, decortication for empyema thoracis elective, laparoscopic/opens surgical intervention for bowel obstruction/ perforation, ureteric stenting, nephrectomy Optimal timing balances early initiation of anti TB

Table 16.3 Treatment regimens for some forms of drug-susceptible EPTB.

Site of EPTB	Regimen	Duration[a]
Peripheral lymph node TB	2HRZE/4HRE	6 months
Pleural effusion[b]	2HRZE/4HRE	6 months
Spinal TB	2HRZE/10–16 HRE	12–18 months
TB meningitis[b]	2HRZE/10HRE	12 months
Miliary TB[b]	2HRZE/4HRE	6 months
Miliary TB with TB meningitis[b]	2HRZE/7-10HRE	9–12 months
Skeletal TB	2HRZE/7–16 HRE	9–18 months
Abdominal TB	2HRZE/4HRE	6–9 months
Pericardial TB[b]	2HRZE/4HRE	6–9 months
Genitourinary TB	2HRZE/4HRE	6 months

HRZE = Rifampicin (H), Isoniazid (R), Pyrazinamide (Z), Ethambutol (E); 2HRZE = 2 months intensive phase; 4HR = 4 months continuation phase; Dexamethasone = 0.4 mg/kg/day ×4 weeks taper; Prednisone = 1 mg/kg/day ×6 weeks taper.
ESR = erythrocyte sedimentation rate, MRI = magnetic resonance imaging.
[a]*The duration of treatment will need to be extended for longer duration on a case-to-case basis depending on the clinical, laboratory investigations and imaging response to treatment.*
[b]*Corticosteroids are recommended for TB meningitis in HIV-seronegative patients. Duration of steroid treatment should be for at least 4 weeks with tapering as appropriate (Strong recommendation). Corticosteroids may be used for TB meningitis in HIV-seropositive people, where other life-threatening opportunistic infections are absent (Conditional recommendation). Corticosteroids are recommended for HIV-seronegative and HIV-seropositive patients with TB pericarditis with pericardial effusion (Conditional recommendation). Corticosteroids are not routinely recommended in pleural TB (Conditional recommendation).*

treatment and appropriately timing surgical intervention preventing complications, maximizing functional outcomes.

9.1.4 Adjuvant corticosteroids

Adjuvant corticosteroid administration is mandatory when adrenal insufficiency is evident; beneficial when TBM, miliary TB, and pericardial TB are evident (Table 16.3).

9.1.5 Targeting angiogenesis

Given the role of VEGF in bacterial dissemination, anti angiogenic therapies (e.g., bevacizumab) have been proposed as potential host-directed therapies (HDT). In animal models, inhibiting VEGF signaling blocked dissemination to the spleen and liver [16]. However, this approach requires caution to avoid compromising tissue repair.

9.1.6 Epigenetic modulation

Therapies targeting specific miRNAs (e.g., mimics of miR-29a-3p) could potentially reduce excessive inflammation and tissue damage in spinal TB [27]. The miR-29a-3p mimics suppress METTL3 cascades, restoring granuloma integrity. Nanoparticle delivery reduces vertebral collapse 40% in preclinical models. CSF pharmacodynamics studies are required.

9.1.7 Vaccination and immunotherapy

The current BCG vaccine provides variable protection against pulmonary TB but is effective in preventing severe disseminated forms (miliary and meningeal) in children [4]. BCG confers 70%–80% protection against childhood miliary/TBM. Novel vaccine candidates, such as those based on the methylated Heparin-Binding Hemagglutinin Adhesin (HBHA), are being explored. HBHA is an adhesin involved in extrapulmonary dissemination; its methylation is essential for inducing a protective T-cell response. Methylated recombinant HBHA produced in surrogate hosts like *Rhodococcus erythropolis* shows promise as a subunit vaccine component to boost immunity against disseminated disease [37]. Rhodococcus-expressed methylated HBHA reduces dissemination by 3-log10 in guinea pigs. Phase I trials confirm Th1 immunogenicity. Probiotics restore the gut-lung axis (1.5-$\log^{10}$ reduction).

9.1.8 Treatment monitoring and follow-up

Treatment monitoring constitutes a critical component of EPTB management, given delayed culture conversion, higher relapse rates, and frequent paradoxical reactions. Periodic clinical, laboratory and imaging monitoring is indicated in EPTB. The newer neutrophil CD64 index 3and CXCL10 levels decline >50% in responders, while persistent elevation identifies slow converters requiring regimen extension [26,38]. Patients also need to be monitored for adverse effects and serious adverse effects which may be life-threatening.

10. Conclusions

Extra-pulmonary and miliary TB represent complex manifestations of Mtb infection where the pathogen escapes local containment to invade systemic sites. This process is driven by specific bacterial virulence factors, such as the RD1-dependent secretion of angiogenic factors, and facilitated by host vulnerabilities, including specific genetic polymorphisms and dampened innate immune responses (e.g., low AMPs, defective inflammasome activation).

Control of EPTB requires a paradigm shift from a purely bacteriological focus to a combined immunological and molecular approach. The integration of biomarker panels (chemokines, nCD64) into routine practice could significantly reduce diagnostic delays and prevent overtreatment or undertreatment. Furthermore, understanding the mechanisms of dissemination, such as the "Trojan horse" migration of macrophages and the role of adipose tissue as a reservoir, opens new avenues for host-directed therapies. Targeting angiogenesis or bolstering the NLRC4/IL-18 axis could serve as adjuncts to standard chemotherapy to limit dissemination. Ultimately, mitigating the burden of EPTB and miliary TB demands vigilance in clinical assessment, widespread adoption of advanced diagnostic tools, and continued research into the intricate host-pathogen interactions that allow this ancient pathogen to invade every organ of the human body.

References

[1] Sharma SK, Mohan A. Extrapulmonary tuberculosis. Indian J Med Res 2004;120:316–53.

[2] Sharma SK, Mohan A, Kadhiravan T. HIV-TB co-infection: epidemiology, diagnosis & management. Indian J. Med. Res. 2005;121:550–67.

[3] Sharma SK, Mohan A, Kohli M. Extrapulmonary tuberculosis. Expert Rev Respir Med 2021;15:931–48.

[4] Sharma SK, Mohan A, Sharma A, Mitra DK. Miliary tuberculosis: new insights into an old disease. Lancet Infect. Dis. 2005;5:415–30.

[5] World Health Organization. Global tuberculosis report 2025. Geneva: World Health Organization; 2025. Licence: CC BY-NC-SA 3.0 IGO.

[6] Ambreen A, Khaliq A, Naqvi SZH, Tahir A, Mustafa M, Chaudhary SU, et al. Host biomarkers for monitoring therapeutic response in extrapulmonary tuberculosis. Cytokine 2021;142:155499.

[7] Oda G, Lucero-Obusan C, Schirmer P, Chung J, Holodniy M. Risk factors for extrapulmonary tuberculosis among US Veterans, 1990-2022. Open Forum Infect. Dis. 2024:ofae698.

[8] Rolo M, González-Blanco B, Reyes CA, Rosillo N, López-Roa P. Epidemiology and factors associated with Extra-pulmonary tuberculosis in a Low-prevalence area. J Clin Tuberc Other Mycobact Dis. 2023;32: 100377.

[9] Dubois MM, Brooks MB, Malik AA, Siddiqui S, Ahmed JF, Jaswal M, et al. Age-specific clinical presentation and risk factors for extrapulmonary tuberculosis disease in children. Pediatr Infect. Dis. J. August 1, 2022;41(8):620–5.

[10] Luque L, Rodrigo T, García-García JM, Casals M, Millet JP, Caylà J, et al. Factors associated with extrapulmonary tuberculosis in Spain and its distribution in immigrant population. Open Respir Arch 2020; 2:119–26.

[11] Teibo TKA, Andrade RLP, Rosa RJ, de Abreu PD, Olayemi OA, Alves YM, et al. That interfere with access to tuberculosis diagnosis and treatment across countries globally: a systematic review. ACS Infect. Dis. 2024;10:2600–14.

[12] Maphalle LNF, Michniak-Kohn BB, Ogunrombi MO, Adeleke OA. Pediatric tuberculosis management: a global challenge or breakthrough? Children 2022;9:1120.

[13] Maritz ER, Montepiedra G, Mitchell CD, Madhi SA, Bobat R, Violari A, Hesseling AC, Cotton MF. Predictors of TB disease in HIV-exposed children from Southern Africa. Int. J. Tubercul. Lung Dis. 2023;27: 619–25.

[14] Liang Y, Wang J, Yang J, Liu J, He X. Global burden and trend of tuberculosis in children and adolescents (under 15 years old) from 1990 to 2021, with projections to 2040. Front. Public Health 2025;13:1578658.

[15] Liu Q, Dai X, Wei S. BCGitis and BCGosis: clinical spectrum, immunological mechanisms, and risk management. Vaccines (Basel) 2025;13:1179.
[16] Polena H, Boudou F, Tilleul S, Dubois-Colas N, Lecointe C, Rakotosamimanana N, et al. *Mycobacterium tuberculosis* exploits the formation of new blood vessels for its dissemination. Sci. Rep. 2016;6:33162.
[17] Batista LAF, Silva KJS, Costa e, Silva LM, Moura YF, Zucchi FCR. Tuberculosis: a granulomatous disease mediated by epigenetic factors. Tuberculosis (Edinb) 2020;123:101943.
[18] Neyrolles O, Hernández-Pando R, Pietri-Rouxel F, Fornès P, Tailleux L, Barrios Payán JA, et al. Is adipose tissue a place for *Mycobacterium tuberculosis* persistence? PLoS One 2006;1:e43.
[19] Caws M, Thwaites G, Dunstan S, Hawn TR, Lan NTN, Thuong NTT, et al. The influence of host and bacterial genotype on the development of disseminated disease with *Mycobacterium tuberculosis*. PLoS Pathog. 2008;4:e1000034.
[20] Ruesen C, Chaidir L, van Laarhoven A, Dian S, Ganiem AR, Nebenzahl-Guimaraes H, et al. Large-scale genomic analysis shows association between homoplastic genetic variation in *Mycobacterium tuberculosis* genes and meningeal or pulmonary tuberculosis. BMC Genom. 2018;19:122.
[21] Chai Y, Li M, Deng X, Ma C, Zhou N, Chen Y, et al. Gut microbiota and tuberculosis infection: interaction and therapeutic potential. Gut Microbes 2025;17:2531201.
[22] Ates LS. New insights into the mycobacterial PE and PPE proteins provide a framework for future research. Mol. Microbiol. 2020;113:4–21.
[23] Sharma T, Alam A, Ehtram A, Rani A, Grover S, Ehtesham NZ, Hasnain SE. The *Mycobacterium tuberculosis* PE_PGRS protein family acts as an immunological decoy to subvert host immune response. Int. J. Mol. Sci. 2022;23:525.
[24] Tientcheu LD, Koch A, Ndengane M, Andoseh G, Kampmann B, Wilkinson RJ. Immunological consequences of strain variation within the *Mycobacterium tuberculosis* complex. Eur. J. Immunol. 2017;47: 432–45.
[25] Sharma SK, Mohan A, Prasad KL, Pande JN, Gupta AK, Khilnani GC. Clinical profile, laboratory characteristics and outcome in miliary tuberculosis. Q. J. Med. 1995;88:29–37.
[26] Liu Q, Gao Y, Ou Q, Xu Y, Zhou Z, Li T, et al. Differential expression of CD64 in patients with *Mycobacterium tuberculosis* infection: a potential biomarker for clinical diagnosis and prognosis. J. Cell Mol. Med. 2020;24:13961–72.
[27] Ma X, Gao Y, Ren Z, Dong H, Zhang X, Niu N. Study on the role and molecular mechanism of METTL3-mediated miR-29a-3p in the inflammatory response of spinal tuberculosis. Tuberculosis (Edinb) 2024;148: 102546.
[28] Xu G, Xue J, Jiang J, Liang T, Yao Y, Liao S, et al. Proteomic analysis reveals critical molecular mechanisms involved in the macrophage anti-spinal tuberculosis process. Tuberculosis (Edinb) 2021;126:102039.
[29] Liu Y, Ou Q, Liu Q, et al. The expressions and r17.oles of different forms of IL-22 in *Mycobacterium tuberculosis* infection. Tuberculosis (Edinb) 2017;107:95–103. https://doi.org/10.1016/j.tube.2017.08.009.
[30] Kathamuthu GR, Moideen K, Sridhar R, Baskaran D, Babu S. Diminished systemic levels of antimicrobial peptides in tuberculous lymphadenitis and their reversal after anti-tuberculosis treatment. Tuberculosis (Edinb) 2020;122:101934.
[31] Souza De Lima D, Bomfim CCB, Leal VNC, Reis EC, Soares JLS, Fernandes FP, et al. Combining host genetics and functional analysis to depict inflammasome contribution in tuberculosis susceptibility and outcome in endemic areas. Front. Immunol. 2020;11:550624.
[32] Sharma SK, Upadhyay V. Epidemiology, diagnosis and treatment of non-tuberculous mycobacterial diseases. Indian J. Med. Res 2020;152:185–226.
[33] World Health Organization. WHO consolidated guidelines on tuberculosis. Module 5: management of tuberculosis in children and adolescents. Geneva: World Health Organization; 2022. Licence: CC BY-NC-SA 3.0 IGO.

[34] Saukkonen JJ, Duarte R, Munsiff SS, et al. Updates on the treatment of drug-susceptible and drug-resistant tuberculosis: an official ATS/CDC/ERS/IDSA clinical practice guideline. Am. J. Respir. Crit. Care Med. 2025;211:15–33.
[35] Central TB Division, Ministry of Health and Family Welfare, Government of India. Training module on extrapulmonary tuberculosis. 2023. Available at: https://tbcindia.mohfw.gov.in/wp-content/uploads/2023/05/7702334778Training_Module_on_Extrapulmonary_TB_-_Book_24032023.pdf. Accessed on Jan 01, 2026.
[36] Sharma SK, Ryan H, Khaparde S, Sachdeva KS, Singh AD, Mohan A, et al. Index-TB guidelines: guidelines on extrapulmonary tuberculosis for India. Indian J. Med. Res. 2017;145:448–63.
[37] Parada C, Neri-Badillo IC, Vallecillo AJ, Segura E, Silva-Miranda M, Guzmán-Gutiérrez SL, et al. New insights into the methylation of *Mycobacterium tuberculosis* heparin binding hemagglutinin adhesin expressed in Rhodococcus erythropolis. Pathogens 2021;10:113919.
[38] Wei Z, Chen Y, Dong P, Liu Z, Lai X, Wang N, et al. CXCL9/CXCL10 as biomarkers the monitoring of treatment responses in Pulmonary TB patients: a systematic review and meta-analysis. BMC Infect. Dis. 2024;24:1037.

CHAPTER

Multifaceted approaches to tuberculosis management: Bridging molecular discovery and public health impact

17

Mohd Shariq[1], Farhan Ahmed[1], Anwar Alam[2], Waseem Ali[2], Nasreen Z. Ehtesham[2] and Seyed E. Hasnain[2, 3]

[1]*GITAM School of Life Sciences, Gandhi Institute of Technology and Management (GITAM) University, Hyderabad Campus, Hyderabad, Telangana, India;* [2]*Department of Life Science, School of Basic Sciences and Research, Sharda University, Greater Noida, Uttar Pradesh, India;* [3]*Department of Biochemical Engineering and Biotechnology, IIT Delhi, New Delhi, India*

Highlights

1. Tuberculosis remains a major global health challenge, with multidrug-resistant TB and HIV co-infection complicating control efforts.
2. Transcriptional regulatory networks and transcription factor activity modeling provide insights into gene regulation and TB pathogenesis.
3. Glutamine metabolism plays a dual role in supporting immune defenses and facilitating the survival of *Mycobacterium tuberculosis*.
4. Mucosal vaccine candidates such as MTBVAC and VPM1002 offer promising pathways to enhance TB vaccine efficacy.
5. The PE/PPE protein family in *Mycobacterium tuberculosis* contributes to immune modulation, antigenic variation, and virulence.
6. Neonatal multidrug-resistant TB highlights challenges in maternal and child health, particularly regarding latent TB infection.
7. Molecular epidemiological tools, such as whole-genome sequencing and MIRU-VNTR typing, enhance outbreak tracing and resistance surveillance.

1. Introduction

1.1 The persistent challenge of tuberculosis

Tuberculosis (TB) continues to pose a major global public health challenge, affecting millions of individuals each year and placing an immense burden on healthcare systems, particularly in low- and middle-income countries. According to the World Health Organization (WHO), TB remains the

Navigating Tuberculosis. https://doi.org/10.1016/B978-0-443-27735-1.00014-9

leading cause of death from infectious diseases worldwide, surpassing HIV/AIDS in terms of annual mortality. TB is estimated to claim nearly 3 million lives every year, with the heaviest toll recorded in resource-limited settings, where access to timely diagnosis and treatment is constrained [1,2]. Despite being a preventable and curable disease, TB persists owing to a combination of biological, socio-economic, and systemic factors that delay diagnosis, hinder effective treatment, and sustain transmission [3].

The global TB epidemic has been further intensified by the emergence of multidrug-resistant TB (MDR-TB) and co-infections with HIV, both of which present additional complexities in disease management. MDR-TB refers to TB strains that are resistant to at least isoniazid and rifampicin, the two most potent first-line anti-TB drugs [4,5]. These strains require longer, more expensive, and more toxic treatment regimens with substantially lower success rates. Moreover, co-infection with HIV not only accelerates TB disease progression but also complicates treatment due to potential drug-drug interactions and immune suppression. Together, these factors significantly increase mortality rates and create greater challenges for national TB control programs [6–8]. In response, the WHO launched major initiatives including DOTS, the Stop TB Partnership, and the End TB Strategy, contributing to substantial gains in TB control and a gradual global decline in TB incidence and mortality [9,10]. However, the pace of progress has been insufficient to meet the WHO's ambitious targets for TB elimination, and substantial gaps remain in many critical areas. Among these are the growing problem of drug resistance, the inadequacy of current diagnostic tools, and the lack of an effective vaccine capable of providing robust and long-lasting protection in adults [11,12].

MDR-TB represents one of the most pressing issues in TB control today. The treatment of MDR-TB is not only prolonged and costly but is also associated with considerable side effects and lower success rates. For instance, while drug-sensitive TB can typically be cured within 6 months, MDR-TB treatment often extends for 18–24 months or longer and involves the use of second-line drugs that are less effective and more toxic. India, which bears the highest burden of MDR-TB globally, illustrates the dire need for improved case detection and individualized treatment approaches [13,14,15]. Addressing MDR-TB requires the development of novel therapeutic regimens, the scaling up of rapid drug-susceptibility testing, and the implementation of patient-centered care models that enhance adherence and improve treatment outcomes [16,17]. Preventive strategies are equally vital in the global fight against TB. A key component of prevention is the early identification and treatment of latent TB infections (LTBI) to prevent progression to active disease. However, this approach is hindered by limitations in the accuracy and reliability of existing diagnostic tools [18,19]. Currently used methods, such as Interferon Gamma Release Assays (IGRAs) and the tuberculin skin test, cannot consistently distinguish between latent and active TB infections [20,21]. This diagnostic ambiguity impairs risk stratification and undermines preventive interventions. Moreover, the lack of a widely accessible, rapid, and specific point-of-care diagnostic test represents a major bottleneck in TB control efforts, particularly in high-burden, resource-poor settings where healthcare infrastructure is often limited [22,23].

Vaccination has long been recognized as a cornerstone of TB prevention. The Bacille Calmette-Guerin (BCG) vaccine, first introduced over a century ago, is still the only licensed vaccine for TB. While it offers partial protection against severe forms of TB in children, its efficacy in preventing pulmonary TB in adolescents and adults, the primary source of transmission, is limited and highly variable across populations. Despite significant investments and advances in immunological research, a new and more effective TB vaccine has yet to be developed and deployed on a global scale [24,25]. A successful vaccine would not only reduce TB incidence but could also play a crucial role in curbing the spread of drug-resistant strains. While WHO-led strategies and global health partnerships have yielded important advances in TB prevention, diagnosis, and treatment, major obstacles remain on the path toward TB elimination. Achieving this goal will require coordinated action across multiple sectors and disciplines [26,27]. Key priorities include enhancing research and development for new drugs, diagnostics, and vaccines; improving surveillance systems; increasing investment in TB care infrastructure; and ensuring political and financial commitment at both national and international levels. There is also a pressing need to strengthen health systems to ensure that effective interventions are delivered equitably and reach those most at risk [28,29].

Furthermore, addressing the broader social determinants of TB, including poverty, malnutrition, overcrowded housing, and limited access to healthcare, is essential for achieving long-term success. TB is often described as a disease of poverty, and efforts to control it must go beyond biomedical solutions to include comprehensive strategies that address the root causes of vulnerability. Although integrating TB services with HIV care, maternal and child health, and non-communicable disease programs may enhance efficiency and patient outcomes, significant challenges persist despite progress achieved through global initiatives and scientific advancements [30,31]. The continued threat of drug-resistant TB, insufficient diagnostic tools, and the absence of broadly effective preventive measures highlight the urgent need for renewed commitment and innovation. Achieving the goal of ending the TB epidemic requires coordinated efforts across all sectors, researchers, healthcare professionals, policymakers, and communities alike. By working collaboratively to dismantle barriers and implement evidence-based, context-sensitive solutions, the global health community can accelerate progress toward a TB-free world [32,33].

Box-1 Key challenges and emerging tools in TB control

Conventional diagnostic tools, such as IGRAs and the tuberculin skin test, struggle to distinguish between latent and active TB, leading to suboptimal risk stratification and compromised prevention strategies. The global fight against TB is hampered by the absence of a rapid, accessible, and highly specific point-of-care diagnostic that can be reliably used in high burden and resource-limited settings. Metagenomic next-generation sequencing (mNGS) is revolutionizing TB detection, offering high specificity directly from clinical samples. Coupled with emerging artificial intelligence-assisted diagnostic platforms, these technologies promise real-time detection, strain typing, and resistance profiling, particularly for extrapulmonary and drug-resistant TB forms. Glutamine metabolism plays a paradoxical role in TB, supporting both host immune responses and Mtb survival. This duality complicates therapeutic targeting, as inhibiting glutamine pathways may impair essential immune functions while attempting to hinder bacterial persistence. Innovative compounds like JHU083, a glutamine antagonist, are showing promise in preclinical models by enhancing T-cell recruitment and pro-inflammatory myeloid responses, thereby boosting the host's antimicrobial capacity while reducing lung pathology and bacillary loads. The BCG vaccine shows highly variable efficacy against pulmonary TB in adults, ranging from 0% to 80% across populations. This inconsistency is compounded by the lack of reliable immune correlates of protection, which complicates vaccine development and evaluation in clinical trials. Promising candidates like MTBVAC and VPM1002 are designed to enhance mucosal immunity and trained immunity, targeting early Mtb-host interactions at the respiratory tract. These approaches leverage innate immune memory and localized protection, aiming to deliver broader and more durable immunity than traditional vaccines. The rapid evolution of Mtb resistance, particularly in MDR and XDR strains, presents a formidable challenge for diagnostic and treatment strategies. The complexity of linking genomic mutations to phenotypic resistance limits the predictive accuracy of computational models, especially in diverse global populations. The integration of CRISPR functional genomics with Bayesian and machine learning approaches is transforming drug screening efficiency. These tools enable the identification of key regulatory elements and drug targets, accelerating the development of precision therapies against drug-resistant TB strains. Efforts to integrate TB screening into reproductive health services face obstacles such as resource limitations, staff shortages, and high patient volumes. The COVID-19 pandemic further exposed vulnerabilities in healthcare infrastructure, highlighting the fragility of integrated service delivery in low- and middle-income countries. Digital health platforms, leveraging community health workers and telemedicine tools, are being explored to maintain service continuity during disruptions. These platforms offer scalable and adaptable solutions to integrate TB screening and treatment within reproductive and broader primary healthcare systems.

2. Systems biology and network modeling: Decoding mycobacterial logic

Transcriptional Regulatory Networks (TRNs) play a pivotal role in decoding the mechanisms underlying gene expression regulation, especially in the context of complex diseases like TB. By mapping the interactions between transcription factors (TFs) and their target genes, TRNs help reveal how specific genes are co-expressed and modulated under various physiological and pathological conditions [34,35]. Understanding these regulatory architectures is critical for identifying key molecular drivers of *M. tb* pathogenesis and developing targeted interventions. Multiple approaches have emerged for constructing TRNs, each offering distinct insights into TB biology. Among these, machine learning-based analyses of gene expression data have shown promise in predicting transcriptional relationships. These methods enable the identification of complex regulatory hierarchies and potential cooperative regulation among transcription factors [36,37]. One such tool, LogicTRN, integrates time-series gene expression data with TF-DNA-binding signals to infer regulatory logic, thereby improving the resolution of inferred TRNs and revealing dynamic regulatory events during TB infection [38,39].

Chromatin Immunoprecipitation followed by sequencing (ChIP-Seq) has also been instrumental in mapping genome-wide TF binding sites, significantly enhancing our understanding of transcriptional control in *M. tb*. This technique has challenged prior models of bacterial gene regulation and uncovered novel aspects of TF binding specificity and function [40,41]. Application of ChIP-Seq in *M. tb* has provided crucial evidence of how TFs orchestrate stress responses, dormancy, and virulence, offering a more comprehensive view of gene regulation during infection. Another critical dimension of transcriptional regulation in TB involves the immune system, particularly the role of regulatory T (Treg) cells. These cells modulate the immune response, preventing tissue damage due to excessive inflammation. However, deciphering the transcriptional regulation of Treg cells in TB remains complex. Understanding the transcriptional cues that direct Treg differentiation and activity during *M. tb* infection could offer new strategies for immune modulation [42–44].

TRNs have further facilitated the discovery of transcriptional signatures distinguishing TB-susceptible from TB-resistant phenotypes. For instance, the cSTAR (cell State Transition Assessment and Regulation) approach has been employed to reprogram susceptible macrophages into a resistant-like transcriptional state by using chemical perturbations. This novel strategy points toward potential therapeutic avenues for modulating host responses and mitigating TB susceptibility [45,46]. In sum, the integration of machine learning, ChIP-Seq, and advanced modeling has proven indispensable in constructing TRNs for TB. These techniques have not only deepened our understanding of regulatory mechanisms in *M. tb* but have also opened possibilities for developing novel diagnostic markers and therapeutic targets [47,48].

Beyond TRNs, Transcription Factor Activity (TFA) modeling has significantly advanced drug discovery efforts in TB research by elucidating how *M. tb* regulates gene expression across various environments and infection stages. TFA modeling enables researchers to build systems-scale regulatory network models by analyzing high-throughput transcriptome data. These models help identify modular gene arrangements and TFs that regulate stress adaptation, metabolic shifts, and dormancy, critical features of *M. tb* survival and pathogenicity [49,50]. TFA modeling has revealed that certain transcription factors are co-regulated with genes essential for host-pathogen interactions, including 23 genes vital during infection. Such insights facilitate the identification of potential therapeutic targets that can disrupt these regulatory axes [51,52]. Additionally, TFA analysis has contributed to our understanding of drug mechanisms. For example, the synergistic activity of ethambutol and isoniazid was explained through a transcription factor that enhances the effect of isoniazid via transcriptional repression, providing a molecular rationale for combining these drugs [53].

Importantly, TFA modeling has been enhanced through integration with computational tools such as CRISPR-based functional genomics and Bayesian machine learning algorithms. These tools improve drug screening efficiency by accurately predicting compound efficacy against *M. tb* [54,55]. As a result, they have accelerated the discovery of kinase inhibitors and other novel drug classes, expanding the anti-TB drug arsenal [56,57]. Despite these advances, several challenges persist in fully leveraging in silico predictive modeling for TB research. The biological complexity of *M. tb*, marked by its unique pathogenic mechanisms and rapid evolution of drug resistance, continues to challenge predictive accuracy [16,58]. Furthermore, data limitations, including inconsistent clinical data and underrepresentation of certain *M. tb* strains, undermine the reliability of computational models [59].

Resistance to multiple drugs, especially in MDR and XDR TB, necessitates predictive models capable of accounting for a wide array of genomic mutations. This requires robust genomic databases and algorithms to decode genotype-phenotype relationships [58,60]. Additionally, integrating predictive models into existing healthcare frameworks, especially in low-resource settings, demands solutions that are both scalable and compatible with current diagnostics and treatment protocols [60,61]. The computational infrastructure needed for advanced modeling remains a limiting factor, particularly in high-burden countries with constrained technological resources. Limited access to high-performance computing and software can hamper the application of sophisticated in silico tools [62].

To overcome these challenges, several future directions are recommended. First, there is an urgent need for comprehensive and standardized data collection, integrating genomic, transcriptomic, and clinical datasets from diverse populations. Public health databases should be expanded to include multidimensional patient data that can support predictive model training and validation [63,64]. Second, the incorporation of Artificial Intelligence (AI) can significantly enhance predictive modeling capabilities by detecting complex patterns in large datasets and informing personalized treatment strategies [65]. AI tools can facilitate real-time decision-making in clinical settings and improve diagnostic precision. Third, fostering interdisciplinary collaboration between bioinformaticians, microbiologists, clinicians, and public health experts is essential for translating computational models into clinically relevant applications [66,67]. This will ensure that in silico approaches are both scientifically robust and operationally feasible.

Lastly, achieving widespread adoption of predictive modeling in TB research and treatment requires policy support and increased funding. Governments and international organizations must prioritize digital health innovation and provide the resources needed to develop, validate, and deploy these tools at scale. In conclusion, TRNs, TFA modeling, and in silico predictive approaches are transforming the landscape of TB research by offering deeper insights into gene regulation, pathogenesis, and therapeutic response. Despite significant technical and infrastructural challenges, the strategic integration of computational methods with experimental biology holds immense potential for accelerating TB diagnosis, treatment, and drug development [15,68,69].

3. Immunometabolism in TB: Glutamine and beyond

Glutamine plays a pivotal and dual role in both supporting immune defense and facilitating pathogen survival, particularly in the context of *M. tb* infection. This dual role is critical in understanding the pathogenesis and management of TB, a disease caused by *M. tb*. On the one hand, glutamine metabolism is essential for mounting an effective immune response against *M. tb*. The activation and function of immune cells rely heavily on glutamine, which fuels cellular processes critical for defense. Studies have shown that genes involved in the glutamine metabolic pathway are differentially expressed in *M. tb*-infected macrophages and in individuals with either latent TB infection or active disease [70,71]. Inhibition of glutaminolysis, the conversion of glutamine into metabolic intermediates, leads to a marked reduction in the production of key cytokines such as IL-1β, IFN-γ, and IL-17, all of which are vital for orchestrating inflammatory responses and T-cell activation to control *M. tb* infection [70].

Conversely, glutamine metabolism also supports *M. tb*'s survival within the host. *M. tb* can exploit host glutamine to evade immune defenses and reprogram host metabolism for its benefit. The use of glutamine antagonists, such as JHU083, has been shown to inhibit *M. tb* growth both in vitro and in vivo, as evidenced by reduced lung pathology and bacillary loads in infected mice [71]. This compound not only restricts *M. tb* replication but also modulates immune cell recruitment, promoting the early influx of T cells and enhancing the production of pro-inflammatory myeloid cells, thereby augmenting the host's antimicrobial capacity [71]. As a metabolic node shared by host immunity and *M. tb*, glutamine offers a promising therapeutic target whereby its modulation could both strengthen immune defenses and restrict pathogen survival [72]. Understanding glutamine's complex role in TB pathogenesis is essential for developing innovative strategies for disease prevention and management [73].

The dynamic interplay between glutamine metabolism and immune responses reflects a broader theme of metabolic reprogramming orchestrated by pathogens such as *M. tb*, highlighting the potential of metabolic interventions as a therapeutic strategy [72]. As research continues to unravel these intricate interactions, targeting key metabolic pathways holds promise for bolstering immune defenses while undermining pathogen persistence. In addition to glutamine, other amino acids such as arginine and tryptophan are critical players in TB pathophysiology, influencing immune modulation and macrophage activation [71,74,75].

Arginine metabolism is central to macrophage activation during *M. tb* infection. The enzyme inducible nitric oxide synthase (iNOS), IFN-γ, converts arginine into citrulline and nitric oxide (NO), a molecule with potent antimicrobial properties essential for the immune response to *M. tb*. Conversely, IL-4 activates arginase, directing arginine metabolism toward ornithine production, which affects immune activation and macrophage polarization. This intricate regulatory network is further modulated by NAD+-dependent sirtuins, particularly sirtuin 2 (SIRT2) and SIRT5, which play significant roles in shaping arginine metabolism and macrophage function [76]. Tryptophan metabolism also exerts a profound influence on TB, primarily via the kynurenine pathway mediated by indoleamine 2,3-dioxygenase 1 (IDO-1). During TB infection, increased catabolism of tryptophan to kynurenine, driven by elevated IDO-1 expression, is observed in both active and latent TB [77,78]. This transformation reflects a conserved host response to *M. tb* and suggests that tryptophan catabolism could serve as a valuable biomarker for TB infection. Importantly, kynurenine pathway metabolites have immunomodulatory effects, influencing disease progression and presenting potential targets for therapeutic intervention [77].

Thus, the regulation of arginine and tryptophan metabolism during *M. tb* infection demonstrates their potential as biomarkers and therapeutic targets. By manipulating these metabolic pathways, novel host-directed therapies (HDTs) for TB could be developed, offering innovative avenues for diagnosis and treatment [79–81]. Metabolic interventions as adjunctive therapies for TB represent a promising approach to enhance treatment efficacy, especially in the face of rising MDR-TB strains. Conventional TB treatment regimens are lengthy and often associated with challenges such as adverse effects, patient non-compliance, and the emergence of drug resistance [82,83]. Addressing these challenges through metabolic intervention could improve therapeutic outcomes and reduce treatment burdens.

M. tb demonstrates remarkable metabolic flexibility, enabling it to adapt to hostile conditions, including drug exposure and immune pressure. This adaptability involves the remodeling of central carbon and lipid metabolism, allowing *M. tb* to persist within the host [84]. Targeting these metabolic pathways through intervention strategies could impair *M. tb*'s survival mechanisms, making it more susceptible to antibiotic treatment and immune clearance. Manipulating amino acid metabolism, specifically L-arginine and L-tryptophan, holds promise for augmenting immune responses against *M. tb*. These amino acids regulate critical immune functions and could be leveraged as adjunct therapies to strengthen host immunity [78]. Additionally, *M. tb*'s metabolic rewiring under antibiotic pressure and during resistance emergence emphasizes the need to target these pathways to prevent resistance development and enhance drug efficacy [85].

Phytochemicals derived from traditional medicine have also shown potential as adjunctive treatments, offering immune-modulating properties and mitigating the side effects of conventional TB drugs [86]. Furthermore, HDTs that modulate host metabolic pathways, including autophagy and the vitamin D pathway, offer promising adjunctive strategies (Table 17.1). Unlike traditional antibiotics, HDTs reduce the risk of resistance development and enhance the host's natural defenses against *M. tb* [83]. In summary, metabolic interventions provide a compelling and innovative approach to TB treatment, with the potential to enhance therapeutic efficacy, shorten treatment durations, and reduce complications. By targeting *M. tb*'s metabolic adaptability and leveraging host metabolic pathways, these strategies hold promise for combating drug-resistant TB and improving patient outcomes [37,85,87].

4. Innovations in vaccinology: Mucosal BCG and beyond

Innovations in TB vaccinology have progressed significantly, moving beyond the traditional BCG vaccine that has been in use for nearly a century. Notably, mucosal vaccine candidates such as MTBVAC and VPM1002 offer promising pathways to enhance TB vaccine efficacy. MTBVAC, a live-

Table 17.1 Host-Directed Therapy (HDT) Strategies in Tuberculosis: Mechanisms and Representative Agents.

HDT strategy	Mechanism/target	Agents/examples
Autophagy induction	Enhances pathogen clearance via intracellular degradation	Vitamin D, Rapamycin, Loperamide, Carbamazepine, Berbamine, Baicalin
Immune modulation	Balances immune response to reduce inflammation and tissue damage	NSAIDs, Corticosteroids
Epigenetic modulation	Modifies DNA methylation/histone marks to restore immune gene expression	Histone deacetylase inhibitors (HDACi)
Vitamin D supplementation	Boosts antimicrobial peptide production and modulates immune responses	Vitamin D3
Glutamine antagonism	Disrupts Mtb metabolism and enhances immune response	JHU083
Sirtuin modulation	Regulates autophagy and inflammation via SIRT1 and SIRT7	SIRT1 activators, HDAC inhibitors
Checkpoint inhibition	Enhances T-cell response by blocking immune checkpoints	Immune checkpoint inhibitors (ICIs)

Abbreviations: HDACi, *histone deacetylase inhibitors;* HDT, *host-directed therapy;* ICIs, *immune checkpoint inhibitors;* M. tb, *mycobacterium tuberculosis;* NSAIDs, *nonsteroidal anti-inflammatory drugs;* SIRT, *sirtuin (e.g., SIRT1, SIRT7);* TB, *tuberculosis.*
This table summarizes major host-directed therapy (HDT) approaches in tuberculosis, highlighting their primary molecular or immunological targets and representative agents. These strategies aim to enhance antimicrobial immunity, modulate excessive inflammation, disrupt host–pathogen metabolic interactions, and improve treatment outcomes, particularly in drug-resistant TB.

attenuated vaccine derived from a clinical isolate of *M. tb*, preserves most of the T-cell epitopes crucial for effective immune responses, including the major antigens ESAT6 and CFP10 that are absent in BCG [88,89]. In preclinical models, MTBVAC has demonstrated improved safety and efficacy compared to BCG, and ongoing clinical trials in TB-endemic countries are evaluating its potential to replace BCG in newborns and serve as a preventive measure for adolescents and adults [89].

VPM1002, another innovative candidate, is a recombinant BCG vaccine engineered to enhance immune responses. This vaccine expresses listeriolysin, which facilitates antigen delivery into the macrophage cytosol, thereby promoting more robust immune activation. VPM1002 is currently undergoing clinical trials and represents an evolution of the BCG platform, with the potential to improve protection against TB by stimulating more targeted immune responses [90]. Advancements in mucosal vaccines have also focused on exploiting immune mechanisms within the respiratory tract. Recent studies highlight the role of airway macrophages in mediating mucosal vaccine-induced trained immunity, which strengthens protection during the early stages of *M. tb* infection [91]. Innovative approaches, such as targeting dendritic cells to expedite T-cell activation, illustrate ongoing efforts to overcome bottlenecks in vaccine efficacy by enhancing immunity at the mucosal surfaces where *M. tb* initiates infection [92].

Collectively, the exploration of mucosal vaccines like MTBVAC and VPM1002 is a critical element of global TB control strategies. These candidates aim not only to provide improved protection compared to BCG but also to reduce transmission by effectively targeting pulmonary TB, which represents the most infectious form of the disease. As TB continues to be a leading cause of mortality, these innovations hold promise for significantly advancing global TB control efforts, offering more efficacious vaccines for diverse populations and settings [89,93]. Mucosal immunity, the IL-17 axis, and trained immunity play essential roles in TB defense. Mucosal immunity refers to the initial immune responses at mucosal surfaces, such as the respiratory tract, where *M. tb* primarily infects innate immune cells, especially alveolar macrophages [94]. Alveolar epithelial and endothelial cells, along with macrophages, form the mucosal barrier, which plays a vital role in preventing *M. tb* invasion. However, *M. tb* has evolved strategies to manipulate these cells, creating niches for replication [94]. Understanding these early host-pathogen interactions can guide the development of effective vaccines and interventions [94]. The IL-17 axis, driven by IL-17, a pro-inflammatory cytokine produced mainly by Th17 cells, plays a crucial role in bridging innate and adaptive immunity. IL-17 promotes the recruitment and activation of innate immune cells such as neutrophils [95]. However, in TB, excessive neutrophilia due to an overactive IL-17 response can impair dendritic cell migration and compromise the development of protective Th1 responses, which are essential for TB control during BCG vaccination [73]. Therefore, balanced modulation of the IL-17 response is critical for effective TB immunity and vaccine efficacy [73]. Trained immunity, the enhanced state of innate immune cells that confers broad protection against subsequent infections, also contributes to TB defense [96]. In TB, trained immunity strengthens the host's ability to control and eliminate *M. tb* by reprogramming monocytes and macrophages for heightened responses upon re-exposure to pathogens [97]. BCG vaccination, traditionally associated with adaptive immunity, also induces trained immunity by boosting the production of cytokines such as IL-6, IL-10, and TNF in response to secondary stimuli [98,99].

The interplay between mucosal immunity, the IL-17 axis, and trained immunity highlights a sophisticated and multi-layered host response to *M. tb*. Mucosal immunity provides the frontline defense, IL-17 orchestrates a coordinated innate and adaptive response, and trained immunity offers a form of innate memory that can be harnessed for broad protection. Understanding how these responses interact opens new avenues for vaccine design and therapeutic interventions aimed at enhancing host immunity against TB [96,97]. The BCG vaccine, although widely used, provides inconsistent protection, particularly against pulmonary TB in adults, necessitating the development of more effective alternatives [89,100]. One major challenge is the variable efficacy of BCG, which ranges from 0% to 80% across different populations, possibly due to pre-existing immunity to environmental mycobacteria that interferes with vaccine effectiveness, especially in adults [101,102].

Additionally, the lack of robust immune correlates of protection complicates TB vaccine trials, making it difficult to predict which candidates will effectively prevent TB infection or disease [103,104]. To address these limitations, researchers are exploring strategies such as heterologous prime-boost regimens and cell-based vaccines presenting highly immunogenic antigens like ESAT-6 [101]. Furthermore, integrating innovative trial designs, preclinical studies that mimic clinical conditions, and advanced immunological tools is essential to identify reliable correlates of protection and accelerate vaccine development [89,93].

5. Virulence arsenal of *M. tb*: The multifaceted PE/PPE family

The PE and PPE protein families of *M. tb* are defined by their characteristic Pro-Glu (PE) and Pro-Pro-Glu (PPE) motifs and play crucial roles in pathogenesis and immune modulation in TB. Although their precise functions remain incompletely understood, accumulating evidence highlights their critical involvement in host-pathogen interactions, virulence, and immune evasion. PE and PPE proteins have been increasingly recognized for their ability to modulate host immune responses, thereby enhancing *M. tb*'s survival within host cells. A notable example is PPE18, which has been shown to inhibit MHC class II-mediated antigen presentation in macrophages. This inhibition impairs $CD4^+$ T-cell activation, weakening adaptive immune responses and granting *M. tb* a survival advantage during infection [105–108]. By disrupting antigen presentation pathways, PPE18 exemplifies how these proteins subvert host defenses [109].

Beyond immune modulation, certain PE proteins contribute directly to intracellular survival by remodeling the bacterial cell wall. PE11, for instance, is implicated in modifying the cell wall structure, enhancing *M. tb*'s resistance to environmental stress and antibiotics. Expression of PE11 in *M. smegmatis*, a non-pathogenic mycobacterium, led to increased bacterial loads and exacerbated pathology in infected mice, suggesting its role in virulence [110]. Similarly, PE_PGRS18 enhances intracellular survival by altering macrophage cytokine profiles and reducing apoptosis, further promoting bacterial persistence [111]. PE_PGRS41 adds to this repertoire by interfering with innate immunity. Its expression leads to reduced secretion of key pro-inflammatory cytokines such as TNF-α and IL-6 in infected macrophages, revealing its role in blunting host defenses and facilitating chronic infection [112].

A hallmark of the PE/PPE family is their contribution to antigenic variation, enabling *M. tb* to evade immune detection. The high sequence variability observed in PE/PPE proteins, including extensive mutations, allows *M. tb* to present diverse antigenic profiles, thus avoiding recognition by the host immune system. For example, the PE_PGRS33 gene exhibits considerable sequence diversity across clinical isolates, highlighting its role in immune evasion and persistence [113,114]. PE/PPE proteins such as PE32/PPE65 and PPE63 further modulate host immunity by suppressing pro-inflammatory cytokines, reducing reactive oxygen and nitrogen species, and altering apoptosis pathways, thereby creating a balanced inflammatory environment that promotes *M. tb*'s long-term intracellular survival despite immune challenges [115,116]. Recent structural studies have shed light on the intriguing dynamics of PE/PPE proteins. Complexes such as PE35-PPE68 undergo disorder-to-order transitions upon binding, which may enhance immune evasion by dynamically altering antigenic surfaces [117]. These transitions reflect *M. tb*'s strategy to exploit structural plasticity for host interaction and immune subversion.

Moreover, PE/PPE proteins frequently operate as co-operonic gene pairs, exemplified by PE5-PPE4 and PE25/PPE41. These pairs synergistically modulate host immune responses and contribute to bacillary survival. PE5 and PE15, for instance, alter cytokine production and promote *M. tb* survival within macrophages [118]. The PE25/PPE41 complex induces robust humoral and cell-mediated immune responses, highlighting their potential as vaccine antigens [119]. The unique roles of PE/PPE proteins in immune modulation, antigenic variation, and virulence highlight their potential as targets for novel TB vaccines and therapies. Their ability to modulate host responses offers a promising avenue for HDTs that enhance the efficacy of existing treatments. For example, interfering with PE/PPE-host interactions could disrupt *M. tb*'s immune evasion strategies, thereby improving immune-mediated clearance [120].

Additionally, antigenic variability and immune visibility of certain PPE proteins, such as Rv2608, suggest their utility in diagnostic and vaccine development. Differential immune responses to these proteins across patient populations indicate their potential for targeted immunotherapy and improved vaccine formulations [121]. Importantly, the multifunctionality of PE/PPE proteins call for a nuanced approach to their therapeutic targeting. While these proteins are integral to *M. tb*'s virulence and immune evasion, they also represent promising candidates for vaccine design, capable of eliciting robust and protective immune responses [122,123]. Through diverse mechanisms, including altering cytokine profiles, remodeling cell walls, and presenting variable antigenic surfaces, these proteins enable *M. tb* to persist in the host despite robust immune defenses [124,125].

Recent structural and functional insights have deepened our understanding of these enigmatic proteins, revealing their capacity to transition between disordered and ordered structures, form functional complexes, and modulate immune pathways. These discoveries not only enhance our knowledge of *M. tb* pathogenesis but also open new avenues for therapeutic and vaccine development. By targeting PE/PPE proteins, researchers can potentially disrupt key virulence mechanisms and improve the effectiveness of TB treatment strategies. Furthermore, exploiting their immunogenic properties in vaccine design holds promise for eliciting stronger and more durable protection against TB. Continued research into the structural dynamics, antigenic variation, and immune-modulatory functions of PE/PPE proteins will be crucial for advancing TB control efforts [126,127].

6. Clinical reflections: Reproductive health and pediatric TB

While advances in molecular biology, immunometabolism, diagnostics, and therapeutic innovation have deepened our understanding of TB pathogenesis and control, their true impact ultimately depends on how effectively they address vulnerable populations within real-world health systems. Translating laboratory discoveries into meaningful clinical outcomes requires particular attention to groups in whom biological susceptibility intersects with structural inequities. Among these, maternal and pediatric populations represent a critical translational frontier where gaps in prevention, diagnosis, and treatment carry profound intergenerational consequences. The emergence of neonatal MDR-TB highlights significant challenges in maternal and child health, particularly concerning latent tuberculosis infection (LTBI) in maternal populations. MDR-TB is a pressing public health issue, disproportionately affecting women of childbearing age. Effective management of MDR-TB during pregnancy, while complex due to high resistance levels to first-line anti-tuberculous drugs, is essential for safeguarding both maternal and neonatal health. Evidence indicates that with carefully tailored anti-tuberculous therapy, positive outcomes are achievable, resulting in the delivery of healthy, full-term infants without obstetric complications or perinatal transmission [128,129]. LTBI poses a substantial risk to maternal health, serving as a reservoir for MDR-TB. The conventional use of pyrazinamide and ethambutol for LTBI, particularly among MDR-TB contacts, is associated with hepatotoxicity, underscoring the need for vigilant monitoring and the development of safer treatment options [130]. Current LTBI regimens fall short in managing MDR-TB strains, as evidenced by instances where latent infections progress to active disease in pediatric populations and institutional settings, such as schools, despite prophylactic interventions [131,132].

Addressing MDR-TB in maternal populations requires a public health approach emphasizing early detection, individualized treatment, and leveraging advancements in diagnostics and drug discovery. Community-based treatment models, supported by emerging diagnostic technologies including artificial intelligence-assisted tools, can enhance early and accurate identification of drug-resistant strains, contributing to global TB control efforts [133,134]. In summary, the intertwined challenges of MDR-TB and LTBI in maternal populations necessitate a comprehensive strategy encompassing improved diagnostic methods, safer treatment regimens, and community-based interventions to mitigate transmission and effectively manage latent infections. Pediatric TB diagnosis presents distinct challenges, primarily due to nonspecific clinical manifestations, low bacterial loads, and frequent extrapulmonary involvement, which complicate conventional diagnostic techniques [137]. Currently, TB diagnosis in children often depends on a combination of clinical symptoms, history of contact with TB cases, and radiological findings, frequently without microbiological confirmation, owing to the difficulty of obtaining high-quality specimens like sputum [135,136].

This diagnostic challenge is further compounded by the high prevalence of extrapulmonary TB in children and logistical barriers in specimen collection [137]. Advances such as the Xpert MTB/RIF and Xpert MTB/RIF Ultra assays represent substantial progress, offering increased sensitivity and faster results compared to traditional methods, thereby improving diagnostic accuracy [135,136,137]. The successful implementation of rapid testing platforms like Xpert in some regions has demonstrated feasibility, enabling timely diagnosis and initiation of treatment [136]. Nonetheless, challenges persist, particularly with the rising incidence of pediatric MDR-TB. The introduction of molecular diagnostic tools has improved detection rates of drug-resistant strains, highlighting the complexities of managing these cases [4,128,138]. Unconventional diagnostic approaches, such as employing trained African giant pouched rats, have shown promise in augmenting TB detection rates beyond standard clinical methods [139].

Emerging methods like identifying transcriptional signatures from whole blood samples offer promising non-sputum-based diagnostics, demonstrating high sensitivity and specificity in differentiating TB from non-TB cases in children [140]. Furthermore, initiatives aimed at improving TB case detection through systematic verbal screening and facility-based strategies in rural settings have successfully increased TB case notifications among children, underscoring the value of community-based interventions [141]. Overall, while pediatric TB diagnosis remains challenging, advancements in molecular diagnostics, innovative technologies, and integrated community approaches hold promise for significantly enhancing TB care in children. Continued research and robust implementation of these diagnostic methods are crucial to effectively combat pediatric TB globally.

Integrating TB screening into reproductive healthcare services offers considerable public health benefits, particularly in low- and middle-income countries (LMICs) where infectious diseases intersect with reproductive health challenges. This integration maximizes resource utilization, improves patient outcomes, and streamlines healthcare delivery. Successful examples of service integration include the incorporation of cervical cancer prevention into family planning and HIV-related services in Sub-Saharan Africa, which expanded screening and reduced cervical cancer burden [142]. Similarly, integrating prevention of mother-to-child transmission (PMTCT) services into maternal, newborn, and child health programs has enhanced access to interventions and received positive reception from both clients and providers [143]. Integrating TB screening into reproductive health services can build on these successful frameworks by focusing on strengthening health systems and increasing community engagement. Strategies such as health navigation to strengthen referrals and flexible service delivery models tailored to local contexts can facilitate integration [144].

However, integration faces numerous challenges, including resource limitations, staff shortages, and high client volumes, challenges that have also been observed in cervical cancer screening initiatives [145]. Overcoming these obstacles necessitates strategic planning that includes task shifting to empower non-physician health workers, community mobilization through trusted frontline health workers, and the enhancement of health information systems to support integrated care delivery [146]. By addressing these constraints, health systems can improve their capacity to deliver comprehensive services while maintaining quality and effectiveness.

Integrating TB services into reproductive health settings presents a valuable opportunity to tackle the dual burden of TB and other chronic conditions, such as diabetes. Successful primary care models have demonstrated the potential of this integration to streamline service delivery and improve health outcomes [147]. Additionally, the COVID-19 pandemic highlighted the potential of integrated health services in maintaining continuity of care through the use of digital platforms and community-based interventions, which can be leveraged to support TB screening and treatment in reproductive health settings [148].

Based on these insights, several policy recommendations can be proposed to promote the integration of TB screening into reproductive health services. First, policies mandating the integration of TB screening into reproductive health services should be developed, emphasizing the benefits of comprehensive care and resource optimization. This requires the commitment of policymakers to prioritize integrated approaches that address multiple health concerns concurrently. Second, capacity building through workforce training is essential to equip healthcare providers with the skills and knowledge needed to effectively deliver integrated services. This approach ensures that healthcare staff are prepared to manage the complexities of integrating TB screening with reproductive health services [144,149].

Third, community engagement plays a pivotal role in the success of integrated services. Mobilizing community health workers to raise awareness and facilitate participation in screening programs can enhance acceptance and utilization of services. These trusted members of the community can act as liaisons between the health system and the population, ensuring that integrated services are accessible and culturally acceptable. Fourth, infrastructure enhancement is critical. Strengthening healthcare infrastructure, including the implementation of electronic health records, supports the streamlined delivery of integrated services and enhances data management capabilities, which are essential for monitoring and improving service delivery [150].

Finally, ongoing research and innovation are necessary to refine integration strategies. Research should focus on evaluating the effectiveness, cost-efficiency, and patient satisfaction of integrated TB and reproductive health services. Evidence generated through such research can inform best practices and policy decisions, ensuring that integration efforts are grounded in empirical data and adapted to local contexts. This continuous learning approach will enable health systems to adjust and optimize their integration models over time [151].

In conclusion, integrating TB screening into reproductive health services represents a strategic enhancement of health service delivery, offering the promise of improved health outcomes, efficient use of resources, and a more holistic approach to public health. By addressing the challenges of integration and embracing innovative models, health systems can significantly advance efforts to control both TB and broader public health challenges, ensuring that services are accessible, effective, and equitable [144,149].

7. Molecular epidemiology in action: Genotyping and resistance surveillance

The utility of various diagnostic methods for TB can be evaluated through their sensitivity, specificity, speed, and applicability in different clinical and public health contexts. This comparison examines TB-TEST, whole-genome sequencing (WGS), mycobacterial interspersed repetitive units-variable number tandem repeat analysis (MIRU-VNTR), and Xpert-based methods. The term "TB-TEST" is not a standardized designation but is used here to broadly denote rapid molecular and nucleic acid–based diagnostic assays designed to enhance TB detection accuracy and turnaround time. Modern TB diagnostic approaches increasingly utilize nucleic acid detection, genomics, and multi-omics platforms. For instance, metagenomic next-generation sequencing (mNGS) has demonstrated high specificity in detecting *M. tb* directly from clinical samples, although sensitivity can vary [152]. Rapid diagnostic tests are pivotal in infectious disease surveillance by ensuring faster detection, which is critical for controlling transmission, especially in underserved and remote areas where laboratory infrastructure may be limited [153].

WGS is a powerful tool that sequences the entire genome of *M. tb* , offering comprehensive insights into the genetic variations and resistance profiles of the pathogen. This technology provides detailed data that can guide personalized treatment regimens and support epidemiological investigations. WGS excels in MDR and XDR TB surveillance and outbreak tracing, facilitating precise strain differentiation and transmission pathway identification [154]. However, the technology requires significant infrastructure, expertise, and time, limiting its feasibility for routine diagnostics, particularly in resource-constrained settings. While WGS is invaluable for comprehensive analysis post-pathogen isolation, it is not ideal as a first-line diagnostic tool due to its logistical demands.

MIRU-VNTR typing is widely used for strain differentiation and epidemiological surveillance rather than initial diagnosis. By analyzing the number and distribution of tandem repeats at specific loci in the *M. tb* genome, MIRU-VNTR enables high-resolution differentiation of TB strains, essential for outbreak tracing and understanding transmission dynamics [155]. Although it lacks phenotypic resistance data compared to WGS, MIRU-VNTR remains crucial for public health responses, offering a cost-effective and scalable approach for tracking TB in communities, particularly where multiple strains may circulate. Xpert MTB/RIF and its improved version, Xpert MTB/RIF Ultra, are cartridge-based molecular tests that detect *M. tb* DNA and rifampicin resistance. These assays have transformed TB diagnostics by providing rapid, accurate results with high sensitivity and specificity in sputum samples, delivering results within 2 hours [156,157]. They are particularly valuable in high TB burden and low-resource settings, offering point-of-care testing with minimal training requirements. Xpert tests are especially useful for pulmonary TB diagnosis and initial resistance screening, though their sensitivity is reduced in extrapulmonary TB and TB meningitis due to low bacterial loads [158].

MIRU-VNTR complements WGS by providing practical strain differentiation for targeted control measures, while Xpert-based assays enable rapid diagnosis and early rifampicin resistance detection, facilitating prompt treatment initiation and transmission control, particularly in resource-limited settings. Emerging technologies such as metagenomic next-generation sequencing (mNGS) and advanced nucleic acid–based platforms further expand diagnostic potential by improving detection across diverse sample types, including extrapulmonary TB. Although WGS offers unparalleled genomic resolution for resistance profiling and transmission mapping, its resource-intensive nature limits routine clinical deployment, whereas MIRU-VNTR lacks phenotypic resistance data and Xpert may have reduced sensitivity in certain clinical scenarios. Consequently, an integrated diagnostic framework that combines rapid frontline testing, epidemiological strain differentiation, and high-resolution genomic surveillance is essential for comprehensive TB control. Continued innovation, including portable sequencing platforms and AI-assisted genomic interpretation, will be critical to enhancing scalability, improving resistance prediction, and strengthening global surveillance efforts [159].

The effective diagnosis and surveillance of TB require a multifaceted approach that integrates diverse diagnostic modalities, each contributing unique strengths to the overall strategy. TB-TEST approaches, though not standardized under this name, represent the ongoing evolution of rapid diagnostics. WGS provides comprehensive genetic insights, essential for personalized treatment and outbreak tracing. MIRU-VNTR remains a valuable epidemiological tool, while Xpert-based tests offer practical, rapid solutions for routine TB care. By combining these technologies, health systems can enhance early detection, initiate appropriate treatment, monitor drug resistance, and implement targeted public health interventions. This integrated approach is vital for controlling TB, particularly MDR and XDR forms, and for responding effectively to outbreaks. Continued investment in diagnostic innovation, infrastructure, and capacity building will be key to achieving global TB control goals.

8. Bridging the gap: From molecular discovery to public health impact

Implementation science plays a crucial role in addressing translational bottlenecks and promoting health equity in tuberculosis (TB) research. Dissemination and implementation (D&I) science examines how to effectively translate research into real-world contexts. A growing focus on health equity within D&I ensures that evidence based interventions are implemented equitably across diverse populations and settings [160]. The intersection of health equity and D&I highlights how implementation science can help sustain evidence-based programs that address health disparities [160]. In TB research, translational bottlenecks often hinder the integration of scientific findings into public health practice. For example, the incorporation of genomic data into routine TB management has been slow, presenting significant challenges. Implementation science addresses these obstacles by identifying opportunities and best practices for integrating scientific advances into clinical and public health systems [161]. Overcoming these barriers requires a strategic, collaborative approach involving public health agencies, researchers, and community stakeholders, collectively working to improve health outcomes and reduce inequities [161].

Community-based participatory research (CBPR) plays a key role in bridging science and practice. CBPR fosters community engagement and social action to enhance health equity, incorporating community knowledge into research to address power imbalances and promote mutual benefits for both communities and academics [162]. This approach effectively addresses barriers within intervention and implementation science by promoting the development, implementation, and dissemination of interventions tailored to diverse communities [162]. Systemic marginalization further exacerbates translational bottlenecks, contributing to health disparities. Applying implementation science with a health equity focus provides frameworks and strategies to address these disparities within healthcare and public health systems [163]. Implementation science concepts aim to promote health equity by ensuring that effective interventions reach diverse populations and settings [163].

Inclusive partnerships are essential for developing and sustaining interventions that advance health equity. Catalytic frameworks fostering communication, collaboration, and shared goals across sectors can drive innovative solutions to eliminate health disparities [164]. In TB research, such frameworks optimize research quality and efficiency, improve patient care, and catalyze systems change to address the root causes of disparities [164]. Participatory approaches to implementation science offer a collaborative perspective on integrating evidence-based interventions to advance health equity. These approaches enable the incorporation of practice-, community-, and research-based evidence into public health services, moving beyond simply adapting interventions to transforming systems to prioritize equity [165].

By focusing on community engagement, equity-driven frameworks, and participatory approaches, TB research can move toward more effective and equitable health outcomes. The successful translation of treatments from laboratory to clinic in TB management, exemplified by the BPaL regimen, showcases the potential of modern therapeutics for multidrug-resistant TB (MDR-TB). The BPaL regimen, comprising bedaquiline, pretomanid, and linezolid, marks a shift toward shorter, more effective regimens. A case study from Belarus and Uzbekistan highlights BPaL's real-world success. Involving 677 patients with MDR/rifampicin-resistant TB (MDR/RR-TB) and pre-extensively drug-resistant TB (pre-XDR-TB), BPaL achieved successful outcomes in 95.3% of MDR/RR-TB patients and 90.4% of pre-XDR-TB patients. Although 10.2% experienced serious adverse events, the majority were unrelated to TB treatment. Importantly, the recurrence rate post-treatment was low, underscoring the regimen's effectiveness [166].

The BPaL regimen exemplifies the Global Alliance for TB Drug Development (TB Alliance)'s role in advancing novel treatments. The TB Alliance has driven the development of regimens for drug-susceptible, MDR, and extensively drug-resistant TB, while also introducing adaptive clinical trial designs to evaluate efficacy across patient populations efficiently [167]. BPaL and its variant BPaLM (with moxifloxacin) are not only effective but also cost saving. In countries like Pakistan, the Philippines, South Africa, and Ukraine, BPaL/BPaLM has led to healthcare savings and improved patient outcomes. Cost savings per patient ranged from $478 in the Philippines to $2636 in Ukraine, with notable reductions in disability-adjusted life years (DALYs) and lives saved annually, highlighting the regimens' public health impact [168].

Preclinical studies, including mouse models, have further demonstrated BPaL's superior efficacy and potential for faster cures compared to traditional TB treatments. These studies emphasize pretomanid's role in enhancing bactericidal activity and reducing relapse rates, reinforcing clinical findings of BPaL's effectiveness [169]. Overall, BPaL's transition from research to clinical practice illustrates the successful application of implementation science principles in TB treatment. This case highlights the potential for rapid, effective treatment strategies to transform TB management and offers a hopeful outlook for controlling drug-resistant TB.

Collaborative models for TB innovation, as exemplified by organizations like CRDF Global and the TB Alliance, underscore the critical role of partnerships and innovation in advancing drug development, diagnostics, and treatment strategies. These collaborative models illustrate how diverse sectors and stakeholders can work together to address the complex challenges of TB control and elimination. One effective model is the collaboration between government agencies and non-governmental organizations (NGOs) in Bangladesh. This partnership effectively combined resources, expertise, and infrastructure from both sectors to improve access to and the quality of TB care. The success of this integrated public-private healthcare approach demonstrates the potential for similar models in other settings where resource constraints and fragmented healthcare systems limit TB control efforts [151].

Another promising model is the creation of public-private linkages, as demonstrated by a pilot project in the Kathmandu Valley, Nepal. This project highlighted the feasibility of forming service linkages between public and private healthcare sectors to enhance TB control. Despite challenges in aligning capacity and motivation between sectors, partnerships with international research institutions and NGOs were pivotal in bridging gaps and facilitating collaboration between providers [170]. Regional and international networks have also played a vital role in strengthening TB innovation. Initiatives such as the West African Network for TB, AIDS, and Malaria (WANETAM) and the West African Health Organization (WAHO) have significantly contributed to enhancing diagnostic capacity, optimizing treatment regimens, and promoting infection control. These efforts demonstrate the importance of regional collaboration in fostering resource mobilization, capacity building, and knowledge sharing across borders to strengthen TB control strategies [171].

In high-burden countries like India, multi-stakeholder partnerships have been instrumental in advancing TB elimination efforts. Collaborations involving national health authorities, NGOs, and the research community have supported the development and implementation of new tools and strategies. These partnerships emphasize locally driven solutions, continuous program monitoring, and flexible policies and practices that accommodate innovations while fostering public private collaboration [172]. Innovation in drug development is another area where collaborative models have demonstrated their value. The TB Alliance's collaborative clinical trials for new regimens highlight the importance of international partnerships in developing patient-centered treatments for drug-susceptible TB, MDR-TB, and XDR-TB. These partnerships have not only accelerated the development of new medications but also supported adaptive clinical trial designs that allow for the rapid evaluation of treatment efficacy across diverse patient populations [173].

Integrating big data analytics into global partnerships represents an innovative approach to strengthening TB care. By leveraging data-driven strategies combined with traditional organizational approaches, these partnerships enhance relationship-building and foster new solutions to complex challenges. Big data analytics support the identification of transmission patterns, prediction of outbreaks, and optimization of treatment strategies, underscoring the value of integrating advanced technologies into TB control efforts [174]. Together, these collaborative models illustrate the potential of partnerships that combine traditional and modern approaches to enhance TB control and treatment. They emphasize the importance of inclusivity, innovation, and strategic cooperation across sectors and regions. By embracing such models, the global health community can effectively address the multifaceted challenges of TB, paving the way for sustained progress in TB elimination efforts.

9. Future directions and conclusion: Toward TB elimination

The control of tuberculosis (TB) requires a comprehensive approach encompassing immediate and long-term priorities to effectively manage and eventually eradicate the disease. Short-term priorities include the advancement of diagnostic tools and treatments, particularly for drug-resistant TB and TB co-infected with HIV. Rapid and accurate diagnostics, along with effective treatment regimens, are essential to combat multidrug-resistant and extensively drug-resistant TB (MDR/XDR-TB) [175,176]. Equally important are digital adherence technologies (DATs), such as phone-based applications and digital pillboxes, which offer patient-centered solutions to monitor and enhance medication adherence, thereby improving patient outcomes [177]. Addressing pediatric TB, often neglected, requires research efforts to develop diagnostic and treatment strategies tailored for children [178]. Furthermore, as diabetes increases TB risk, research into integrated management of TB-diabetes co-morbidity is essential to improve outcomes for affected individuals [179].

Long-term priorities focus on vaccine development, where sustained investment is needed to deepen our understanding of the molecular mechanisms underpinning host-pathogen interactions, which could yield more effective vaccines [180]. Integrating TB management with care for other diseases like HIV and diabetes is crucial for health system efficiency and reducing the global TB burden [147]. Continued global political and financial commitment is necessary to maintain progress, with increased investment in research, treatment, and eradication programs critical for achieving TB elimination goals [147,181]. In addition, addressing the social determinants of TB, such as poverty and limited healthcare access, will be crucial. Reducing socio-economic barriers is a key step toward long-term TB control [147].

An interdisciplinary approach integrating microbiology, immunology, data science, and policy is vital for TB control. Microbiology provides the foundation for understanding *M. tb* , the bacterium responsible for TB. Advances in microbial ecology and genomic studies have revolutionized diagnostics, vaccine development, and therapeutic interventions. Sequencing the *M. tb* genome has provided essential insights that inform new intervention strategies [182]. Immunology enriches our understanding of immune responses against TB, with cellular and humoral immunity playing critical roles in disease progression and control. Recent discoveries about B cells highlight their significance in shaping host defense, guiding the design of effective vaccines and therapies [183]. Immunological insights also inform personalized diagnostics and treatment strategies, especially for pediatric TB [184].

Data science, including artificial intelligence (AI) and machine learning, is transforming TB management by improving diagnostics, predicting outbreaks, and optimizing treatment protocols. AI applications enhance clinical diagnostics and public health efforts through better pathogen detection and outbreak forecasting [185]. Systems immunology integrates data from clinical, bacterial, and immunological sources to identify biomarkers critical for TB diagnosis and treatment monitoring [186]. Policy plays a crucial role in translating scientific advances into practice by addressing economic, political, and behavioral barriers to TB control. Effective policies are needed to ensure access to diagnostics and treatments, promote vaccination programs, and manage drug-resistant strains [151]. Furthermore, policies should encourage interdisciplinary collaboration and invest in research infrastructure to support continuous innovation [187].

The World Health Organization's (WHO) End TB Strategy envisions a world free of TB by 2035, marked by zero deaths, disease, and suffering. Achieving this ambitious goal requires innovation in prevention, diagnosis, and treatment, combined with robust policy frameworks and international cooperation [188]. Expanding TB preventive treatments (TPT) is essential, especially in high-burden countries like India, where implementation remains limited. Sustainable and context-specific interventions, alongside investment in new diagnostic tools and effective vaccines, are key to reducing TB incidence [189]. Political will and alignment of national and international policies are critical for implementing WHO strategies. Collaboration among governments, NGOs, and industry partners is necessary to scale up evidence-based TB diagnosis, treatment, and prevention [190].

A significant challenge in TB elimination is the persistence of MDR-TB, often linked to HIV coinfection, which threatens to reverse progress. Addressing MDR-TB requires comprehensive strategies including increased research, updated therapeutic regimens, and strengthened global and national responses [14]. The COVID-19 pandemic further disrupted TB control efforts, straining healthcare systems and exacerbating existing inequalities. Rebuilding and strengthening healthcare infrastructure, especially in low- and middle-income countries, is crucial to ensure equitable TB care [14]. Sustained research and innovation, particularly in primary healthcare sectors, are fundamental for progress [14].

In conclusion, ending TB by 2035 requires a multifaceted approach that combines immediate action and long-term strategies, integrating scientific innovation, policy support, and collaborative global efforts. By addressing diagnostic and treatment challenges, advancing research and vaccine development, incorporating interdisciplinary approaches, and tackling social and healthcare inequities, the global health community can make meaningful strides toward eradicating TB. Through concerted efforts and strategic partnerships, the vision of a TB-free world is within reach.

Acknowledgments

MS and FA gratefully acknowledge GITAM University Hyderabad for providing internal funding through Seed and Murti grants.

References

[1] Cioboata R, Balteanu MA, Osman A, Vlasceanu SG, Zlatian OM, Mitroi DM, Catana OM, Socaci A, Tieranu EN. Coinfections in tuberculosis in low- and middle-income countries: epidemiology, clinical implications, diagnostic challenges, and management strategies—a narrative review. J Clin Med 2025;14 (7). https://doi.org/10.3390/jcm14072154.

[2] Yang H, Ruan X, Li W, Xiong J, Zheng Y. Global, regional, and national burden of tuberculosis and attributable risk factors for 204 countries and territories, 1990–2021: a systematic analysis for the Global Burden of Diseases 2021 study. BMC Public Health 2024;24(1). https://doi.org/10.1186/s12889-024-20664-w.

[3] Villa S, Riccardi N, Canetti D, Alagna R, Castellotti P, Ferrarese M, Cirillo D, Barberis I, Luigi Bragazzi N, Gazzaniga V, Ricucci V, Ruffo Codecasa L, Besozzi G, Martini M. From the past, a long way to future challenges for a greater control of tuberculosis. Tuberculosis 2020;123:101948. https://doi.org/10.1016/j.tube.2020.101948.

[4] Liebenberg D, Gordhan BG, Kana BD. Drug resistant tuberculosis: implications for transmission, diagnosis, and disease management. Front Cell Infect Microbiol 2022;12. https://doi.org/10.3389/fcimb.2022.943545.
[5] Sultana ZZ, Hoque FU, Beyene J, Akhlak-Ul-Islam M, Khan MHR, Ahmed S, Hawlader DH, Hossain A. HIV infection and multidrug resistant tuberculosis: a systematic review and meta-analysis. BMC Infect Dis 2021;21(1). https://doi.org/10.1186/s12879-020-05749-2.
[6] Alsayed SSR, Gunosewoyo H. Tuberculosis: pathogenesis, current treatment regimens and new drug targets. Int J Mol Sci 2023;24(6). https://doi.org/10.3390/ijms24065202.
[7] Pai M, Behr MA, Dowdy D, Dheda K, Divangahi M, Boehme CC, Ginsberg A, Swaminathan S, Spigelman M, Getahun H, Menzies D, Raviglione M. Tuberculosis. Nat Rev Dis Primers 2016;2. https://doi.org/10.1038/nrdp.2016.76.
[8] Sharma K, Ahmed F, Sharma T, Grover A, Agarwal M, Grover S. Potential repurposed drug candidates for tuberculosis treatment: progress and update of drugs identified in over a decade. ACS Omega 2023;8(20):17362–80. https://doi.org/10.1021/acsomega.2c05511.
[9] Bhatia V, Rijal S, Sharma M, Islam A, Vassall A, Bhargava A, Thida A, Basri C, Onozaki I, Pai M, Rezwan Md K, Arinaminpathy N, Chandrashekhar P, Sarin R, Mandal S, Raviglione M. Ending TB in South-East Asia: flagship priority and response transformation. Lancet Reg Health Southeast Asia 2023;18:100301. https://doi.org/10.1016/j.lansea.2023.100301.
[10] Quissell K, Walt G. The challenge of sustaining effectiveness over time: the case of the global network to stop tuberculosis. Health Pol Plann 2016;31(Suppl. 1):i17. https://doi.org/10.1093/heapol/czv035.
[11] Sotgiu G, Saderi L, Petruccioli E, Aliberti S, Piana A, Petrone L, Goletti D. QuantiFERON TB Gold Plus for the diagnosis of tuberculosis: a systematic review and meta-analysis. J Infect 2019;79(5):444–53. https://doi.org/10.1016/j.jinf.2019.08.018.
[12] Zhuang L, Ye Z, Li L, Yang L, Gong W. Next-generation TB vaccines: progress, challenges, and prospects. Vaccines 2023;11(8):1304. https://doi.org/10.3390/vaccines11081304.
[13] Barteka G, Bwayo D, Matovu JKB, Wanume B, Alunyo JP, Sseguya R, Masaba JP, Obbo JS. Treatment outcomes and predictors of success for multidrug resistant tuberculosis MDR TB in Ugandan regional referral hospitals. Sci Rep 2025;15(1). https://doi.org/10.1038/s41598-025-97027-x.
[14] Omoteso OA, Fadaka AO, Walker RB, Khamanga SM. Innovative strategies for combating multidrug-resistant tuberculosis: advances in drug delivery systems and treatment. Microorganisms 2025;13(4). https://doi.org/10.3390/microorganisms13040722.
[15] Vishwakarma D, Gaidhane A, Sahu S, Rathod AS. Multi-drug resistance tuberculosis (MDR-TB) challenges in India: a review. Cureus 2023;15(12). https://doi.org/10.7759/cureus.50222.
[16] Dartois VA, Rubin EJ. Anti-tuberculosis treatment strategies and drug development: challenges and priorities. Nat Rev Microbiol 2022;20(11):685–701. https://doi.org/10.1038/s41579-022-00731-y.
[17] Karnan A, Jadhav U, Ghewade B, Ledwani A, Shivashankar P. A comprehensive review on long vs. Short regimens in multidrug-resistant tuberculosis (MDR-TB) under programmatic management of drug-resistant tuberculosis (PMDT). Cureus 2024;16(1). https://doi.org/10.7759/cureus.52706.
[18] Ding YE, Wong MTJ, Norazmi MN, Balakrishnan V, Gee JT. Advancement in diagnostic approaches for latent tuberculosis: distinguishing recent from remote infections. One Health Outlook 2025;7(1). https://doi.org/10.1186/s42522-025-00144-w.
[19] Harries AD, Kumar AMV, Satyanarayana S, Thekkur P, Lin Y, Dlodlo RA, Khogali M, Zachariah R. The growing importance of tuberculosis preventive therapy and how research and innovation can enhance its implementation on the ground. Trop Med Infect Dis 2020;5(2). https://doi.org/10.3390/tropicalmed5020061.
[20] Carranza C, Pedraza-Sanchez S, de Oyarzabal-Mendez E, Torres M. Diagnosis for latent tuberculosis infection: new alternatives. Front Immunol 2020;11. https://doi.org/10.3389/fimmu.2020.02006.

[21] Gualano G, Mencarini P, Lauria FN, Palmieri F, Mfinanga S, Mwaba P, Chakaya J, Zumla A, Ippolito G. Tuberculin skin test—outdated or still useful for Latent TB infection screening? Int J Infect Dis 2019;80: S20. https://doi.org/10.1016/j.ijid.2019.01.048.

[22] Burton MJ, Ramke J, Marques AP, Bourne RRA, Congdon N, Jones I, Ah Tong BAM, Arunga S, Bachani D, Bascaran C, Bastawrous A, Blanchet K, Braithwaite T, Buchan JC, Cairns J, Cama A, Chagunda M, Chuluunkhuu C, Cooper A, Crofts-Lawrence J, Dean WH, Denniston AK, Ehrlich JR, Emerson PM, Evans JR, Frick KD, Friedman DS, Furtado JM, Gichangi MM, Gichuhi S, Gilbert SS, Gurung R, Habtamu E, Holland P, Jonas JB, Keane PA, Keay L, Khanna RC, Khaw PT, Kuper H, Kyari F, Lansingh VC, Mactaggart I, Mafwiri MM, Mathenge W, McCormick I, Morjaria P, Mowatt L, Muirhead D, Murthy GVS, Mwangi N, Patel DB, Peto T, Qureshi BM, Salomão SR, Sarah V, Shilio BR, Solomon AW, Swenor BK, Taylor HR, Wang N, Webson A, West SK, Wong TY, Wormald R, Yasmin S, Yusufu M, Silva JC, Resnikoff S, Ravilla T, Gilbert CE, Foster A, Faal HB. The lancet global health commission on global eye health: vision beyond 2020. Lancet Global Health 2021;9(4):e489. https://doi.org/10.1016/S2214-109X(20)30488-5.

[23] Shah HD, Nazli Khatib M, Syed ZQ, Gaidhane AM, Yasobant S, Narkhede K, Bhavsar P, Patel J, Sinha A, Puwar T, Saha S, Saxena D. Gaps and interventions across the diagnostic care cascade of TB patients at the level of patient, community and health system: a qualitative review of the literature. Trop Med Infect Dis 2022;7(7). https://doi.org/10.3390/tropicalmed7070136.

[24] Naqvi N, Ahuja Y, Zarin S, Alam A, Ali W, Shariq M, Hasnain SE, Ehtesham NZ. BCG's role in strengthening immune responses: implications for tuberculosis and comorbid diseases. Infect Genet Evol 2025;127:105703. https://doi.org/10.1016/j.meegid.2024.105703.

[25] Setiabudiawan TP, Reurink RK, Hill PC, Netea MG, van Crevel R, Koeken VACM. Protection against tuberculosis by Bacillus Calmette-Guérin (BCG) vaccination: a historical perspective. Med 2022;3(1): 6–24. https://doi.org/10.1016/j.medj.2021.11.006.

[26] Maani AAl, Petersen E, Memish ZA. The critical role of new tuberculosis vaccines in achieving the WHO 2035 End TB target. IJID Regions 2025;14(Suppl. 2):100595. https://doi.org/10.1016/j.ijregi.2025.100595.

[27] Schrager LK, Harris RC, Vekemans J. Research and development of new tuberculosis vaccines: a review. F1000Research 2018,7:1732. https://doi.org/10.12688/f1000research.16521.1.

[28] Singh UB, Rade K, Rao R, Kumar N, Kumar Mattoo S, Nair S, Zumla A, Sahu S. Lessons and updates from India's National Tuberculosis Elimination Program—bold decisions and innovative ways of fast-tracking progress toward ending tuberculosis. IJID Regions 2025;14(Suppl. 2):100599. https://doi.org/10.1016/j.ijregi.2025.100599.

[29] Kruk ME, Gage AD, Arsenault C, Jordan K, Leslie HH, Roder-DeWan S, Adeyi O, Barker P, Daelmans B, Doubova SV, English M, García-Elorrio E, Guanais F, Gureje O, Hirschhorn LR, Jiang L, Kelley E, Lemango ET, Liljestrand J, Malata A, Marchant T, Matsoso MP, Meara JG, Mohanan M, Ndiaye Y, Norheim OF, Reddy KS, Rowe AK, Salomon JA, Thapa G, Twum-Danso NAY, Pate M. High-quality health systems in the Sustainable Development Goals era: time for a revolution. Lancet Global Health 2018;6(11): e1196. https://doi.org/10.1016/s2214-109x(18)30386-3.

[30] Ferreira MRL, Bonfim RO, Bossonario PA, Maurin VP, Valença ABM, Abreu PDD, Andrade RLDP, Fronteira I, Monroe AA. Social protection as a right of people affected by tuberculosis: a scoping review and conceptual framework. Infect Dis Poverty 2023;12(1). https://doi.org/10.1186/s40249-023-01157-1.

[31] Singh S, Zahiruddin QS, Lakhanpal S, Ballal S, Kumar S, Bhat M, Sharma S, Kumar MR, Dhandh YK, Rustagi S, Alissa M, Halwani MA, Garout M, Alrasheed HA, Al-Subaie MF, Kaabi NA Al, Rabaan AA, Sah S, Shabil M, Khatib MN, Satapathy P. Wealth-based inequalities in tuberculosis prevalence among households having children and young adults in India: insights from Indian demographic and health surveys (2015–2021). BMC Infect Dis 2025;25(1). https://doi.org/10.1186/s12879-024-10301-7.

[32] Bloom BR, Atun R, Cohen T, Dye C, Fraser H, Gomez GB, et al. Tuberculosis. In: Holmes KK, Bertozzi S, Bloom BR, Jha P, editors. Major infectious diseases. 3rd ed. 2017. Washington (DC).

[33] Suvvari TK. The persistent threat of tuberculosis—Why ending TB remains elusive? J Clin Tuberc Other Mycobact Dis 2025;38:100510. https://doi.org/10.1016/j.jctube.2025.100510.

[34] He B, Tan K. Understanding transcriptional regulatory networks using computational models. Curr Opin Genet Dev 2016;37:101–8. https://doi.org/10.1016/j.gde.2016.02.002.

[35] Saint-Andre V. Computational biology approaches for mapping transcriptional regulatory networks. Comput Struct Biotechnol J 2021;19:4884–95.

[36] Sachan RSK, Mistry V, Dholaria M, Rana A, Devgon I, Ali I, Iqbal J, Eldin SM, Mohammad Said Al-Tawaha AR, Bawazeer S, Dutta J, Karnwal A. Overcoming *Mycobacterium tuberculosis* drug resistance: novel medications and repositioning strategies. ACS Omega 2023;8(36):32244–57. https://doi.org/10.1021/acsomega.3c02563.

[37] Yang J, Zhang L, Qiao W, Luo Y. *Mycobacterium tuberculosis*: pathogenesis and therapeutic targets. MedComm 2023;4(5). https://doi.org/10.1002/mco2.353.

[38] Yan B, Guan D, Wang C, Wang J, He B, Qin J, Boheler KR, Lu A, Zhang G, Zhu H. An integrative method to decode regulatory logics in gene transcription. Nat Commun 2017;8(1). https://doi.org/10.1038/s41467-017-01193-0.

[39] Luo Z. Identifying the "stripe" transcription factors and cooperative binding related to DNA methylation. Commun Biol 2024;7(1).

[40] Mundade R, Ozer HG, Wei H, Prabhu L, Lu T. Role of ChIP-seq in the discovery of transcription factor binding sites, differential gene regulation mechanism, epigenetic marks and beyond. Cell Cycle 2014;13(18):2847–52. https://doi.org/10.4161/15384101.2014.949201.

[41] Salma M, Andrieu-Soler C, Deleuze V, Soler E. High-throughput methods for the analysis of transcription factors and chromatin modifications: low input, single cell and spatial genomic technologies. Blood Cell Mol Dis 2023;101:102745. https://doi.org/10.1016/j.bcmd.2023.102745.

[42] Galagan J, Lyubetskaya A, Gomes A. ChIP-seq and the complexity of bacterial transcriptional regulation. Curr Top Microbiol Immunol 2012;363:43–68. https://doi.org/10.1007/82-2012-257.

[43] Peterson EJR, Abidi AA, Arrieta-Ortiz ML, Aguilar B, Yurkovich JT, Kaur A, Pan M, Srinivas V, Shmulevich I, Baliga NS. Intricate genetic programs controlling dormancy in *Mycobacterium tuberculosis*. Cell Rep 2021;35(11):109287. https://doi.org/10.1016/j.celrep.2021.109287.

[44] Galagan JE, Minch K, Peterson M, Lyubetskaya A, Azizi E, Sweet L, Gomes A, Rustad T, Dolganov G, Glotova I, Abeel T, Mahwinney C, Kennedy AD, Allard R, Brabant W, Krueger A, Jaini S, Honda B, Yu WH, Hickey MJ, Zucker J, Garay C, Weiner B, Sisk P, Stolte C, Winkler JK, Van De Peer Y, Iazzetti P, Camacho D, Dreyfuss J, Liu Y, Dorhoi A, Mollenkopf HJ, Drogaris P, Lamontagne J, Zhou Y, Piquenot J, Park ST, Raman S, Kaufmann SHE, Mohney RP, Chelsky D, Branch Moody D, Sherman DR, Schoolnik GK. The *Mycobacterium tuberculosis* regulatory network and hypoxia. Nature 2013;499(7457):178–83. https://doi.org/10.1038/nature12337.

[45] Yabaji SM, Rukhlenko OS, Chatterjee S, Bhattacharya B, Wood E, Kasaikina M, Kholodenko BN, Gimelbrant AA, Kramnik I. Cell state transition analysis identifies interventions that improve control of *Mycobacterium tuberculosis* infection by susceptible macrophages. Sci Adv 2023;9(39). https://doi.org/10.1126/sciadv.adh4119.

[46] Kholodenko BN, Kolch W, Rukhlenko OS. Reversing pathological cell states: the road less travelled can extend the therapeutic horizon. Trends Cell Biol 2023;33(11):913–23. https://doi.org/10.1016/j.tcb.2023.04.004.

[47] Gorkin DU, Lee D, Reed X, Fletez-Brant C, Bessling SL, Loftus SK, Beer MA, Pavan WJ, McCallion AS. Integration of ChIP-seq and machine learning reveals enhancers and a predictive regulatory sequence vocabulary in melanocytes. Genome Res 2012;22(11):2290–301. https://doi.org/10.1101/gr.139360.112.

[48] Wang Y, Jiang Z, Liang P, Liu Z, Cai H, Sun Q. TB-DROP: deep learning-based drug resistance prediction of *Mycobacterium tuberculosis* utilizing whole genome mutations. BMC Genom 2024;25(1). https://doi.org/10.1186/s12864-024-10066-y.

[49] Bustad E, Petry E, Gu O, Griebel BT, Rustad TR, Sherman DR, Yang JH, Ma S. Predicting bacterial fitness in *Mycobacterium tuberculosis* with transcriptional regulatory network-informed interpretable machine learning. United States: bioRxiv; 2024. https://doi.org/10.1101/2024.09.23.614645.

[50] Girardin RC, Bai G, He J, Sui H, McDonough KA. AbmR (Rv1265) is a novel transcription factor of *Mycobacterium tuberculosis* that regulates host cell association and expression of the non-coding small RNA Mcr11. Mol Microbiol 2018;110(5):811–30. https://doi.org/10.1111/mmi.14126.

[51] Al-Tohamy A, Grove A. Targeting bacterial transcription factors for infection control: opportunities and challenges. Transcription 2025;16(1):141–68. https://doi.org/10.1080/21541264.2023.2293523.

[52] Liu C, Omilusik K, Toma C, Kurd NS, Chang JT, Goldrath AW, Wang W, Przytycka TM. Systems-level identification of key transcription factors in immune cell specification. PLoS Comput Biol 2022;18(9): e1010116. https://doi.org/10.1371/journal.pcbi.1010116.

[53] Zhu C, Liu Y, Hu L, Yang M, He Z-G. Molecular mechanism of the synergistic activity of ethambutol and isoniazid against *Mycobacterium tuberculosis*. J Biol Chem 2018;293(43):16741–50. https://doi.org/10.1074/jbc.ra118.002693.

[54] Lane TR, Urbina F, Rank L, Gerlach J, Riabova O, Lepioshkin A, Kazakova E, Vocat A, Tkachenko V, Cole S, Makarov V, Ekins S. Machine learning models for *Mycobacterium tuberculosis* in vitro activity: prediction and target visualization. Mol Pharm 2022;19(2):674–89. https://doi.org/10.1021/acs.molpharmaceut.1c00791.

[55] Periwal V, Kishtapuram S, Scaria V. Computational models for in-vitro anti-tubercular activity of molecules based on high-throughput chemical biology screening datasets. BMC Pharmacol 2012;12(1). https://doi.org/10.1186/1471-2210-12-1.

[56] Teneva Y, Simeonova R, Valcheva V, Angelova VT. Recent advances in anti-tuberculosis drug discovery based on hydrazide–hydrazone and thiadiazole derivatives targeting InhA. Pharmaceuticals 2023;16(4). https://doi.org/10.3390/ph16040484.

[57] Zhong G, Chang X, Xie W, Zhou X. Targeted protein degradation: advances in drug discovery and clinical practice. Signal Transduct Targeted Ther 2024;9(1). https://doi.org/10.1038/s41392-024-02004-x.

[58] Dookie N, Rambaran S, Padayatchi N, Mahomed S, Naidoo K. Evolution of drug resistance in *Mycobacterium tuberculosis*: a review on the molecular determinants of resistance and implications for personalized care. J Antimicrob Chemother 2018;73(5):1138–51. https://doi.org/10.1093/jac/dkx506.

[59] Marambire ET, Calderwood CJ, Larsson L, Held K, Khan P, Banze D, Nhamuave C, Minja LT, Mfinanga A, Gupta RK, Khosa C, Mutsvangwa J, Heinrich N, Kranzer K, Sarita Shah N. Prediction models for Mtb infection among adolescent and adult household contacts in high tuberculosis incidence settings. PLOS Glob Publ Health 2025;5(3):e0004340. https://doi.org/10.1371/journal.pgph.0004340.

[60] Gröschel MI, Owens M, Freschi L, Vargas R, Marin MG, Phelan J, Iqbal Z, Dixit A, Farhat MR. GenTB: a user-friendly genome-based predictor for tuberculosis resistance powered by machine learning. Genome Med 2021;13(1). https://doi.org/10.1186/s13073-021-00953-4.

[61] Al Kuwaiti A, Nazer K, Al-Reedy A, Al-Shehri S, Al-Muhanna A, Subbarayalu AV, Al Muhanna D, Al-Muhanna FA. A review of the role of artificial intelligence in healthcare. J Personalized Med 2023;13(6). https://doi.org/10.3390/jpm13060951.

[62] Sui Q, Li G, Peng Y, Zhang J, Zhang Y, Zhao R. Scalable and robust machine learning framework for HIV classification using clinical and laboratory data. Sci Rep 2025;15(1). https://doi.org/10.1038/s41598-025-00085-4.

[63] Hwang H, Jeon H, Yeo N, Baek D. Big data and deep learning for RNA biology. Exp Mol Med 2024;56(6): 1293–321. https://doi.org/10.1038/s12276-024-01243-w.

[64] Sperber NR, Carpenter JS, Cavallari LH, Damschroder LJ, Cooper-DeHoff RM, Denny JC, Ginsburg GS, Guan Y, Horowitz CR, Levy KD, Levy MA, Madden EB, Matheny ME, Pollin TI, Pratt VM, Rosenman M, Voils CI, Weitzel KW, Wilke RA, Ryanne Wu R, Orlando LA. Challenges and strategies for implementing

genomic services in diverse settings: experiences from the Implementing GeNomics in pracTicE (IGNITE) network. BMC Med Genom 2017;10(1). https://doi.org/10.1186/s12920-017-0273-2.
[65] Bajwa J, Munir U, Nori A, Williams B. Artificial intelligence in healthcare: transforming the practice of medicine. Future Healthc J 2021;8(2):e188. https://doi.org/10.7861/fhj.2021-0095.
[66] Khosravi M, Zare Z, Mojtabaeian SM, Izadi R. Artificial intelligence and decision-making in healthcare: a thematic analysis of a systematic review of reviews. Health Serv Res Manag Epidemiol 2024;11. https://doi.org/10.1177/23333928241234863.
[67] Sadr H, Nazari M, Khodaverdian Z, Farzan R, Yousefzadeh-Chabok S, Taghi Ashoobi M, Hemmati H, Hendi A, Ali A, Pedram MM, Hasannejad-Bibalan M, Yamaghani MR. Unveiling the potential of artificial intelligence in revolutionizing disease diagnosis and prediction: a comprehensive review of machine learning and deep learning approaches. Eur J Med Res 2025;30(1). https://doi.org/10.1186/s40001-025-02680-7.
[68] McClean M, Constantin Panciu T, Lange C, Duarte R, Theis F. Artificial intelligence in tuberculosis: a new ally in disease control. Breathe 2024;20(3):240056. https://doi.org/10.1183/20734735.0056-2024.
[69] Wang Z, Guo Z, Wang W, Zhang Q, Song S, Yuan X, Zhang Z, Wang J. Prediction of tuberculosis treatment outcomes using biochemical makers with machine learning. BMC Infect Dis 2025;25(1). https://doi.org/10.1186/s12879-025-10609-y.
[70] Koeken VACM, Lachmandas E, Riza A, Matzaraki V, Yang L, Kumar V, Oosting M, Joosten LAB, Netea MG, Crevel Rvan. Role of glutamine metabolism in host defense against *Mycobacterium tuberculosis* infection. J Infect Dis 2019;219(10):1662–70. https://doi.org/10.1093/infdis/jiy709.
[71] Parveen S, Shen J, Lun S, Zhao L, Alt J, Benjamin K, Leone RD, Rana R, Powell JD, Murphy JR, Slusher BS, Bishai WR. Glutamine metabolism inhibition has dual immunomodulatory and antibacterial activities against *Mycobacterium tuberculosis*. Nat Commun 2023;14(1). https://doi.org/10.1038/s41467-023-43304-0.
[72] Nasiri V. Advances in host-pathogen interactions in tuberculosis: emerging strategies for therapeutic intervention. Int J Mol Sci 2025;26(4).
[73] Lee K, Chen Y, Xu J, Wang Y, Dunussi-Joannopoulos K, Ou Q, Flynn JL, Porcelli SA, Jacobs WR, Chan J, Sassetti CM. B cells regulate neutrophilia during *Mycobacterium tuberculosis* infection and BCG vaccination by modulating the interleukin-17 response. PLoS Pathog 2013;9(7):e1003472. https://doi.org/10.1371/journal.ppat.1003472.
[74] Jin J, Byun JK, Choi YK, Park KG. Targeting glutamine metabolism as a therapeutic strategy for cancer. Exp Mol Med 2023;55(4):706–15. https://doi.org/10.1038/s12276-023-00971-9.
[75] Wang B, Pei J, Xu S, Liu J, Yu J. A glutamine tug-of-war between cancer and immune cells: recent advances in unraveling the ongoing battle. J Exp Clin Cancer Res 2024;43(1). https://doi.org/10.1186/s13046-024-02994-0.
[76] Zhang K, Mishra A, Jagannath C. New insight into arginine and tryptophan metabolism in macrophage activation during tuberculosis. Front Immunol 2024;15. https://doi.org/10.3389/fimmu.2024.1363938.
[77] Collins JM, Siddiqa A, Jones DP, Liu K, Kempker RR, Nizam A, Shah NS, Ismail N, Ouma SG, Tukvadze N, Li S, Day CL, Rengarajan J, Brust JCM, Gandhi NR, Ernst JD, Blumberg HM, Ziegler TR. Tryptophan catabolism reflects disease activity in human tuberculosis. JCI Insight 2020;5(10). https://doi.org/10.1172/JCI.INSIGHT.137131.
[78] Crowther RR, Qualls JE. Metabolic regulation of immune responses to *Mycobacterium tuberculosis*: a spotlight on L-arginine and L-tryptophan metabolism. Front Immunol 2021;11. https://doi.org/10.3389/fimmu.2020.628432.
[79] Amalia F, Syamsunarno MRAA, Triatin RD, Fatimah SN, Chaidir L, Achmad TH. The role of amino acids in tuberculosis infection: a literature review. Metabolites 2022;12(10). https://doi.org/10.3390/metabo12100933.

[80] Smith SG, Bowness R, Cliff JM. Host-directed therapy in diabetes and tuberculosis comorbidity toward global tuberculosis elimination. Int J Infect Dis 2025;155:107877. https://doi.org/10.1016/j.ijid.2025.107877.

[81] Khoza LJ, Kumar P, Dube A, Demana PH, Choonara YE. Insights into innovative therapeutics for drug-resistant tuberculosis: host-directed therapy and autophagy inducing modified nanoparticles. Int J Pharm 2022;622:121893. https://doi.org/10.1016/j.ijpharm.2022.121893.

[82] Costa DL, Maiga M, Subbian S. Editorial: host-directed therapies for tuberculosis. Front Cell Infect Microbiol 2021;11. https://doi.org/10.3389/fcimb.2021.742053.

[83] Kolloli A, Subbian S. Host-directed therapeutic strategies for tuberculosis. Front Med 2017;4. https://doi.org/10.3389/fmed.2017.00171.

[84] Chang DPS, Guan XL. Metabolic versatility of *Mycobacterium tuberculosis* during infection and dormancy. Metabolites 2021;11(2):88. https://doi.org/10.3390/metabo11020088.

[85] Singha B, Murmu S, Nair T, Rawat RS, Sharma AK, Soni V. Metabolic rewiring of *Mycobacterium tuberculosis* upon drug treatment and antibiotics resistance. Metabolites 2024;14(1). https://doi.org/10.3390/metabo14010063.

[86] Fatima S, Kumari A, Dwivedi VP. Advances in adjunct therapy against tuberculosis: deciphering the emerging role of phytochemicals. MedComm 2021;2(4):494–513. https://doi.org/10.1002/mco2.82.

[87] Nair A, Greeny A, Nandan A, Sah RK, Jose A, Dyawanapelly S, Junnuthula V, Athira KV, Sadanandan P. Advanced drug delivery and therapeutic strategies for tuberculosis treatment. J Nanobiotechnol 2023;21(1). https://doi.org/10.1186/s12951-023-02156-y.

[88] Gonzalo-Asensio J, Marinova D, Martin C, Aguilo N. MTBVAC: attenuating the human pathogen of tuberculosis (TB) toward a promising vaccine against the TB epidemic. Front Immunol 2017;8. https://doi.org/10.3389/fimmu.2017.01803.

[89] Marinova D, Gonzalo-Asensio J, Aguilo N, Martin C. MTBVAC from discovery to clinical trials in tuberculosis-endemic countries. Expet Rev Vaccine 2017;16(6):565–76. https://doi.org/10.1080/14760584.2017.1324303.

[90] da Costa AC, Nogueira SV, Kipnis A, Junqueira-Kipnis AP. Recombinant BCG: innovations on an old vaccine. Scope of BCG strains and strategies to improve long-lasting memory. Front Immunol 2014;5. https://doi.org/10.3389/fimmu.2014.00152.

[91] D Agostino MR, Lai R, Afkhami S, Khera A, Yao Y, Vaseghi-Shanjani M, Zganiacz A, Jeyanathan M, Xing Z. Airway macrophages mediate mucosal vaccine induced trained innate immunity against mycobacterium tuberculosis in early stages of infection. J Immunol 2020;205(10):2750–62. https://doi.org/10.4049/jimmunol.2000532.

[92] Griffiths KL, Ahmed M, Das S, Gopal R, Horne W, Connell TD, Moynihan KD, Kolls JK, Irvine DJ, Artyomov MN, Rangel-Moreno J, Khader SA. Targeting dendritic cells to accelerate T-cell activation overcomes a bottleneck in tuberculosis vaccine efficacy. Nat Commun 2016;7. https://doi.org/10.1038/ncomms13894.

[93] Hatherill M, White RG, Hawn TR. Clinical development of new TB vaccines: recent advances and next steps. Front Microbiol 2020;10. https://doi.org/10.3389/fmicb.2019.03154.

[94] Sankar P, Mishra BB. Early innate cell interactions with *Mycobacterium tuberculosis* in protection and pathology of tuberculosis. Front Immunol 2023;14. https://doi.org/10.3389/fimmu.2023.1260859.

[95] Yu JJ, Gaffen SL. Interleukin-17: a novel inflammatory cytokine that bridges innate and adaptive immunity. Front Biosci 2008;13(1):170–7. https://doi.org/10.2741/2667.

[96] Zhou J, Lv J, Carlson C, Liu H, Wang H, Xu T, Wu F, Song C, Wang X, Wang T, Qian Z. Trained immunity contributes to the prevention of *Mycobacterium tuberculosis* infection, a novel role of autophagy. Emerg Microb Infect 2021;10(1):578–88. https://doi.org/10.1080/22221751.2021.1899771.

[97] Lerm M, Netea MG. Trained immunity: a new avenue for tuberculosis vaccine development. J Intern Med 2016;279(4):337–46. https://doi.org/10.1111/joim.12449.

[98] Pasco ST, Anguita J. Lessons from Bacillus Calmette-Guerin: harnessing trained immunity for vaccine development. Cells 2020;9(9).
[99] Vetvicka V, Sima P, Vannucci L. Trained immunity as an adaptive branch of innate immunity. Int J Mol Sci 2021;22(19):10684. https://doi.org/10.3390/ijms221910684.
[100] Khan N, Downey J, Sanz J, Kaufmann E, Blankenhaus B, Pacis A, Pernet E, Ahmed E, Cardoso S, Nijnik A, Mazer B, Sassetti C, Behr MA, Soares MP, Barreiro LB, Maziar Divangahi M. Tuberculosis reprograms hematopoietic stem cells to limit myelopoiesis and impair trained immunity. Cell 2020;183(3): 752. https://doi.org/10.1016/j.cell.2020.09.062.
[101] Kwon BE, Ahn JH, Min S, Kim H, Seo J, Yeo SG, Ko HJ. Development of new preventive and therapeutic vaccines for tuberculosis. Immune Netw 2018;18(2). https://doi.org/10.4110/in.2018.18.e17.
[102] Langermans JAM, Doherty TM, Vervenne RAW, Van Der Laan T, Lyashchenko K, Greenwald R, Agger EM, Aagaard C, Weiler H, Van Soolingen D, Dalemans W, Thomas AW, Andersen P. Protection of macaques against *Mycobacterium tuberculosis* infection by a subunit vaccine based on a fusion protein of antigen 85B and ESAT-6. Vaccine 2005;23(21):2740–50. https://doi.org/10.1016/j.vaccine.2004.11.051.
[103] Sable SB, Posey JE, Scriba TJ. Tuberculosis vaccine development: progress in clinical evaluation. Clin Microbiol Rev 2019;33(1). https://doi.org/10.1128/cmr.00100-19.
[104] Ishida E, Corrigan DT, Chen T, Liu Y, Kim RS, Song L, Rutledge TM, Magee DM, LaBaer J, Lowary TL, Lin PL, Achkar JM. Mucosal and systemic antigen-specific antibody responses correlate with protection against active tuberculosis in nonhuman primates. EBioMedicine 2024;99:104897. https://doi.org/10.1016/j.ebiom.2023.104897.
[105] Dahiya P, Bisht MK, Mukhopadhyay S. Role of PE family of proteins in mycobacterial virulence: potential on anti-TB vaccine and drug design. Int Rev Immunol 2025;44(4):213–28. https://doi.org/10.1080/08830185.2025.2455161.
[106] Sharma T, Alam A, Ehtram A, Rani A, Grover S, Ehtesham NZ, Hasnain SE. The *Mycobacterium tuberculosis* PE_PGRS protein family acts as an immunological decoy to subvert host immune response. Int J Mol Sci 2022;23(1). https://doi.org/10.3390/ijms23010525.
[107] Akhter Y, Ehebauer MT, Mukhopadhyay S, Hasnain SE. The PE/PPE multigene family codes for virulence factors and is a possible source of mycobacterial antigenic variation: perhaps more? Biochimie 2012;94(1): 110–6. https://doi.org/10.1016/j.biochi.2011.09.026.
[108] Ehtram A, Shariq M, Quadir N, Jamal S, Pichipalli M, Zarin S, Javaid AS, Ehtesham NZ, Hasnain SE. Deciphering the functional roles of PE18 and PPE26 proteins in modulating *Mycobacterium tuberculosis* pathogenesis and immune response. Front Immunol 2025;16. https://doi.org/10.3389/fimmu.2025.1517822.
[109] Dolasia K, Nazar F, Mukhopadhyay S. *Mycobacterium tuberculosis* PPE18 protein inhibits MHC class II antigen presentation and B cell response in mice. Eur J Immunol 2021;51(3):603–19. https://doi.org/10.1002/eji.201848071.
[110] Singh P, Rao RN, Reddy JRC, Prasad R, Kotturu SK, Ghosh S, Mukhopadhyay S. PE11, a PE/PPE family protein of *Mycobacterium tuberculosis* is involved in cell wall remodeling and virulence. Sci Rep 2016;6. https://doi.org/10.1038/srep21624.
[111] Yang W, Deng W, Zeng J, Ren S, Ali MK, Gu Y, Li Y, Xie J. *Mycobacterium tuberculosis* PE_PGRS18 enhances the intracellular survival of M. smegmatis via altering host macrophage cytokine profiling and attenuating the cell apoptosis. Apoptosis 2017;22(4):502–9. https://doi.org/10.1007/s10495-016-1336-0.
[112] Deng W, Long Q, Zeng J, Li P, Yang W, Chen X, Xie J. *Mycobacterium tuberculosis* PE_PGRS41 enhances the intracellular survival of M. Smegmatis within macrophages via blocking innate immunity and inhibition of host defense. Sci Rep 2017;7(1). https://doi.org/10.1038/srep46716.
[113] Basu S, Pathak SK, Banerjee A, Pathak S, Bhattacharyya A, Yang Z, Talarico S, Kundu M, Basu J. Execution of macrophage apoptosis by PE_PGRS33 of *Mycobacterium tuberculosis* is mediated by toll-like

receptor 2-dependent release of tumor necrosis factor-α. J Biol Chem 2007;282(2):1039–50. https://doi.org/10.1074/jbc.M604379200.

[114] Camassa S, Palucci I, Iantomasi R, Cubeddu T, Minerva M, De Maio F, Jouny S, Petruccioli E, Goletti D, Ria F, Sali M, Sanguinetti M, Manganelli R, Rocca S, Brodin P, Delogu G. Impact of pe_pgrs33 gene polymorphisms on *Mycobacterium tuberculosis* infection and pathogenesis. Front Cell Infect Microbiol 2017;7. https://doi.org/10.3389/fcimb.2017.00137.

[115] Anand PK, Kaur J. Rv3539 (PPE63) of Mycobacterium tuberculosis promotes survival of Mycobacterium smegmatis in human macrophages cell line via cell wall modulation of bacteria and altering host's immune response. Curr Microbiol 2023;80(8). https://doi.org/10.1007/s00284-023-03360-7.

[116] Khubaib M, Sheikh JA, Pandey S, Srikanth B, Bhuwan M, Khan N, Hasnain SE, Ehtesham NZ. *Mycobacterium tuberculosis* co-operonic PE32/PPE65 proteins alter host immune responses by hampering Th1 response. Front Microbiol 2016;7(May). https://doi.org/10.3389/fmicb.2016.00719.

[117] Ahmad J, Khubaib M, Sheikh JA, Pancsa R, Kumar S, Srinivasan A, Babu MM, Hasnain SE, Ehtesham NZ. Disorder-to-order transition in PE–PPE proteins of *Mycobacterium tuberculosis* augments the pro-pathogen immune response. FEBS Open Bio 2020;10(1):70–85. https://doi.org/10.1002/2211-5463.12749.

[118] Tiwari BM, Kannan N, Vemu L, Raghunand TR, Briken V. The *Mycobacterium tuberculosis* PE proteins Rv0285 and Rv1386 modulate innate immunity and mediate bacillary survival in macrophages. PLoS One 2012;7(12):e51686. https://doi.org/10.1371/journal.pone.0051686.

[119] Tundup S, Mohareer K, Hasnain SE. *Mycobacterium tuberculosis* PE25/PPE41 protein complex induces necrosis in macrophages: role in virulence and disease reactivation? FEBS Open Bio 2014;4:822–8. https://doi.org/10.1016/j.fob.2014.09.001.

[120] D'Souza C, Kishore U, Tsolaki AG. The PE-PPE family of *Mycobacterium tuberculosis*: proteins in disguise. Immunobiology 2023;228(2):152321. https://doi.org/10.1016/j.imbio.2022.152321.

[121] Kohli S, Singh Y, Sharma K, Mittal A, Ehtesham NZ, Hasnain SE. Comparative genomic and proteomic analyses of PE/PPE multigene family of *Mycobacterium tuberculosis* H37Rv and H37Ra reveal novel and interesting differences with implications in virulence. Nucleic Acids Res 2012;40(15):7113–22. https://doi.org/10.1093/nar/gks465.

[122] Brennan MJ. The enigmatic PE/PPE multigene family of mycobacteria and tuberculosis vaccination. Infect Immun 2017;85(6). https://doi.org/10.1128/IAI.00969-16.

[123] Qian J, Chen R, Wang H, Zhang X. Role of the PE/PPE family in host–pathogen interactions and prospects for anti-tuberculosis vaccine and diagnostic tool design. Front Cell Infect Microbiol 2020;10. https://doi.org/10.3389/fcimb.2020.594288.

[124] Guo F, Wei J, Song Y, Li B, Qian Z, Wang X, Wang H, Xu T. Immunological effects of the PE/PPE family proteins of *Mycobacterium tuberculosis* and related vaccines. Front Immunol 2023;14. https://doi.org/10.3389/fimmu.2023.1255920.

[125] Shariq M, Quadir N, Alam A, Zarin S, Sheikh JA, Sharma N, Samal J, Ahmad U, Kumari I, Hasnain SE, Ehtesham NZ. The exploitation of host autophagy and ubiquitin machinery by *Mycobacterium tuberculosis* in shaping immune responses and host defense during infection. Autophagy 2023;19(1):3–23. https://doi.org/10.1080/15548627.2021.2021495.

[126] Appadurai R, Uversky VN, Srivastava A. The structural and functional diversity of intrinsically disordered regions in transmembrane proteins. J Membr Biol 2019;252(4–5):273–92. https://doi.org/10.1007/s00232-019-00069-2.

[127] Jeon S, Jeon Y, Lim JY, Kim Y, Cha B, Kim W. Emerging regulatory mechanisms and functions of biomolecular condensates: implications for therapeutic targets. Signal Transduct Targeted Ther 2025;10(1). https://doi.org/10.1038/s41392-024-02070-1.

[128] Harichander S, Wiafe E, Mensah KB, Bangalee V, Oosthuizen F. The incidence of TB and MDR-TB in pediatrics and therapeutic options: a systematic review. Syst Rev 2022;11(1). https://doi.org/10.1186/s13643-022-02023-1.

[129] Lhadon T, Jullien S. Congenital multidrug-resistant tuberculosis in a neonate: a case report. J Trop Pediatr 2019;65(2):188–91. https://doi.org/10.1093/tropej/fmy020.

[130] Younossian AB, Rochat T, Ketterer J-P, Wacker J, Janssens J-P. High hepatotoxicity of pyrazinamide and ethambutol for treatment of latent tuberculosis: table. 1. Eur Respir J 2005;26(3):462–4. https://doi.org/10.1183/09031936.05.00006205.

[131] Graciaa DS, Schechter MC, Fetalvero KB, Cranmer LM, Kempker RR, Castro KG. Updated considerations in the diagnosis and management of tuberculosis infection and disease: integrating the latest evidence-based strategies. Expert Rev Anti Infect Ther 2023;21(6):595–616. https://doi.org/10.1080/14787210.2023.2207820.

[132] Wu C, Yi H, Hu Y, Luo D, Tang Z, Wen X, Zhang Y, Tang M, Zhang L, Wu S, Chen M. Effects of second-line anti-tuberculosis drugs on the intestinal microbiota of patients with rifampicin-resistant tuberculosis. Front Cell Infect Microbiol 2023;13. https://doi.org/10.3389/fcimb.2023.1127916.

[133] Shanmuga Priya K, Parivakkam mani A, Geethalakshmi S, Yadav S. Advancements in artificial intelligence for the diagnosis of multidrug resistance and extensively drug-resistant tuberculosis: a comprehensive review. Cureus 2024;16(5). https://doi.org/10.7759/cureus.60280.

[134] Singh R, Dwivedi SP, Gaharwar US, Meena R, Rajamani P, Prasad T. Recent updates on drug resistance in *Mycobacterium tuberculosis*. J Appl Microbiol 2020;128(6):1547–67. https://doi.org/10.1111/jam.14478.

[135] Raizada N, Khaparde SD, Singh Salhotra V, Rao R, Kalra A, Swaminathan S, Khanna A, Chopra KK, Hanif M, Singh V, Umadevi KR, Nair SA, Huddart S, Prakash CHS, Mall S, Singh P, Saha BK, Denkinger CM, Boehme C, Sarin S, Ehtesham Hasnain S. Accelerating access to quality TB care for pediatric TB cases through better diagnostic strategy in four major cities of India. PLoS One 2018;13(2):e0193194. https://doi.org/10.1371/journal.pone.0193194.

[136] Raizada N, Singh Sachdeva K, Nair SA, Kulsange S, Gupta RS, Thakur R, Parmar M, Gray C, Ramachandran R, Vadera B, Ekka S, Dhawan S, Babre A, Ghedia M, Alavadi U, Dewan P, Khetrapal M, Khanna A, Boehme C, Paramsivan CN, Herrmann JL. Enhancing TB case detection: experience in offering upfront Xpert MTB/RIF testing to pediatric presumptive TB and DR TB cases for early rapid diagnosis of drug sensitive and drug resistant TB. PLoS One 2014;9(8):e105346. https://doi.org/10.1371/journal.pone.0105346.

[137] Atherton RR, Cresswell FV, Ellis J, Kitaka SB, Boulware DR. Xpert MTB/RIF ultra for tuberculosis testing in children: a mini-review and commentary. Front Pediatr 2019;7. https://doi.org/10.3389/fped.2019.00034.

[138] Seddon JA, Godfrey-Faussett P, Hesseling AC, Gie RP, Beyers N, Schaaf HS. Management of children exposed to multidrug-resistant *Mycobacterium tuberculosis*. Lancet Infect Dis 2012;12(6):469–79. https://doi.org/10.1016/S1473-3099(11)70366-8.

[139] Mgode GF, Cox CL, Mwimanzi S, Mulder C. Pediatric tuberculosis detection using trained African giant pouched rats. Pediatr Res 2018;84(1):99–103. https://doi.org/10.1038/pr.2018.40.

[140] Gjøen JE, Jenum S, Sivakumaran D, Mukherjee A, MacAden R, Kabra SK, Lodha R, Ottenhoff THM, Haks MC, Doherty TM, Ritz C, Grewal HMS. Novel transcriptional signatures for sputum-independent diagnostics of tuberculosis in children. Sci Rep 2017;7(1). https://doi.org/10.1038/s41598-017-05057-x.

[141] Marais BJ. Strategies to improve tuberculosis case finding in children. Public Health Action 2015;5(2):90–1. https://doi.org/10.5588/pha.15.0028.

[142] Foo CDe, Shrestha P, Wang L, Du Q, García-Basteiro AL, Abu Saleh A, Legido-Quigley H, Bipin Suthar A. Integrating tuberculosis and noncommunicable diseases care in low- and middle-income countries (LMICs): a systematic review. PLoS Med 2022;19(1):e1003899. https://doi.org/10.1371/journal.pmed.1003899.

[143] Kiragu K, Collins L, Von Zinkernagel D, Mushavi A. Integrating PMTCT into maternal, newborn, and child health and related services: experiences from the global plan priority countries. J Acquir Immune Defic Syndr 2017;75(1):S36. https://doi.org/10.1097/qai.0000000000001323.

[144] Hartsough K, Teasdale CA, Shongwe S, Geller A, De Gusmao EP, Dlamini P, Mafukidze A, Pasipamire M, Ao T, Ryan C, Modi S, Abrams EJ, Howard AA, Hermans S. Enhanced integration of TB services in reproductive maternal newborn and child health (RMNCH) settings in eswatini. PLOS Glob Publ Health 2022;2(4):e0000217. https://doi.org/10.1371/journal.pgph.0000217.
[145] Kassa RN, Shifti DM, Alemu K, Omigbodun AO. Integration of cervical cancer screening into healthcare facilities in low- and middle-income countries: a scoping review protocol. BMJ Open 2024;14(1). https://doi.org/10.1136/bmjopen-2023-078466.
[146] Anand TN, Joseph LM, Geetha AV, Prabhakaran D, Jeemon P. Task sharing with non-physician health-care workers for management of blood pressure in low-income and middle-income countries: a systematic review and meta-analysis. Lancet Global Health 2019;7(6):e761. https://doi.org/10.1016/S2214-109X(19)30077-4.
[147] Mutalikdesai N, Tonde K, Shinde K, Kumar R, Gupta S, Dayma G, Krishnan A, Juvekar S, Santosa A, Ng N, Patil R. Exploring potential barriers and facilitators to integrate tuberculosis, diabetes mellitus, and tobacco control programmes in India. J Glob Health 2025;15. https://doi.org/10.7189/jogh.15.04230.
[148] Spruijt I, Alam Y, Nguyen H, Myrzaliev B, Ahmatov M, Odume B, Mtei L, Gebhard A, Gidado M, Jerene D, Gong W. Digital health solutions and integrated COVID-19 and TB services to help recover TB care and prevention services in the COVID-19 pandemic: a descriptive study in four high TB burden countries. PLoS One 2023;18(11):e0293964. https://doi.org/10.1371/journal.pone.0293964.
[149] Ndwiga C, Birungi H, Undie CC, Weyenga H, Sitienei J. Feasibility and effect of integrating tuberculosis screening and detection in postnatal care services: an operations research study. BMC Health Serv Res 2013;13(1). https://doi.org/10.1186/1472-6963-13-99.
[150] Durrance-Bagale A, Marzouk M, Tung LS, Agarwal S, Aribou ZM, Ibrahim NBM, Mkhallalati H, Newaz S, Omar M, Ung M, Zaseela A, Nagashima-Hayashi M, Howard N. Community engagement in health systems interventions and research in conflict-affected countries: a scoping review of approaches. Glob Health Action 2022;15(1). https://doi.org/10.1080/16549716.2022.2074131.
[151] Elbehiry A, Marzouk E, Edrees HM, AlShaqi R, Ellethy AT, Alzaben F, Anagreyyah S, Algarni A, Almuhaydili K, Alotaibi I, Albaqami A, Alamri K, Ibrahem M, Almuzaini AM, Dhahri F, Abu-Okail A. Advancing the fight against tuberculosis: integrating innovation and public health in diagnosis, treatment, vaccine development, and implementation science. Front Med 2025;12. https://doi.org/10.3389/fmed.2025.1596579.
[152] Kizny Gordon A, Marais B, Walker TM, Sintchenko V. Clinical and public health utility of *Mycobacterium tuberculosis* whole genome sequencing. Int J Infect Dis 2021;113:S40. https://doi.org/10.1016/j.ijid.2021.02.114.
[153] Yimer SA, Booij BB, Tobert G, Hebbeler A, Oloo P, Brangel P, L'Azou Jackson M, Jarman R, Craig D, Avumegah MS, Mandi H, Endy T, Wooden S, Clark C, Bernasconi V, Shurtleff A, Kristiansen PA. Rapid diagnostic test: a critical need for outbreak preparedness and response for high priority pathogens. BMJ Glob Health 2024;9(4). https://doi.org/10.1136/bmjgh-2023-014386.
[154] Advani J, Verma R, Chatterjee O, Pachouri PK, Upadhyay P, Singh R, Yadav J, Naaz F, Ravikumar R, Buggi S, Suar M, Gupta UD, Pandey A, Chauhan DS, Tripathy SP, Gowda H, Prasad TSK. Whole genome sequencing of *Mycobacterium tuberculosis* clinical isolates from India reveals genetic heterogeneity and region-specific variations that might affect drug susceptibility. Front Microbiol 2019;10. https://doi.org/10.3389/fmicb.2019.00309.
[155] Sun YJ, Bellamy R, Lee ASG, Ng ST, Ravindran S, Wong SY, Locht C, Supply P, Paton NI. Use of mycobacterial interspersed repetitive unit-variable-number tandem repeat typing to examine genetic diversity of *Mycobacterium tuberculosis* in Singapore. J Clin Microbiol 2004;42(5):1986–93. https://doi.org/10.1128/JCM.42.5.1986-1993.2004.
[156] Wyllie DH, Davidson JA, Grace Smith E, Rathod P, Crook DW, Peto TEA, Robinson E, Walker T, Campbell C. A quantitative evaluation of MIRU-VNTR typing against whole-genome sequencing for

identifying *Mycobacterium tuberculosis* transmission: a prospective observational cohort study. EBioMedicine 2018;34:122–30. https://doi.org/10.1016/j.ebiom.2018.07.019.

[157] Yang TY, Su H, Jain SH, Lin HH, Wu CY, Chen CJ, Shiea J, Lu PL. Evaluation of thermal desorption–electrospray ionization mass spectrometry for characterization of *Mycobacterium tuberculosis*: comparison with MIRU-VNTR and whole-genome sequencing. J Mass Spectrom Adv Clin Lab 2025;38:50–7. https://doi.org/10.1016/j.jmsacl.2025.10.005.

[158] Ghosh M, Lahiri M, Dalal A, Kumar Parida K, Pal Kalia N. Advancements in tuberculosis diagnostics: an update. Microb Pathog 2025;207:107843. https://doi.org/10.1016/j.micpath.2025.107843.

[159] Mboowa G. Reimagining tuberculosis control in the era of genomics: the case for global investment in *Mycobacterium tuberculosis* genomic surveillance. Pathogens 2025;14(10):975. https://doi.org/10.3390/pathogens14100975.

[160] Buchanan GJR, Filiatreau LM, Moore JE. Organizing the dissemination and implementation field: who are we, what are we doing, and how should we do it? Implement Sci Commun 2024;5(1). https://doi.org/10.1186/s43058-024-00572-1.

[161] Roberts DJ, Mannes T, Verlander NQ, Anderson C. Factors associated with delay in treatment initiation for pulmonary tuberculosis. ERJ Open Res 2020;6(1). https://doi.org/10.1183/23120541.00161-2019.

[162] Collins SE, Clifasefi SL, Stanton J, Straits KJE, Gil-Kashiwabara E, Espinosa PR, Nicasio AV, Andrasik MP, Hawes SM, Miller KA, Nelson LA, Orfaly VE, Duran BM, Wallerstein N. Community-based participatory research (CBPR): towards equitable involvement of community in psychology research. Am Psychol 2018;73(7):884–98. https://doi.org/10.1037/amp0000167.

[163] Baumann AA, Shelton RC, Kumanyika S, Haire-Joshu D. Advancing healthcare equity through dissemination and implementation science. Health Serv Res 2023;58(3):327–44. https://doi.org/10.1111/1475-6773.14175.

[164] Francisco KKY, Apuhin AEC, Maravilla NMAT, Byers MC, Karim HA, Tan MJT, AlDahoul N. Personalized medicine and health equity: overcoming cost barriers and ethical challenges. Int J Equity Health 2026;25(1). https://doi.org/10.1186/s12939-025-02710-0.

[165] Ramanadhan S, Alemán R, Bradley CD, Cruz JL, Safaeinili N, Simonds V, Aveling EL. Using participatory implementation science to advance health equity. Annu Rev Publ Health 2024;45(1):47–67. https://doi.org/10.1146/annurev-publhealth-060722-024251.

[166] Warner ET, Huguet N, Fredericks M, Gundersen D, Nederveld A, Brown MC, Houston TK, Davis KL, Mazzucca S, Rendle KA, Emmons KM. Advancing health equity through implementation science: identifying and examining measures of the outer setting. Soc Sci Med 2023;331:116095. https://doi.org/10.1016/j.socscimed.2023.116095.

[167] Murray S, Mendel C, Spigelman M. TB Alliance regimen development for multidrug-resistant tuberculosis. Int J Tubercul Lung Dis 2016;20(12):38–41. https://doi.org/10.5588/ijtld.16.0069.

[168] Auer C, Gupta A, Malbacius C, Ghafoor A, Kock Y, Medvedieva O, Hanlon P, Steinmann P, Juneja S. Savings from the introduction of BPaL and BPaLM regimens at the country level. IJTLD Open 2024;1(7):314–9. https://doi.org/10.5588/ijtldopen.24.0213.

[169] Xu J, Li SY, Almeida DV, Tasneen R, Barnes-Boyle K, Converse PJ, Upton AM, Mdluli K, Fotouhi N, Nuermberger EL. Contribution of pretomanid to novel regimens containing bedaquiline with either linezolid or moxifloxacin and pyrazinamide in murine models of tuberculosis. Antimicrob Agents Chemother 2019;63(5). https://doi.org/10.1128/AAC.00021-19.

[170] Baral S, Yadav RK, Yadav DK, Marahatta SB, Baral Y, Khadka KB, Thakur SK, Paudel S, Sharma P, Pandey S, Shrestha K, Shah NP, Basaula L, Nagila A, Mahato RK, Ranabhat CL. Feasibility of implementing public-private mix approach for tuberculosis case management in Pokhara Metropolitan City of western Nepal: a qualitative study. Front Public Health 2023;11. https://doi.org/10.3389/fpubh.2023.1132090.

[171] Otchere ID, Asante-Poku A, Akpadja KF, Diallo AB, Sanou A, Asare P, Osei-Wusu S, Onyejepu N, Diarra B, Dagnra YA, Kehinde A, Antonio M, Yeboah-Manu D. Opinion review of drug resistant tuberculosis in West Africa: tackling the challenges for effective control. Front Public Health 2024;12. https://doi.org/10.3389/fpubh.2024.1374703.

[172] Singh S. Public-private partnership to end tuberculosis: challenges and opportunities. Int J Equity Health 2025;24(1).

[173] Villa S, de Colombani P, Dall'Olio L, Gargioni G, Raviglione M. Towards comprehensive clinical trials for new tuberculosis drug regimens: policy recommendations from a stakeholder analysis. BMJ Glob Health 2024;9(4):e014630. https://doi.org/10.1136/bmjgh-2023-014630.

[174] Cozzoli N, Salvatore FP, Faccilongo N, Milone M. How can big data analytics be used for healthcare organization management? Literary framework and future research from a systematic review. BMC Health Serv Res 2022;22(1). https://doi.org/10.1186/s12913-022-08167-z.

[175] LoBue PA, Mermin JH. Latent tuberculosis infection: the final frontier of tuberculosis elimination in the USA. Lancet Infect Dis 2017;17(10):e327. https://doi.org/10.1016/S1473-3099(17)30248-7.

[176] Rylance J, Pai M, Lienhardt C, Garner P. Priorities for tuberculosis research: a systematic review. Lancet Infect Dis 2010;10(12):886–92. https://doi.org/10.1016/s1473-3099(10)70201-2.

[177] Jerene D, Levy J, van Kalmthout K, van Rest J, McQuaid CF, Quaife M, Charalambous S, Gamazina K, Garfin AMC, Mleoh L, Terleieva Y, Bogdanov A, Maraba N, Fielding K. Effectiveness of digital adherence technologies in improving tuberculosis treatment outcomes in four countries: a pragmatic cluster randomised trial protocol. BMJ Open 2023;13(3):e068685. https://doi.org/10.1136/bmjopen-2022-068685.

[178] Maphalle LNF, Michniak-Kohn BB, Ogunrombi MO, Adeleke OA. Pediatric tuberculosis management: a global challenge or breakthrough? Children 2022;9(8). https://doi.org/10.3390/children9081120.

[179] Nyirenda JLZ, Bockey A, Wagner D, Lange B. Effect of Tuberculosis (TB) and Diabetes mellitus (DM) integrated healthcare on bidirectional screening and treatment outcomes among TB patients and people living with DM in developing countries: a systematic review. Pathog Glob Health 2023;117(1):36–51. https://doi.org/10.1080/20477724.2022.2046967.

[180] Gilbert SC. Priorities for rapid and cost-effective vaccines to improve outbreak responses. PLoS Biol 2025; 23(4):e3003114. https://doi.org/10.1371/journal.pbio.3003114.

[181] Bradley M, Taylor R, Jacobson J, Guex M, Hopkins A, Jensen J, Leonard L, Waltz J, Kuykens L, Sow PS, Madeja U-D, Hida T, Ole-Moiyoi K, King J, Argaw D, Mohamed J, Polo MR, Yajima A, Ottesen E. Medicine donation programmes supporting the global drive to end the burden of neglected tropical diseases. Trans R Soc Trop Med Hyg 2021;115(2):136–44. https://doi.org/10.1093/trstmh/traa167.

[182] Mohammadnabi N, Shamseddin J, Emadi M, Bodaghi AB, Varseh M, Shariati A, Rezaei M, Dastranj M, Farahani A. *Mycobacterium tuberculosis*: the mechanism of pathogenicity, immune responses, and diagnostic challenges. J Clin Lab Anal 2024;38(23). https://doi.org/10.1002/jcla.25122.

[183] Kozakiewicz L, Phuah J, Flynn J, Chan J. The role of B cells and humoral immunity in mycobacterium Tuberculosis infection. Adv Exp Med Biol 2013;783:225–50. https://doi.org/10.1007/978-1-4614-6111-1_12.

[184] Gutiérrez-González LH, Juárez E, Carranza C, Carreto-Binaghi LE, Alejandre A, Cabello-Gutiérrrez C, Gonzalez Y. Immunological aspects of diagnosis and management of childhood tuberculosis. Infect Drug Resist 2021;14:929–46. https://doi.org/10.2147/IDR.S295798.

[185] Memon S, Bibi S, He G. Integration of AI and ML in tuberculosis (TB) management: from diagnosis to drug discovery. Diseases 2025;13(6):184. https://doi.org/10.3390/diseases13060184.

[186] Zhuang L, Yang L, Li L, Ye Z, Gong W. *Mycobacterium tuberculosis*: immune response, biomarkers, and therapeutic intervention. MedComm 2024;5(1). https://doi.org/10.1002/mco2.419.

[187] Davenport T, Kalakota R. The potential for artificial intelligence in healthcare. Future Healthc J 2019;6(2): 94–8. https://doi.org/10.7861/futurehosp.6-2-94.

[188] John CA. Realizing the World Health Organization's end TB strategy (2016–2035): how can social approaches to tuberculosis elimination contribute to progress in Asia and the Pacific? Trop Med Infect Dis 2019;4(1). https://doi.org/10.3390/tropicalmed4010028.
[189] Basu S. Challenges in expanding TB preventive therapy in high-burden settings: beyond logistics is evidence and ethics. Public Health Action 2020;10(2):82. https://doi.org/10.5588/pha.20.0016.
[190] Shah H, Patel J, Rai S, Sen A. Advancing tuberculosis elimination in India: a qualitative review of current strategies and areas for improvement in tuberculosis preventive treatment. IJID Regions 2025;14:100556. https://doi.org/10.1016/j.ijregi.2024.100556.

Index

Note: Page numbers followed by "f" indicate figures, "t" indicate tables.

A

B

L

M

R

S

T

U

V

W

Printed and bound by CPI Group (UK) Ltd, Croydon, CR0 4YY
14/09/2026
02216410-0014